RECENT DEVELOPMENTS IN QUANTUM OPTICS

RECENT DEVELOPMENTS IN QUANTUM OPTICS

Edited by
Ramarao Inguva
The University of Wyoming
Laramie, Wyoming

SPRINGER SCIENCE+BUSINESS MEDIA, LLC

Library of Congress Cataloging-in-Publication Data

Recent developments in quantum optics / edited by Ramarao Inguva.
p. cm.
"Proceedings of the International Conference on Quantum Optics, held January 5-10, 1991, in Hyderabad, India"--T.p. verso.
Includes bibliographical references and index.
ISBN 978-1-4613-6275-3
1. Quantum optics--Congresses. I. Inguva, Ramarao, 1941- . II. International Conference on Quantum Optics (1991 : Hyderabad, India)
QC446.15.R43 1993
535--dc20 92-46382
CIP

Proceedings of the International Conference on Quantum Optics, held January 5–10, 1991, in Hyderabad, India

ISBN 978-1-4613-6275-3 ISBN 978-1-4615-2936-1 (eBook)
DOI 10.1007/978-1-4615-2936-1

PREFACE

This volume is composed of papers (invited and contributed) presented at the International Conference on Coherence and Quantum Optics held at the University of Hyderabad January 5-January 10, 1991. It has been organized by Professor Girish Agarwal and his colleagues at the School of Physics, University of Hyderabad, Hyderabad, India under partial support from the Department of Science and Technology, Government of India, International Center for Theoretical Physics, Trieste, Italy and the National Science Foundation, USA. Without the untiring efforts of Prof. Girish Agarwal and the members of his quantum office group, the Conference and the present volume would not have been possible. Some extraordinary circumstances resulted in a delay of the publication of the present volume. Our sincere apologies to all the authors. We deeply regret the inconvenience caused due to the delay.

A debt of gratitude is due to Ms. Kim Bella for the excellent typing job of the different versions and the final version of the manuscript. It is a pleasure to acknowledge the efforts of Ms. Pat Vann, Mr. Greg Safford and Mr. Eric Katz of the Plenum Publishing, without whose interest and persistence this volume would not have been possible.

CONTENTS

QUANTUM OPTICS: THEORY

SQUEEZED AND NON-CLASSICAL LIGHT

CAVITY QED - LASERS AND MASERS

COHERENCE THEORY

PHOTON STATISTICS AND INTENSE FIELD OPTICAL RESONANCE

OPTICAL BISTABILITY, FOUR WAVE MIXING AND PHASE CONJUGATION

THE QUANTUM MECHANICS OF PARTICLES IN TIME-DEPENDENT QUADRUPOLE FIELDS

Roy J. Glauber

Lyman Laboratory of Physics
Harvard University
Cambridge, MA 02138

ABSTRACT

We solve the quantum mechanical problem of the motion of charged particle in a quadrupole field that varies arbitrarily with time. The time-dependent wave functions are shown to be in a simple one-to-one correspondence with the wave functions of an ordinary static-field harmonic oscillator. We define and generate a set of coherent states for the charged particle that provide Gaussian localization about the classical particle trajectories. When the field variation is periodic in time, these minimum uncertainty states are shown to pulsate periodically in size and in phase. Under the action of external fields, coherent states retain their coherence. We discuss the way in which the Kapitza pseudopotential approach approximates the present exact results.

1. INTRODUCTION

The possibility of trapping particles in time-dependent electromagnetic fields suggests that they may be slowed down substantially by those fields, or brought nearly to rest. Minimizing the effects of particle motion in this way could lead to significant improvements in the precision of spectroscopy. Most theoretical discussions of trapping have been undertaken in classical terms to date and the resulting insights have played a formative role in the development of the technique. It is nonetheless clear that truly minimizing particle energies implies a need to analyze their motions quantum mechanically. Only by appealing to quantum mechanics can we hope to understand in detail the influence of residual particle motions on radiative processes, and to deal with those influences quantitatively.

Since there is no static field configuration in which charged particles can be permanently trapped, we must deal with their motion in explicitly time-dependent fields. The quantum states we must find are therefore not stationary states, or energy eigenstates of any kind. Finding them thus presents somewhat novel problems in the context of quantum mechanics, and developing simple means for doing that possesses a certain methodological interest in its own right.

The Paul trap [1] offers a particularly simple example of a trapping problem since its basic equations of motion, even though they contain a time-dependent coefficient,

Recent Developments in Quantum Optics, Edited
by R. Inguva, Plenum Press, New York, 1993

remain fundamentally linear in structure. Those equations can easily be solved by a simple technique [2] for generating a constant of the motion. The result, we shall show, is that even though the particle motion has no stationary states, we can construct for it a complete set of non-stationary states that are in an elementary one-to-one correspondence with the energy eigenstates of ordinary (static-field) harmonic oscillators. We can combine these states, furthermore, to form analogues of the coherent states of the static-field oscillators. These are correspondingly useful wave-packet states that follow, in their mean behavior, all the details of the classical motion of a particle in the time-dependent field.

Many years ago P. Kapitza [3] suggested an interesting approximation for treating classically the motion of a charged particle in a rapidly oscillating field. He assumed that the motion can be separated into two components, one of large amplitude that varies more slowly than the applied field, and the other much smaller in amplitude but varying with the same frequency as the applied field. The approximation suggested that the slow component of the motion can be regarded as taking place in a certain effective static potential. We shall show, by formulating an exact solution for the Paul trap, that the Kapitza approximation can hold quite accurately in quantum mechanical as well as classical terms.

2. FORMULATION OF CONSTANTS OF THE MOTION

The Paul trap is based on an oscillating quadrupole field, that is to say, an oscillating potential quadratic in each of the three cartesian coordinates of a charged particle. Since the problem is thus separable, it suffices to consider one of them for the present, and call it $q(t)$. The time dependence of the quadratic potential, on the other hand, need not be restricted to simple oscillations. We shall see that the problem of finding the quantum mechanical motion can be solved in large measure without specifying the time dependence of the field at all. Let us consider therefore the motion that follows from the Hamiltonian

$$H_o(q,p,t) = \frac{p^2}{2m} + \frac{m}{2}W(t)q^2, \tag{2.1}$$

in which p and q are momentum and coordinate operators respectively and $W(t)$ expresses the time dependence of the externally applied field.

The Hamiltonian equations of motion

$$\dot{q} = \frac{p}{m} \tag{2.2}$$

and

$$\dot{p} = -mW(t)q \tag{2.3}$$

can be combined into the second order equation

$$\ddot{q} + W(t)q = 0. \tag{2.4}$$

If $q_1(t)$ and $q_2(t)$ are any two solutions of this equation, they must obey the Wronskian relation

$$q_1\dot{q}_2 - \dot{q}_1 q_2 = \text{constant}. \tag{2.5}$$

Our strategy will be to use this Wronskian relation to generate a particularly useful constant of the motion of the quantum mechanical system. We note, for this purpose, that q_1 and q_2 may be *any* solutions of the differential equation (2.4). We are free, for example, to let $q_1(t)$ to be an ordinary c-number solution of Eq. (2.4), and to let $q_2(t)$ be the general operator solution $q(t)$ to Eqs. (2.2) and (2.3), regarded as Heisenberg equations of motion.

We will find it useful therefore to introduce the complex c-number function $u(t)$, which we define to satisfy the differential equation

$$\ddot{u} + W(t)u = 0, \tag{2.6}$$

and the initial condition

$$u(0) = 1. \tag{2.7}$$

A second initial condition is necessary for the specification of $u(t)$ and we shall write that as

$$\dot{u}(0) = i\omega, \tag{2.8}$$

where ω is a real positive parameter that we are free to choose quite arbitrarily, at least for the present. A useful property of $u(t)$ is that it too obeys the Wronskian identity

$$u^*(t)\dot{u}(t) - u(t)\dot{u}^*(t) = 2i\omega. \tag{2.9}$$

When this function $u(t)$ and the coordinate $q(t)$ are substituted into Eq. (2.5), we see that the combination $u(t)\dot{q}(t) - \dot{u}(t)q(t)$ must be constant in time. By including a factor of the mass m in this expression, and giving it a convenient normalizing factor we can define the function

$$C(t) = \frac{i}{\sqrt{2m\hbar\omega}}\left\{u(t)p(t) - m\dot{u}(t)q(t)\right\}, \tag{2.10}$$

which is evidently an operator constant of the motion. It is therefore equal at all times to its initial value

$$C(t) = C(0) = \frac{i}{\sqrt{2m\hbar\omega}}(p - im\omega q), \tag{2.11}$$

an expression that we can recognize as the annihilation operator for excitations of an ordinary harmonic oscillator of frequency ω. It will be convenient then to use the quantum states of just such a static-field oscillator as a basis for describing our time-dependent system.

The operators p and q, which in Eq. (2.11) stand for the initial values of $p(t)$ and $q(t)$, can also be taken to be the corresponding time-independent operators characteristic of the Schrödinger picture. If we define the annihilation operator

$$a = \frac{i}{\sqrt{2m\hbar\omega}}(p - im\omega q) \tag{2.12}$$

for the excitations of a static-field oscillator of frequency ω, its ground state $|n = 0 >_\omega$ must obey the relation

$$a|n = 0 >_\omega = 0. \tag{2.13}$$

The constancy of $C(t)$, expressed by Eq. (2.11), then assures us that

$$C(t)|n = 0 >_\omega = 0 \tag{2.14}$$

at all times. This fact furnishes us with a simple means of determining the state at any time for our time-dependent system. The operator $C(t)$ is defined in Eq. (2.10) within the Heisenberg picture; that is, it is a linear combination of the Heisenberg operators $p(t)$ and $q(t)$. It is the time dependence of the linear combination coefficients that compensates for the time dependence of $p(t)$ and $q(t)$ and reduces $C(t)$ to constancy. The unitary transformation from the Schrödinger to the Heisenberg picture takes the characteristic form

$$q(t) = U^{-1}(t)qU(t), \tag{2.15}$$

in which the operator U obeys the time-dependent Schrödinger equation and has the initial value $U(0) = 1$. We shall not, however, need to find $U(t)$ at all. It will suffice simply to observe that $C(t)$ is related to its Schrödinger-picture counterpart $C_s(t)$ by the relation

$$C(t) = U^{-1}(t)C_s(t)U(t), \tag{2.16}$$

and that $C_s(t)$ is given by

$$C_s(t) = \frac{i}{\sqrt{2m\hbar\omega}}\left\{u(t)p - m\dot{u}(t)q\right\}, \tag{2.17}$$

which is a time-dependent combination of time-independent operators.

3. SOLUTION FOR THE WAVE FUNCTIONS

To find the time-dependent states we note that we can rewrite Eq. (2.13) or (2.14) in the form

$$C_s(t)U(t)|n = 0 >_\omega = 0, \tag{3.1}$$

that follows from Eq. (2.16). The unitarily transformed state vector $U(t)|n = 0 >_\omega$ is the Schrödinger state of the time-dependent system that evolves from the initial $(t = 0)$ static-field oscillator state $|n = 0 >_\omega$. If we write it as

$$|n = 0, t > \equiv U(t)|n = 0 >_\omega, \tag{3.2}$$

then we see that it obeys the equation

$$C_s(t)|n = 0, t >= 0, \tag{3.3}$$

or

$$\left\{u(t)p - m\dot{u}(t)q\right\}|n = 0, t >= 0. \tag{3.4}$$

Expressed in coordinate space, this is the first-order differential equation

$$\left\{u(t)\frac{\hbar}{i}\frac{\partial}{\partial q'} - m\dot{u}(t)q'\right\} < q'|n = 0, t >= 0, \tag{3.5}$$

which has the normalized solution

$$< q'|n = 0, t >= \left(\frac{m\omega}{\pi\hbar}\right)^{\frac{1}{4}} \frac{1}{\{u(t)\}^{\frac{1}{2}}} exp\left\{\frac{im}{2\hbar}\frac{\dot{u}(t)}{u(t)}q'^2\right\}. \tag{3.6}$$

This is at all times a Gaussian wave function similar in several ways to the ground state wave function for the static-field oscillator, and it reduces to the latter wave function for the constant field specified by $W(t) = \omega^2$ and $u(t) = exp(i\omega t)$. In the more general case it differs from the static wave function only by a time-dependent complex scale factor.

It is worth emphasizing that, because we have not needed to specify the time dependence of the field, the wave function (3.6) applies as well to the motion of unbound particles as it does to ones that are trapped by the time-varying field. The rapid decrease of the wave function as $q'^2 \to \infty$, and its square integrability are, in fact, not related to the question of trapping. They only require that $Im\{\dot{u}(t)/u(t)\}$ be non-negative, and that condition is assured, for arbitrary time dependence of the field, by writing the Wronskian condition (2.9) in the form

$$Im\left\{\frac{\dot{u}(t)}{u(t)}\right\} = \frac{\omega}{|u(t)|^2} \geq 0. \tag{3.7}$$

The density function that characterizes this state simplifies, when use is made of the Wronskian relation (2.9), to the form

$$| < q'|n = 0, t > |^2 = \left\{\frac{m\omega}{\pi\hbar|u(t)|^2}\right\}^{\frac{1}{2}} exp\left\{-\frac{m\omega}{\hbar}\frac{q'^2}{|u(t)|^2}\right\}. \tag{3.8}$$

It shows that the expectation value of $q^2(t)$ is given by

$$< q^2(t) >_{n=0} = \frac{\hbar}{2m\omega}|u(t)|^2. \tag{3.9}$$

We could, alternatively, have started our time-dependent system out in any of the n-quantum states of the static-field oscillator

$$|n>_\omega = \frac{a^{\dagger n}}{\sqrt{n!}}|0>_\omega . \tag{3.10}$$

The time-dependent Schrödinger state vector would then be

$$|n,t>=U(t)|n>_{\omega}=\frac{[C_s^{\dagger}(t)]^n}{\sqrt{n!}}|0,t>. \tag{3.11}$$

When this relation is recast in the coordinate representation, it generates the Hermite polynomials much as the corresponding relation (3.10) does for the static-field oscillator. In this way we find the entire sequence of normalized coordinate-space wave functions

$$<q'|nt>=\frac{1}{\sqrt{n!}}\left(\frac{m\omega}{\pi\hbar}\right)^{\frac{1}{4}}\frac{1}{[u(t)]^{\frac{1}{2}}}\left\{\frac{u^*(t)}{2u(t)}\right\}^{\frac{n}{2}}H_n\left(\left\{\frac{m\omega}{\hbar|u(t)|^2}\right\}^{\frac{1}{2}}q'\right)exp\left\{\frac{im}{2\hbar}\frac{\dot{u}(t)}{u(t)}q'^2\right\}. \tag{3.12}$$

Wave functions that fall into this form for the case in which the function $W(t)$ has a cosine dependence on the time have been derived by M. Combescure [4]. The present approach leaves the time-dependent spring constant $W(t)$ unrestricted.

It is equally easy to return to Eq. (3.4) and use it, by similar steps, to solve for the entire sequence of wave functions in momentum space. They are found to be

$$<p'|nt>=\frac{1}{\sqrt{n!}}\left(\frac{\omega}{\pi m\hbar}\right)^{\frac{1}{4}}\frac{1}{[\dot{u}(t)]^{\frac{1}{2}}}\left\{\frac{\dot{u}^*(t)}{2\dot{u}(t)}\right\}^{\frac{n}{2}}H_n\left(\left\{\frac{\omega}{m\hbar|\dot{u}(t)|^2}\right\}^{\frac{1}{2}}p'\right)exp\left\{\frac{-i}{2m\hbar}\frac{u(t)}{\dot{u}(t)}p'^2\right\}. \tag{3.13}$$

In particular the momentum space density for the $n=0$ state is

$$|<p'|0t>|^2=\left\{\frac{\omega}{\pi m\hbar|\dot{u}(t)|^2}\right\}^{\frac{1}{2}}exp\left\{-\frac{\omega}{m\hbar}\frac{p'^2}{|\dot{u}(t)|^2}\right\}. \tag{3.14}$$

and the expectation value of $p^2(t)$ is

$$<p^2(t)>_{n=0}=\frac{m\hbar}{2\omega}|\dot{u}(t)|^2. \tag{3.15}$$

4. THE COHERENT STATES

The n-quantum states for the ordinary harmonic oscillator of frequency ω form a complete set. We have found the time-dependent wave function that evolves from each of them when it is chosen as the initial state. The wave function that evolves from an arbitrary choice of the initial state may therefore be written as a linear combination of these wave functions.

A particularly interesting motion of the time-dependent system is one that follows from an initially coherent state. We take the Schrödinger state vector at time $t=0$, in this case, to be the coherent state $|\alpha>_{\omega}$, which satisfies the relation

$$C(t)|\alpha>_{\omega}=a|\alpha>_{\omega}=\alpha|\alpha>_{\omega} \tag{4.1}$$

for an arbitrary complex value of α. The time dependent state $|\alpha,t>$ that evolves from this initial state then obeys the relation

$$C_s(t)|\alpha, t> = \alpha|\alpha, t> . \qquad (4.2)$$

To find the corresponding configuration space wave function we can either solve this equation, as we solved the previous ones, or form the appropriate linear combination of the solutions given by Eq. (3.12). The result in either case may be written as the Gaussian wave function

$$< q'|\alpha, t> = (\frac{m\omega}{\pi\hbar})^{\frac{1}{4}} \frac{1}{[u(t)]^{\frac{1}{2}}} exp\{\frac{im}{2\hbar}\frac{\dot{u}(t)}{u(t)}q'^2 + \frac{\alpha q'}{u(t)}\sqrt{\frac{2m\omega}{\hbar}} - \frac{\alpha^2 u^*(t)}{2u(t)} - \frac{1}{2}|\alpha|^2\}. \quad (4.3)$$

It is a form that shares and considerably generalizes some of the remarkable properties of the coherent state wave function for an ordinary harmonic oscillator. The configuration space density function, for example, takes the form

$$| < q'|\alpha, t > |^2 = \{\frac{m\omega}{\pi\hbar|u(t)|^2}\}^{\frac{1}{2}} exp\{\frac{-m\omega}{\hbar|u(t)|^2}[q' - (\frac{2\hbar}{m\omega})^{\frac{1}{2}} Re\{\alpha u^*(t)\}]^2\}. \qquad (4.4)$$

It is a Gaussian function of the same form as the ground state density given by Eq. (3.8), but is centered on the mean coordinate

$$< q(t) >_\alpha = \sqrt{\frac{2\hbar}{m\omega}} Re\{\alpha u^*(t)\}. \qquad (4.5)$$

The width of the Gaussian density distribution, in other words, varies with time precisely as the ground state distribution does, while its center is carried along an essentially classical trajectory.

5. UNCERTAINTY PRODUCTS

The coherent states of an ordinary harmonic oscillator are well known to yield the minimal uncertainty product

$$< (\delta q)^2 >< (\delta p)^2 > = \frac{\hbar^2}{4}, \qquad (5.1)$$

where $\delta q = q - < q >$, and $\delta p = p - < p >$. For the generalized coherent states defined by Eqs. (4.2) and (4.3), on the other hand, the uncertainty product is the same as that in the $n = 0$ state. It is given, in other words, by the product of the expressions in Eqs. (3.9) and (3.15).

$$< (\delta q)^2 >< (\delta p)^2 > = \frac{\hbar^2}{4\omega}|u(t)|^2|\dot{u}(t)|^2. \qquad (5.2)$$

This expression indeed takes the value $\hbar^2/4$ at time $t = 0$, but at later times, as we shall see, it is in general larger.

There is nonetheless a well-defined sense in which the uncertainty product (5.2) is minimal. The combination of the Schwarz inequality and the canonical commutation rule provides that

$$< (\delta q)^2 >< (\delta p)^2 > \geq \frac{1}{4} < \delta q \delta p + \delta p \delta q >^2 + \frac{1}{4}\hbar^2. \tag{5.3}$$

The Hamiltonian of an ordinary harmonic oscillator possesses a kind of rotational symmetry in the variables p and $m\omega q$ that makes the expectation value of the anticommutator $\{\delta q, \delta p\}$ vanish in any stationary state. That symmetry is in general broken, however, by the time dependence of the function $W(t)$ in the Hamiltonian stated in Eq. (2.1). The expectation value of the anticommutator need not vanish, and in fact we find

$$< \{\delta q(t), \delta p(t)\} >^2 = \hbar^2 \{|u(t)|^2 |\frac{\dot{u}(t)}{\omega}|^2 - 1\}, \tag{5.4}$$

an expression that can never be negative, and so it follows that the uncertainty product, though minimal, must in general exceed $(1/4)\hbar^2$. In more physical terms the uncertainty relation (5.2) differs from that for an ordinary static-field oscillator by the inevitable presence of correlations between position and momentum in the generalized coherent states.

6. EFFECT OF EXTERNALLY APPLIED FIELDS

The simplest way to alter the state of a trapped particle is to apply an external forcing field. Let us assume that the particle described by the Hamiltonian H_o of Eq. (2.1) is subjected to an additional uniform time-dependent force field of magnitude $mF(t)$. Its Hamiltonian is then

$$H = H_o - mF(t)q, \tag{6.1}$$

and the coordinate $q(t)$ must satisfy the inhomogeneous differential equation

$$\ddot{q}(t) + W(t)q(t) = F(t). \tag{6.2}$$

The presence of the inhomogeneous term $F(t)$ means that the expression $C(t)$ that we constructed in Eq. (2.10) is no longer a constant of the motion. In fact we find, from the Heisenberg equations of motion, that

$$\frac{d}{dt}C(t) = i\sqrt{\frac{m}{2\hbar\omega}}u(t)F(t), \tag{6.3}$$

but that relation shows us that we can construct a new constant of the motion in the form

$$C'(t) = C(t) - i\sqrt{\frac{m}{2\hbar\omega}} \int_o^t u(t')F(t')dt'. \tag{6.4}$$

This new constant of the motion, when evaluated at the initial time again equals the annihilation operator a, and so must retain that value at all times,

$$C'(t) = a. \tag{6.5}$$

If we assume, for example, that the particle is initially in the state $|n = 0>_\omega$, then we find that its state $|t>$ at time t obeys the relation

$$C'_S(t)|t>= 0, \tag{6.6}$$

where C'_S is the Schrödinger-picture version of $C'(t)$. This relation can also be written in the form

$$C_S(t)|t>= \alpha(t)|t>, \tag{6.7}$$

with

$$\alpha(t) = i\sqrt{\frac{m}{2\hbar\omega}}\int_o^t u(t')F(t')dt'. \tag{6.8}$$

By comparing this equation with Eq. (4.2), that we solved earlier, we see that the new state generated by the forcing field is the coherent state

$$|t>= |\alpha(t), t> . \tag{6.9}$$

The effect of such a forcing field can always be represented by a unitary displacement transformation in the complex α-plane. It therefore always carries an initially coherent state, such as the $n = 0$ state, into another coherent state.

7. TRAPPING BY PERIODIC FIELDS

It is clear from the structure of the wave functions (3.6) and (3.12) and from their corresponding densities, such as that of Eq. (3.8), that the condition for trapping is the essentially classical requirement that $|u(t)|^2$ remain bounded. The most obvious example of trapping is the static-field oscillator for which W is a positive constant. If we then choose $\omega = \sqrt{W}$, we have $u(t) = e^{i\omega t}$ and $|u(t)|^2 = 1$. An equally obvious example of unstable behavior is one in which the potential is constant in time and repulsive, i.e., W is a negative constant. Then if we choose $\omega = \sqrt{-W}$, we find $|u(t)|^2 = 1 + 2sinh^2\omega t$, and the density distributions spread out explosively and without bound.

The quadrupole field in the Paul traps is made to vary periodically in time. If its period is T, then we have

$$W(t+T) = W(t), \tag{7.1}$$

and the differential equation (2.6) for the function $u(t)$ becomes the well-known equation of Hill, a second order equation with a periodic coefficient [5]. It is always possible to find two linearly independent solutions u_1 and u_2 to this equation with the property that increasing t by one period multiplies the solutions by two constants, say σ_1 and σ_2 respectively,

$$u_j(t+T) = \sigma_j u_j(t) \quad j = 1, 2. \tag{7.2}$$

If either of the constants σ_1 and σ_2 should have an absolute value greater than unity, the corresponding solution $u_j(t)$ will clearly increase in modulus without bound over many periods.

The two constants σ_1 and σ_2 are in fact linked together in value by the constancy of the Wronskian. If we evaluate that expression both at t and at $t+T$, we find

$$\sigma_1\sigma_2 = 1, \tag{7.3}$$

so if one solution increases in modulus with time, the other must decrease. Now the initial condition on the derivative of $u(t)$ contains the parameter ω, which we are still free to choose arbitrarily; its value only specifies a set of basis functions. If the particle we are describing is truly trapped, $|u(t)|^2$ must remain bounded for any and all real, positive values of ω. Since both solutions $u_1(t)$ and $u_2(t)$ must be combined in general to meet this initial condition, neither of them can be allowed to grow in modulus without bound. It follows then that we must have

$$|\sigma_1| = |\sigma_2| = 1. \tag{7.4}$$

If we define the characteristic exponent μ by the relations

$$\sigma_1 = e^{i\mu T}, \quad \sigma_2 = e^{-i\mu T}, \tag{7.5}$$

then we see that μ must be real if trapping is to occur, and that we may take it to be positive as well. We can furthermore define the functions

$$\varphi_1(t) = e^{-i\mu t}u_1(t) \tag{7.6}$$

$$\varphi_2(t) = e^{i\mu t}u_2(t) \tag{7.7}$$

which are both periodic with period T

$$\varphi_j(t+T) = \varphi_j(t) \quad j = 1,2. \tag{7.8}$$

The solutions $u_j(t)$ may therefore be expressed in the characteristic Floquet form,

$$u_1(t) = e^{i\mu t}\varphi_1(t), \tag{7.9}$$

$$u_2(t) = e^{-i\mu t}\varphi_2(t), \tag{7.10}$$

in which the exponent μ is real and the periodic functions φ_j can be taken to have the initial values

$$\varphi_j(0) = 1 \qquad j = 1,2. \tag{7.11}$$

Let us assume, for simplicity, that the function $W(t)$ is even-valued

$$W(-t) = W(t). \tag{7.12}$$

Then the solution $u_2(t)$ can be taken simply to be $u_1(-t)$. The most general solution of the differential equation (2.6) can thus be written in the form

$$u(t) = Au_1(t) + Bu_1(-t), \tag{7.13}$$

where the constants A and B are to be determined by the initial conditions

$$u(0) = 1 = A + B, \tag{7.14}$$

$$\dot{u}(0) = i\omega = (A - B)\dot{u}_1(0). \tag{7.15}$$

Whatever values these equations give to A and B, the value of $|u(t)|^2$ remains bounded for the solution (7.13).

It is worth remembering once more at this point that the parameter ω can be chosen arbitrarily and that there may be considerable simplification to be gained in the description of the motion by making an appropriate choice. If we give to ω, for example, the value

$$\omega = -i\dot{u}_1(0), \tag{7.16}$$

then we see that the solution of the equations (7.14) and (7.15) reduce to

$$A = 1, \qquad B = 0, \tag{7.17}$$

and the solution for $u(t)$ becomes just

$$u(t) = u_1(t) = e^{i\mu t}\varphi_1(t). \tag{7.18}$$

Our special choice of the basis states, in other words, has greatly simplified the way we see the motion.

What this isolated Floquet solution for $u(t)$ does is to separate the motion of the particle, in effect, into two components, very much as Kapitza suggested [3]. The exponential $e^{i\mu t}$ best describes the relatively slow part of the motion while the periodic function $\varphi_1(t)$ describes the rapid oscillations at the frequency of the driving field (and its harmonics) that make up what is often called the "micromotion." Indeed, when the Floquet solution (7.18) is substituted into the wave functions we have constructed, they all simplify considerably in their behavior. That is evident because the function $|u(t)|^2$ which figures in them becomes periodic, as does the logarithmic derivative

$$\frac{\dot{u}(t)}{u(t)} = i\mu + \frac{\dot{\varphi}_1(t)}{\varphi_1(t)}, \quad \text{etc.} \tag{7.19}$$

The density function (3.8) for the $n = 0$ state, for example, can be seen to pulsate in size about the normal value for an oscillator of the frequency specified by Eq. (7.16). The coherent state wave packets undergo the same periodic pulsations in size while they follow the classical trajectories of trapped particles. In this sense they are rather like the coherent states of ordinary oscillators. They undergo no secular spread or diffusion. Their new feature is the throbbing that represents "micromotion," in addition to their following a more intricate classical path.

To be a bit more specific about the correspondence of our results with the Kapitza approximation, let us take the function $W(t)$ to have the form

$$W(t) = \omega_1^2 cos\Omega t, \tag{7.20}$$

where Ω is the angular frequency of the driving field and ω_1^2 is a measure of its strength. Then, according to the Kapitza approximation, as long as we have $\Omega >> \omega_1$, the slow component of the motion oscillates with frequency $\omega_1^2/\Omega\sqrt{2}$ in an effectively static harmonic oscillator potential [3].

When the function $W(t)$ has the cosine time dependence indicated in Eq. (7.19), the solution $u_1(t)$ formulated in Eq. (7.9) becomes a Mathieu function [5]. For the case in which $\omega_1/\Omega << 1$, it is possible to expand both the characteristic exponent μ and the logarithmic derivative $\dot{u}_1(0)/u_1(0)$ in powers of the ratio ω_1/Ω [6]. We then find for the characteristic exponent the value

$$\mu = \frac{\omega_1^2}{\Omega\sqrt{2}}\{1 + \frac{25}{32}(\frac{\omega_1}{\Omega})^4 + ...\}, \tag{7.21}$$

while our parameter ω, determined by Eq. (7.16), is

$$\omega = \frac{\omega_1^2}{\Omega\sqrt{2}}\{1 + 2(\frac{\omega_1}{\Omega})^2 + ...\}. \tag{7.22}$$

In the lowest approximation then we have $\omega \approx \mu$, and both of these values agree with the Kapitza frequency. The analysis we have carried out clearly extends the Kapitza approximation to higher orders in ω_1/Ω.

The two frequencies ω and μ may more generally be quite different in numerical value, so it is interesting to ask about their separate physical meanings. The Fourier analysis of the motion specified by Eq. (7.18) contains components of frequencies $\mu + j\Omega$ for all integer values of j, positive and negative. Our particle, when coupled to the radiation field, should then emit spontaneously photons of all frequencies $|\mu + j\Omega|$, and it is easy to show that it does that, for example, when it begins in the quantum states given by Eqs. (3.11) or (3.12). It does not, on the other hand, emit any photons of frequency ω. That frequency which we have chosen as the best for a set of oscillator basis states, emerges in several senses we shall discuss in future work as a mean frequency of oscillation for the complex motion described by $u(t)$.

I am greatly indebted to Professor W. Paul for demonstrating a model of his trap for me, to Jene Golovchenko for asking many questions about trapping, and to Vladimir Man'ko for telling me many new things about harmonic oscillators. This work was supported in part by a grant from the Department of Energy.

REFERENCES

1. For a review see: W. Paul, *Rev. Mod. Phys.* **62**, 531 (1990).
2. V. V. Dodonov and V. I. Man'ko, in *Invariants and the Evolution of Nonstationary Quantum Systems*, ed. M. A. Markov (Nova Science Publishers, Commack, N.Y., 1989), p. 103.

3. Summarized in L. D. Landau and E. M. Lifschitz, *Mechanics*, (Pergamon Press, Oxford, 1969).

4. M. Combescure, *Ann. Inst. Henri Poincaré* **44**, 293 (1986).

5. E. T. Whittaker and G. N. Watson, *Modern Analysis* (Macmillan, New York 1943).

6. J. Meixner and F. W. Schäfke, *Mathieusche Funktionen und Sphäroidfunktionen* (Springer Verlag, Berlin 1954) p. 119.

LOCALIZATION OF PHOTONS IN RANDOM AND QUASIPERIODIC MEDIA

S. Dutta Gupta

School of Physics
University of Hyderabad
Hyderabad-500 134, India

1. INTRODUCTION

Localization[1] of photons in random and quasiperiodic structures has been the subject of intense theoretical and experimental investigations for the past few years. Photon localization studies were motivated by the discovery of Anderson localization[2] in disordered solid state systems. A profound alteration in the electronic wave function in a random potential was first noticed by Anderson if randomness is sufficiently strong. The traditional view on scattering from random potentials was as follows: One would think that there would be loss of phase coherence of the Bloch waves in the length scale of the mean free path. Nevertheless, the wavefunction would remain extended. Anderson discovered that for strong disorder the envelope of the wave function shows an exponential decay from some site i.e., ψ(r) behaves as

$$|\psi(\gamma)| \sim exp(-|\vec{r} - \vec{r}_o|/\xi) \tag{1.1}$$

where, ξ is the localization length. The model considered by Anderson is essentially a tight binding model where a single band is formed from s-like atomic orbitals with energies $\varepsilon_{\vec{\ell}}$ corresponding to site $\vec{\ell}$. The bandwidth B is given by B=2VZ, where V is the overlap energy integral and Z is the coordination number. $\varepsilon_{\vec{\ell}}$'s were assumed to be random with a common distribution with width Γ. It was shown that there exists a critical value of Γ, namely $\Gamma_c \sim$ B lnz which determines the nature of the wave function. For example, for $\Gamma > \Gamma_c$, states at the middle of the band (by inference all states) are localized, whereas, for $\Gamma < \Gamma_c$ states at the middle are extended. Instead of critical randomness parameter Γ_c, Mott[3] introduced the concept of critical energy E_c (mobility edge) separating localized and extended states. In the language of Mott the Anderson transition (i.e. the disappearance of regions of extended states as Γ crosses Γ_c) was interpreted as the disappearance of regions of extended states caused by coincidence of two adjacent E_c's. Somewhat later, after the discovery of Anderson, the scaling theory of localization[4] was developed. It was shown that in one dimension, all the states are localized even for weak randomness. In two dimensions all the states are localized, but there is a change in the character as a scaling parameter is changed from low to high values. In three dimensions, for sufficient disorder the existence of the mobility edge was shown. It is understood that all the above cited results

Recent Developments in Quantum Optics, Edited
by R. Inguva, Plenum Press, New York, 1993

hold for infinite systems at zero temperature. Finite size and finite temperature corrections have to be incorporated for system size $L \neq \infty$ and temperature $T \neq 0$. A very important consequence of the scaling theory was the predicted deviation from classical Ohmic rule. Resistance is no more linear in system size, rather, it grows exponentially. The deviation from classical Ohmic rule was demonstrated in simple and clear terms earlier by Landauer.[5] Landauer considered electrons incident on an array of obstacles. The following assumptions were made regarding the nature of the obstacles: a) they are non overlapping, b) distances between the obstacles are independent, c) phase shift of the free electron wave between adjacent obstacles are varied uniformly over integral multiples of 2π. It was shown that the total resistance Ω of the one dimensional array is proportional to the ensemble average $< R/(1-R) >$ where R is the reflection probability of the array. It was also shown that

$$< R/(1-R) >= (1/2)\Big[exp\Big\{n\ ln\Big(\frac{1-r}{1+r}\Big)\Big\} - 1\Big] \tag{1.2}$$

where, r is the individual obstacle reflection probability. It is clear from Eq. (1.2) that the quantity $< R/(1-R) >$ (later known as Landauer resistivity) is not linear in the system size n, rather it grows exponentially. If instead we take a sequence of classical obstacles which just intercepts a portion r of the incident beam then $< R/(1-R) >= nr/(1-r)$ implying a linear growth with the system size. Thus wave character of the incident beam is crucial for non Ohmic behavior of the resistance which is a manifestation of the localized states. In fact, any wave equation in a random medium can lead to localized solutions. In this context, optics can play a significant role to bring out the essential features of the physics of localization. The positive and negative aspects of optics compared to solid state physics will emerge from a simple comparison of the basic units, photons and electrons. In solid state physics, electron-electron and electron-phonon interactions are inevitable. This has been a real threat for direct tests for the scaling theory predictions. On the other hand, optical experiments are more pure since photons are noninteracting. Moreover, the large velocity of light renders any random medium practically static. In addition, the polarization of the photons adds a new feature to the localization problem. The fact that the polarization of light can play a crucial role was recently understood. The scalar theory for 3d periodic structures predicts the existence of gaps, whereas vector theory leads to pseudogaps characterized by nonzero density of states.[6] Finally, in optics one can easily incorporate nonlinear effects via the field dependence of the medium parameters. This opens up new questions whether nonlinearity inhibits localization or assists it. However, there are practical difficulties in setting up experiments to see strong localization. For strong localization the following inequality should hold[7]

$$kl^* \leq 1, \tag{1.3}$$

where k is the wave vector and l^* is the momentum exchange mean free path. For optical systems typically $kl^* \geq 100$. It remains an experimental challenge to prepare a random medium with $kl^* \sim 1$. Since it is difficult to observe Anderson localization, one can investigate the precursor to Anderson localization, often referred to as weak localization. Weak localization manifests itself in enhanced back scattering from the random sample. A qualitative picture of weak localization[8,9] and a few experimental[10–12] results will be discussed.

It was pointed out that in one dimensional random systems all the states are localized. Thus such systems cannot show critical (i.e. power law bounded wave

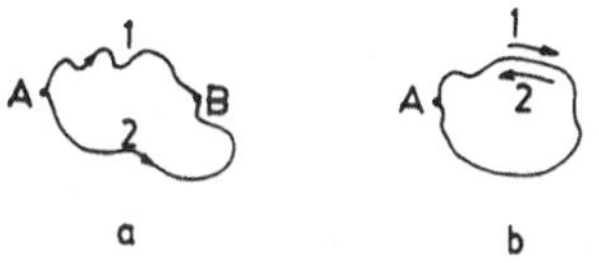

Figure 1. Transport from point A to point B along ray tubes 1 and 2 when a) A and B differ, and b) when A and B coincide.

function) behavior. Quasiperiodic structures with two incommensurate periods are intermediate between periodic and random media. Recently, there has been a lot of interest in such structures[13–15] because of the very rich properties of spectra for such systems. Some of the important results on one dimensional quasiperiodic Fibonacci system in optical realization[16] will be cited. Finally, the effect of intensity dependent refractive index on the transmission properties as well as on the states of the Fibonacci layered medium will be considered.[17–19] It will be shown that nonlinearity leads to multistable output[17,18] and inhibits localization.[19] Moreover, the nonlinear structure will be shown to lead to new bulk localized states[18,19] which do not have any linear analogue. These bulk localized states will be shown to be self similar.

The organization of the paper is as follows: In Section 2, a qualitative explanation of weak localization resulting in enhanced back scattering will be given, some experiments on weak photon localization will be discussed. In Section 3, properties of linear and nonlinear Fibonacci stack of dielectric layers will be described.

2. WEAK LOCALIZATION: A PRECURSOR TO ANDERSON LOCALIZATION

A simple physical picture of the back scattering phenomenon associated with weak localization was given by Khmelnitsky.[8] Under the assumptions that a) the electron wavelength $\lambda << 1$ (mean free path) and b) temperature T is so small that times of all inelastic processes are much longer than the elastic collision time, Khmelnitsky considered the transport from point A to point B. This may be conceived as propagation of wave packets along ray tubes 1 and 2 (see Fig. 1). The tube width is proportional to λ and its form reflects a diffusional character of electron motion in a medium with impurities. Let A_i be the probability amplitude corresponding to the ith tube, and let W be the probability of reaching B from A. W can be expressed in terms of the probability amplitudes A_i as follows:

$$W = |\Sigma_i A_i|^2 = \sum_i |A_i|^2 + \sum_{i \neq j} A_i A_j^* \tag{2.1}$$

The first term on the right hand side of (2.1) is the sum of probabilities corresponding to separate trajectories, whereas the second term represents the interference effects. If interference is neglected, one would recover the standard Boltzmann description. In most cases interference is not important since the trajectories i and j differ significantly in length. As a result the interference term represents a sum of quickly oscillating terms and turns out to be negligible. The only exception is the case when A and B coincide, i.e., when we compute the probability of return. Then the two amplitudes A_1 and A_2 are coherent and their interference is important. Thus, the wave effects allowed for, the probability of return becomes larger than that given by the classical probability of return. Thus a quantum particle may be less mobile in a random potential than given by the kinetic equation.

A more definitive description of the enhanced back scattering phenomenon was given by Bergman,[9] who considered an electron with momentum $\vec{k}$ having the wave function $exp(ikr)$ undergoing elastic scattering. The electron in state $\vec{k}$ is scattered after time τ (elastic scattering time) into a state $\vec{k}'_1$, after 2τ into state $\vec{k}'_2$ etc. There is a finite probability, that the electron will be scattered into the vicinity of the state $-\vec{k}$, for example, after n scattering events

$$\vec{k} \rightarrow \vec{k}'_1 \rightarrow \vec{k}'_2 \rightarrow \ldots\ldots \rightarrow \vec{k}'_n = -\vec{k}$$

Let the associated momentum transfer be given by $\vec{g}_1, \vec{g}_2, \ldots .\vec{g}_n$. There is an equal probability for the following process

$$\vec{k} \rightarrow \vec{k}''_1 \rightarrow \vec{k}''_2 \rightarrow \ldots\ldots \rightarrow \vec{k}''_n = -\vec{k}$$

with momentum transfers $\vec{g}_n$, $\vec{g}_{n-1}$, $\vec{g}_{n-2}, \ldots .\vec{g}_1$. This complimentary scattering series has the same change of momentum in the opposite sequence and the amplitude in the final state $-\vec{k}$ for both the scattering processes are the same. Since the final amplitudes A' and A'' are phase coherent and equal $A' = A'' = A$, the total intensity is given by $|A' + A''|^2 = 4|A|^2$. If the two amplitudes were not coherent then the total scattering intensity would be $2|A|^2$. Thus the scattering intensity into the state $-\vec{k}$ is by a factor two larger compared to the case of incoherent scattering. In fact, there is an increase in the scattering intensity inside a cone of angular width of the order λ/ℓ (ℓ is the elastic mean free path) centered in the back scattering direction.

Two similar experiments[10,11] on coherent back scattering were reported almost simultaneously. The essential features and the results are basically the same. In one of them[10] the multiple scattering of visible light from an Argon ion laser by aqueous suspensions of submicron size monodisperse polystyrene spheres was studied. Beads of diameters d=0.109, 0.35, 0.46, and 0.8 μm, covering the whole range from pure Rayleigh ($d << \lambda$) to Rayleigh-Gans ($d > \lambda$) were used. Starting from a volume fraction of 10% the concentration of the beads was stepwise reduced. Two polarizers with parallel and crossed orientations were used at the input and output faces to study the polarization characteristics of the scattered light. A sharp peak centered at the back scattering direction was observed. The effect of the variation of the density n on the angular width of the back scattered peak was also studied. It was found that the width depends linearly on density confirming the theoretical predictions. Note that for negligible interparticle interaction n is proportional to λ/ℓ and thus angular width W $\sim \lambda/\ell$. The studies on the polarization aspects of the back scattered light revealed that for parallel orientation of the polarizers, the back scattered intensity is larger compared to the case when the polarizers are crossed. This difference was explained in terms of the angular dependence of the light scattered from each individual sphere.

Time domain experiments[12] involving pulse propagation in random media were suggested for the search of photon localization. It was pointed out that the back scattering method does not probe the medium beyond a few mean free path lengths. It was claimed that the time domain method a) permits probing from a few to a few million mean free path lengths b) can serve as a quantitative means for separating elastic scattering and absorptive processes and c) has the abililty to distinguish between those systems which exhibit simple diffusive behavior and those which do not. Explicit results on the measurements in the time domain on strong multiple scattering events leading to diffusive transfer of photons were reported. The possibility of meeting the Ioffe-Regel condition for strong localization $kl^* \leq 1$ was also discussed.

3. PHOTON LOCALIZATION IN ONE DIMENSIONAL QUASIPERIODIC STRUCTURES

The literature on localization in one or higher dimensional quasiperiodic structures is rather extensive.[13] The interest in quasiperiodic structures was further stimulated by the work of Levine and Steinhardt who proposed the concept of quasicrystals with a different kind of symmetry. These theoretical studies received an impetus from the experimental discovery of the quasicrystal phase in metallic alloys,[20] which was later followed by another realization of quasiperiodic superlattice structure by Merlin[15] et al. The optical realization of a one dimensional quasiperiodic superlattice involving a stack of dielectric layers arranged in a Fibonacci way was proposed and studied theoretically by Kohmoto et al.[16] A Fibonacci multilayer is constructed recursively by slabs A and B (with linear refractive indices n_a and n_b and widths d_a and d_b respectively) as $S_{j+1} = S_{j-1}S_j$ with $S_o = (\mathrm{B})$ and $S_1 = (A)$. Thus $S_2 = (BA)$, $S_3 = (ABA)$, $S_4 = (BAABA)$, etc. In this section we discuss in detail the transmission properties of such structures and the localization features when the slabs are made of linear/nonlinear dielectric.

To begin with, we summarize the results pertaining to a linear Fibonacci multilayer[16] where the constituent slabs were assumed to have the same optical thickness δ. It was shown that the system can be described by a (2x2) matrix (belonging to SL(2,R)) map, which can be subsequently reduced to a trace map. The most interesting consequence of the analysis was the self similarity (see Fig. 2) and multifractal character of the transmission coefficient as a function of δ for various generations. Moreover, it was shown that, the allowed states form a Cantor set with Lebesgue measure zero. Later a detailed treatment of the finite Fibonacci structure elaborating the findings of Kohmoto et al was presented.[21] Along with transmission, the Landauer resistivity $\bar{R} = R/T$ (reflection/transmission coefficients) was studied. It is clear that $\bar{R}$ becomes exponential or power-law bounded as a function of material length for localized and critical states, respectively. For an extended state the transmission is given by a constant or a bounded function of material length. The nature of the field distribution for various values of δ was also investigated. The major findings can be summarized as follows: In the limit when the number of layers becomes large, all the observable states are exponentially bounded surface states. However, for a smaller number of layers some states appear to be critical but cross over to the exponentially localized as the number of layers increases. The exponential (critical) character of the states was tested by looking directly at the field distribution or by inspecting the linearity of the dependence of $\log(\bar{R})$ on the number of layers j (on $\log(j)$).

Very recently, the theory for a nonlinear Fibonacci multilayer[17–19] was developed using the nonlinear characteristic matrix formalism[22] and fully incorporating the nonlinearity of the boundary conditions.[23] It was assumed that the pair of slabs, possess a nonlinearity given by the displacement

$$\vec{D}_{\mathrm{NL}} = \varepsilon\chi\left[A\vec{E}(\vec{E}.\vec{E}^*) + B\vec{E}^*(\vec{E}.\vec{E})\right], \tag{3.1}$$

where ε and χ are the linear dielectric constant and the constant of nonlinear interaction, respectively, $\vec{E}$ is the electric field vector and the constants A and B define the strength and type of nonlinear interaction. It was shown that the nonlinearity can lead to a very rich multivalued character (see Fig. 3) so far as the power dependence of the transmission is considered.[17] In other words, high intensities can render a system transparent which is opaque in the zero intensity (linear) limit. States, corresponding to the total transmission of the nonlinear structure (for example point

P1 in Fig. 3) were investigated and found to be bulk localized described by a 'sech' distribution.[18] This is shown in Fig. 4a where we have plotted the sum of the forward and backward wave intensities, corresponding to point P1 in Fig. 3 as a function of layer number j. The distribution corresponding to points P2 and P3 in Fig. 3 (not shown) are characterized respectively, by two, three humps instead of one for point P1. These states for various generations were shown to be self similar. A detailed

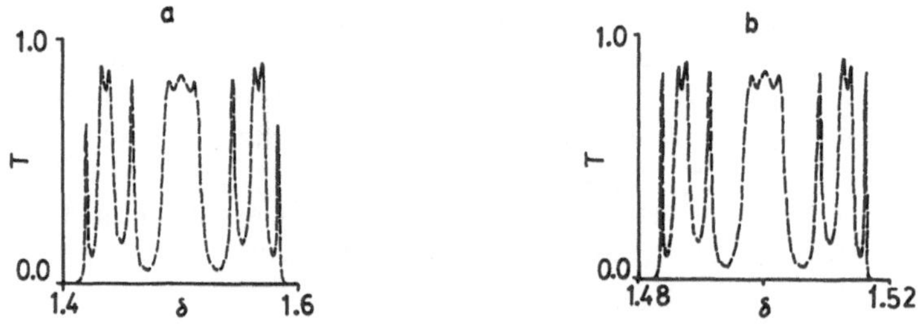

Figure 2. Self similarity of the transmission coefficient as a function of optical path $\delta = (\omega/c)n_a d_a = (\omega/c)n_b d_b$ for Fibonacci multilayers a) S_9 (55 layers) and b) S_{12} (233 layers). Note the difference in the horizontal scale. The parameters are as follows: $n_a = 2$, $n_b = 3$. The layered structure is embedded in a medium with refractive index n_a.

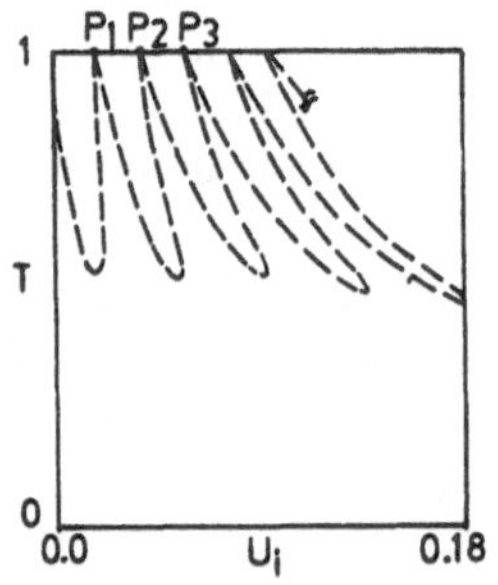

Figure 3. Transmission coefficient T as a function of incident power U_i for S_9 (55 layers) and $\delta = 2\pi$. Note the multivalued character of the transmission. Points P1, P2 and P3 etc. correspond to total transmission states of the nonlinear structure.

study of the effects of nonlinearity on localized, extended and critical states was later presented.[19] It was shown that strong surface localized states which is observed for the forbidden states ($T \sim O$) in the linear theory gets more and more extended till it occupies the whole structure as the power is increased. Thus, nonlinearity leads to delocalization. However, the extended states corresponding to the allowed regions ($T \sim 1$) in the linear theory remain almost unaffected by the nonlinearity (except for a scaling of the amplitude). The evidence of critical-like states in the nonlinear structure was also shown. The most important result was the evidence of bulk localization which persisted even if one increases the number of layers. This is in sharp contrast to linear theory and is an indication of the delicate interplay between nonlinearity and dispersion. Such bulk localized states were also reported in nonlinear periodic structures.[24]

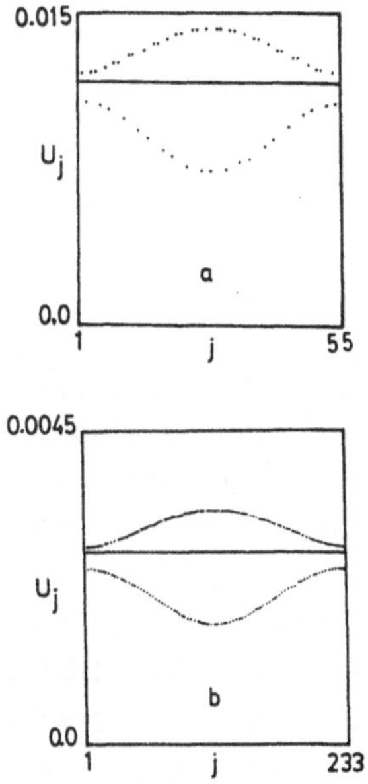

Figure 4. Self similarity of the bulk localized states for (a) S_9 and (b) S_{12}. Both can be described by sech distribution. The upper (lower) set of points gives the sum intensity in layers of type A (B).

4. CONCLUSIONS

Thus we have presented a very brief survey of the achievements of solid state physics in the context of localization problem and pointed out the role optics can play to explore the essential features of this interesting phenomenon. We have discussed some very first experiments on weak localization. We have discussed another important candidate for localization i.e., Fibonacci superlattices, and presented results pertaining to linear and nonlinear structures.

REFERENCES

1. For a review of localization in Solid State Physics see, for example P. Lee and T. V. Ramakrishnan, *Rev. Mod. Phys.* **57** 287 (1985); For a more recent review, see R. Merlin, *IEEE. J. Quantum Electron.* **24**, 1791 (1988); S. A. Gradeskur and V. D. Freilikher, *Sov. Phys. Usp.* **33**(2), 134 (1990).
2. P. W. Anderson, *Phys. Rev.* **109**, 1492 (1958).
3. N. F. Mott, *Adv. Phys.* **16**, 49 (1967), and *Philos. Mag.* **17**, 1259 (1968).
4. E. Abrahams, P. W. Anderson, D. C. Licciardello and T. V. Ramakrishnan, *Phys. Rev. Lett.* **42**, 673 (1979); P. W. Anderson, D. J. Thouless, E. Abrahams and D. S. Fisher, *Phys. Rev.* **B22**, 3519 (1980).

5. R. Landauer, *Philos. Mag.* **21**, 863 (1970).
6. K. M. Leung and Y. F. Lin, *Phys. Rev. Lett.* **65**, 2646 (1990); Z. Zhang and S. Satpathy, *Phys. Rev. Lett.* **65**, 2650 (1990).
7. A. F. Ioffe and A. R. Regel, *Prog. Semicond.* **4**, 237 (1960); In the context of Optics see, S. John, *Phys. Rev. Lett.* **58**, 2486 (1987).
8. D. E. Khmelnitsky, *Physica* **126B**, 235 (1984).
9. G. Bergmann, *Physica* **126B**, 229 (1984).
10. P. W. Wolf and G. Marset, *Phys. Rev. Lett.* **55**, 2696 (1985).
11. M. P. Van Albada and Ad. Lagendijk, *Phys. Rev. Lett.* **55**, 2692 (1985).
12. G. H. Watson, P. A. Fleury and S. L. McHall, *Phys. Rev. Lett.* **58**, 945 (1987).
13. See, for example, R. Merlin in Ref. 1; M. Kohmoto, B. Sutherland and C. Tang, *Phys. Rev.* **B35**, 1020 (1987); M. Kohmoto and J. R. Banavar, *Phys. Rev.* **B34**, 563 (1986); M. Kohmoto, L. P. Kadanoff and C. Tang, *Phys. Rev. Lett.* **50**, 1870 (1983); P. Hawrylak and J. J. Quinn *Phys. Rev. Lett.* **57**, 380 (1986).
14. D. Levine and P. J. Steinhardt, *Phys. Rev. Lett.* **53**, 2477 (1984).
15. R. Merlin, K. Bajima, R. Clarke, F. T. Juang and P. K. Bhattacharya, *Phys. Rev. Lett.* **55**, 1768 (1986).
16. M. Kohmoto, B. Sutherland and K. Iguchi, *Phys. Rev. Lett.*
17. S. Dutta Gupta and D. S. Ray, *Phys. Rev.* **B38**, 3628 (1988).
18. S. Dutta Gupta and D. S. Ray, *Phys. Rev.* **B40**, 10604 (1989).
19. S. Dutta Gupta and D. S. Ray, *Phys. Rev.* **B41**, 8047 (1990).
20. D. Schechtman, I. Black, D. Gratias and J. W. Cahn, *Phys. Rev. Lett.* **53**, 1951 (1984).
21. E. Sendler and D. G. Steel, *J. Opt. Soc. Am.* **B5**, 1636 (1988).
22. S. Dutta Gupta and G. S. Agarwal, *J. Opt. Soc. Am.* **B4**, 691 (1987).
23. G. S. Agarwal and S. Dutta Gupta, *Opt. Lett.* **12**, 829 (1987).
24. S. Dutta Gupta, *J. Opt. Soc. Am.* **B6**, 1927 (1989); W. Chen and D. L. Mills, *Phys. Rev. Lett.* **58**, 160 (1987).

ENHANCED FUNDAMENTAL LINEWIDTH OF A LASER DUE TO OUTCOUPLING

W.A. Hamel*, M.P. van Exter and J.P. Woerdman

Huygens Laboratory, University of Leiden
P.O. Box 9504, 2300 RA Leiden
The Netherlands

INTRODUCTION

A laser can be thought of as an amplitude-stabilized oscillator. Such an oscillator has a finite linewidth, due to phase-changing events. In most cases these events have a "technical" origin, such as fluctuations in the cavity length due to acoustic perturbations. However, even in a perfectly stable environment there is still phase diffusion due to spontaneous emission; this leads to the quantum-limited or fundamental linewidth as first discussed by Schawlow and Townes [1]. In recent years it has been shown by others [2,3,4,5] and by us [6,7,8,9] that the standard (Schawlow-Townes) formula for the fundamental linewidth must be modified if the outcoupling through the mirrors is large. In this paper we review our work [6,7,8,9] in this field, starting in section 2 with theory. In section 3 we report on the diagnostics of the semiconductor lasers used in the experiments. The technique of linewidth measurement is discussed in section 4 and actual results are given in section 5, together with a comparison with theory.

THEORY

We first discuss diffusion of the phase of the laser field due to spontaneous emission, described by a diffusion coefficient D. The latter is related to the laser linewidth $\delta\nu$ (FWHM) through $\delta\nu = D/2\pi$. Consider a cavity with two mirrors with intensity reflectivities R_1 and R_2. The light in the cavity is split into the part moving to the left and the part moving to the right (see Fig. 1). Large outcoupling clearly leads to nonuniformity of the traveling-wave intensities. At each point in the cavity the time evolution of either the leftward or rightward component of the field can be determined. We assume that the laser is single-frequency. Suppose that we could travel along with the light on its roundtrip in the cavity and observe how the optical field evolves from one position to the other. Spontaneous emission "in the proper direction and frequency range" will disturb the phase of the traveling wave, leading to a local phase diffusion proportional to the rate of

*current address: PTT Telecom b.v., P.O. Box 30150, 2500 GA's Gravenhage, the Netherlands

Recent Developments in Quantum Optics, Edited
by R. Inguva, Plenum Press, New York, 1993

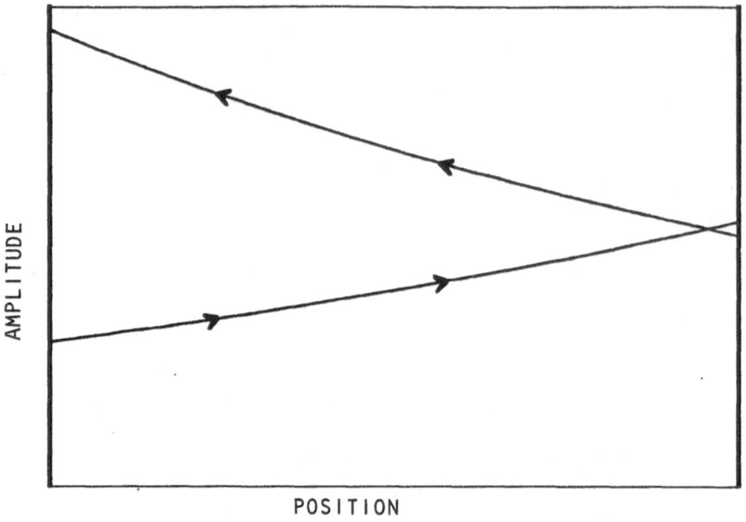

Figure 1. The optical field inside the cavity is split into the component moving to the right, $I^+(z)$, and the component moving to the left, $I^-(z)$. In this example the intensity reflection coefficients of the mirrors have been chosen as $R_1 = 10\%$ and $R_2 = 90\%$. (From ref. 9, with permission)

spontaneous emission and inversely proportional to the local intensity of the traveling wave. If one assumes that the optical phase remains unchanged on reflection, the phase diffusion averaged over one roundtrip is found to be

$$D = C_1 \frac{1}{2L} \int_0^L dz N_2(z) \{ \frac{1}{I+(z)} + \frac{1}{I-(z)} \} \tag{1}$$

where N_2(z) is the density of the excited-state population and I^+(z) and I^-(z) is the intensity of the light moving to the right and to the left, respectively. The constant C_1 is a measure for the spontaneous emission rate "in the proper direction and frequency range" and includes both the oscillator strength of the lasing transition and the geometry of the cavity. To eliminate this geometrical factor we require that Eq. (1) reduces to the standard expression $\delta\nu = \gamma_c/4\pi n$ in the limit of low outcoupling. Here γ_c is the cavity loss rate and n the number of photons in the laser mode. This number is given by

$$n = C_2 \frac{1}{2L} \int_0^L dz (I^+(z) + I^-(z)) \tag{2}$$

where the constant $C_2 = 2V/(v_g h\nu)$, with V the volume of the optical field and v_g the group velocity. Combining Eqs. (1) and (2) and comparing with the standard result for a four-level laser with a spatially uniform excited-state population (i.e. N_2 is constant) one finds

$$\delta\nu = \frac{\gamma_c}{4\pi n} K \tag{3}$$

$$\gamma_c = r_{sp} = 2C_1C_2N_2 \tag{4}$$

$$K = \left(\frac{1}{2L}\int_0^L dz\{\frac{1}{I^+(z)} + \frac{1}{I^-(z)}\}\right)\left(\frac{1}{2L}\int_0^L dz\{I^+(z) + I^-(z)\}\right) \quad (5)$$

where r_{sp} is the spontaneous emission rate and K the so-called excess-noise factor.

It is easily verified that $K = 1$ for a uniform intracavity intensity and that any non-uniformity leads to $K \geq 1$. In other words: at the start of a roundtrip the field amplitude is low and spontaneous emission will have a strong disturbing effect on the optical phase. During the roundtrip the field amplitude is amplified and the disturbing effect of spontaneous emission becomes less. Integrated over the length of the cavity the increase in phase diffusion in the low-field regions is larger than the decrease in the high-field regions and a net enhancement in the linewidth remains. This is similar to the fact that the average speed during a bicycle ride on hilly terrain is less than in flat country: the increased speed on the downhill slopes is not big enough to compensate for the slowdown experienced during the climbs.

In the absence of saturation and for a uniform excited-state population the excess-noise factor can be easily obtained by substituting in Eq. (5) exponential functions for $I^+(z)$ and $I^-(z)$ and by using the proper boundary conditions (i.e. mirror reflectivities)

$$K = \left[\frac{(\sqrt{R_1} + \sqrt{R_2})(1 - \sqrt{R_1 R_2})}{\sqrt{R_1 R_2} ln R_1 R_2}\right]^2 \quad (6)$$

Numerical simulation shows that saturation may indeed be neglected for the conditions of the experiments discussed in section 5. [6,10]

An intriguing consequence of Eq. (6) is that two lasers, having the same cavity roundtrip loss (depending on the product $R_1 R_2$), but different R_1 and R_2, have different fundamental linewidths solely due to a difference in intensity distribution inside the cavity. The laser with the most asymmetric mirror set will have the largest linewidth [7]. In section 5 we review experimental evidence for this effect [8].

An alternative theory to derive the fundamental limit of the laser linewidth in the case of large outcoupling makes use of the concept of non-orthogonal eigenmodes, where the non-orthogonality results from the non-Hermicity associated with the losses. This non-orthogonality leads to an excess spontaneous power in the laser mode and thus to an increase in linewidth by a factor [7,11]

$$K = \left|\frac{\int_0^L dz|E(z)|^2}{\int_0^L dz E(z)^2}\right|^2 \quad (7)$$

where $E(z)$ is the total optical field inside the cavity i.e. the sum of the leftward and rightward traveling fields.

In the absence of a mechanism that could break the inversion symmetry (such as a magnetic field), the gain for the leftward and rightward traveling light is the same, making the product $I^+(z)I^-(z)$ independent of position [12]. Under this general condition one can show that Eq. (7) is identical to Eq. (5) and thus a one-to-one relation is found between the theory of non-orthogonal eigenmodes and the model for traveling-wave phase diffusion presented above.

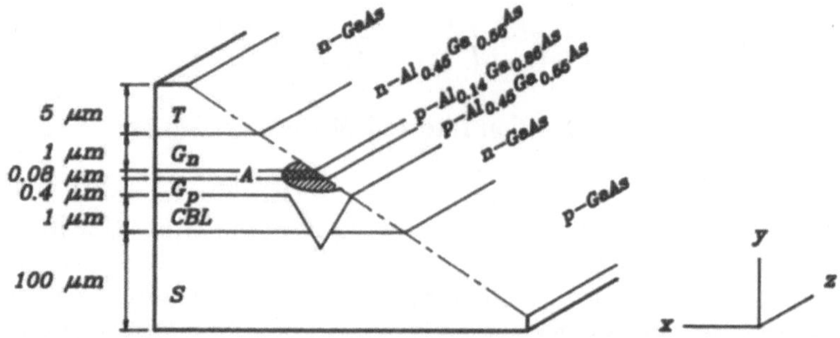

Figure 2. Schematic illustration of VSIS laser structure. The V-shaped geometry is typical for weakly index-guided lasers. The index-guiding is provided by the refractive-index difference between the current-blocking layer (CBL), consisting of GaAs, and the cladding layers (G_p and G_n), consisting of AlGaAs.

For semiconductor lasers the situation is more complicated than described by the above theories and the usual expression for the fundamental limit to the laser linewidth is given by [3,9]

$$\delta\nu = \delta\nu_0 + \frac{h\nu\gamma_c^2}{4\pi P_{out}}\frac{\gamma_{sat}}{\gamma_c}\eta_{opt}n_{sp}(1+\alpha^2)K \qquad (8)$$

where $\delta\nu_0$ is a power-independent contribution to the linewidth, η_{opt} is the optical power extraction efficiency, n_{sp} is the spontaneous emission factor [11] and α is the linewidth enhancement factor [13]. The factor γ_{sat}/γ_c is close to one and can usually be neglected. It arises from the fact that saturation leads to a redistribution of the intensity over the cavity and thus slightly affects the outcoupling efficiency. The additional factors appearing in Eq. (8) can be measured (see section 3), however with insufficient accuracy if one wants to deduce a value of K from a measurement of $\delta\nu$. Our strategy has therefore been to compare the linewidths of two semiconductor lasers grown on the same wafer; in this way the (hopefully identical) device-dependent factors η_{opt}, n_{sp} and α cancel in the comparison.

LASER STRUCTURE AND DIAGNOSTICS

The semiconductor lasers used in the experiment were V-grooved Substrate Inner Stripe (VSIS) lasers fabricated at Philips Research Laboratories.† They are weakly index-guided AlGaAs lasers with an Al content of ~ 14%. This results in an emission wavelength of about 790 nm at room temperature. The structure of these lasers is given schematically in Fig. (2). The lasers are fabricated using Liquid Phase Epitaxy (LPE). The index-guiding is provided as a side-effect of the presence of the V-grooved current-blocking layer (CBL). The primary function of this layer is to

†The lasers were made available to us by G.A. Acket of Philips Research Laboratories. The laser facets were coated by W.E. van Es. The lasers were mounted in such a way that the full output of both facets was available; they were tested by R.R. Drenten and H.J. den Blanken.

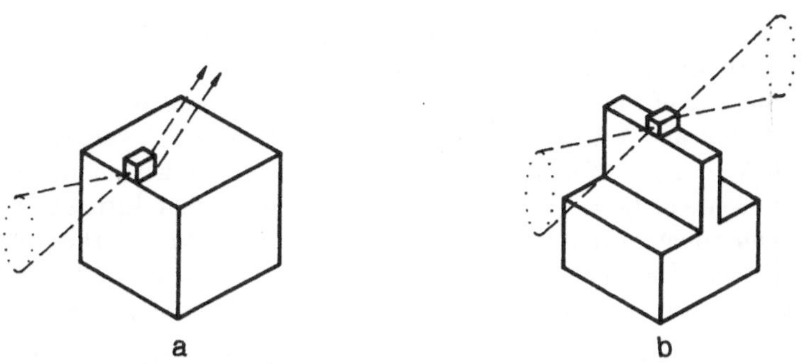

Figure 3: Comparison between standard mounting (a) and our mounting (b) of the VSIS lasers on an aluminum heatsink. Whereas the output of the rear facet is not or only partly accessible in the conventional mounting, our mounting permits the collection of all light emitted from both sides of the laser. (From ref. 14, with permission)

confine the current to a small area in the active layer in order to get high carrier densities. However, as the optical field distribution penetrates somewhat into the CBL and since the refractive index in the CBL (consisting of GaAs) is lower than that of the AlGaAs cladding layer, the field is weakly guided in the transverse (x) direction. This weakly index-guided character makes these lasers oscillate in a single longitudinal mode which is required for our experiment. The fact that the CBL is made of GaAs that absorbs radiation at 790 nm makes the internal losses of these lasers rather high, resulting in a threshold current of around 40- 50 mA at room temperature.

We used six lasers for the experiments; they were all grown on the same wafer. This is very important as many device-dependent parameters differ widely from wafer to wafer, mainly due to poor growth reproduction. The lasers were mounted on a non-standard aluminum heatsink, sketched in Fig. (3), so that the full output of both sides was accessible. The facets of three lasers were coated with multi-layer (HfO_2, SiO_2) dielectric stacks with nominal reflection coefficients of 10% for R_1 and 90% for R_2 respectively; these lasers will be referred to as 10-90 lasers from now on. Three other lasers were coated with a $\lambda/2$ protective layer so that their reflection coefficients were $\simeq$ 32% on both sides (30-30 lasers), determined by the Fresnel reflection. In this way the product of the reflectivities was the same ($\simeq$ 0.09) for both sets of lasers. From Eq. (6), the K-factors of these two sets of lasers are predicted to differ by a factor of 1.33. After the linewidth measurements, described in the next section, all lasers were characterized using the diagnostics described in Refs [6,14]. The results are given in Table 1. We notice that there is a wide spread in threshold current among the lasers. The threshold current is determined by more factors than just the total loss (i.e. transparency current, confinement factor etc.) so that a certain spread may be expected. However, the measured differential quantum efficiency η is almost the same for all lasers. The latter is given by $\eta = \eta_i \eta_{opt}$, with η_i the internal quantum efficiency, $\eta_{opt}(= \alpha_m/(\alpha_m + \alpha_i))$ the optical-extraction efficiency, α_m the mirror-loss coefficient and α_i the internal-loss coefficient. If we assume η_i and α_i to be equal for

Table 1. Measured parameters for the six lasers used in the experiment. dP_1/dP_2 is the ratio of the output powers of both facets, R_1 and R_2 are the reflection coefficients of the facets, I_{th} is the threshold current, α_i is the internal-loss coefficient and λ is the lasing wavelength. Lasers 1,2,3 were standard lasers with $R_1 = R_2 \simeq 32\%$; lasers 4,5,6 had coated facets with nominal reflection coefficients of $R_1 = 10\%$ and $R_2 = 90\%$. Laser 4 died during the diagnostic measurement, so no experimental data are available on the reflectivity of its facets. The main cause for the error margins in the reflectivities is the experimental error in the measurement of the output power.

laser	η %	dP_1/dP_2 -	α_i cm^{-1}	R_1R_2 %	R_1 %	R_2 %	I_{th} mA	λ nm
1	39(4)	1	72(7)	10(3)	32(4)	32(4)	43.9	788
2	41(4)	1	73(7)	8(2)	28(4)	28(4)	44.4	788
3	39(4)	1	66(7)	12(3)	35 (4)	35(4)	41.8	782
4	37(4)	17.3	-	-	-	-	48.2	781
5	38(4)	21.7	83(8)	5.6(1.4)	6.6(1.5)	84(2)	52.1	784
6	39(4)	25.8	71(7)	7.7(1.9)	8.7(2.0)	89(2)	44.1	790

all lasers, a reasonable assumption since all lasers have been grown on the same wafer, we can conclude that α_m, and thus the product of R_1R_2, is approximately equal for all lasers used. If we look at the measurements of the individual facet reflectivities we see that the theoretical value of 32% for the uncoated lasers is reproduced with the experimental error. For the coated lasers the agreement with the nominal reflection coefficients of 10% and 90% is somewhat less.[‡] Calculating the ratio of the expected K-factors based on the measured reflectivities (averaged per set of three lasers) yields the value of 1.5 ± 0.2, which still contains the nominally expected value of 1.33 with the error margin.

EXPERIMENTAL SET-UP FOR LINEWIDTH MEASUREMENT

As the frequency of a semiconductor laser depends both on temperature and injection current [11], these parameters were highly stabilized to avoid frequency jitter during a linewidth measurement. The current source (Seastar LD-200) had a specified stability of better than 10 nA over a timescale of seconds which, taking into account the known dependence of the frequency on the injection current ($\simeq$ -3 GHz/mA), translates into a frequency stability of $\sim$ 30 kHz. We used a home-built computer-controlled two-stage Peltier temperature stabilization system. In both stages the temperature was monitored by measuring the voltage across a NTC resistor. After subtracting a bias voltage, which served to set the temperature (0-50 °C), the remaining error signal was amplified and subsequently sampled by a 12-bit AD converter at a sampling rate of 2.5 Hz. The thermal relaxation times of both stages of the laser mount were much larger than on one second, so this sampling rate was large enough. The error signal was further processed in a microcomputer, programmed to give a PI-type feedback with controllable gain and integration time. The processed error signal was then converted by a DA converter and sent to a current source to regulate the current through the Peltier elements. This computer-controlled PI-control was

[‡]These nominal values were obtained by measuring the reflectivity of a glass substrate which was coated simultaneously with the lasers and by recalculating the measured reflection coefficients to a GaAs substrate.

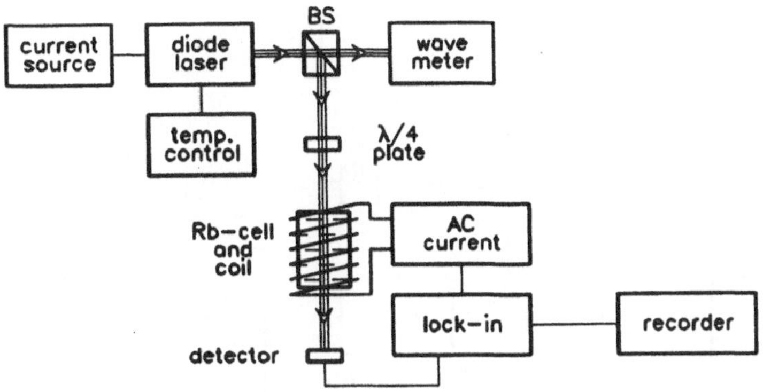

Figure 4. Experimental set-up used to determine the frequency stability of the lasers.

necessary as the relaxation times of two loops were so large that analog processing was unpractical.

The frequency stability of a laser, (passively) stabilized by the above-mentioned current source and temperature controller, was measured as follows [15]. The output beam of the laser was circularly polarized and sent through a room-temperature Rb vapour cell (fig. 4). The wavelength of the laser was tuned to the centre of the $^{85}RbD_1, F = 2 \rightarrow \{F = i\}$ transition at 795 nm. The cell was placed in a sinusoidally modulated magnetic field. The Rb spectrum is then modulated by the Faraday effect so that one can detect the deviation of the laser frequency from line centre in a phase-sensitive way. Modulating the absorption line instead of the laser frequency has the advantage that the stability of the laser frequency is not affected. From the measured output signal of the lock-in amplifier the frequency stability was estimated to be better than 10 MHz over a period of one second and better than 30 MHz over a period of one minute. These fluctuations are apparently due to temperature fluctuations, since the frequency jitter due to injection-current fluctuations is much smaller. Using the known dependence of the frequency on the temperature ($\simeq$ -30 GHz/K) we find that the temperature stability was better than 1 mK over a period of one minute and 300 μK over a period of one second.

When collecting light from the laser in order to perform the linewidth measurements great care was taken to avoid optical feedback into the laser since this is known to drastically influence the linewidth (see e.g. ref. [16]). For that purpose we used an antireflection-coated collimating lens (Melles Griot 06LXP005) with an aperture in front, resulting in a very small numerical aperture (NA $\simeq$ 0.02). In order to further reduce feedback, the lens was tilted with respect to the laser axis and followed by a Faraday isolator (Isowave I-7590-T, isolation > 30 dB). The effective absence of feedback was checked using a highly sensitive method [17] that is based on the fact that, if the feedback is influencing the linewidth, the feedback-induced linewidth depends on the feedback phase. If one scans the laser frequency, e.g. by changing the injection current linearly as a function of time, the observed linewidth will change periodically, with a period equal to the inverse of the time that it takes the light to travel to

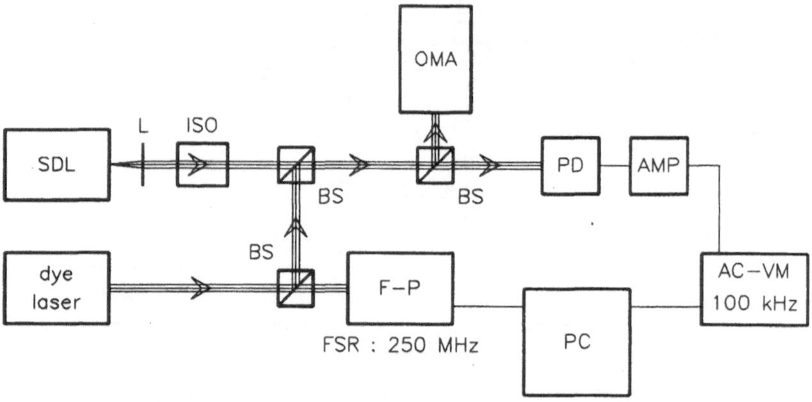

Figure 5. Experimental set-up for the linewidth measurements. The output of the semiconductor laser is heterodyned with that of the dye laser, which acts as a local oscillator. The beat signal of the two lasers is detected by an AC voltmeter tuned to a frequency of 100 kHz. The linewidth of the dye laser ($\simeq$1 MHz) is much smaller than that of the semiconductor laser. A (non-scanning) confocal Fabry-Perot interferometer with a 250 MHz free spectral range was used for the frequency calibration.

the feedback element and back. For this purpose the linewidth was monitored by a scanning Fabry-Perot interferometer§ which was sufficiently optically isolated from the laser. This method has the advantage that, apart from testing if there is any feedback, it also determines the distance from the laser to the (unintentional) feedback element and thus helps to identify the latter.

The spectral lineshape was measured using a heterdyne method [18]. The experimental set-up is shown in Fig. (5). A tunable cw single-mode ring dye laser (Spectra Physics 380D with LD700 dye, linewidth $\simeq$ 1MHz, pumped by a Kr^+ ion laser) was used as a local oscillator. Its frequency was tuned across the semiconductor laser frequency spectrum and the beat spectrum was recorded with a photodiode and a fixed-frequency (100 kHz) bandpass filter. Figure 6 shows a typical experimental lineshape. Note that the heterodyne experiment measures the amplitude spectrum of the semiconductor laser, and not the power spectrum as a FP interferometer would do. The power spectrum corresponding to the fundamental linewidth is expected to be Lorentzian and thus the amplitude spectrum $\tilde{E}(\nu)$ is expected to be the square-root of a Lorentzian,

$$\tilde{E}(\nu) \propto \sqrt{\mathcal{L}(\nu)} \propto \frac{\frac{1}{2}\Delta\nu}{\sqrt{(\nu-\nu_0)^2 + (\frac{1}{2}\Delta\nu)^2}}, \tag{9}$$

having a FWHM width of

$$\delta\nu_E = \sqrt{3}\Delta\nu \tag{10}$$

§One may ask why this FP interferometer was not used for the linewidth measurements. The main reason for this is the range of linewidths to be measured, from 25 to 125 MHz. This would require a FP with a free spectral range of at least 1 GHz and a finesse much larger than 100, in order to avoid complicated de-convolution problems. Such a FP was not available. Furthermore, the described heterodyne method gives a superior signal to noise ratio.

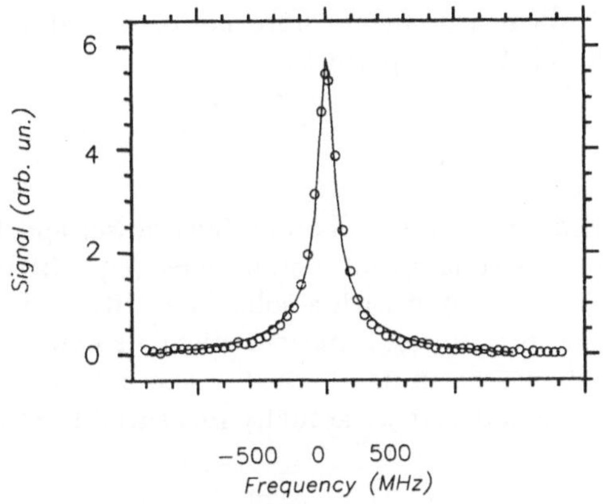

Figure 6. Typical lineshape measurement. The circles refer to the experiment and the solid line is a fit to the square-root of a Lorentzian (see text). In this example the linewidth (FWHM) of the power spectrum was determined to be 66 MHz. (From ref. 8, with permission)

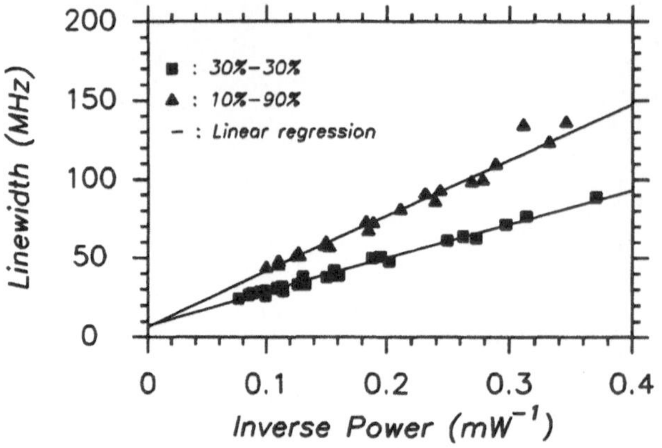

Figure 7. Measured linewidth versus inverse total output power for the three 30-30 lasers and the three 10-90 lasers. The total output power is the sum of the output powers of both facets. Note that it is impossible to distinguish the experimental points of individual lasers for both the 10-90 and the 30-30 categories. All measurements were performed at a laser temperature of 20°C. (From ref. 8, with permission)

where $\Delta\nu$ is the FWHM width of the (Lorentzian) power spectrum. We have fitted the observed lineshapes to square-root Lorentzians (eq. (9)); the best fit for the measurement in Fig. 6 is indicated with a solid line. Jitter, due to temperature and injection-current fluctuations during a measurement, is expected to add a Gaussian component to the lineshape. From its absence in the data we deduce that the lasers were sufficiently stabilized and that we actually measured the fundamental linewidth.

RESULTS AND DISCUSSION

In Fig. 7 we have plotted the measured linewidths of all six lasers as a function of the inverse of the total output power (i.e., from both facets).¶ When extrapolating the data to zero inverse output power, we observe a power-independent contribution to the linewidth. Such a power-independent contribution has also been observed by many others (see e.g. ref. [19]) and several theoretical explanations have been proposed such as occupation fluctuation noise [20], 1/f noise [21], injection-current fluctuations [22], intensity dependence of the linewidth-enhancement factor [23] and influence of sidemodes [19,24,25]. Also, outcoupling-induced non-uniform saturation has been predicted to result in a power-independent contribution to linewidth by us [10] and others [26]. As the measured power-independent linewidth is the same for both sets of lasers (i.e. 30-30 and 10-90) we conclude that the latter mechanism does not contribute significantly.

The data for the three 30-30 lasers lie on a straight line. It is not possible to distinguish between individual lasers; this proves that the device-dependent factors which, apart from R_1 and R_2, co-determine the linewidth were indeed equal for these lasers. The same applies to the data for the three 10-90 lasers. The fact that the

¶The scale of the vertical axis differs from that of Fig. 2 in [8] by a factor of $\sqrt{3}$. This is due to the fact that the lineshape measurements in ref. [8] have been fitted to Lorentzians instead of square-root Lorentzians (cf. eq(9)). However, this has no effect on the final result obtained in that reference since only ratios of linewidths are relevant.

power-dependent part of the linewidth is observed to vary linearly with the inverse output power again proves that we did observe the fundamental linewidth. The slopes of these lines, i.e. the linewidth-power products (see eq. (8)), are 215 and 352 MHz·mW, respectively. The ratio of these values is 1.64, i.e. clearly larger than unity. We ascribe this to the effect of outcoupling. For a quantitive comparison, this measured ratio of 1.64 has to be compared with the calculated value for the ratio of the excess-noise factors which is 1.33 when based on the nominal reflection coefficients of the coated facets and 1.5 ± 0.2 when based on the results of our measurements of the reflection coefficients (see section 3). In the latter case we have quantitative agreement between experiment and theory. The linewidth-power products of the 30-30 and 10-90 lasers, when considered separately, are in agreement with eq. (8) when reasonable values for the device-dependent factors are used ($\alpha = 4.3$ and $n_{sp} = 1.5$). Note, however, that α and n_{sp} are known with insufficient accuracy to draw conclusions from the experimental data on either category of lasers separately. This illustrates that our approach, taking the ratio of the linewidths of two sets of lasers, is crucial in providing experimental evidence of the effect of outcoupling on the fundamental linewidth of a laser.

Next we discuss the possibility that the measured difference in linewidth-power product arises from some unwanted side-effect. As all lasers come from the same wafer and show so little difference in linewidth within one set we may assume that the device-dependent factors are sufficiently equal. We therefore look for effects caused by the difference in reflection coefficients as this seems to be the only real difference of the two sets. Different reflection coefficients lead to different axial intensity distributions in the lasers. Most consequences of a non-uniformity of the axial intensity distribution have been discussed in ref. [10] in the context of saturation; effects thereof have been shown to be negligible for the conditions of our experiments. However, a possibility not mentioned up to now is that of thermal effects. Due to the internal losses the non-uniform intensity distribution will cause a non-uniform absorbtion and hence a non-uniform axial temperature distribution. The temperature determines the bandgap and thus the gain spectrum. Therefore, a non-uniformity of the temperature will lead to a non-uniformity of the gain spectrum and thus, possibly, to a different gain distribution at the lasing wavelength which, in turn, would lead to a different distribution for the intensity, the linewidth-enhancement factor and all other factors mentioned in section 2, leading to a different linewidth. Calculations, done at Philips Research Laboratories [27] show that, in a VSIS laser, the temperature difference between the aluminum heatsink and the active layer is typically 10 °C; it is determined by the power input into the active layer on the one hand and the thermal flow resistance to the heatsink on the other hand. The major part of this 10 °C difference is caused by electrical dissipation and only a minor part by optical dissipation. It is easily calculated that the intensity variations inside a 10-90 laser are never more than ≃50% (see e.g. Fig. 1). This means that the axial temperature variations in our lasers due to the non-uniformity of axial optical intensity are at most a few °C. Using a simple gain model, the gain (at fixed wavelength) can be estimated to vary by about 1% or 1 cm^{-1} per °C. The resulting gain variation (i.e. $<$ a few %) is small compared to the gain variation (typically 10%) due to the effect of saturation [10]. Therefore we conclude that temperature variations along the laser axis are not expected to have a significant influence on the results presented here.

All theory presented here is based on a one-dimensional description of the laser cavity and of the optical field, thus neglecting possible transverse effects. This is correct as long as the laser oscillates in one transverse mode with a transverse intensity distribution that is always the same. However, as the intensity in the centre of this distribution (on the laser axis) will be higher than away from this centre (off-axis), we might expect that, through the effect of saturation, the carrier density is more depleted on-axis than off-axis, resulting in a change of the transverse intensity distribution when the injection current is changed. This might lead, for example, to a saturation-induced transverse excess-noise factor (also called Petermann factor) for an index-guided semiconductor laser. However, numerical calculations of the transverse distributions of the intensity and carrier density in VSIS lasers [27] show that this effect is small as compared to the axial saturation and, hence, can be neglected. It is clear after this first experimental demonstration of enhanced fundamental linewidth much room is left for further, more quantitative experiments. One might think, for example, of doing an experiment like that described in section 4 on only one laser, first measuring its linewidth when the facets are uncoated, subsequently coating the facets, and then measuring the linewidth again. As the diagnostics described in section 3 and ref. [14] can then be done twice for the same laser, first with known (uncoated) and then with unknown (coated) facet reflectivities, the latter reflectivities can be determined more accurately than when using different lasers, thus allowing a more accurate comparison of experiment and theory. It would also be interesting to do the experiment on lasers with more pronounced outcoupling asymmetries and therefore larger excess-noise factors, e.g. a 1-100 laser which has $K = 4.6$. In principle, the gain in a semiconductor laser can be made high enough to let such a device oscillate, although it is difficult to predict whether it would do so in a single longitudinal mode.

REFERENCES

1. A.L. Schawlow and C.H. Townes, "Infrared and optical masers," *Phys. Rev.* **112**, 1940-1949 (1958).
2. K. Ujihara, "Phase noise in a laser with output coupling," *IEEE J. Quantum Electron.* **QE-20**, 814-818 (1984).
3. C.H. Henry, "Theory of spontaneous emission in open resonators and its application to lasers and optical amplifiers," *IEEE J. Lightwave Technol.* **LT-4**, 288-297 (1986).
4. J. Arnaud, "Natural linewidth of semiconductor lasers," *Electron. Lett.* **22**, 538-540 (1986).
5. S. Prasad and B.S. Abbott, "Mirror transmission and laser phase diffusion in the quantum regime," *Phys. Rev.* A **38**, 3551-3555 (1988).
6. W.A. Hamel, "Effect of outcoupling on the quantum-limited linewidth of a semiconductor laser," Ph.D. thesis, Leiden University, February 1991.
7. W.A. Hamel and J.P. Woerdman, "Nonorthogonality eigenmodes of a laser," *Phys. Rev.* A **40**, 2785-2787 (1989).
8. W.A. Hamel and J.P. Woerdman, "Observation of enhanced fundamental linewidth of a laser due to nonorthogonality of its eigenmodes," *Phys. Rev. Lett.* **64**, 1506-1509 (1990).
9. M.P. van Exter, W.A. Hamel and J.P. Woerdman, "Non-uniform phase diffusion in a laser," *Phys. Rev.* A **43** (June 1991).
10. W.A. Hamel, M.P. van Exter, A. Shore and J.P. Woerdman, "Numerical study of the effect of saturation on the linewidth of a semiconductor laser," submitted to IEEE J. Quantum Electron.
11. K. Petermann, "Laser diode modulation and noise", Kluwer Academic Publishers (Dordrecht, 1988).

12. A.E. Siegman, "Lasers," University Science Books (Mill Valley, 1986).
13. C.H. Henry, "Theory of the linewidth of semiconductor lasers," *IEEE J. Quantum Electron.* **QE-18**, 259-264 (1982).
14. W.A. Hamel, M. Babeliowsky, J.P. Woerdman and G.A. Acket, "Diagnostics of asymmetrically coated semiconductor lasers," accepted for publication in IEEE Photon. Technol. Lett.
15. R.A. Valenzuela, L.J. Cimini, R.W. Wilson, K.C. Reichmann and A. Grot, "Frequency stabilization of AlGaAs lasers to the absorbtion of Rubidium using the Zeeman effect", *Electron. Lett.* **24**, 725-726 (1988).
16. G.P Agarwal, "Line narrowing in a single-mode injection laser due to external optical feedback", *IEEE J. Quantum Electron.* **QE-20**, 468-471 (1984).
17. J. Käppel and W. Heinlein, "Experimental determination of paracitic optical feedback by diode laser linewidth measurements", *Electron. Lett.* **25**, 447-448 (1989).
18. K. Kikuchi, "Lineshape measurement of semiconductor lasers below threshold", *IEEE J. Quantum Electron.* **QE-24**, 1814-1817 (1988).
19. W. Elsässer and E.O Göbel, "Multimode effects in the spectral linewidth of semiconductor lasers," *IEEE J. Quantum. Electron.* **QE-21**, 687-692 (1985).
20. K. Vahala and A. Yariv, Appl. *Phys Lett.* **43**, 140-142 (1983).
21. K. Kikuchi, "Effect of 1/f type FM noise on semiconductor-laser linewidth residual in high-power limit," *IEEE J. Quantum Electron.* **QE-25**, 684-688 (1989).
22. G.P. Agrawal and R. Roy, "Effect of injection-current fluctuations on the spectral linewidth of semiconductor lasers," *Phys. Rev.* A **37**, 2495-2501 (1988).
23. G.P Agrawal, "Intensity dependence of the linewidth enhancement factor and its implications for semiconductor lasers," *IEEE Photon. Technol. Lett.* **1**, 212-214 (1989).
24. S.E. Miller, "The influence of power level on injection laser linewidth and intensity fluctuations including side-mode contributions," *IEEE J. Quantum Electron.* **QE-24**, 1873-1876 (1988).
25. U. Krüger and K. Petermann, "The semiconductor laser linewidth due to the presence of side modes," *IEEE J. Quantum Electron.* **QE-24**, 2355-2358 (1988).
26. S. Prasad, private communication.
27. H.J. den Blanken, private communication.

EINSTEIN-PODOLSKY-ROSEN CORRELATIONS[1]

Virendra Singh

Tata Institute of Fundamental Research
Homi Bhabha Road
Bombay 400 005, India

We review here some recent progress in the study of Einstein-Podolsky-Rosen (EPR) correlations (Bell (1964), Bell (1987), Bohm (1951), Einstein, Podolsky and Rosen (1935), Selleri (1988), Singh (1988)) and their generalizations. The topics included are (a) two particle interferometry (b) multiparticle correlations and (c) theories with signal locality i.e. with no faster than light signals.

TWO PARTICLE INTERFEROMETRY

The EPR correlations arise from the existence of two particle entangled states in quantum mechanics. Irrespective of the spatial separation between the two particles, the wavefunction of the total two particle system for these states cannot be written as a product of wavefunctions of the individual one particle subsystems. Thus despite the physical interaction between the two particles being negligible the subsystems cannot be assigned separate wavefunctions and this is rather counterintuitive. Schrödinger (1935) was so struck by this aspect of quantum mechanics that he wrote "I would not call that one but rather the characteristic of quantum mechanics."

If we produce two beams of "two particle entangled states," such that these beams have definite phase relationships, and make them interfere, we obtain phenomena which come under the novel field of two particle interferometry. This is a significant enrichment of the field of interferometry. The earlier interference experiments, such as Young's slit experiments etc., involved interference between single particle beams (single particle interferometry). The quantum mechanics owes a lot to these single particle interferometry experiments and it is hoped that the two particle interferometry would also provide significant inputs for the study of quantum phenomenon. We shall follow the treatment given by Horne, Shimony and Zeilinger (1989, 1990).

Let the operator $A_1(A_2)$ refer to some physical observables for the first (second) particle. Let $j = (1,2)$

[1] Invited talk delivered at the International Conference on Quantum Optics (Jan. 5-10, 1991) held at Hyderabad.

Recent Developments in Quantum Optics, Edited
by R. Inguva, Plenum Press, New York, 1993

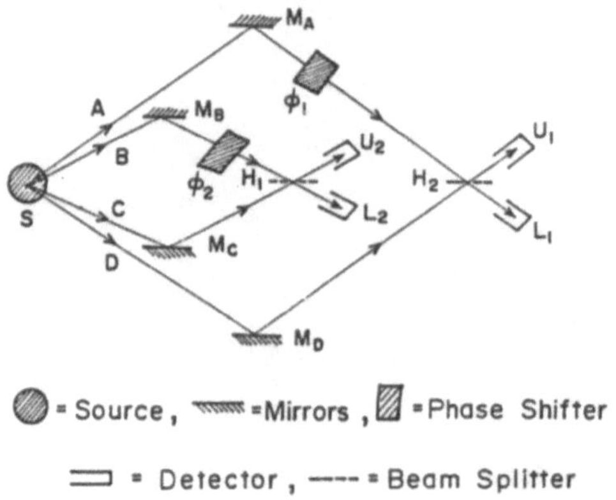

Figure 1

$$A_1|\alpha>_1=\alpha|\alpha>_1,\ \ A_1|\delta>_1=\delta|\delta>_{1,1}<\alpha|\delta>_1=0$$

$$A_2|\beta>_2=\beta|\beta>_2,A_2|\gamma>_2=\gamma|\gamma>_{2,2}<\beta|\gamma>_2=0$$

we may try to prepare the entangled two particle state

$$|\psi>=\frac{1}{\sqrt{2}}[|\alpha>_1|\gamma>_2+|\delta>_1|\beta>_2]$$

Here the physical observables A_1 and A_2 need not refer to spin variables as is normally the case for a typical experiment checking Bell's inequalities. In fact any variable, e.g. linear momentum, such that $A_1+A_2=A_T$ is a conserved quantity, can be used to prepare such two particle states by using different partitions of A_T. We can now prepare such states and then carry out two particle interferometry experiments by directing, introducing known phaseshifts between them and finally recombining them to produce the interference pattern.

A typical arrangement is given in figure 1. A source S emits an ensemble of pairs of particles into the beams A, B, C, D with momenta $\vec{k}_A$, $\vec{k}_B$, $\vec{k}_C$ and $\vec{k}_D$ satisfying $|\vec{k}_A|=|\vec{k}_D|$, $|\vec{k}_B|=|\vec{k}_C|$ such that each pair is described by

$$|\psi>=\frac{1}{\sqrt{2}}[|\vec{k}_A>_1|\vec{k}_C>_2+|\vec{k}_D>_1|\vec{k}_B>_2].$$

For the first component in this coherent superposition the first (second) particle after being reflected by the mirror $M_A(M_C)$ and phase shifted by phase $\phi_1(\phi_2)$ and acted on by the beam splitter $H_1(H_2)$ is detected by the detector U_1 or $L_1(U_2$ or $L_2)$; for the second component the first (second) particle after being reflected by the mirror $M_D(M_B)$ and acted on by the beam splitter $H_1(H_2)$ is detected by the detector U_1 or $L_1(U_2$ or $L_2)$.

Let the joint detection probability by the two detectors D_1 (which is either U_1 or L_1) and D_2 (which is either U_2 or L_2) be denoted by $p(D_1, D_2; \phi_1, \phi_2)$. We then have

$$p(U_1, U_2; \phi_1, \phi_2) = p(L_1, L_2; \phi_1, \phi_2) = \eta^2(1 + cos(\phi_2 - \phi_1 + \theta))/4$$

$$p(U_1, L_2; \phi_1, \phi_2) = p(U_2, L_1; \phi_1, \phi_2) = \eta^2(1 - cos(\phi_2 - \phi + \theta))/4,$$

where η is the detector efficiency and θ is the phase introduced between the two interfering two-particle beams as a result of the exact arrangement of the mirrors and beam splitters. The angle θ is independent of ϕ_1 and ϕ_2.

The cosine factors in these joint probabilities is a factor typical of quantum mechanical interference. Note however that in the arrangement used there is no single particle interference as

$$p(U_1; \phi_1, \phi_2) = p(U_2; \phi_1, \phi_2) = p(L_1; \phi_1, \phi_2) = p(L_2; \phi_1, \phi_2) = \frac{1}{2}\eta^2.$$

The arrangement thus is quintessentially a two-particle interference one.

The dichotomic nature of spin variables, used in derivations of Bell's inequalities in the usual EPR context, is here replaced by the dichotomic nature of the detector U_1 or L_1 for the first particle and U_2 or L_2 for the second. We can then derive the analogue of the usual Bell's inequalities for this experimental setups and these are given by

$$2 \geq |E_\theta(\phi_1', \phi_2') + E_\theta(\phi_1', \phi_2'') + E_\theta(\phi_1'', \phi_2') - E_\theta(\phi_1'', \phi_2'')|.$$

The earliest two-particle interferometry arrangement was by Ghosh and Mandal (1987). More recently J. Rarity and P. Tapster (1990) provided a violation of Bell's inequality which does not use polarisation measurements. Franson (1989) had earlier shown that the quantum mechanical uncertainties in position or the emission time of a particle can lead to effects which are in violation of local realism and proposed experiments using two-particle interferometry. These have been experimentally realised by P. Kwiat et. al (1990) and by Z. Ou et. al (1990).

MULTIPARTICLE CORRELATIONS

John Bell derived in 1964 within the framework of local realism, which was the basis of E.P.R.'s program of a possible acceptable completion of quantum mechanism, his celebrated inequalities which showed that such a completion is not compatible with statistical predictions of quantum mechanics. Recently it was shown by Greenberger, Horne and Zeilinger (GNZ) (1989) that, using three or more particle correlations, the same incompatibility between local realism and quantum mechanics can be demonstrated using perfect correlations, rather than statistical ones, in quantum mechanics. See also Greenberger et. al (1990).

We present the GNZ argument in the Mermin (1990a) version. Let three spin-1/2 particles be emitted by a source, and let the wavefunction of the three particle system by given by

$$\psi = \frac{1}{\sqrt{2}}[|\uparrow,\uparrow,\uparrow> - |\downarrow,\downarrow,\downarrow>]$$

where $\uparrow$ ($\downarrow$) specifies the spin up (down) wavefunction of a particle along it's direction of motion. All the particles are taken to be coplaner ($y-z$ plane) and their z-axis is taken along the direction of their motion. Denoting the spin $\vec{S}^j = 1/2\vec{\sigma}^j$ for the jth particle (j = 1,2,3) it is easy to check that the wavefunction ψ is simultaneous eigen state of the operators $A_1 = \sigma_x^{(1)}\sigma_y^{(2)}\sigma_y^{(3)}, A_2 = \sigma_y^{(1)}\sigma_x^{(2)}\sigma_y^{(3)}$ and $A_3 = \sigma_y^{(1)}\sigma_y^{(2)}\sigma_x^{(3)}$ with eigenvalue +1.

In view of this circumstance, if we measure the y-spin component of any two particles we can predict with certainty the x-spin component of the third one. Thus the x-component of the spin of a particle is an element of reality using EPR definition. Similarly a simultaneous measurement of the x-spin component of one particle and the y-spin component of the second one allows us to infer with certainty the y-spin component of the third particle. Thus the y-spin component is equally an element of reality. Let us therefore associate the numbers $m_x^{(j)}, m_y^{(j)}$ which can only take the values +1 or -1 with $\sigma_x^{(j)}$ and $\sigma_y^{(j)}$. We must have $M_1 = m_x^{(1)}m_y^{(2)}m_y^{(3)} = 1, M_2 = m_y^{(1)}m_x^{(2)}m_y^{(3)} = 1, M_3 = m_y^{(1)}m_y^{(2)}m_x^{(3)} = 1$ for our system.

We now note that $A = A_1A_2A_3 = \sigma_x^{(1)}\sigma_x^{(2)}\sigma_x^{(3)}$. If we measure A then local realism would demand that the result of measurement be given by $m_x^{(1)}m_x^{(2)}m_x^{(3)}$ i.e. it should be equal to +1 as it is a product of $M_1M_2M_3$ which are all equal to +1. Quantum mechanically however $A|\psi> = (-1)|\psi>$ i.e. a measurement of A must lead to a value (-1) with certainty. The incompatibility between local realism and quantum mechanics is thus sharply brought out.

In a real experimental situation, where one has to deal with imperfect detectors etc., it is better to depend on correlation inequalities such as those of Clauser et. al (1969). Mermin (1990b) has shown that for the n-particle state

$$|\Phi> = \frac{1}{\sqrt{2}}[|\uparrow_1,\uparrow_2,\cdots,\uparrow_n> - |\downarrow_1,\downarrow_2,\cdots,\downarrow_n>]$$

and the operator B given by

$$B = \frac{1}{2i}\Big[\prod_{j=1}^{n}(\sigma_x^{(j)} + i\sigma_y^{(j)}) - \prod_{j=1}^{n}(\sigma_x^{(j)} - i\sigma_y^{(j)})\Big],$$

while we have quantum mechanically $<\Phi|B|\Phi> = 2^{n-1}$, the corresponding linear combination of correlation functions, in local realistic formulation, is bounded by $2^{n/2}$ (n even) or $2^{(n-1)/2}$ (n odd). Thus the violation is by an exponentially large factor $2^{(n-2)/2}$ (n even) and $2^{(n-1)/2}$ (n odd). Roy and Singh (1991) have shown that the violation can be as large as $2^{(n-1)/2}$ for both odd and even n if one chooses the state Φ and the operator B a little differently. Further one can also obtain many Bell-type inequalities for multiparticle correlations using a method given earlier for two particle correlations (Roy and Singh (1978, 1979)).

SIGNAL LOCALITY

Subject to some caveats, realting to detector efficiencies or extra assumptions involved in checking Bell's inequalities, the present experimental evidence overwhelmingly favours quantum mechanics over local realism. It is therefore worthwhile to investigate weaker versions of locality. In particular we would certainly like our theories to obey signal locality i.e. there should not be any faster than light propagation of signals.

Roy and Singh (1989) have recently made the notion of signal locality precise within the framework of hidden variables. Thus in the EPR-Bohm context we have the expectation values, denoted by angular brackets.

$$< A >= \int d\lambda \rho(\lambda, a, b) A(\lambda, a, b)$$

$$< B >= \int d\lambda \rho(\lambda, a, b) B(\lambda, a, b)$$

$$< AB >= \int d\lambda \rho(\lambda, a, b) A(\lambda, a, b) B(\lambda, a, b)$$

where λ refer to all the hidden variables, a and b refer to the detector orientation measuring the physical observables A and B respectively and $\rho(\lambda, a, b)$ is the normalised positive probability density. Unlike Bell (1964), but in accordance with the view of Bohr (1935), we allow $A(\lambda, a, b)$ to depend not only on a but on b as well and similarly for $B(\lambda, a, b)$ to depend not only on b but a as well. Signal locality is formulated by requiring $< A >= A(a), < B >= B(b)$. Thus $< A >$ does not depend on the distant detector setting b and similarly $< B >$ does not depend on a.

Signal local theories then lead to the experimentally testable inequalities (with $\eta_A^2 = \eta_B^2 = 1$)

$$1 + \eta_A A(a) + \eta_B B(b) + \eta_A \eta_B P(a, b) \geq 0$$

where $P(a, b) \equiv < AB >$. The proof involves only

$$\rho(\lambda, a, b) \geq 0, \quad 1 \geq |A(\lambda, a, b)|, \quad 1 \geq |B(\lambda, a, b)|.$$

Singh and Roy (1991) also derive a complete set of multiparticle correlation inequalities.

It is possible to show that quantum mechanics satisfies the requirement imposed by us on signal-local theories.

REFERENCES

Bell, J.S., 1964, *Physics* 1:195.

Bell, J.S., 1987, *Speakable and Unspeakable in Quantum Mechanics*, Cambridge.

Bohm, D., 1951, *Quantum Theory*, Prentice Hall.

Bohr, N., 1935, *Phys. Rev.* **48**:696.

Clauser, J.F., Horne, M.A., Shimony, A. and Holt, R.A., 1969, *Phys. Rev. Lett* **23**:880.

Einstein, A., Podolsky, B. and Rosen, N., 1935, *Phys. Rev.* **47**:777.

Franson, J.D., 1989, *Phys. Rev. Lett.* **62**:2205.

Ghosh, R. and Mandel, L., 1987, *Phys. Rev. Lett.* **59**:1903.

Greenberger, D.M., Horne, M.A. and Zeilinger A., 1989, "Going beyond Bell's theorem" in *Bell's Theorem, Quantum Theory and Conception of the Universe* edited by M. Kafatos, Kluwer Academic, Dordrecht, pp. 73-76.

Greenberger, D.M., Horne, M.A., Shimony, A. and Zeilinger, A., 1990, *Am. J. Phys.* **58**:1131.

Horne, M., Shimony, A. and Zeilinger, A., 1989, *Phys. Rev. Lett.* **62**:2209.

Horne, M., Shimony, A. and Zeilinger, A., 1990, "Introduction to Two Particle interferometry," in Sixty-Two Years of Uncertainty, edited by A. Miller, Plenum, New York.

Kwiat, P.G., Vareka, W.A., Hong, C.K., Nathel, H. and Chiao, R.Y., 1990, *Phys. Rev.* **A41**:2910.

Mermin, N.D., 1990a, *Am. J. Phys.* **58**:731.

Mermin, N.D., 1990b, *Phys. Rev. Lett.* **65**:1838.

Ou, Z.Y. and Zou, X.Y., Wang, L.J. and Mandel, L., 1990, Phys. Rev. Lett. **65**:321.

Rarity J. and Tapster, P.R., 1990, *Phys. Rev. Lett.* **64**:2495.

Roy, S.M. and Singh, V., 1978, *Jour. Phys.* **A11**, L167.

Roy, S.M. and Singh, V., 1979, *Jour. Phys.* **A12**, 1003.

Roy, S.M. and Singh, V., 1989, *Phys. Lett.* **A139**:437.

Roy, S.M. and Singh, V., 1991, *Phys. Rev. Lett.* **67**, 2761.

Selleri, F., 1988, (ed.): *Quantum Mechanics Versus Local Realism: The Einstein, Podolsky and Rosen Paradox*, Plenum, New York.

Schrödinger, E., 1935, Proc. Cambridge Phil. Soc. 31:55.

Singh, V., 1988, "Quantum Physics: Some Fundamental Aspects" in *Schrödinger Centenary Surveys in Physics* edited by V. Singh and S. Lal, World Scientific.

Singh, V. and Roy, S.M., 1991, "Theories with Signal Locality and their Experimental Tests" in *M.A.B. Bég Memorial Volume* edited by A. Ali and P. Hoodbhoy, World Scientific.

SIMULTANEOUSLY SHARP WAVE AND PARTICLE-LIKE PROPERTY OF SINGLE PHOTON STATES IN A TWO-PRISM EXPERIMENT

Partha Ghose

S.N. Bose National Centre for Basic Sciences
DB-17, Sector-1, Salt Lake
Calcutta 700 064, India

Dipankar Home

Department of Physics
Bose Institute
Calcutta 700 009, India

ABSTRACT

We discuss a two-prism experiment proposed by Ghose, Home and Agarwal [1] for which the formalism of quantum optics predicts anticoincidences for 'Single Photon States'. This implies simultaneous particle and wave-like propagation in contradiction with the complementarity principle.

If one wants to use classical pictures to describe quantum phenomena, it is well known that incompatible descriptions arise on using concepts such as particles or waves. Niels Bohr tried to resolve this problem through his complementarity principle which expresses the impossibility of simultaneously performing experiments corresponding to incompatible classical descriptions. This mutual exclusiveness between complete 'particle-knowledge' and complete 'wave-knowledge' ensures the inner consistency of using classical pictures to interpret quantum phenomena. Recently, however, certain experiments [2] have revealed variable degrees of sharpness of wave and particle-like behaviour, showing that it is possible to obtain partial wave-knowledge and partial particle-knowledge from the same experimental arrangement (unsharp particle and wave-like properties) in terms of the "which path" ("Welcher Weg") information and the corresponding contrast of the interference pattern. In this paper we shall discuss a new experiment with a single-photon source which is different from such "Welcher Weg" experiments and in which simultaneously sharp particle and wave-like properties should be seen in contradiction with the complementarity principle, although there is no ambiguity in the quantum mechanical mathematical description of the experiment.

The classical analogue of the proposed experiment [1] was performed by J.C. Bose [3] in 1897 as reported in Sommerfeld's "Optics" [4]. Bose took two asphalt prisms

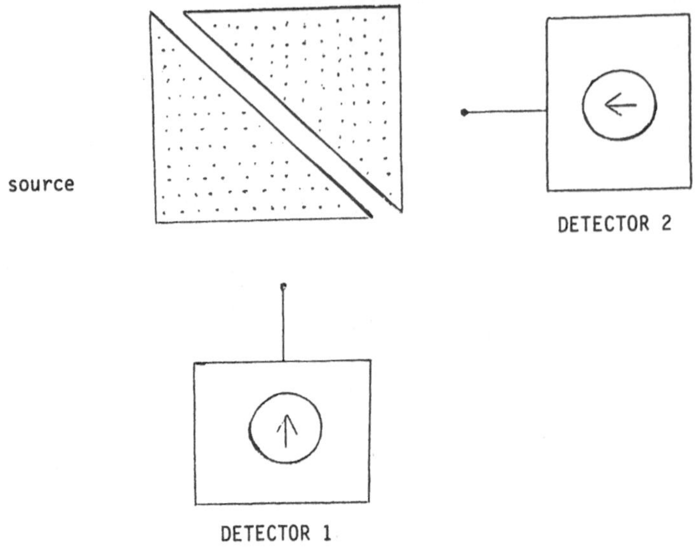

Figure 1

and placed them opposite each other with a large air gap between them (Fig. 1). When microwaves with $\lambda = 20$ cm were incident on the first prism, they were found to be totally internally reflected by it. As he decreased the air gap and made it of the order of several centimeters, Bose found that the waves could tunnel through the gap. This was a striking confirmation of the wave nature of microwaves. Similar experiments can also be done with visible light. Feynman [5] has given a detailed explanation of this effect based on the theory of classical electrodynamics.

The question that arises is: what would happen if this experiment is performed with "single photon states?" Let the single photon state be described by the state vector

$$\Psi = a_t \Psi_t + a_r \Psi_r \tag{1}$$

where Ψ_t is the state that optically tunnels through the gap between the prisms and Ψ_r the state that is internally reflected by the first prism. Let the final states of the two identical detectors (Fig. 1) be D_1 and D_2. The total state vector of the combined system after registrations by the detectors is given by (assuming ideal 100% efficient detectors)

$$\Psi' = a_t \Psi_t D_2 + a_r \Psi_r D_1 \tag{2}$$

For multi-photon states and classical light pulses $< D_1 | D_2 > \neq 0$ and coincidence counts are predicted. However, for single photon states $< D_1 | D_2 > = 0$ (anticoincidence) is the only possibility, and Ψ collapses to a mixed state comprising Ψ_t and Ψ_r with weight factors $|a_t|^2$ and $|a_r|^2$ respectively. One can therefore label each registered photon as coming either after tunneling through the gap or after internal reflection from the first prism. Tunneling through the gap is a clear-cut evidence of the wave-like propagation (to use a classical picture) of a single photon (it should disappear on making the gap larger than the wavelength), whereas perfect anticoincidence of the counts is clear-cut evidence of its particle-like propagation (again to use a classical picture).

To elucidate the difference from a double-slit interference experiment, let us recall that the state vector Ψ of a single photon in such an experiment can be written as

$$\Psi = a_1\Psi_1 + a_2\Psi_2 \tag{3}$$

where Ψ_1 and Ψ_2 are the states emerging from the two slits. The interference between Ψ_1 and Ψ_2 is usually interpreted classically as evidence of its wave-like property. If one places detectors near the two slits, the total wave function of the combined system after detection will be (assuming ideal 100% efficient detectors)

$$\Psi' = a_1\Psi_1 D_1 + a_2\Psi_2 D_2 \tag{4}$$

Since $< D_1|D_2 >= 0$ for single particle states, Ψ collapses to an incoherent mixture of Ψ_1 and Ψ_2 and the interference disappears. Here the anticoincidence between the two detector counts (particle-like propagation) invariably destroys the interference pattern. This is the genesis of Bohr's complementarity principle. In the 'two-prism' experiment, however, anticoincidence is concurrent with optical tunneling. Using classical language, one is therefore forced to use wave and particle pictures to describe the outcome of a single experimental arrangement. Nevertheless, the experiment is consistent with both the Einstein-de Broglie version of wave-particle dualism [6] and the viewpoint advocated by Heisenberg [7] who wrote in 1959: ". . . the concept of complementarity introduced by Bohr into the interpretation of quantum theory has encouraged the physicists to use an ambiguous rather than an unambiguous language, to use the classical concepts in a somewhat vague manner in conformity with the principle of uncertainty . . . When this vague and unsystematic use of the language leads into difficulties, the physicist has to withdraw into the mathematical scheme and its unambiguous correlation with the experimental facts."

A crucial feature of the proposed experiment is the use of genuine "single photon states" and not attenuated classical light pulses even though the average energy per pulse may be smaller than that of a photon of the same frequency. The latter type of source is known not to show any non-classical effect [8]. A very interesting experiment with such weak pulses and "single photon states" incident on a beam-splitter has already been done by Aspect et al. [9] which corroborates the complementarity principle. It shows coincidences with weak pulses but anticoincidences with single photon states exhibiting their particle-like propagation. Nevertheless, if the two channels of the beam splitter are re-combined, interference is observed ("a photon interferes with itself"). The 'two-prism' experiment proposed by us uses tunneling rather than interference and can confront the complementarity principle. It is already under way at the Central Research Laboratory, Hamamatsu Photonics K.K. Japan (Y. Mizobuchi et al.) [10].

ACKNOWLEDGEMENT

The authors gratefully acknowledge collaboration with Prof. G.S. Agarwal. One of the authors (DH) thanks the Department of Science and Technology (Government of India) for supporting his research (Project No. SP.S2/K-42/88).

REFERENCES

1. P. Ghose, D. Home and G. S. Agarwal, *Phys. Lett.* **A 153**, 403 (1991).
2. W. K. Wootters and W. H. Zurek, *Phys. Rev.* **D 19**, 473 (1979); L. S. Bartell, *Phys. Rev.* **D 21**, 1968 (1980); P. Mittelstaedt et al., *Found. Phys.* **17**, 891 (1987); D. M. Greenberger and A. Yasin, *Phys. Lett.* **128A**, 391 (1988); D. Home and P. N. Kaloyerou, *J. Phys.* **A 22**, 3253 (1989); H. Rauch, in: Proc. 3rd Int. Symp. on Foundations of Quantum Mechanics in the Light of New Technology, eds. S. Kobayashi et al. (The Physical Society of Japan, Tokyo, 1990).
3. J. C. Bose, in: Collected Physical Papers (Longmans, Green & Co., London, 1927), pp. 44-49.
4. A. Sommerfeld, Optics (Academic Press, New York, 1964), pp. 32-33.
5. R. P. Feynman, R. B. Leighton and M. Sands,The Feynman Lectures on Physics, Vol. II (Addison-Wesley, Reading, Mass, 1964) pp. 33-12.
6. F. Selleri, Quantum Paradoxes and Physical Reality (Kluwer, Dordrecht, 1990), Chapter 3.
7. W. Heisenberg, Physics and Philosophy (Harper and Row, New York, 1959), p. 179.
8. R. Loudon, *Rep. Progr. Phys.* **43**, 913 (1980).
9. P. Grangier, G. Roger and A. Aspect, *Europhysics Lett. 1*, **173** (1986); A. Aspect and P. Grangier, Hyperfine Interactions 37, 3 (1987).
10. Y. Mizobuchi, private communication.

WIGNER FUNCTION DESCRIPTION OF NONLOCAL FEATURES OF QUANTUM FIELDS GENERATED IN NONLINEAR OPTICAL PROCESSES

A. Venugopalan and R. Ghosh

School of Physical Sciences
Jawaharlal Nehru University
New Dehli - 100 067, India

ABSTRACT

Fields generated in a large number of nonlinear optical processes (including those with losses) have a Wigner distribution which is Gaussian centered around the mean value of the field.[4] We show[9] that the Bell inequality for an optical correlation experiment with two coupled modes generated in a nonlinear process can be expressed as an inequality relating to the parameters of the underlying Wigner distribution function. The example of parametric down-conversion, which has already been used in experiments[2] to demonstrate the quantum nonlocality, is considered.

INTRODUCTION

There has been considerable recent interest in nonclassical features of the electromagnetic field, in the form of antibunching, sub-Poissonian photon statistics, squeezing, quantum interference and nonlocality. Each of these quantum phenomena is normally expressed by the violation of a particular inequality associated with the quantum state under consideration.

For a quantum electromagnetic field, complete information is contained in the density matrix $\hat{\rho}$ and information about the statistical properties of the field can be obtained from moments of the field operators. Establishment of the quantum condition puts restrictions on these moments which are the parameters of the phase-space distribution function corresponding to the density matrix $\hat{\rho}$. In the diagonal coherent state ($|\{v\}>$) representation of the density operator $\hat{\rho}$,

$$\hat{\rho} = \int P(\{v\})\ |\{v\}><\{v\}|\ d\{v\}, \tag{1}$$

where P($\{v\}$) is some weight functional or phase-space density (the Glauber-Sudarshan P function) and the integral is to be taken over all values of the set of complex amplitudes $\{v\}$. If $P(\{v\})$ is <u>not</u> a probability density, then the state is nonclassical. In general $P(\{v\})$ can be negative and highly singular. In the case of a quantum field, where no well-behaved $P(\{v\})$ function exists, other quasiprobability distributions are often used, as for example the Q and Wigner functions. All known nonclassical and nonlocal effects can then be expressed in terms of the parameters of the respective distribution function.

THE WIGNER FUNCTION

The Wigner function corresponding to the density matrix $\hat{\rho}$ of a single mode of the radiation field is defined[6] as:

$$W(z,z^*) = \pi^{-4} Tr\Big[\hat{\rho}\int d^2p\ exp\{-[p(z^* - \hat{a}^+) - p^*(z - \hat{a})]\}\Big], \tag{2}$$

The boson operators $\hat{a}, \hat{a}^+$ characterize the field and the integral is over the entire complex p-plane.

It has been shown [4] that the Wigner function for fields generated in a large class of nonlinear optical processes is a Gaussian which is centered around the mean value of the field. These studies have established the quantum conditions required for squeezing, and quantum statistics exhibited by the generated fields in terms of such distribution function. Consider a field with Gaussian Wigner function:

$$W(z,z^*) = [\pi(t^2 - 4|\mu|^2)^{\frac{1}{2}}]^{-1} exp[-\{\mu(z-z_o)^2 + \mu^*(z^* - z_o^*)^2 + t|z - z_o^*|^2\}/(t^2 - 4|\mu|^2)], \tag{3}$$

where

$$< \hat{a} >= z_o, \tag{4}$$

$$< \hat{a}^2 >= -2\mu^* + z_o^2, \tag{5}$$

$$< \hat{a}^{+2} >= -2\mu + z_o^{*2}, \tag{6}$$

$$< \hat{a}^+\hat{a} >= t - \frac{1}{2} + |z_o|^2, \tag{7}$$

and

$$t > \mu + \mu^*. \tag{8}$$

Fluctuations in the photon-number $\hat{n} \equiv \hat{a}^+\hat{a}$ can be written in terms of the Wigner parameters as

$$< (\Delta\hat{n})^2 > \equiv < \hat{n}^2 > - < \hat{n} >^2 = t^2 + 2|z_o|^2 t - \frac{1}{4} - 2z_o^{*2}\mu^* - 2z_o^2\mu + 4\mu\mu^*. \tag{9}$$

For sub-Poissonian photon statistics,

$$< (\Delta\hat{n})^2 > - < \hat{n} > < 0, \tag{10}$$

which gives

$$t^2 + 2|z_o|^2 t + \frac{1}{4} - 2z_o^{*2}\mu^* - 2z_o^2\mu + 4|\mu|^2 - |z_o|^2 - t < 0. \tag{11}$$

PARAMETRIC DOWN-CONVERSION IN A CAVITY

In the case of two-photon squeezed laser[5] and degenerate parametric down-conversion in a cavity[8], the dynamical equation for the Wigner function is known to have the form of a linearized Fokker-Planck equation. The resulting solution is a Gaussian, given by Eq. (3). We quote the results for the down-conversion process which has been used in a number of recent correlation experiments demonstrating nonclassical[1] and nonlocal interference.[2] Let the pump mode of frequency $\omega_c = 2\omega_a$ generate two modes of frequencies ω_a each. Let $\hat{c}(\hat{c}^+)$ and $\hat{a}(\hat{a}^+)$ be the annihilation (creation) operators associated with the pump and generated modes respectively. The Hamiltonian describing the down-conversion process can be written as:

$$\hat{H} = \hbar\omega_a \hat{a}^+\hat{a} + 2\hbar\omega_a \hat{c}^+\hat{c} + i\hbar/2[G\hat{a}^{+2}\hat{c} - H.c.] + i\hbar(\varepsilon_c \hat{c}^+ e^{-2i\omega_a t} - H.c.), \tag{12}$$

G gives the magnitude of the parametric coupling and is proportional to $\chi^{(2)}$, and ε_c denotes the coherent field driving the mode $\hat{c}$. γ_a and γ_c are the decay rates associated with the two modes. If $\gamma_c >> \gamma_a$, then linearization around the steady-state value $< \hat{a} >$ (setting $\hat{a} =< \hat{a} > +\hat{A}$, $\hat{A}/ < \hat{a} > << 1$) gives the Fokker-Planck equation for the Wigner function for mode A:

$$\frac{\partial W_A}{\partial t} = \frac{\partial}{\partial z}[(Kz - < \hat{c} > Gz^*)W_A] + \frac{\partial^2}{\partial z \partial z^*}\frac{(KW_A)}{2} + c.c. \tag{13}$$

where

$$K = \gamma_a + |G|^2| < \hat{a} > |^2/\gamma_c,$$

$$\gamma_c < \hat{c} >= \varepsilon_c - G^*| < \hat{a} > |^2/2,$$

$$\gamma_a < \hat{a} >= G < \hat{a} >^* < \hat{c} > . \tag{14}$$

Again $W_A(z, z^*)$ is a Gaussian given by (3). The values of the (Wigner) parameters in (3) are thus found to be

$$t = (1 - |G < \hat{c} > /K|^2]^{-1}/2, \tag{15a}$$

$$\mu = -(G < \hat{c} >)^*[1 - |G < \hat{c} > /K|^2]^{-1}/4K, \tag{15b}$$

$$z_o =< \hat{a} >, \tag{15c}$$

$$\gamma_a\gamma_c < \hat{a} >= G < \hat{a} >^* [\varepsilon_c - G^* < \hat{a} >^2 /2]. \tag{15d}$$

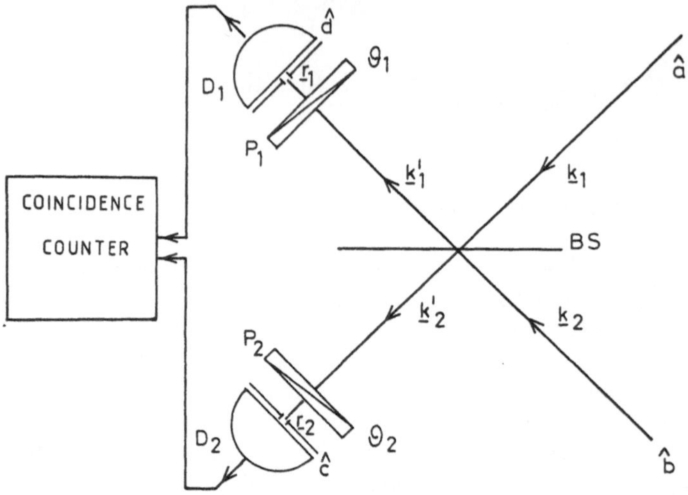

Figure 1. Schematic diagram of a photon correlation experiment for testing the Bell inequality.

Denoting $\varepsilon \equiv G\varepsilon_c/\gamma_a\gamma_c$, for operation above threshold ($\varepsilon > 1$), one obtains from (15d)

$$|z_o|^2 = 2|\gamma_a\gamma_c/G^2|(\varepsilon - 1), \tag{16a}$$

$$|G < \hat{c} > /K| = 1/(2\varepsilon - 1). \tag{16b}$$

The phase of z_o is given by

$$arg[G\varepsilon_c(z_o^*)^2] = 0. \tag{16c}$$

Of course z_o is zero below threshold ($\varepsilon < 1$).

THE BELL INEQUALITY FOR PHOTON CORRELATION EXPERIMENTS

Consider a typical set up for a polarization correlation experiment (see Fig. 1)[9]. Let $\hat{a}$ and $\hat{b}$ be two correlated modes with wave vectors $\underline{k}_1$ and $\underline{k}_2$ coming out of a nonlinear material. These are made to fall from opposite sides on a beam-splitter (BS). $\hat{d}$ and $\hat{c}$ are the mixed beams which arrive at the detectors (D_1 and D_2) placed at points $\underline{r}_1$ and $\underline{r}_2$ with two polarizers (P_1 and P_2) set at variable angles θ_1 and θ_2 in front of them respectively. The Bell inequality in this case has the following well known form[3]:

$$S \equiv P(\theta_1,\theta_2) - P(\theta_1,\theta_2') + P(\theta_1',\theta_2) + P(\theta_1',\theta_2') - P(\theta_1',-) - P(-,\theta_2) \leq 0. \quad (17)$$

$P(\theta_1,\theta_2)$ is the joint probability density of detecting two photons for polarizer settings of θ_1 and θ_2 measured by the coincidence counter. $P(\theta_1,-)$ stands for the probability when the second polarizer is removed. The joint probability density of detection of two photons is given as:

$$P(\theta_1,\theta_2) = K < \hat{d}^+\hat{c}^+\hat{c}\hat{d} >, \quad (18)$$

where K is a constant characterizing the detectors. For correlation experiments[2] with the down-converted signal and idler beams, where $\hat{a}$ and $\hat{b}$ are x- and y-polarized respectively, the polarized (scalar) fields at the detectors are

$$\hat{d}(\underline{r}_1,\theta_1) = X_a\hat{a} + X_b\hat{b}, \quad (19a)$$

$$\hat{c}(\underline{r}_2,\theta_2) = Y_a\hat{a} + Y_b\hat{b}, \quad (19b)$$

with

$$X_a = icos\theta_1\sqrt{R_x}exp(i\underline{k}_1' \cdot \underline{r}_1),$$

$$X_b = sin\theta_1\sqrt{T_y}exp(i\underline{k}_2 \cdot \underline{r}_1),$$

$$Y_a = cos\theta_2\sqrt{T_x}exp(i\underline{k}_1 \cdot \underline{r}_2),$$

$$Y_b = -isin\theta_2\sqrt{R_y}exp(i\underline{k}_2' \cdot \underline{r}_2). \quad (20)$$

Without explicitly specifying the state of the incident field at this stage, we assume that expectations of unpaired operators vanish. For $\underline{k}_1$ and $\underline{k}_2$ parallel to $\underline{k}_2'$ and $\underline{k}_1'$ respectively, from (18) we get

$$P(\theta_1,\theta_2) = K[R_xT_xcos^2\theta_1cos^2\theta_2 < \hat{n}_a(\hat{n}_a-1) > +R_yT_ysin^2\theta_1sin^2\theta_2 < \hat{n}_b(\hat{n}_b-1) > +$$

$$(\sqrt{T_x}\sqrt{T_y}sin\theta_1cos\theta_2 + \sqrt{R_x}\sqrt{R_y}cos\theta_1sin\theta_2)^2 < \hat{n}_a\hat{n}_b >], \quad (21)$$

where $\hat{n}_a \equiv \hat{a}^+\hat{a}$ and $\hat{n}_b \equiv \hat{b}^+\hat{b}$ are the photon-number operators for the two beams incident on the beam-splitter. The probability density when the second polarizer is removed is calculated using unitarity:

$$P(\theta_1,-) = P(\theta_1,\theta_2) + P(\theta_1,\theta_2+\pi/2)$$

$$= K[R_xT_xcos^2\theta_1 < \hat{n}_a(\hat{n}_a-1) > +R_yT_ysin^2\theta_1 < \hat{n}_b(\hat{n}_b-1) > +$$

$$(T_xT_y sin^2\theta_1 + R_xR_y cos^2\theta_1) < \hat{n}_a\hat{n}_b >]. \quad (22)$$

For comparison of the different probabilities, all of them should be scaled by the joint probability density when both polarizers are removed:

$$P(-,-) = P(\theta_1,-) + P(\theta_1+\pi/2,-) = K[R_xT_x < \hat{n}_a(\hat{n}_a-1) > +$$

$$R_yT_y < \hat{n}_b(\hat{n}_b-1) > +(T_xT_y + R_xR_y) < \hat{n}_a\hat{n}_b >]. \quad (23)$$

Let $R_x = T_x = R_y = T_y = \frac{1}{2}$, and choose angles $\theta_1 = \pi/8,\ \theta_1' = 3\pi/8,\ \theta_2 = \pi/4,\ \theta_2' = 0$. Then from (17), we get

$$4S/K = -0.85[< \hat{n}_a(\hat{n}_a-1) > + < \hat{n}_b(\hat{n}_b-1) >] + 0.41 < \hat{n}_a\hat{n}_b > . \quad (24)$$

The Bell inequality is violated whenever $4S/K > 0$, i.e.,

$$\frac{< \hat{n}_a(\hat{n}_a-1) > + < \hat{n}_b(\hat{n}_b-1) >}{< \hat{n}_a\hat{n}_b >} < 0.48. \quad (25)$$

For optimum choice of angles, the right-hand-side of (25) can be made equal to 0.5. It was shown by Chubarov and Nikolayev[7] (CN) that the Bell inequality could be violated in polarization correlation experiments provided photons of each polarization component are sufficiently sub-Poissonian in their statistics [condition (10) above]. In the case treated by CN,

$$< \hat{n}_a(\hat{n}_a-1) >=< \hat{n}_b(\hat{n}_b-1) >\equiv< \hat{n}(\hat{n}-1) >, \quad (26a)$$

$$< \hat{n}_a\hat{n}_b >=< \hat{n} >^2 . \quad (26b)$$

Hence the condition (25) for violation of the Bell inequality gives

$$\frac{< (\Delta\hat{n})^2 > - < \hat{n} >}{< \hat{n} >^2} < -0.76, \quad (27)$$

implying that the photon statistics have to be sufficiently sub-Poissonian in agreement with CN. For the parametric down-conversion process, the photon statistics is nearly Poissonian with mean $< \hat{n} >$:

$$< \hat{n}_a(\hat{n}_a-1) >=< \hat{n}_b(\hat{n}_b-1) >=< \hat{n} >^2, \quad (28a)$$

$$< \hat{n}_a\hat{n}_b >=< \hat{n} > + < \hat{n} >^2 . \quad (28b)$$

Hence from (25) we get

$$\frac{< \hat{n} >^2}{< \hat{n} > + < \hat{n} >^2} < 0.24, \quad (29a)$$

or

$$< \hat{n} > < 0.32. \tag{29b}$$

From (7), in terms of the Wigner parameters of the field given by (15), the condition for violation of the Bell inequality can be expressed as

$$t + |z_o|^2 < 0.82. \tag{30}$$

Once this inequality holds, the fields are necessarily nonlocal.

SUMMARY

Several optical correlation experiments investigating the quantum mechanical paradox relating to Einstein locality exhibit an explicit violation of the Bell inequality. It is our contention that instead of having different inequalities to describe different quantum features of the electromagnetic field, one can identify a generalized description pertaining to the basic definition of a quantum field, namely in terms of the nonclassical distribution function of the field. We have used the results of Agarwal and Adam[4] regarding the Wigner distribution function generated in a large class of nonlinear processes producing correlated emissions, and have given a description[9] of the quantum nonlocality (expressed by the violation of the Bell inequality) in terms of the corresponding Wigner parameters.

ACKNOWLEDGEMENT

AV acknowledges financial support from the University Grants Commission, India.

REFERENCES

1. R. Ghosh, C. K. Hong, Z. Y. Ou and L. Mandel, Phys. Rev. A 34, 3962 (1986); R. Ghosh and L. Mandel, Phys. Rev. Lett. 59, 1903 (1987).
2. Z. Y. Ou and L. Mandel, Phys. Rev. Lett. 61, 50 (1988); C. O. Alley and Y. H. Shih, Phys. Rev. Lett. 61, 2921 (1988).
3. J. S. Bell, Physics 1, 195 (1965). For a review, see J. F. Clauser and A. Shimony, Rep. Prog. Phys. 41, 1881 (1978).
4. G. S. Agarwal and G. Adam, Phys. Rev. A38, 750 (1988).
5. R. Ghosh and G. S. Agarwal, Phys. Rev. A39 (Rapid Comm.), 1582 (1989); Phys. Rev. A42, 665 (1990).
6. See, for example, W. H. Louisell, "Quantum Statistical Properties of Radiation" (Wiley, New York, 1973), p. 173.
7. M. S. Chubarov and E. P. Nikolayev, Phys. Lett. 110A, 199 (1985).
8. G. S. Agarwal and S. Dutta Gupta, Phys. Rev. A39, 2961 (1989). Given their Eq. (2.2), we find that the first relation in Eq. (2.6) should read D=K/2, instead of D=K.
9. A. Venugopalan and R. Ghosh, *Phys. Rev.* A 44, 6109 (1991).

NEAR DIPOLE–DIPOLE INTERACTION EFFECTS IN QUANTUM AND NONLINEAR OPTICS

Charles M. Bowden

Weapons Sciences Directorate, AMSMI-RD-WS
Research, Development, and Engineering Center
U. S. Army Missile Command
Redstone Arsenal, Alabama 35898–5248

INTRODUCTION

It is shown that electromagnetic field propagation in dense nonlinear media (many atoms within a cubic wavelength) results in a nonlinear modified form of the Bloch–Maxwell formulation for two–level atoms. In particular, microscopic near dipole–dipole interaction causes an atomic excitation–dependent resonance frequency renormalization in the macroscopic Bloch equations, describing the medium, which are coupled to the macroscopic field equation. For optically thin media, of thickness less than a resonance wavelength, and in the mean field limit, the interaction produces, in addition, subradiant and superradiant decays. The resulting renormalized Bloch–Maxwell equations are rich with respect to predictions of new phenomena obtained from them.

It is well known that the microscopic, local static electric field, $\underline{E}_L$, which couples with an atomic or molecular dipole moment in a simple cubic lattice, is related to the macroscopic field $\underline{E}$ and volume polarization, $\underline{P}$, in the dielectric medium, by the Lorentz–Lorenz relation[1],

$$\underline{E}_L = \underline{E} + \left(\frac{4\pi}{3} + s\right)\underline{P} \ , \tag{1}$$

where s is a structure factor to account for crystal symmetry environment less than cubic. It is generally assumed that the relation, Eq. (1), is also valid for isotropic, homogeneous media, with $s = 0$, as well. An important consequence of Eq. (1) in condensed matter, as well as for liquids, is that it, together with Maxwell's equations, leads to the well known Clausius–Mossotti equation relating the microscopic atomic or molecular polarizability, α, to the macroscopic dielectric function, ϵ, of the medium,

$$\alpha = \frac{3}{4\pi N}\left(\frac{\epsilon - 1}{\epsilon + 2}\right) , \tag{2}$$

where N is the volume density.

Recent Developments in Quantum Optics, Edited
by R. Inguva, Plenum Press, New York, 1993

The corresponding problem for time varying, propagating, fields is fundamentally different, since causality and retardation effects must be explicitly taken into account. It has been shown[2], however, based upon the extinction theorem[3], that for a stationary, monochromatic, time-dependent, propagating field in a linear isotropic, homogeneous medium, the relation (1) (with s = 0) is valid and the Clausius–Mossotti relation (2), follows. This leaves open the question concerning propagating fields in nonlinear media.

The purpose of this paper is to present the development of the Maxwell–Bloch formulation for electromagnetic field propagation in dense, nonlinear media, having many atoms within the cubic resonance wavelength. The model consists of a medium of saturable two–level atoms, coupled to the electromagnetic field in semiclassical approximation. We consider a homogeneously broadened system, although inhomogeneous broadening can be readily addressed, and the effects of induced dipole-dipole interaction upon the macroscopic equations of motion and propagation.

MODEL

The model consists of a collection of homogeneously broadened, two–level atoms, coupled to the electromagnetic field in the dipole and semiclassical approximations. The system is described by the optical Bloch equations[4], for the level population difference, W , and the slowly–varying, off–diagonal matrix element, R_{ab} , which is proportional to the slowly–varying polarization envelope, $\mathcal{P}$,

$$\frac{\partial W}{\partial t} = -\gamma_L (W+1) - \frac{u}{\hbar}\left[\varepsilon_L^* R_{ab} + \text{c. c.}\right], \tag{3}$$

$$\frac{\partial R_{ab}}{\partial t} = -\left(i\Delta + \gamma_T\right) R_{ab} + \frac{u}{2\hbar}\varepsilon_L W , \tag{4}$$

where $|a\rangle$ represents the upper level and $|b\rangle$ the lower level, separated, at zero field strength, by the atomic energy level spacing, $\hbar\omega_0$. Furthermore, $\Delta = \omega - \omega_0$ is the detuning of the field central frequency from atomic resonance, u is the matrix element of the transition dipole moment, and γ_L and γ_T are the homogeneous relaxation and dephasing rates, respectively. The microscopic slowly–varying electric field envelope, ε_L , is the electric field which drives an atom of polarization amplitude, $\mathcal{P}$, in the dipole interaction approximation, $H_I = -\underline{E}_L \bullet \underline{P}$.

The macroscopic electric field amplitude, $\underline{E}$, is related to the macroscopic polarization in the medium, P, by Maxwell's wave equation,

$$\nabla^2 \underline{E} - \frac{1}{c^2}\frac{\partial^2 \underline{E}}{\partial t^2} = \frac{4\pi}{c^2}\frac{\partial^2 \underline{P}}{\partial t^2} , \tag{5}$$

where c is the speed of light in vacuum. In terms of slowly–varying amplitudes, ε, and $\mathcal{P}$,

$$E = \frac{1}{2}\left[\varepsilon\, e^{-i(\omega t - kz)} + \text{c. c.}\right] \qquad (6)$$

$$P = \frac{1}{2}\left[\mathcal{P} e^{-i(\omega t - kz)} + \text{c. c.}\right] \qquad (7)$$

$$\mathcal{P} = i\, u\, n\, R_{ab} \qquad (8)$$

where n is the density of atoms in the medium.

It is to be noted that the microscopic field, ε_L in Eqs. (3) and (4), is not, in general, equivalent to the macroscopic field, ε, of Maxwell's equation, Eqs. (5) and (6). Fundamentally, at least, ε_L, the field that drives an atom, does not contain the atom's self–field, whereas the macroscopic Maxwell field, Eqs. (5) and (6), contain all local fields present[5]. The fundamental problem for dense media, media having many atoms within a cubic resonance wavelength on the average, is to establish the relationship between the microscopic field, ε_L, which drives an atom, and the corresponding macroscopic Maxwell field, ε.

In an earlier derivation[6], we developed the set of Bloch–like equations for dense media, coupled to the Maxwell wave equation in the slowly–varying envelope approximation (SVEA), within the framework of quantum electrodynamics. This was done for a long pencil–like medium, and we derived a set of extended Maxwell–Bloch (EMB) equations for a homogeneous medium having many atoms within a cubic resonance wavelength. Here, we consider a similar medium, but with large Fresnel number.

For this purpose, we consider the microscopic field $\underline{E}_L(\underline{r}_i,t)$, which interacts with an atom of the medium located at position, $\underline{r}_i$, and consider all atoms stationary. The field is taken as plane wave and linearly polarized. The field can, in terms of the reaction field due to the induced dipoles in the medium[6], be written in the form,

$$\underline{E}_L(\underline{r}_i,t) = E_{ex}(\underline{r}_i,t) + \sum_{j=1}^{N} B_{ij}\, e^{-i\underline{k}_o \bullet (\underline{r}_i - \underline{r}_j)}\, R_{ab}\left(t - \frac{r_{ij}}{c}\right) \qquad (9)$$

where[6]

$$B_{ij} = \frac{3}{2}\beta\left\{\left[\hat{\underline{p}}_i \bullet \hat{\underline{p}}_j - \left(\hat{\underline{p}}_i \bullet \hat{\underline{r}}_{ij}\right)\left(\hat{\underline{p}}_j \bullet \hat{\underline{r}}_{ij}\right)\right] F_I\left(k r_{ij}\right) + \left(\hat{\underline{p}}_i \bullet \hat{\underline{r}}_{ij}\right)\left(\hat{\underline{p}}_j \bullet \hat{\underline{r}}_{ij}\right) F_{II}\left(k r_{ij}\right)\right\} \qquad (10)$$

and

$$F_I(k\,r_{ij}) = e^{ik r_{ij}} \left(\frac{1}{k^2 r_{ij}^2} + \frac{i}{k^3 r_{ij}^3} - \frac{i}{k\,r_{ij}} \right) , \tag{11a}$$

$$F_{II}(k\,r_{ij}) = e^{ik r_{ij}} \left(- \frac{2i}{k^3 r_{ij}^3} - \frac{2}{k^2 r_{ij}^2} \right) , \tag{11b}$$

and $E_{ex}(\underline{r}_i,t)$ is an externally applied field. Here, $k = \omega/c$, $\underline{r}_{ij} = \underline{r}_i - \underline{r}_j$, and $\hat{\underline{r}}_{ij}$ and $\hat{\underline{p}}_i$ are unit vectors for the relative spacial separation of the dipole moments and the moments, respectively, and

$$\beta = \frac{2\,|u|^2 k^3}{3\hbar} . \tag{11c}$$

Referring to Figure 1, the field, Eq. (9), can be written in terms of contributions from three distinct zones: (a) the near zone, $E'(\underline{r}_i,t)$, $r_{ij} << \lambda$; (b) the far, or radiation, zone, $E_R(\underline{r}_i,t)$, $\underline{r}_{ij} >> \lambda$; (c) and the intermediate region, $E_{\mathcal{I}}(\underline{r}_i,t)$, $r_{ij} \approx \lambda$, where λ is an atomic resonance wavelength. Thus,

$$E_L(\underline{r}_i,t) = E_{ex}(\underline{r}_i,t) + E(\underline{r}_i,t) + E'(\underline{r}_i,t). \tag{12}$$

The near field, $E'(\underline{r}_i,t)$, due to the near dipoles (near in the z–coordinate) is taken as the contribution from all the dipoles contained in the slab of thickness Δz, at z, containing the atom located at $\underline{r}_i$, Figure 1. Thus, E' is the microscopic contribution, and the remainder will be taken as the macroscopic field. Contrary to the static field analog, the microscopic field, E' , does not vanish for an isotropic, homogeneous medium in the plane wave approximation. Also, for propagating, time–varying, fields, retardation and causality effects must be explicitly taken into account. We proceed to calculate the contributions of the fields in the respective zones.

A. Near Zone – Microscopic Field

We use cylindrical geometry, and consider a plane wave field propagating in the positive z–direction. For simplicity, the atom at position r_i is located on the z–axis, as illustrated in Figure 2. All the dipole moments in the medium are considered plane–polarized and oriented along the x–axis. We proceed to calculate the field at position, r_i , due to all the atomic dipoles contained in the cylinder of thickness $d = \Delta z$, Figure 1, at z containing the atom located at r_i . Considering all the atoms identical within the cylindrical volume, we have, from (9), after replacing the summation with an integration in the continuum limit,

$$E'(z,t) = R_{ab}(z,t)\,\Gamma(r,z,\rho) \quad ,$$

$$\Gamma = \frac{3}{2}\beta n \int_0^{2\pi} d\phi \int_{-d/2}^{d/2} dz \int_{\rho_{min}}^{\rho_{max}} \rho\, d\rho\, B \ , \tag{13}$$

$$B = \sin^2\phi\ F_I(kr) + \cos^2\phi\ F_{II}(kr)$$

$$+ \frac{|z - z_i|^2}{r^2} \cos^2\phi \left[F_I(kr) - F_{II}(kr)\right], \qquad (14)$$

where $r = [\rho^2 + |z - z_i|^2]^{1/2}$, and we must exclude a small volume about the atom at $z = z_i$, $\frac{4}{3}\pi\delta^3 = \frac{1}{n}$, and $r_{ij} \rightarrow r$. The integrals indicated in Eqs. (13) and (14) are easily performed, and in the limit $\delta \rightarrow 0$, $d << \rho_{max}$, the result is

$$\Gamma = \frac{2 i \pi \beta n}{k^3} + \frac{3 \pi \beta n}{k^2} d, \qquad (15)$$

yielding two terms, the first, which is purely imaginary, is independent of the thickness of the cylinder, d, and the other, which is purely real, is dependent upon d. Substituting the value for β, Eq. (11c) into the above result, gives

$$\Gamma = i \frac{4\pi |u|^2 n}{3\hbar} + \frac{2\pi |u|^2 nk}{\hbar} d. \qquad (16)$$

When the local, microscopic field is eliminated from Eqs. (3) and (4) in terms of the macroscopic field, the first term in (16) leads to a population difference, W, dependent dynamical shift in the atomic resonance frequency in Eq. (4), and the second term contributes W–dependent cooperative relaxation and dephasing in Eqs. (3) and (4), i.e., superradiant and subradiant decays. This would be the result, then, for a thin film of thickness d, governed by the dynamical equations,

$$\frac{dW}{dt} = -\gamma_L (W+1) - \frac{u}{\hbar}\left[\varepsilon^* R_{ab} + c.\,c.\right] - \frac{1}{\tau_R} \left|R_{ab}\right|^2, \qquad (17)$$

$$\frac{d R_{ab}}{dt} = - i\left(\Delta - \bar{\epsilon} W\right)R_{ab} + \frac{u}{2\hbar}\varepsilon W - \left(\gamma_T - \frac{1}{2\tau_R} W\right) R_{ab}. \qquad (18)$$

Here,

$$\tau_R^{-1} = \frac{4\pi k |u|^2}{\hbar} nd, \quad \bar{\epsilon} = \frac{4\pi}{3} \frac{|u|^2}{\hbar} n. \qquad (19)$$

These equations, therefore, constitute the EMB equations for a thin film of atoms driven by a field ε, in the limit of a density such that there are many atoms within a cubic resonance wavelength. These results were derived in our previous work[6] under the same conditions, except for the value of ϵ, which, in reference 6, differs from Eq. (19) by a factor of 1.5, which is due to an error in the calculation in Appendix B of reference 6. Similar results were obtained recently by Benedict, et al.[7], who discussed several applications to reflection, transmission and switching.

In the limit, $kd = (2\pi d/\lambda) << 1$, the second term in Eq. (16) may be neglected, and we consider the contribution of the near dipoles in the limit that $\Delta z \rightarrow 0$. From Eqs. (12) and (16), we have

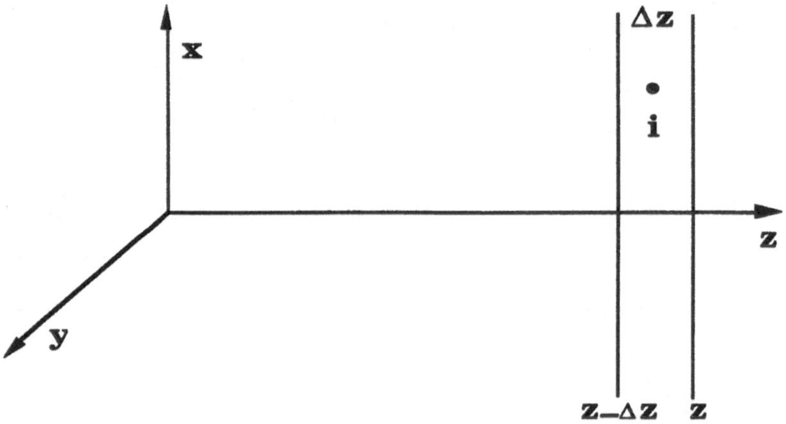

Figure 1. The plane wave, plane–polarized, field propagates in the positive z–direction. Geometry for calculating the microscopic field at the atom located at $r = r_i$: the atom at $r = r_i$ is sandwiched within a cylinder of thickness, Δz, at $z = z_i$.

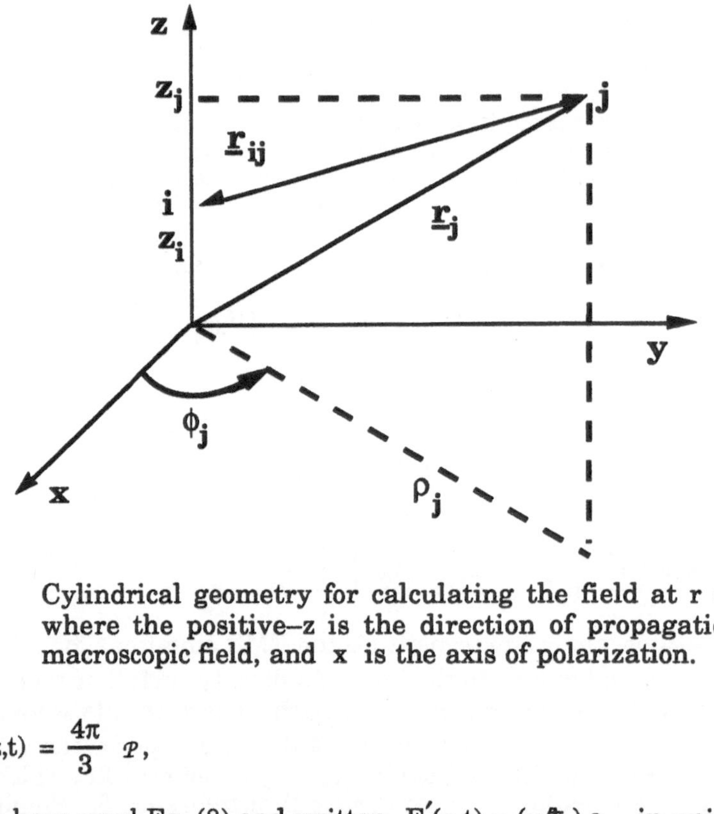

Figure 2. Cylindrical geometry for calculating the field at $r = r_i = z_i$, where the positive–z is the direction of propagation of the macroscopic field, and x is the axis of polarization.

$$E'(z,t) = \frac{4\pi}{3}\, \mathcal{P}, \tag{20}$$

where we have used Eq. (8) and written $E'(z,t) = (u/\hbar)\, \varepsilon_L$ in units of Rabi frequency.

B. Far and Intermediate Zones – Macroscopic Field

To compute the remaining field contributions, we use Eq. (9) in the continuum limit, and the geometry of Fig. 1, to write,

$$E(\mathbf{r},t) = \int_{[\Delta z]} d^3r' \, R_{ab}\left(\underline{r}', t - \frac{|\underline{r}-\underline{r}'|}{c}\right) e^{-i\underline{k}_o \cdot (\underline{r}-\underline{r}')} B\left(|\underline{r}-\underline{r}'|\right), \qquad (21)$$

where $[\Delta z]$ indicates that the region contained in the slab of length Δz at z , in Figure 1, has been excluded from the integration. For plane waves, we assume that the argument of R_{ab} in Eq. (21) is independent of the x and y coordinates,

$$R_{ab}\left(\underline{r}', t - \frac{|\underline{r}-\underline{r}'|}{c}\right) \approx R_{ab}\left(z', t - \frac{|z-z'|}{c}\right). \qquad (22)$$

With this condition, and using the cylindrical geometry depicted in Figures 1 and 2, Eq. (21) can be written explicitly in the form,

$$E(z_i,t) = n \int_0^{2\pi} d\phi \int_0^{z_i-d} dz' \int_0^{\rho_{max}} \rho \, d\rho \, B\left(|z'-z_i|, \rho, \phi\right)$$

$$\times \, e^{-ik_o(z_i-z')} R_{ab}\left(z', t - \frac{|z_i-z'|}{c}\right), \qquad (23)$$

with

$$B\left(|z_i-z|, \phi, r\right) = \frac{3}{2}\beta \left\{ \sin^2\phi \, F_I(kr) + \cos^2\phi \, F_{II}(kr) \right.$$

$$\left. + \frac{|z-z_i|^2}{r^2} \cos\phi \left[F_I(kr) - F_{II}(kr) \right] \right\}, \qquad (24)$$

where, from Figure 2, $\rho^2 = r^2 - |z-z_i|^2$, and F_I and F_{II} are given by Eqs. (11) with $r_{ij} \to r$. The integrations in ϕ and ρ can be taken immediately, resulting in the relation,

$$E(z_i,t) = \frac{3\pi\beta n}{k^2} \int_0^{z_i-d} dz' \, R_{ab}\left(z', t - \frac{|z_i-z'|}{c}\right)$$

$$- \frac{3}{2} \frac{\pi\beta n}{k^2} \int_0^{z_i-d} dz' \, e^{-ik_o(z_i-z')} R_{ab}\left(z', t - \frac{|z_i-z'|}{c}\right)$$

$$\times \, e^{iR_{max}} \left[1 + i k^2 \frac{|z'-z_i|^2}{R_{max}^3} \right], \qquad (25)$$

$$R_{max}^2 = k^2 \rho_{max}^2 + k^2 |z-z_i|^2 .$$

For $k\,\rho_{max} >> 1$, the second term in Eq. (25) vanishes, giving

$$E\left(z_i,t\right) = \frac{3\pi\beta n}{k^2} \int_0^{z_i-d} dz'\, R_{ab}\left(z',t - \frac{\left|z_i-z'\right|}{c}\right) . \qquad (26)$$

It is to be noted that the result, Eq. (26), is identical to the result which is obtained using the radiation zone approximation[6] , i.e.,

$$B \approx -\frac{3i\beta}{2}\frac{e^{ikr}}{kr}\sin^2\alpha \quad , \quad \cos\alpha = \frac{\hat{x}\bullet r}{|r|} . \qquad (27)$$

The intermediate zone contribution is estimated by taking $\Delta z \rightarrow 0$ and $z_i = d$ in Eq. (26) to yield exactly the result corresponding to the second term in Eq. (16) for the thin film. Thus, if we take $d\rightarrow 0$, and $z_i\rightarrow z$ in Eq. (26), we include both the far and intermediate zone contributions to the macroscopic field,

$$E(z,t) = \frac{3\pi\beta n}{k^2}\int_0^z dz'\, R_{ab}\left(z',t-\frac{\left|z-z'\right|}{c}\right) . \qquad (28)$$

It is noted, here, that only the retarded dipoles contribute to the field at z , satisfying causality requirements.

By taking the appropriate partial derivatives of both sides of Eq. (28), we arrive at Maxwell's wave equation in the SVEA,

$$\frac{\partial E}{\partial t} + c\,\frac{\partial E}{\partial z} = \frac{3\pi\beta c u}{k^2} R_{ab} . \qquad (29)$$

Combining Eq. (20) with Eq. (12),

$$E_L(z,t) = E_{ex}(z,t) + E(z,t) + \frac{4\pi}{3}\mathcal{P}(z,t) , \qquad (30)$$

where $E(z,t)$ is the macroscopic field which is governed by Maxwell's equation, Eq. (29). It is noted that Eq. (30) is identical in form to the "local field correction", for static fields[1], the well–known Lorentz–Lorenz correction, although for different reasons and conditions.

If Eq. (30) is used in the Bloch equations, Eqs. (3) and (4), to eliminate ε_L , the result is $(E_L \equiv u\varepsilon_L/\hbar)$

$$\frac{dW}{dt} = -\gamma_L(W+1) - \frac{u}{h}\left[\varepsilon^* R_{ab} + c.\,c.\right] \qquad (31)$$

$$\frac{dR_{ab}}{dt} = -i\left[\Delta - \epsilon W\right]R_{ab} + \frac{u}{2\hbar}\varepsilon W - \gamma_T R_{ab} , \qquad (32)$$

where

$$\bar{\epsilon} = \frac{4\pi}{3\hbar}|u|^2 n , \qquad (33)$$

and, from Eq. (29), the associated Maxwell's equation in SVEA, is given by

$$\frac{1}{c}\frac{\partial\varepsilon}{\partial t} + \frac{\partial\varepsilon}{\partial z} = \frac{3\pi\beta}{k^2}R_{ab} . \qquad (34)$$

Thus, Eqs. (31) – (34) form a self-consistent set of EMB equations which govern temporal and spacial evolution of the electric field and atomic exciton for isotropic, homogeneous dense media modeled by a homogeneously broadened, saturable, two-level system.

CONCLUSIONS

It is noted that the transformation, Eq. (30), leaves the form of Eq. (3) invariant, as indicated by Eq. (31). However, Eq. (4) now becomes nonlinear in the atomic variables, Eq. (32), resulting in an atomic excitation, W , dependent shift in the atomic resonance frequency, which is dependent upon the strength of the near dipole–dipole interaction, Eq. (33). This aspect renders the EMB equations, Eqs. (31) – (34) sufficiently rich in new optical phenomena to open an entire field of investigation.

Already, we have analyzed intrinsic optical bistability[6] (IOB) in the steady-state solutions to Eqs. (31)–(34), as well as quantum noise effects, bimodality and first passage time effects in IOB[8,9]. Propagational asymptotic conditions and analytic solutions for self-induced transparency (SIT) have been derived[10,11], and clear and unique signatures of the near dipole–dipole contribution identified. The spacial and temporal evolution of the first-order phase transition in IOB has been treated[12,13]. Others have examined dynamical effects and induced cooperativity in transmission and reflection from thin films[7] and linear and nonlinear spectral shifts[14,15].

REFERENCES

1. J. D. Jackson, "Classical Electrodynamics," second edition (John Wiley, New York, 1962), Ch. 4.
2. M. Born and E. Wolf, "Principles of Optics," (MacMillian Co., New York, 1964), Ch. 2.
3. P. P. Ewald, **Ann. D. Physik.** 49:1 (1916). C. W. Oscen, **Ann. D. Physik.** 48:1 (1915).
4. L. Allen and J. H. Eberly, "Optical Resonance and Two-level Atoms," (John Wiley, New York, 1975).
5. J. Van Kranendonk and J. E. Sipe, in "Progress in Optics XV," edited by E. Wolf (North-Holland, Amsterdam, 1977), p. 245.
6. Y. Ben-Aryeh, C. M. Bowden, and J. C. Englund, **Phys. Rev. A** 34: 3917 (1986).
7. M. G. Benedict, V. A. Malyshev, E. D. Trifonov, and A. I. Zaitsev, **Phys. Rev. A** 43:3845 (1991).
8. Y. Ben-Aryeh and C. M. Bowden, **Opt. Comm.** 72:335 (1989).
9. Y. Ben-Aryeh and C. M. Bowden, **J. Opt. Soc. Am. B** 8:1168 (1988).
10. C. R. Stroud, C. M. Bowden, and L. Allen, **Opt. Comm.** 67:387 (1988).
11. C. M. Bowden, A. Postan, and R. Inguva, **J. Opt. Soc. Am. B** 8:1081 (1991).
12. R. Inguva and C. M. Bowden, **Phys. Rev. A** 41:1670 (1990).
13. Y. Ben-Aryeh, C. M. Bowden, and J. C. Englund, **Opt. Comm.** 61:147 (1987).
14. F. Friedberg, S. R. Hartmann, and J. T. Manasseh, **Phys. Rev. A** 39:3444 (1989).
15. F. Friedberg, S. R. Hartmann, and J. T. Manasseh, **Phys. Rev. A** 40:2446 (1989).

QUANTUM FIELD THEORY OF DIELECTRICS AND SOLITONS

P.D. Drummond

NTT Basic Research Laboratories
Musashino-Shi, Tokyo 180
Japan

ABSTRACT

Recent experiments with laser generated pulses in optical fibers have resulted in the first direct observation of quantum effects in solitons. This paper reviews the theoretical background to this discovery. The theory is derived using the method of macroscopic quantization of Maxwell's equations. It is necessary to include the effects of both nonlinearity and dispersion in the equations. This results in a relatively simple quantum field theory, namely the quantum nonlinear Schrödinger equation. As well as describing propagation, the field theory is able to treat the nonclassical nature of the radiation produced. The resulting experimental effect is quantum phase diffusion, resulting in quadrature squeezing.

1. INTRODUCTION

In agreement with theoretical predictions[1,2,3,4,5], recent experiments[6] at IBM Research Laboratories have provided the first *direct* evidence for the quantum evolution of a soliton. By this I mean that the experiments have given results which can only be explained by quantizing the relevant equations of motion. In this case, the quantum field theory turns out to be the well-known quantum nonlinear Schrödinger equation. By comparison, while certain solid-state transport processes are soliton enhanced, thermal activation appears to hide the uniquely quantum-mechanical properties. The interesting thing that stands out in the IBM experiments (and in more recent experiments at NTT Laboratories[10]), is that these experiments treat solitons in the most direct possible way. Pulses from a laser are first injected into the nonlinear medium. They then propagate for a period of time, are extracted, and a quantum measurement is performed.

While many quantum field theories have been investigated theoretically, only a few are both realistic and soluble. Most quantum field theories are either tractable but unphysical, or else physical but intractable. The exceptions, of course, are quantum electrodynamics[7] and quantum chromodynamics[8], which have had amazing successes. Despite this, many quantum field theories found to have interesting properties[9], do not yet have any experimental verification. However, it is only when a theoretical idea is testable in experiment, that it really becomes interesting to most physicists. It is now possible to duplicate the properties of simple quantum field theories in nonlinear dielectrics, and carry out quantum-limited experiments. Thus, theoretical and experimental techniques developed in laser physics and quantum optics, are opening a new window onto quantum field theory.

In the IBM experiments, the initial injected state corresponds to a linear superposition of different photon numbers. Quantum effects are observed that originate in quantum interference between the underlying fundamental quantum solitons. These nonlinear bound states are the eigenstates of the quantum field theory Hamiltonian. In this sense, the IBM experiments are sensitive to the spectroscopy – the level structure – of the solitons. This seems to be the first observation of this type. In the recent NTT experiments, quantum properties of soliton collisions[11] are observed.

To the high energy physicist, this might come as a surprise. The interaction energies involved are infinitesimal, and the experiments require subtle phase measurements. Despite this, however, the IBM experiments are now revealing a surprisingly good agreement with theoretical predictions. This is perhaps all the more surprising when the number of particles involved is as large as it is: more than 10^9 photons are bound together in each soliton. Thus, as well as testing quantum field theory, the experiments test how well quantum mechanics works at large particle number. These measurements complement the more usual phase-insensitive measurements performed in traditional scattering experiments.

In this review of the theoretical background to these experiments, I shall focus on the issue of deriving the relevant quantum field theory. Since the binding energies are low, the underlying quantum theory is actually quantum electrodynamics, or QED. However, while QED is well-studied, it is not at all straightforward to derive the nonlinear quantum theory of electromagnetic propagation in a real dielectric, from first principles. This is particularly true when it is realized that the experiments take place in silica fibers. Not only are the fibres inhomogeneous, but silica has a disordered, covalently bonded, lattice structure.

For this reason, it is useful to start from the viewpoint of macroscopic quantization - although many of the results obtained can be also treated in simplified microscopic models. In macroscopic quantization, the usual Dirac quantization procedure[12] is implemented at the level of the average electromagnetic fields[13], rather than at the microscopic level of the lattice structure. This allows the simpler theory of the nonlinear quantum Schrödinger equation to be extracted from the complexities of the full interacting field-matter system. We can think of this as having the same relation to QED as this well established theory has to quantum chromodynamics - or perhaps as QCD has to even deeper field theories. In other words, it is a low-energy approximation with a restricted, but well-defined region of validity. The extent that this procedure is valid is of course tested in the experiments.

An outline of this review is as follows. In Section (2), the relevant Maxwell equations for a nonlinear dielectric are presented in a form suitable for later quantization. In Section (3), the dispersionless, nonlinear quantization method of Hillery and Mlodinow is reviewed. In Section (4), the Lagrangian is extended to include dispersion. A suitable modal expansion is presented in Section (5), to allow quantization of modes inside a restricted bandwidth. In Section (6), the resulting photon-polariton annihilation and creation operators are used to define continuum fields, whose equation of motion is the nonlinear quantum field theory of interest. In Section (7), the results are summarized, together with a comparison between this approach and some others.

2. CLASSICAL NONLINEAR DIELECTRIC THEORY

The starting point of this theoretical investigation will be the classical field equations. For simplicity, the derivation here will be just one-dimensional, in terms of fields scaled by a nominal transverse scale length R. This is equivalent to the physical situation of waves propagating in a wave-guide, where R determines a transverse scale-length. Single-mode optical fibers, as used in communications systems, can be described to an excellent approximation in this way. A full spatial theory can be worked out using similar techniques[14].

In one dimension, Maxwell's equations are simply

$$\begin{aligned} \partial_z E_x &= -\mu \partial_t H_y \\ \partial_z H_y &= -\partial_t D_x. \end{aligned} \tag{1}$$

Here, I suppose that all electric fields are x-polarized, and all magnetic fields are y-polarized. From now on, the polarization indices will be therefore omitted. The magnetic susceptibility is μ, and the nonlinearity is provided through the relationship of D to E. I will take the viewpoint that D is a canonical coordinate, and E is a function of D. This is similar to the mechanical problem of a position dependent force, with D analogous to the position. In fact, the analogy goes deeper than this. Just as with position, it is possible to specify the displacement field via the deposition of charges along the x-coordinate, and measure the resulting force via the potential difference.

This experiment is difficult, since the frequency range of interest is around 10^{14} Hz. It does show, however, that it is operationally meaningful to define a causal dielectric response function as a functional Taylor series, which gives the electric field E in terms of the displacement field D at earlier times:

$$E(t) = E[D] = \sum_{n>0} \int_0^\infty \cdots \int_0^\infty \zeta^{(n)}(\tau_1, \cdots \tau_n) D(t-\tau_1) \cdots D(t-\tau_n) d^n \vec{\tau}. \tag{2}$$

This equation is the inverse of the Bloembergen[15] expansion of nonlinear optics, which is why the expansion coefficients might look unfamiliar. The relationship between the two is essentially a canonical transformation, which is necessary since Bloembergen's work started with the microscopic Hamiltonian in the **p.A** form. A simple variable change or canonical transformation, first introduced by Goeppert-Meyer[16] gives the **r.D** or multipolar[17] interaction Hamiltonian, which is often used in modern calculations[18]. This leads to the above expansion. Many texts write the dipole coupling using the *total* **E** field –– and then claim to obtain the polarization in terms of this field. However, the form of the response function given above is simpler to justify microscopically. Accordingly, equations (1) and (2) combine to define the equations of motion I will use.

As one might expect, the electromagnetic energy is simply composed of magnetic plus electrostatic terms[15]. The nonlinear electrostatic term arises from integrating the force (E) with respect to the displacement (D). The total energy in length L is therefore given, provided $E(-\infty) = 0$, by

$$W = \int_0^L \left[\frac{1}{2}\mu H^2 + \int_{-\infty}^t E(\tau)\dot{D}(\tau) d\tau \right] dz. \tag{3}$$

With these results for the equations of motion and the energy, it is now possible to attempt a canonical quantization of the theory.

3. QUANTUM FIELD THEORY

The essential point of macroscopic canonical quantization is that the equations of motion in the classical limit must come from a Lagrangian whose corresponding canonical Hamiltonian is the system energy. Both requirements - the correct equations of motion, and the correct energy - are necessary. It is easy to demonstrate examples of a Hamiltonian with the right energy, but the wrong equations of motion[13] - or vice versa. To fulfill both these requirements in one Lagrangian takes care, since the equations we are interested in are the nonlinear, dispersive version of Maxwell's equations that hold for a real dielectric. I will deliberately neglect, for simplicity, the dielectric internal degrees of freedom – apart from those that are already included as the polarizability term in Maxwell's equations.

The problem, then, is to find a Lagrangian that generates (1) and (2), while defining a canonical Hamiltonian equal to (3). In the dispersionless case of response functions $\zeta^{(n)}$ that are instantaneous, the answer is known from an insightful paper[19] of Hillery and Mlodinow. It is simplest to present their results for a charge-free medium, using a dual potential. This is just like the usual vector potential, except that the roles of the magnetic and electric fields are interchanged. The dual potential Λ is defined by:

$$\begin{aligned} \partial_t \Lambda &= H \\ \partial_z \Lambda &= -D. \end{aligned} \tag{4}$$

The Lagrangian is then:

$$\mathcal{L} = \int_0^L \left[\frac{1}{2}\mu \dot{\Lambda}^2 - U(\partial_z \Lambda) \right] dz, \tag{5}$$

where:

$$U(D) \equiv \int_0^D E[D] \cdot dD. \tag{6}$$

It is easy to verify that $\mathcal{L}$ has all the required properties. It is also local, involving only products of fields at the same location, which is usually regarded as a desirable physical property. In addition, it is not even necessary for U to be a homogeneous (spatially uniform) function. The theory is still correct[14] if all the linear and nonlinear dielectric properties are spatially varying. The corresponding Hamiltonian is

$$\mathcal{H} = \int_0^L \left[\frac{1}{2\mu} \Pi^2 + U(\partial_z \Lambda) \right] dz. \tag{7}$$

After quantizing, and taking the large L limit, the canonical momentum and equal-time commutators are:

$$\begin{aligned} \hat{\Pi} &= \mu \dot{\hat{\Lambda}} \\ \left[\hat{\Lambda}(t, z_1), \hat{\Pi}(t, z_2) \right] &= i\hbar \delta(z_1 - z_2) \end{aligned} \tag{8}$$

Accordingly, these equations define a one-dimensional quantum field theory which might describe a nonlinear dielectric waveguide, like an optical fiber.

The instantaneous response approximation leads to a difficulty which will be resolved in the next section. This is that while the commutator of the electric displacement field and the magnetic field is the usual one, the electric field commutation relation with the magnetic field is modified from its free field value[20]. Even in the linear case,

$$\left[\hat{E}(t,z_1),\hat{B}(t,z_2)\right] = -\frac{i\hbar}{\varepsilon}\frac{\partial}{\partial z_1}\delta(z_1-z_2) \tag{9}$$

which clearly involves the dielectric permittivity ε, rather than the free field permittivity ε_0. However, an interaction is usually thought not to change equal-time commutation relations. What has happened here? Surely, equal-time electromagnetic commutation relations must be invariant when interactions are introduced.

Apart from this, the whole approach is very simple, with some enlightening results. In particular, there are obvious advantages in using the displacement field as a fundamental canonical field, as opposed to the electric field. These advantages are also well-known in the multipolar theory of microscopic quantum electrodynamics. It is also possible to use the vector potential, both microscopically and macroscopically. However, this leads to nonlinear functions of field time-derivatives, which are not so easily treated as the above theory. Nevertheless, the vector potential approach is certainly possible. What does not seem possible is to use the electric field as a canonical variable. While there appears to be no fundamental reason why this should be the case, all attempts along these lines have so far led either to incorrect equations of motion, or else to an incorrect Hamiltonian.

4. DISPERSION

The instantaneous nonlinear field theory omits an important physical property. This is the dispersion in the dielectric, which is due to the causal time-delay between changes in D and corresponding changes in E. Because the electrons have a finite mass, the polarization – and hence the electric field – is unable to respond instantaneously. This causes the group velocity in a dielectric to become frequency-dependent, which is essential to soliton formation.

This physical property was not included in the above Lagrangian. In fact, the previous theory is really only able to treat monochromatic fields. Even then, the effects of dispersion are felt through the Hamiltonian. Because the dielectric can store energy too, it's time-dependent response changes the total energy. Solving this dispersion problem, as we shall see, gives us not only the correct energy, but also a better understanding of the commutation relations.

Dispersion is often attributed to the delayed response of the dielectric polarization to the electric field. However, this is really an inappropriate description. The total electric field of a macroscopic system *includes* the field due to the polarization. It therefore seems more natural to regard the displacement field as inducing the dielectric response of the medium. We also know[16,17] that at the microscopic level the multipolar canonical Hamiltonian has a fundamental coupling of the polarization field to the displacement field. Other possible forms of the microscopic Hamiltonian include the minimal coupling to the vector potential (as used in Bloembergen's nonlinear optics calculations[15]), or even hybrid combinations[14]. However, there is *no* known exact microscopic Hamiltonian involving the total electric field. For this reason, it seems preferable to keep using the displacement field as the fundamental canonical variable.

At the microscopic level, it is transparent how to evaluate the dielectric response function in the form used here. One has to specify the displacement field, use the

Hamiltonian time-evolution to compute the polarization, and hence obtain the total electric field. Of course, the use of macroscopic fields would require local-field corrections in order to be related to calculated microscopic quantities. In this paper I will take the pragmatic approach of minimal quantization, using the measured permittivities and frequency-dependent refractive indices to evaluate the response functions. This is less accurate than a complete microscopic theory, but can at least be expected to reproduce the observable macroscopic behaviour. Hence, define

$$\zeta^{(n)}(\omega;\omega_1,\cdots\omega_n)=\delta_{-\omega,\omega_1+\cdots+\omega_n}\int_0^\infty\cdots\int_0^\infty e^{i(\omega_1 t_1+\cdots+\omega_n t_n)}\zeta^{(n)}(t_1,\cdots t_n)d^n\vec{t}, \tag{10}$$

so that, compared to the usual dielectric permittivity,

$$\zeta^{(1)}(\omega;-\omega)=[\varepsilon(\omega)]^{-1}. \tag{11}$$

The commutation relation difficulty can now be resolved. Clearly, the polarization response vanishes at t=0, so that

$$\zeta^{(1)}[t=0]=\varepsilon_0^{-1}, \tag{12}$$

while for nonlinear response functions with $n>1$,

$$\zeta^{(n)}[t_1=0,\cdots t_n=0]=0. \tag{13}$$

With these definitions, the electric field commutators at equal times are related to the displacement field commutators in the same way as they are for the free field case. The problem of the dispersionless theory is simply that it cannot be valid over the infinitely wide wavelength range necessary to derive the complete equal-time commutator.

Obtaining a Lagrangian for the dispersive equations is straightforward. I shall give a very simple technique for doing this, which is still in the spirit of macroscopic quantization. This involves writing the equations of motion in terms of band limited complex fields, just as one does in classical dispersive calculations. Here, the relevant band is chosen so that the absorption is low, which will result in real response functions in frequency space. The dual potential Λ is therefore expanded as:

$$\Lambda(t,z)=\Lambda^{\omega}(t,z)+\Lambda^{\omega*}(t,z) \tag{14}$$

where

$$\Lambda^{\omega}(t,z)=\mathcal{A}(t,z)e^{-i\omega t}. \tag{15}$$

The amplitude $\mathcal{A}$ is assumed to be slowly varying in time.

However, it is easier to work with Λ^ω directly, which therefore becomes a narrow-band field near frequency ω. Next, a Taylor series expansion of $\zeta^{(1)}(\omega;-\omega)$ in frequencies near ω is employed to simplify the wave-equation. This technique is common in classical dispersion theory, and was first introduced into macroscopic field quantization by Kennedy and Wright [22]. Defining $\Lambda^{-\omega}=\Lambda^{\omega*}$, the resulting wave-equation in the rotating-wave approximation is, from Eq. (1):

$$\begin{aligned}\mu\partial_t^2\Lambda^\omega &= \partial_z\Bigg[\sum_{n>1}\sum_{\omega_j=\pm\omega}-\zeta^{(n)}[-\omega;\omega_1,\cdots\omega_n](-\partial_z\Lambda^{\omega_1})\cdots(-\partial_z\Lambda^{\omega_n}) \\ &\quad +[\zeta+i\zeta'\partial_t-\frac{1}{2}\zeta''\partial_t^2+\cdots](\partial_z\Lambda^\omega)\Bigg]. \end{aligned}\tag{16}$$

Here the terms ζ, ζ', ζ'' represent a quadratic Taylor series expansion, valid near ω, so that:

$$\zeta^{(1)}(-\omega;\omega) \equiv \zeta + \omega\zeta' + \frac{1}{2}\omega^2\zeta''. \tag{17}$$

The careful reader will have recognized that all the higher order nonlinear terms have a similar expansion. I will neglect these nonlinear dispersion terms, as they are normally very small in dielectrics unless near an absorption band.

A straightforward modification of the earlier Lagrangian to allow for the new wave-equation, still local in the fields, is:

$$\begin{aligned}\mathcal{L} &= \int_0^L \Big\{\mu|\dot{\Lambda}^\omega|^2 - \zeta|\partial_z\Lambda^\omega|^2 - U^N(\partial_z\Lambda^\omega, \partial_z\Lambda^{\omega *}) \\ &\quad - \frac{i}{2}\zeta'\left[\partial_z\dot{\Lambda}^\omega\partial_z\Lambda^{\omega *} - \partial_z\dot{\Lambda}^{\omega *}\partial_z\Lambda^\omega\right] - \frac{1}{2}\zeta''|\partial_z\dot{\Lambda}^\omega|^2\Big\}\, dz \end{aligned} \tag{18}$$

where

$$U^N(\partial_z\Lambda^\omega, \partial_z\Lambda^{\omega *}) \equiv \frac{3}{2}\zeta^{(3)}(-\omega;\omega,\omega,-\omega)|\partial_z\Lambda^\omega|^4. \tag{19}$$

At this point I have retained only the terms in $\zeta^{(3)}$, equivalent to terms in $\chi^{(3)}$ in the Bloembergen notation. This is due to the choice of only one carrier frequency, together with the rotating-wave approximation and the neglect of higher order nonlinearities.

In order to complete the canonical theory, a Hamiltonian must be obtained. The canonical momentum fields are:

$$\Pi^{\pm\omega} = \frac{\delta\mathcal{L}}{\delta\dot{\Lambda}^{\pm\omega}} = \mu\dot{\Lambda}^{\mp\omega} \pm \frac{i}{2}\zeta'\partial_z^2\Lambda^{\mp\omega} + \frac{1}{2}\zeta''\partial_z^2\dot{\Lambda}^{\mp\omega}. \tag{20}$$

The corresponding Hamiltonian is therefore:

$$\mathcal{H} = \int_0^L \left\{\mu|\dot{\Lambda}^\omega|^2 - \frac{1}{2}\zeta''|\partial_z\dot{\Lambda}^\omega|^2 + \zeta|\partial_z\Lambda^\omega|^2 + U^N(\partial_z\Lambda^\omega, \partial_z\Lambda^{\omega *})\right\} dz. \tag{21}$$

The Hamiltonian is left here in terms of the field derivatives, rather than being expressed as a function of the canonical momenta. This allows a direct comparison with the classical energy of a dielectric, given a monochromatic excitation at frequency ω.

In the case of a linear dielectric, the total energy – including the energy of the polarized medium – has a known expression. This is, on averaging over a cycle

$$< W >_{cycle} = \int_0^L \left[\mathcal{E}^*(x)\frac{\partial}{\partial\omega}\left[\omega\varepsilon(\omega)\right]\mathcal{E}(x) + \frac{1}{\mu}|\mathcal{B}(x)|^2\right] dx. \tag{22}$$

Here $\mathcal{E}(x)$, $\mathcal{B}(x)$ are envelope functions at angular frequency ω, so that

$$< W >_{cycle} = \int_0^L \left[\mu|\dot{\Lambda}^\omega|^2 + \zeta(\omega)^2\frac{\partial}{\partial\omega}\left[\omega\varepsilon(\omega)\right]|\partial_z\Lambda^\omega|^2\right] dz. \tag{23}$$

It is straightforward to verify that this expression is identical to that of our Hamiltonian in the linear case, since:

$$\zeta^2(\omega)\frac{\partial}{\partial\omega}\left[\omega\varepsilon(\omega)\right] \equiv \zeta - \frac{1}{2}\omega^2\zeta''. \tag{24}$$

In summary, this local dispersive Lagrangian generates both the correct Maxwell equations and the correct energy, provided nonlinear dispersion is negligible, as is usually the case.

5. DISPERSIVE QUANTIZATION

Now that a Hamiltonian that is correct inside a band of frequencies near ω is known, the theory can be quantized. While there are many ways to do this, the most useful is to quantize the spatial Fourier transform of $\Lambda^{\omega}(t,z)$. This method is simple to extend to three dimensions, where transversality is required, as it allows the spatial derivatives in the Lagrangian to be readily handled. Using spatial modes also permits the limitation of the range of frequencies that are excited, so that $< \hat{\Lambda}^{\omega}(t,z) >\sim e^{-i\omega t}$, as required.

In fact, the Hamiltonian can have many possible excitation frequencies, only some of which are inside the required frequency band. Accordingly, the Hilbert space must be restricted. For this approximate theory to be valid, modes which result in envelopes varying as $< \Lambda^{\omega}(t,z) >\sim e^{-i\omega' t}$ with ω' outside the required range, must remain in the vacuum state. This is a relatively mild restriction, which corresponds physically to the requirement that there is negligible inelastic scattering of photons into absorption bands.

For simplicity, the modes here will be the solutions that diagonalize the linear Hamiltonian, as these have the structure (in a uniform medium) of:

$$\hat{\Lambda}^{\omega}(t,z) = \sum_k e^{ikz}\left[\hat{a}_k\lambda_k + \hat{b}_k^{\dagger}\mu_k\right]. \tag{25}$$

Terms like $\hat{b}_k^{\dagger}$ must remain in the vacuum state, since they vary as $\sim e^{i\omega t}$, and are therefore not in the required frequency band. This leaves a restricted set of modes of interest, with an expansion of:

$$\hat{\Lambda}^{\omega}(t,z) = \sum_k e^{ikz}\lambda_k\hat{a}_k. \tag{26}$$

Calculation of λ_k shows that in order to obtain the usual form of

$$\hat{\mathcal{H}}_0 = \sum \hbar\omega_k\hat{a}_k^{\dagger}\hat{a}_k. \tag{27}$$

it is necessary that:

$$\lambda_k = i\left[\frac{\hbar\partial\omega/\partial k}{2Lk\zeta^{(1)}(-\omega_k,\omega_k)}\right]^{1/2} \tag{28}$$

where $\hat{a}_k^{\dagger}$, $\hat{a}_{k'}$ have the standard commutators:

$$\left[\hat{a}_k,\hat{a}_{k'}^{\dagger}\right] = \delta_{k,k'}. \tag{29}$$

The final quantum Hamiltonian is therefore written, choosing normal ordering for the operators, as:

$$\hat{\mathcal{H}} = \sum_k \hbar\omega_k\hat{a}_k^{\dagger}\hat{a}_k + \int_0^L : U^N\left(\partial_z\hat{\Lambda}^{\omega},\partial_z\hat{\Lambda}^{\omega *}\right) : dz \tag{30}$$

where:

$$\omega_k = k\sqrt{\zeta(-\omega_k,\omega_k)/\mu}. \tag{31}$$

This is very straightforward to understand physically. The operators $\hat{a}_k^{\dagger}$, $\hat{a}_k$ generate the quantum excitations of the coupled matter-field system, for co-rotating frequencies near ω. These are photon-polaritons, propagating along the waveguide at a velocity $\partial\omega/\partial k$, which is the group velocity corresponding to a wave vector k. If we wish to study inhomogeneous waveguides with boundaries, then it is necessary to find modal solutions to the corresponding inhomogeneous Maxwell equations.

When there is a nonlinear refractive-index, or $\zeta^{(3)}$ term, these quanta interact via the Hamiltonian nonlinearity. It is this coupling that leads to soliton formation. The above Hamiltonian was originally used to obtain predictions of quantum phase-diffusion and quadrature squeezing, which have now been verified in optical fiber soliton experiments. I note here that, at the expense of losing strict locality, a Lagrangian function of a_k can be readily defined to generate the above Hamiltonian directly. This eliminates the problem of nonresonant modes, but also some physical intuition. In fact, the extra bosons are an indication that a local dispersive theory must possess additional quanta. These correspond physically to the addition of matter variables. The reason why they are omitted is simply that the refractive index data is not sufficient to characterize them properly, and they usually do not have the frequency and phase matching necessary for them to become excited.

6. CONTINUUM FIELDS

Having obtained the quantum Hamiltonian, expanded in terms of the quanta of the coupled matter-field system, it is useful to define a photon-polariton amplitude field[2], which has a direct interpretation in photodetection. It should be emphasized that this field, and its annihilation and creation operators, is only equivalent to the usual non-interacting photon field when propagating in a free-space region with no dielectric. A slowly varying envelope field is defined in the Heisenberg picture, for propagation in the $+z$ direction, as:

$$\hat{\Psi}(t,z) = \sqrt{\frac{1}{L}} \sum_k e^{[i(k-k_0)z+i\omega t]} \hat{a}_k(t). \tag{32}$$

This has an equal-time commutator of:

$$\left[\hat{\Psi}(t,z_1), \hat{\Psi}^\dagger(t,z_2)\right] = \tilde{\delta}(z_1 - z_2), \tag{33}$$

where $\tilde{\delta}$ is defined as a tempered version of the usual Dirac delta-function:

$$\tilde{\delta}(z) \equiv \frac{1}{L} \sum_k e^{i(k-k_0)z}. \tag{34}$$

Here, $< \hat{\Psi}^\dagger(t,z)\hat{\Psi}(t,z) >$ is the photon-polariton density. The summation is over momenta k with corresponding frequencies inside the relevant band, and with the usual spacing of $\Delta k = 2\pi/L$. I define $k_0 = k(w)$ as the central wavenumber in the frequency band.

In terms of this field, the nonlinear interaction can be approximated with the replacement:

$$U^N(\partial_z \hat{\Lambda}, \partial_z \hat{\Lambda}^\dagger) \simeq -\frac{1}{2}\chi\hbar |\hat{\Psi}(t,z)|^4. \tag{35}$$

Here $\chi = 3\hbar\chi^{(3)}\omega^2 v^2/4\epsilon R^2 c^2$ has the sign of the well-known Bloembergen coefficient, $\chi^{(3)}$. Also, $v = \partial\omega/\partial k|_{k=k_0}$ is the group velocity at the carrier frequency ω. This replacement involves the neglect of frequency dependent coefficients in the nonlinear terms only, which is an excellent approximation when the excitation is narrow-band compared to the carrier frequency. The resulting interaction Hamiltonian, relative to the carrier frequency ω, is obtained on expanding ω_k around k_0:

$$\begin{aligned} \hat{\mathcal{H}}_I \simeq & \left(\frac{\hbar}{2}\right) \int \left\{ iv \left[\frac{\partial}{\partial z}\hat{\Psi}^\dagger \cdot \hat{\Psi} - \hat{\Psi}^\dagger \frac{\partial}{\partial z}\hat{\Psi} \right] \right. \\ & \left. + \omega'' \frac{\partial}{\partial z}\hat{\Psi}^\dagger \cdot \frac{\partial}{\partial z}\hat{\Psi} - \chi \hat{\Psi}^{\dagger 2}\hat{\Psi}^2 \right\} dz. \end{aligned} \tag{36}$$

This leads to the following equation of motion, where $\omega'' = \partial^2\omega/\partial k^2|_{k=k_0}$ is the group velocity dispersion:

$$\left(v\frac{\partial}{\partial z} + \frac{\partial}{\partial t}\right)\hat{\Psi}(t,z) = \left[\frac{i\omega''}{2}\frac{\partial^2}{\partial z^2} + i\chi\hat{\Psi}^\dagger\hat{\Psi}\right]\hat{\Psi}(t,z). \tag{37}$$

In a frame in which $z' = z - vt$, this equation reduces to the quantum nonlinear Schrödinger equation, in the form:

$$i\frac{\partial}{\partial t}\hat{\Psi}(t,z') = \left[-\frac{\omega''}{2}\frac{\partial^2}{\partial z'^2} - \chi\hat{\Psi}^\dagger\hat{\Psi}\right]\hat{\Psi}(t,z'). \tag{38}$$

This is one of the simplest known nonlinear quantum field theories[23]. The discrete energy eigenstates, or quantum solitons, can be found using techniques originally due to Bethe[24]. Physically, the theory is equivalent to spinless particles interacting with a delta function potential. The interaction is attractive, giving rise to bound quantum solitons, when $\chi > 0$. This is the case in silica fibers. Soliton formation only takes place when $\omega'' > 0$, a condition known as anomalous dispersion. This occurs, by coincidence, in the wavelength region around $1.5\mu m$ where silica is the most transparent solid dielectric known, making it ideal for quantum soliton experiments[25].

In typical laser experiments, the initial excitation is a coherent state with a photon number of 10^9. Hence, it has proved more useful to use coherent-state expansions to calculate observable quantities like quadrature variances and phase fluctuations. These methods are reviewed elsewhere in this volume. Using these techniques, and assuming vanishing boundary terms, an equivalent c-number stochastic equation that represents normally ordered correlation functions can be found to be:

$$\frac{\partial}{\partial t}\Psi(t,z') = \left[i\frac{\omega''}{2}\frac{\partial^2}{\partial z'^2} + i\chi\Psi^\dagger\Psi + (i\chi)^{\frac{1}{2}}\eta(t,z')\right]\Psi(t,z') \tag{39}$$

where:

$$< \eta(t_1,z_1')\eta(t_2,z_2') >= \delta(t_1-t_2)\tilde{\delta}(z_1'-z_2') \tag{40}$$

together with a similar equation for the complex c-number field $\Psi^\dagger$, representing the conjugate field $\hat{\Psi}^\dagger$.

Finally, it is often convenient to work in a frame in which the role of z and t variables are reversed. This can be achieved by changing variables to a comoving time variable defined as $\tau = t - z/v$, and neglecting terms that involve higher z derivatives and cross-derivatives[2]. The result is only approximately valid, under conditions of slow spatial variation relative to the pulse-length. The equations are best expressed in terms of the photon flux amplitude, which I define as $\Phi(\tau,z) = v^{\frac{1}{2}}\Psi(t,z)$, together with a flux nonlinearity $\chi_\Phi = \chi/v^2$ and a dispersion term $k'' = -\omega''/v^3$, so that:

$$\frac{\partial}{\partial z}\Phi(\tau,z) = \left[-i\frac{k''}{2}\frac{\partial^2}{\partial\tau^2} + i\chi_\Phi\Phi^\dagger\Phi + (i\chi_\Phi)^{\frac{1}{2}}\eta(\tau,z)\right]\Phi(\tau,z). \tag{41}$$

These equations are equivalent to a suitably modified Hamiltonian, which must be regarded as causing translation in position, not in time. The corresponding commutation relations are equal-space commutators, and they are only approximately valid; although for small τ, equal space (z) is obviously equivalent to equal time (t). It is strictly the *equal-time* version of the commutation relations which are fundamental, and stochastic or operator equations to first order in $\partial/\partial z$ are only approximate.

There have also been intriguing attempts to reformulate quantum mechanics using equal-space commutators. I will not go into the details here, except to note that there are difficult questions of causality involved when specifying time-dependent boundary conditions for nonlinear systems of interacting particles. For this reason, I prefer to use equal-time Dirac quantization.

Using the stochastic techniques given above, the observable phase diffusion and squeezing have been worked out for coherent inputs[4]. A strong practical limitation on experiment originates from the refractive-index fluctuations and Raman scattering in real media. This has led to the prediction of a phase-noise window in the vicinity of $1ps$ pulse durations, which is now experimentally verified[6].

CONCLUSION

A quantization of macroscopic equations must be able to generate both the correct equations of motion, and the Hamiltonian corresponding to the classical energy. This has been carried out here for a dispersive, nonlinear dielectric. The method is one of the simplest that uses a local Lagrangian, and can be extended to higher dimensions if necessary. This type of theory gives quantitative predictions about quantum correlations, which can be tested experimentally. These experimental tests are of a new type in quantum field theory, and allow quantum soliton properties to be investigated. The experimental results are in accordance with theory.

There have been a number of alternative quantization proposals, which often have incorrect classical equations, or do not have a Hamiltonian corresponding to the classical energy. Having incorrect classical equations generates obvious problems -- for example, there may be no solitons predicted at all! Without the restriction on the Hamiltonian, there are severe non-uniqueness problems. The quantum energies ($\hbar\omega$) are intrinsically undefined, because this type of theory can always be transformed to one with different quantum energies, just by rescaling the Lagrangian. To provide a dimensional reference for the size of Planck's constant, it is essential to equate the scale of the classical energies with those of the corresponding quantum theory.

Quantization with equal-space commutation relations has also been attempted. I have shown that this procedure is able to give approximately valid results, for certain types of physical situation. It also has computational advantages when there are well-defined boundary conditions. In the case of nonlinear dispersive equations, the method is restricted to a co-moving time coordinate, with nearly coherent propagation and slow variation in space. This is also the region of validity of the corresponding slowly-varying classical equations which have only first-order spatial derivatives.

It is clear that laser techniques can be used to test quantum field theory predictions in ways that are not possible using traditional scattering methods. This opens up a new area of theoretical and experimental investigation; obvious cases that are of interest include topological and higher-dimensional solitons. It is worth noting that as solitons have been recently suggested for use in communications systems[26], these purely scientific investigations could have technological implications.

REFERENCES

1. S. J. Carter, P. D. Drummond, M. D. Reid and R. M. Shelby, *Phys. Rev. Lett.* **58**, 1841 (1987).

2. P. D. Drummond and S. J. Carter, *J. Opt. Soc. Am.* B**4**, 1565 (1987).
3. P. D. Drummond, S. J. Carter and R. M. Shelby, *Opt. Lett.* **14**, 373 (1989).
4. R. M. Shelby, P. D. Drummond and S. J. Carter, *Phys. Rev.* A**42**, 2966 (1990).
5. H. A. Haus and Y. Lai, *J. Opt. Soc. Am.* B**7**, 386 (1990).
6. M. Rosenbluh and R. M. Shelby, *Phys. Rev. Lett.* **66**, 153 (1991).
7. P. A. M. Dirac, Proc. Roy. Soc. (London), Ser. A. **114**, 243 (1927).
8. An excellent example of this, is the discovery of quark-antiquark resonance states, predicted using the standard QCD arguments.
9. For an instructive and clear account of progress in quantum field theory, see: S. Coleman, *Aspects of symmetry*, (Cambridge University Press, Cambridge, 1985).
10. S. R. Friberg, Technical Digest, OSA Annual Meeting, Paper MF1 (1991).
11. H. A. Haus, K. Watanabe, Y. Yamamoto, *J. Opt. Soc. Am.* B**6**, 1138 (1989).
12. P. A. M. Dirac, *The Principles of Quantum Mechanics* (Oxford, Clarendon, 1958); *Lectures on Quantum Mechanics* (Belfer Graduate School of Science, New York, 1964).
13. Y. R. Shen, *Phys. Rev.* **155**, 921 (1967).
14. P. D. Drummond, *Phys. Rev.* A**42**, 6845 (1990).
15. N. Bloembergen, *Nonlinear Optics* (Benjamin, New York, 1965).
16. M. Goeppert-Meyer, *Ann. Physik.* **9**, 273 (1931).
17. E. Power, S. Zienau, Phil. Trans. Roy. Soc. Lord., A**251**, 427 (1959); R. Loudon, *The Quantum Theory of Light* (Clarendon Press, Oxford, 1983).
18. S. Geltmann, *Phys. Lett.* **4**, 168 (1963).
19. M. Hillery and L. D. Mlodinow, Phys. Rev. A**30**, 1860 (1984).
20. S. T. Ho and P. Kumar, unpublished.
21. P.D. Drummond, *Phys. Rev.* A**39**, 2718 (1989).
22. T. A. B. Kennedy and E. M. Wright, *Phys. Rev.* A**38**, 212 (1988).
23. S. Takeno (Ed.), *Dynamical Problems in Soliton Systems*, (Springer, Berlin, 1984).
24. H. A. Bethe, *Z. Physik* **71**, 205 (1931).
25. A. Hasegawa and F. Tappert, *Appl. Phys. Lett.* **23**, 142 (1973).
26. L. F. Mollenauer, Philos. Trans. Roy. Soc. Lon. A**315**, 435 (1985).

TEN YEARS OF THE POSITIVE P-REPRESENTATION

C.W. Gardiner

Physics Department
University of Waikato
New Zealand

P.D. Drummond

Physics Department
University of Queensland
Queensland 4072, Australia

The positive P-representation is an example of a new class of phase-space representations, which utilize a nonclassical phase-space. These have many applications, and have been widely used in quantum optics in the last decade, since their introduction. The utility of the positive P-representation is that it retains many of the advantages of the diagonal P-representation, while still remaining nonsingular, even for nonlinear Hamiltonians. However, attention to boundary terms and to the use of stable integration methods are important.

INTRODUCTION: WHAT IS THE POSITIVE P-REPRESENTATION?

The development of quantum mechanics was strongly influenced by the recognition that there is a clear formal similarity between classical and quantum theories. This, in fact, led Schrödinger to his discovery of the coherent state of the harmonic oscillator, as a nearly classical state. In this case, the coherent state is labelled by a complex parameter $\alpha = x + ip$. The state is a minimum uncertainty state in the sense that is minimizes the Heisenberg uncertainty product. In addition to this desirable property, the coherent state remains a coherent state under time evolution, with the classical paths for the variables x and p. More technically, a coherent state is usually defined today as an eigenstate of the annihilation operator a:

$$a|\alpha >= \alpha|\alpha > \tag{1}$$

Despite the formal similarity of quantum and classical mechanics, there is no practical way of representing quantum uncertainties using a positive distribution on a classical phase space. This fact was discovered by von Neumann [1], who showed that no classical, positive phase-space distribution could exist for all quantum states, with correct marginal distributions in both x and p simultaneously. This is perhaps the

Recent Developments in Quantum Optics, Edited
by R. Inguva, Plenum Press, New York, 1993

earliest "no hidden variables theorem" in quantum mechanics, and can be regarded as the precursor of more powerful results of Einstein, Podolsky and Rosen, and the famous Bell inequalities. In a fundamental sense, this theorem shows that there is an intrinsic difference between classical and quantum uncertainties in measurement.

There are many reasons why classical-like phase space distribution techniques are popular in theoretical quantum physics. These techniques take advantage of the well-developed theory of Fokker-Planck equations and stochastic equations, and are often able to give results where other methods fail. It is clear, however, that quantum phase space distributions must have properties differing from those of classical phase-space distributions. Probably the most widely known phase-space method is that of the Wigner distribution function. [2] This has correct marginal distributions in x and in p, at the expense of developing negative values for certain quantum states.

An alternative that is sometimes used is the Q-distribution, defined by: $Q(\alpha) = <\alpha|\rho|\alpha>$. This is always positive, and has the physical interpretation that is corresponds to the probability of measuring x and p simultaneously, giving $\alpha = x + ip$. Simultaneous measurement of non-commuting operators like x and p is possible, but entails a larger variance than any single measurement of x or p by itself. Accordingly, the Q-distribution pays for its positivity by having larger marginal variances than the Wigner function, for isolated measurements on the canonical variables. This property is to be expected in view of von Neumann's fundamental result, and can be corrected for when necessary.

Another approach to this problem was that of Glauber and Sudarshan, [3] who independently achieved a significant breakthrough in the early 1960s. These workers recognized that the classical evolution of coherent states made them the natural basis for the study of radiation fields with nearly classical coherence properties. A single mode radiation field is characterized by just one set of harmonic oscillator operators. Accordingly, it was proposed that an arbitrary density operator ρ be represented by:

$$\rho = \int d^2\alpha|\alpha><\alpha|P(\alpha) \tag{2}$$

Unfortunately, the function $P(\alpha)$ often does not exist, except as a highly singular generalized function. The reason for this is not hard to find, since it is possible to show that $P(\alpha)$ can be convolved with a two-dimensional Gaussian distribution, to give the Wigner function. Thus, the Wigner function always has larger marginal variances than the P-function. This leads to substantial difficulties if the quantum state is a position eigenstate, having a Wigner function of zero marginal variance on the x axis: and hence a P-function of negative variance! Despite this, the technique proved to have many computational advantages in laser physics, and was widely used in this field.

That certain types of nonlinear interactions would cause problems was not widely understood until the late 1970s, when techniques were developed to obtain the corresponding nonclassical states of the radiation field in the laboratory. In these nonclassical states, the Glauber-Sudarshan P-representation is negative or singular. The more general nonlinear Hamiltonians used in these cases have Fokker-Planck equations which have a non-positive-semidefinite diffusion, and hence no corresponding stochastic process. In fact, this even occurs in the laser Fokker-Planck equations, although this fact was generally ignored - or overcome by using a scaling technique to remove the offending terms.

It might seem that this could be overcome by returning to the use of the Q-function, which has a classical phase space, and is always positive. In fact, the Q-function - and the Wigner function - can always be calculated from the P-function. However, these distributions on a classical phase space generally do *not* have positive-semidefinite diffusion in their Fokker-Planck equations, and frequently have time-evolution equations with higher than second order derivatives. In such cases, there is no stochastic interpretation of the time evolution. This leads to substantial difficulties in actual calculations, even when the distribution is well defined.

An often more useful resolution of the problem was provided by the observation that, in cases of nonpositive-definite diffusion, a naive application of standard techniques would result in the following generic stochastic equation:

$$\dot{\alpha} = -i\omega\alpha + \xi_1(t), \quad \dot{\alpha}^* = -i\omega\alpha^* + \xi_2(t) \tag{3}$$

Here the two noise terms $\xi_1(t)$ and $\xi_2(t)$ are uncorrelated random noise terms, describing a departure from the classical path of Schrödinger. They are present because the Hamiltonian of these nonlinear systems has extra terms beyond the normal harmonic oscillator terms, meaning that the quantum coherent state neither remains a coherent state, nor follows a classical path.

Since the extra stochastic terms are uncorrelated, they are not conjugate, which leads to an inconsistency in (3). However it was realized that the resulting equations could be re-interpreted as applying to two *independent* complex variables, which were only complex conjugate in the mean. This required the introduction of a normalized representation, having off-diagonal terms in the coherent state expansion of the density operator. The new representation, although on an enlarged phase space, can be shown to have at least one positive distribution function for all quantum states or density operators. This new representation — one of a class of generalized P-representations — is called the positive P-representation. The detailed properties and development of the new representation are given in the following sections.

THE GLAUBER-SUDARSHAN P-REPRESENTATION

The Glauber-Sudarshan P-representation is defined by

$$\rho = \int d^2\alpha P(\alpha, \alpha^*)|\alpha><\alpha|. \tag{4}$$

Glauber and *Sudarshan* [3] both showed that such a representation of the density operator did exist for a large class of ρ, and indeed *Klauder* et al. [4] showed that, provided $P(\alpha, \alpha^*)$ is permitted to be a sufficiently singular generalized function, such a representation exists for any density operator ρ. From the fact that $Tr\{\rho\} = 1$ it is easily deduced that

$$\int d^2\alpha P(\alpha, \alpha^*) = 1. \tag{5}$$

Further, for any normal product of creation and destruction operators we can write

$$<(a^\dagger)^r a^s> = \int d^2\alpha (\alpha^*)^r \alpha^s P(\alpha, \alpha^*). \tag{6}$$

Thus the quantity $P(\alpha,\alpha^*)$ plays the role of a kind of probability density for the variables α and α^*, in that the means of *normally ordered* products of quantum operators are simple moments of $P(\alpha,\alpha^*)$.

The usefulness of this representation can be shown by the example of the driven damped harmonic oscillator, for which the master equation may be written

$$\begin{aligned}\dot{\rho} = & -i\omega[a^\dagger a,\rho] \\ & +\tfrac{1}{2}K(\bar{N}+1)(2a\rho a^\dagger - a^\dagger a\rho - \rho a^\dagger a) \\ & +\tfrac{1}{2}K\bar{N}(2a^\dagger\rho a - aa^\dagger\rho - \rho aa^\dagger).\end{aligned} \tag{7}$$

The procedure for using the P-representation to solve this equation is as follows. We first note that the action of the creation and destruction operators on $|\alpha >$ is

$$a|\alpha >= \alpha|\alpha >, \qquad a^\dagger|\alpha >= (\alpha^* + \partial/\partial\alpha)|\alpha > \tag{8}$$

with the corresponding Hermitian conjugate. We then insert the expansion (5) into the master equation, and use the correspondences (8). We integrate by parts, and *discard any surface terms*, provided the distribution is sufficiently rapidly vanishing at the boundaries. This leads to a correspondence of the form

$$\begin{array}{ll} a\rho \leftrightarrow \alpha P(\boldsymbol{\alpha}\ ), & a^\dagger\rho \leftrightarrow (\alpha^* - \partial/\partial\alpha)P(\boldsymbol{\alpha}\ ) \\ \rho a^\dagger \leftrightarrow \alpha^* P(\boldsymbol{\alpha}\ ), & \rho a \leftrightarrow (\alpha - \partial/\partial\alpha^*)P(\boldsymbol{\alpha}\ ). \end{array} \tag{9}$$

Thus, wherever a creation or destruction operator occurs in the master equation, one can translate this into the corresponding operation on $P(\alpha)$. (NOTE: the order of the operators which appear to the right of ρ in the master equation is reversed by making this translation process.) The equation which results is a Fokker-Planck equation:

$$\frac{\partial P}{\partial t} = \left\{\frac{1}{2}K\left(\frac{\partial}{\partial\alpha}\alpha + \frac{\partial}{\partial\alpha^*}\alpha^*\right) - i\omega\left(\frac{\partial}{\partial\alpha}\alpha - \frac{\partial}{\partial\alpha^*}\alpha^*\right) + K\bar{N}\frac{\partial^2}{\partial\alpha\partial\alpha^*}\right\}P. \tag{10}$$

Classical stochastic theory [5] shows that this Fokker-Planck equation is equivalent to the stochastic differential equation in the form

$$\frac{d\alpha}{dt} = -(\frac{1}{2}K - i\omega)\alpha + \sqrt{\bar{N}}\ \xi(t), \tag{11}$$

and the corresponding complex conjugate equation. The quantity $\xi(t)$ is a white noise fluctuating force with the correlation properties

$$< \xi(t) >= 0, \qquad < \xi(t)\xi^*(t') >= \delta(t-t') \tag{12}$$

$$< \xi(t)\xi(t') >=< \xi^*(t)\xi^*(t') >= 0 \tag{13}$$

The equivalence to a classical SDE is not always so straightforward. It is not unusual to find an equation which appears to be of the Fokker-Planck form, but which has a diffusion matrix with some negative eigenvalues. Add to this the independent problem that the Glauber-Sudarshan P-representation does not always exist, and it becomes apparent that there are severe limitations on its practical use. It was this fact that motivated *Drummond* and *Gardiner* [6] to introduce a class of generalized P-representations, by expanding in non-diagonal coherent state projection operators. The generalized P-representations are defined as follows. We set

$$\rho = \int_{\mathcal{D}} \Lambda(\alpha, \alpha^+) P(\alpha, \alpha^+) d\mu(\alpha, \alpha^+), \tag{14}$$

where

$$\Lambda(\alpha, \alpha^+) = \frac{|\alpha >< \alpha^{+*}|}{< \alpha^{+*}|\alpha >}, \tag{15}$$

$d\mu(\alpha, \alpha^+)$ is the integration measure which may be chosen to define different classes of possible representations and $\mathcal{D}$ is the domain of integration. The projection operator $\Lambda(\alpha, \alpha^+)$ is analytic in (α, α^+). In the case of the Positive P-Representation we choose $d\mu(\alpha, \alpha^+) = d^2\alpha d^2\alpha^+$. This representation allows (α, α^+) to vary independently over the whole complex plane. In this case the normally ordered moments are given by

$$< (a^\dagger)^m a^n >= \int_{\mathcal{D}} d\mu(\alpha, \alpha^+) \alpha^{+m} \alpha^n P(\alpha, \alpha^+). \tag{16}$$

The positive P-representation has quite strong existence properties. NB: for brevity, we shall use the notation $\boldsymbol{\alpha} = (\alpha, \alpha^+)$.

Theorem (Drummond-Gardiner): A positive P-representation exists for any quantum density operator ρ, with

$$P(\boldsymbol{\alpha}) = (1/4\pi^2) exp(-|\alpha - \alpha^{+*}|^2/4) < \frac{1}{2}(\alpha + \alpha^{+*})|\rho|\frac{1}{2}(\alpha + \alpha^{+*}) > . \tag{17}$$

The direct definition of the Positive P-representation as an expansion in operators makes the examination of convergence and the neglect of boundary terms very difficult. A better definition is given by saying that $P(\boldsymbol{\alpha})$ is a positive P-function corresponding to a density operator ρ provided

$$\chi_P(\lambda, \lambda^*) \equiv \int\int P(\boldsymbol{\alpha}) exp(\lambda\alpha^+ - \lambda^*\alpha) d^2\alpha d^2\alpha^+ \tag{18}$$

is identical with the quantum characteristic function of the density operator,

$$\chi(\lambda, \lambda^*) \equiv Tr\{\rho e^{\lambda a^\dagger} e^{-\lambda^* a}\}. \tag{19}$$

Convergence Conditions: It can be shown [7] that the integral (18) always converges provided $P(\boldsymbol{\alpha})$ is normalizable. To do this requires a careful definition of the four dimensional integral, which must be understood as the limit of an integral over disks of radii R, R' in the α, α^+ planes as $R, R' \to \infty$.

This is much milder than one might expect from (18), which would lead one to expect that $P(\boldsymbol{\alpha})$ would need to drop off exponentially as $|\alpha|, |\alpha^+| \to \infty$ in the direction of positive $\lambda\alpha^+, -\lambda^*\alpha$.

OPERATOR IDENTITIES

From the definitions (15) of the nondiagonal coherent state projection operators, the following identities can be obtained. Again, α is used to denote (α, α^+):

$$\begin{aligned} a\Lambda(\boldsymbol{\alpha}) &= \alpha\Lambda(\boldsymbol{\alpha}) \\ a^\dagger\Lambda(\boldsymbol{\alpha}) &= (\alpha^+ + \partial/\partial\alpha)\Lambda(\boldsymbol{\alpha}) \\ \Lambda(\boldsymbol{\alpha})a^\dagger &= \alpha^+\Lambda(\boldsymbol{\alpha}) \\ \Lambda(\boldsymbol{\alpha})a &= (\partial/\partial\alpha^+ + \alpha)\Lambda(\boldsymbol{\alpha}). \end{aligned} \tag{20}$$

By substituting the above identities into a master equation like (7), using (14), which defines the generalized P-representation, and using partial integration (provided the boundary terms vanish), these identities can be used to generate operations on the P-function depending on the representation. For the Positive P-Representation we now use the analyticity of $\Lambda(\boldsymbol{\alpha})$ and note that if

$$\alpha = \alpha_x + i\alpha_y, \qquad a^+ = a_x^+ + i\alpha_y^+, \tag{21}$$

then

$$(\partial/\partial\alpha)\Lambda(\boldsymbol{\alpha}) = (\partial/\partial\alpha_x)\Lambda(\boldsymbol{\alpha}) = (-i\partial/\partial\alpha_y)\Lambda(\boldsymbol{\alpha}) \tag{22}$$

and

$$(\partial/\partial\alpha^+)\Lambda(\boldsymbol{\alpha}) = (\partial/\partial\alpha_x^+)\Lambda(\boldsymbol{\alpha}) = (-i\partial/\partial\alpha_y^+)\Lambda(\boldsymbol{\alpha}) \tag{23}$$

so that as well as all of (20) being true in this case, we also have

$$a^\dagger\rho \leftrightarrow (\alpha^+ - \partial/\partial\alpha_x)P(\boldsymbol{\alpha}) \leftrightarrow (\alpha^+ + i\partial/\partial\alpha_y)P(\boldsymbol{\alpha})$$

$$\rho a \leftrightarrow (\alpha - \partial/\partial\alpha_x^+)P(\boldsymbol{\alpha}) \leftrightarrow (\alpha + i\partial/\partial\alpha_y^+)P(\boldsymbol{\alpha}). \tag{24}$$

All these correspondences can now be used to derive Fokker-Planck equations when appropriate.

TIME-DEVELOPMENT EQUATIONS

A time development equation for the density operator in the form of a master equation can be reduced to a c-number Fokker-Planck equation by using the mappings (20). The basic procedure is as follows

i) By use of the operator identities (20) write the master equation in the form [where $(\alpha, \alpha^+) = \boldsymbol{\alpha} \equiv (\alpha^{(1)}, \alpha^{(2)}; \mu = 1, 2]$:

$$\begin{aligned} \frac{\partial \rho}{\partial t} &= \int_{C,C'} \Lambda(\boldsymbol{\alpha}) \frac{\partial P(\boldsymbol{\alpha})}{\partial t} d\alpha d\alpha^+ \\ &= \int_{C,C'} \left\{ \left[A^\mu(\boldsymbol{\alpha}) \frac{\partial}{\partial \alpha^\mu} + \tfrac{1}{2} D^{\mu\nu}(\boldsymbol{\alpha}) \frac{\partial}{\partial \alpha^\mu} \frac{\partial}{\partial \alpha^\nu} \right] \Lambda(\boldsymbol{\alpha}) \right\} P(\boldsymbol{\alpha}) d\alpha d\alpha^+. \end{aligned} \tag{25}$$

ii) Now integrate by parts, and drop surface terms, provided that the distribution allows this. *This may not always be possible*, but if it is at least one solution is obtained by equating the coefficients of $\Lambda(\boldsymbol{\alpha})$:

$$\frac{\partial P(\boldsymbol{\alpha})}{\partial t} = \left[-\frac{\partial}{\partial \alpha^\mu} A^\mu(\boldsymbol{\alpha}) + \frac{1}{2} \frac{\partial}{\partial \alpha^\mu} \frac{\partial}{\partial \alpha^\nu} D^{\mu\nu}(\boldsymbol{\alpha}) \right] P(\boldsymbol{\alpha}). \tag{26}$$

iii) Alternatively we can write an equation which is completely real. The symmetric matrix $\boldsymbol{D}(\boldsymbol{\alpha})$ can always be factorized into the form

$$\boldsymbol{D}(\boldsymbol{\alpha}) = \boldsymbol{B}(\boldsymbol{\alpha}) \boldsymbol{B}^T(\boldsymbol{\alpha}) . \tag{27}$$

We now write

$$\boldsymbol{A}(\boldsymbol{\alpha}) = \boldsymbol{A}_x(\boldsymbol{\alpha}) + \boldsymbol{i} \boldsymbol{A}_y(\boldsymbol{\alpha}) , \qquad \boldsymbol{B}(\boldsymbol{\alpha}) = \boldsymbol{B}_x(\boldsymbol{\alpha}) + \boldsymbol{i} \boldsymbol{B}_y(\boldsymbol{\alpha}) \tag{28}$$

where $\mathbf{A}_x$, $\mathbf{A}_y$, $\mathbf{B}_x$, $\mathbf{B}_y$ are real.

iv) We define a four dimensional drift vector by

$$\mathcal{A}(\boldsymbol{\alpha}) \equiv (A_x^{(1)}(\boldsymbol{\alpha}), A_x^{(2)}(\boldsymbol{\alpha}), A_y^{(1)}(\boldsymbol{\alpha}), A_y^{(2)}(\boldsymbol{\alpha})) \tag{29}$$

and a corresponding diffusion matrix by

$$\mathcal{D}(\boldsymbol{\alpha}) = \begin{pmatrix} \mathbf{B}_x \mathbf{B}_x^T, & \mathbf{B}_x \mathbf{B}_y^T \\ \mathbf{B}_y \mathbf{B}_x^T, & \mathbf{B}_y \mathbf{B}_y^T \end{pmatrix} (\boldsymbol{\alpha}) \equiv \mathcal{B}(\boldsymbol{\alpha}) \mathcal{B}^T(\boldsymbol{\alpha}), \tag{30}$$

where

$$\mathcal{B}(\boldsymbol{\alpha}) = \begin{pmatrix} \mathbf{B}_x, & 0 \\ \mathbf{B}_y, & 0 \end{pmatrix} (\boldsymbol{\alpha}) \tag{31}$$

and $\mathcal{D}$ is thus explicitly positive semidefinite (and not positive definite).

v) The Master equation then yields (after partial integration, and neglect of surface terms), a Fokker-Planck equation which is equivalent to the Ito stochastic differential equations

$$\frac{d}{dt} \begin{pmatrix} \boldsymbol{\alpha}_x \\ \boldsymbol{\alpha}_y \end{pmatrix} = \begin{pmatrix} A_x(\boldsymbol{\alpha}) \\ A_y(\boldsymbol{\alpha}) \end{pmatrix} + \begin{pmatrix} \mathbf{B}_x(\boldsymbol{\alpha}) \boldsymbol{\xi}(t) \\ \mathbf{B}_y(\boldsymbol{\alpha}) \boldsymbol{\xi}(t), \end{pmatrix} \tag{32}$$

or recombining real and imaginary parts

$$d\boldsymbol{\alpha} / dt = A(\boldsymbol{\alpha}) + \boldsymbol{B}(\boldsymbol{\alpha}) \boldsymbol{\xi}(t). \tag{33}$$

Apart from the substitution $\alpha^* \to \alpha^+$, (33) is just the stochastic differential equation which would be obtained by using the Glauber-Sudarshan representation and naively converting the Fokker-Planck equation with a non-positive-definite diffusion matrix into an Ito stochastic differential equation.

In our derivation, the two formal variables (α, α^*) have been replaced by variables in the complex plane (α, β) that are allowed to fluctuate independently. The positive P-representation as defined here thus appears as a mathematical justification of this procedure.

NEGLECT OF BOUNDARY TERMS

In the derivation of the positive P-representation Fokker-Planck equation in the previous section, partial integration is used, and it was assumed that boundary terms at infinity could be neglected. The direct derivation in terms of projection operators $\Lambda(\boldsymbol{\alpha})$ makes it very difficult to assess the magnitude of these. However, defining the positive P-representation through the characteristic function makes the problem merely one of calculus. The equation of motion for the characteristic function is obtained by the rules

$$\rho a^\dagger \leftrightarrow \frac{\partial \chi}{\partial \lambda}, \qquad a^\dagger \rho \leftrightarrow \Big(-\lambda^* + \frac{\partial}{\partial \lambda}\Big)\chi$$

$$a\rho \leftrightarrow -\frac{\partial \chi}{\partial \lambda^*}, \qquad \rho a \leftrightarrow \Big(\lambda - \frac{\partial}{\partial \lambda^*}\Big)\chi \tag{34}$$

which easily gives the correspondence (24) provided the boundary terms can be neglected. We can estimate these boundary terms for any particular case, provided we know the behaviour of $\boldsymbol{A}(\boldsymbol{\alpha})$, $\boldsymbol{B}(\boldsymbol{\alpha})$ as $\alpha, \alpha^+ \to \infty$. In most cases of interest, these coefficients are low order polynomials in α, α^+, and we will find that we require a slightly faster drop off as $r, r' \to \infty$ than $(rr')^{-2-\epsilon}$.

The neglect of surface terms which is necessary to derive all the Fokker-Planck equations equivalent to (32) is a deceptively simple step, but there is as yet no general criterion for determining when this is justifiable. In all cases of linear Hamiltonians, the requirement is trivially satisfied. Thus, for example, the harmonic oscillator has no boundary terms. This means that the treatment of the damping in the harmonic oscillator can be easily extended to cover any nonclassical quantum state, including those for which there is no well-behaved Glauber-Sudarshan P-representation. At zero temperature, the damping equations are deterministic, which allows the effects of damping on non-classical fields to be calculated in a very straightforward way. Similarly, the linear or linearized equations often used in treating squeezed or correlated states in a cavity, have no boundary corrections.

A nonlinear example [8] where the boundary terms vanish is the case of subharmonic generation in a cavity, with the driving field adiabatically eliminated. This has a finite phase-space manifold at zero temperature, and it is straightforward to show that the boundary terms vanish. It is known in this case that the steady-state distribution has an exact potential solution. All quantum moments are calculable in the steady-state. In addition, the dynamics can be readily simulated numerically, and even the above-threshold tunnelling [9] can be analytically calculated. This problem can also be treated, with some difficulty, using a numerical solution of the master equation.

Exact agreement is obtained between these number-state results, and the Positive-P results. This, of course, is expected, since there are no boundary terms, and no other approximations like system-size expansions are used.

It is useful to compare this situation of subharmonic generation in a cavity, with that for the more traditional distributions, like the Wigner, Q, or Glauber-Sudarshan representations. In none of these is it possible to even define an exact stochastic process, let alone find the steady-state quantum distribution. This appears to be generic when there is a nonlinearity in these classical-like representations, due to the occurrence of higher-order or non-positive definite diffusion terms. In the case of the Wigner distribution, it is often argued that the higher-order derivatives can be neglected at large photon number, leading to a semi-classical theory. However, this procedure does not give the correct tunnelling rates for this problem. [9] Depending on exactly which stage in the calculation the higher derivatives are dropped, the tunnelling rate may be many orders of magnitude too high or too low. It is also impossible to obtain the steady-state quantum distribution, without approximations - nor is there any proof that the boundary terms vanish in the Wigner case.

Another case of some interest is the anharmonic oscillator Hamiltonian, it just leads to a second order diffusion in the case of the positive P-representation. There are no deterministic trajectories that can escape to infinity in a finite time, so the surface terms are presumably zero, provided the initial distribution is bounded. In fact, the stochastic equations are trivially soluble. It is useful, again, to compare this situation with the more traditional representations. None of these have stochastic equations. The P and Q representations have non-positive-definite Fokker-Planck equations, while the Wigner equation is of third order - which automatically prevents any possible probabilistic interpretation of the propagator.

There is one example in which, under rather extreme circumstances, the boundary terms are not negligible. Smith and Gardiner [10] investigated a model of a harmonic oscillator damped by both one photon and two photon pabsorption. In an interaction picture the master equation is

$$\frac{\partial\rho}{\partial t} = \frac{1}{2}\kappa(2a^2\rho a^{\dagger 2} - a^{\dagger 2}a^2\rho - \rho a^2 a^{\dagger 2}) + \frac{1}{2}\gamma\kappa(2a\rho a^\dagger - a^\dagger a\rho - \rho a a^\dagger). \tag{35}$$

Carrying our the positive-P procedures, we arrive at the Ito SDEs

$$\left.\begin{aligned} \frac{d\alpha}{dt} &= -\kappa(\tfrac{1}{2}\gamma\alpha + a^+a^2) + i\sqrt{\kappa}\alpha\xi_1(t) \\ \frac{d\alpha}{dt} &= -\kappa(\tfrac{1}{2}\gamma\alpha^+ + \alpha\alpha^{+2}) - i\sqrt{\kappa}\alpha\xi_2(t). \end{aligned}\right\} \tag{36}$$

By defining $N = \alpha^+\alpha$ and using Ito rules [7], we obtain a single equation

$$\frac{dN}{d\tau} = -(\frac{1}{2}\gamma N + N^2) + iN\xi(\tau) \tag{37}$$

where

$$\tau = 2\kappa t, \qquad \xi(\tau) = \frac{\xi_1(t) + \xi_2(t)}{\sqrt{2\kappa}} \tag{38}$$

and using $v = 1/N$ we can transform this into a linear equation for v

$$\frac{dv}{d\tau} = [1 + (\frac{1}{2}\gamma - 1)v] - iv\xi(\tau) \tag{39}$$

Because the equation is linear, it is readily solved numerically. However, even without a computer, it is clear that the origin has a set of measure zero of trajectories that pass through it - that is, the deterministic drift term does not vanish at the origin. For this reason, we conclude that the distribution is nearly constant near $v = 0$. This implies that $P(\alpha)$ varies as N^{-3} asymptotically. While this allows the distribution to be normalized, it is not sufficiently rapid to eliminate the boundary terms exactly. Despite this problem, the origin is not an attractor — in other words, no trajectories will ever reach this point in a practical simulation. In addition, for the case that the one-photon decay is much greater than the two-photon decay rate, with one photon present on average, then $\gamma >> 1$. In this typical physical situation, the origin in v-space has exponentially decreasing probability densities, so that the boundary terms are negligible. In fact, the ratio of these rates is typically many orders of magnitude. In computer simulations, the only discrepancy that is known occurs in the rather unphysical situation of $\gamma < 1$, although we clearly cannot discount the possibility of exponentially small corrections to the transient results in the typical regime of $\gamma >> 1$.

To complete the picture, we note that again, as before, there are no exact stochastic processes that represent this problem in the standard classical phase-space (P, Q, Wigner) representations. In each case these have either a non-positive-definite or higher-order diffusion term. It is possible to truncate the higher-order terms to obtain approximate results in some cases. However, there has been no systematic treatment to justify this. We conclude that, in this case, there is simply no exact stochastic theory available as yet. Despite this, the positive P-representation is able to provide an approximate theory, with rather good agreement to number state calculations, when the damping ratios have typical values corresponding to experiment. In cases of extremely large relative nonlinearities, it appears preferable to turn to either number-state methods or to complex P-representations.

REFERENCES

1. J. von Neumann: *Mathematical Foundations of Quantum Mechanics* (Princeton University Press, Princeton 1955).
2. E. P. Wigner: *Phys. Rev.* **40**, 749 (1932).
3. R. J. Glauber: *Phys. Rev.* **131**, 2766 (1963); E. C. G. Sudarshan: *Phys. Rev. Lett.* **10**, 277 (1963).
4. J. R. Klauder, J. McKenna, D. G. Currie: *J. Math. Phys.* **6**, 734 (1965).
5. C. W. Gardiner: *Handbook of Stochastic Methods (2nd ed.)* (Springer, Heidelberg 1985, 1990).
6. P. D. Drummond, C. W. Gardiner: *J. Phys.* A**13**, 2353 (1980).
7. C. W. Gardiner: *Quantum Noise* (Springer, Heidelberg 1991).
8. M. Wolinsky, H. J. Carmichael: *Phys. Rev. Lett.* **60**, 1836 (1988).
9. P. Kinsler, P. D. Drummond: *Phys. Rev. Lett.* **64**, 236 (1990); P. D. Drummond, P. Kinsler: *Phys. Rev.* A**40**, 4813 (1989).
10. A. M. Smith, C. W. Gardiner: *Phys. Rev.* A**39**, 3511 (1989).

THERMOFIELD DYNAMICS AND ITS APPLICATIONS TO QUANTUM OPTICS

S. Chaturvedi

School of Physics
University of Hyderabad
Hyderabad 500134, India

INTRODUCTION

Thermofield dynamics (TFD) arose out of efforts to define operators at finite temperatures in the theory of superconductivity.[1] Since then it has developed into a powerful formalism for dealing with quantum field theories at finite temperatures[2-4] and has found numerous applications in non-equilibrium phenomena[5] and in quantum optics.[6-15] The central idea in TFD is to represent a density operator as a vector in an extended Hilbert space. The same idea, in fact, also forms the basis of another formalism known as the Liouville space representation.[16-19] The difference between the two formalisms is as follows. Consider a density operator ρ on a Hilbert space $\mathcal{H}$. Any operator on $\mathcal{H}$ and, in particular, the density operator ρ can be expanded in terms of the operators $|M><N|$ where $|N>$ constitutes a complete orthonormal basis in $\mathcal{H}$. In TFD the operators $|M><N|$ are viewed as basis vectors in the Hilbert space $\mathcal{H} \otimes \mathcal{H}^*$, obtained by taking the direct product of $\mathcal{H}$ with its dual, and are denoted by $|M, N>$. In Liouville space representation,[16-19] on the other hand, $|M><N|$ are regarded as the basis vectors in the Hilbert space $\bar{\mathcal{H}}$ of linear operators on $\mathcal{H}$. In what follows, we shall confine ourselves to the TFD representation and consider its application to quantum optics. In particular, we shall highlight its usefulness for exact or approximate solution of master equations encountered in quantum optics.

A BRIEF SUMMARY OF TFD

In TFD one associates, with a density operator ρ acting on a Hilbert space $\mathcal{H}$, a state vector $|\rho^\alpha>, 0 \leq \alpha \leq 1$ in the extended Hilbert space $\mathcal{H} \otimes \mathcal{H}^*$ so that averages of operators with respect to ρ acquire the appearance of an expectation value.

$$< A >= TrA\rho =< \rho^{1-\alpha}|A|\rho^\alpha > . \tag{1}$$

The state $|\rho^\alpha>$ is given by

$$|\rho^\alpha >= \hat{\rho}^\alpha |I >, \tag{2}$$

Recent Developments in Quantum Optics, Edited
by R. Inguva, Plenum Press, New York, 1993

where

$$\hat{\rho}^{\alpha} = \rho^{\alpha} \otimes I, \tag{3}$$

and

$$|I> = \sum_N |N> \otimes |N> \equiv \sum_N |N, N>, \tag{4}$$

where $|N>$ constitutes any complete orthonormal set in $\mathcal{H}$. The state $|I>$ is simply the counterpart of the resolution of the identity

$$I = \sum_N |N><N|, \tag{5}$$

in terms of a complete orthonormal set $|N>$ in $\mathcal{H}$. In particular, if

$$\rho|\mathrm{N}> = \mathrm{p_N}|\mathrm{N}>. \tag{6}$$

$$|\rho^{\alpha}> = \sum_N p_N^{\alpha}|N, N> . \tag{7}$$

In dealing with bosonic systems it is natural to use, for $|N>$, the number states $|n>$ and to introduce creation and annihilation operators $a^{\dagger}, \tilde{a}^{\dagger}, a$, and $\tilde{a}$ as follows

$$a|n,m> = \sqrt{n}|n-1,m>, \quad \tilde{a}|n,m> = \sqrt{m}|n,m-1>, \tag{8a}$$

$$a^{\dagger}|n,m> = \sqrt{m+1}|n+1,m>, \quad \tilde{a}^{\dagger}|n,m> = \sqrt{m+1}|n,m+1> . \tag{8b}$$

The operators a and $a^{\dagger}$ commute with $\tilde{a}$ and $\tilde{a}^{\dagger}$. It is easily seen that the operators $\tilde{a}$ and $\tilde{a}^{\dagger}$ respectively simulate the action of $a^{\dagger}$ and a on $|n><m|$ from the right.

From the expression for $|I>$ in terms of the number states

$$|I> = \sum_n |n,n>,$$

it follows that

$$a|I> = \tilde{a}^{\dagger}|I>, \quad a^{\dagger}|I> = \tilde{a}|I> . \tag{10}$$

and hence for any operator

$$A(a^{\dagger}, a) = \sum_{p,q} \alpha_{p,q} a^{\dagger p} a^{q}, \tag{11}$$

$$A|I> = \tilde{A}^{\dagger}|I>, \tag{12}$$

where $\tilde{A}$ is obtained from A by making the replacements (tilde conjugation rules) $a \to \tilde{a}, a^{\dagger} \to \tilde{a}^{\dagger}, \alpha \to \alpha^{*}$. Given the evolution equation for ρ^{α}, the relations (10) enable one to transcribe it into a Schrödinger like equation for $|\rho^{\alpha}>$ associated to the density operator ρ. For dissipative systems, as we shall see later this is only possible for $\alpha = 1$.

TFD REPRESENTATIONS FOR SOME FAMILIAR DENSITY OPERATORS

[1] Consider the density operator for a harmonic oscillator at a finite temperature

$$\rho_o = (1 - e^{-\beta}) exp(-\beta a^\dagger a). \tag{13}$$

Since ρ_o is diagonal in the number representation, one obtains using (7), the following expression for the corresponding $|\rho_o^\alpha>$

$$|\rho_o^\alpha >= (1-f)^\alpha \sum_{n=o}^{\infty} (f^\alpha)^n |n,n>, \quad f \equiv e^{-\beta} \tag{14}$$

This expression, after some simple manipulations may be put in the following form

$$|\rho_o^\alpha >= \frac{(1-f)^\alpha}{\sqrt{(1-f^{2\alpha})}} exp[\theta(a^\dagger \tilde{a}^\dagger - a\tilde{a})]|0,0>, tanh\theta = f^\alpha. \tag{15}$$

Note that $|\rho_o^\alpha >$ is unchanged under the replacements $a \to \tilde{a}, a^\dagger \to \tilde{a}^\dagger, \alpha \to \alpha^*$ and one says that $|\rho_o^\alpha >$ is tildian. The tildian property is, in fact, a direct consequence of the hermiticity property of the density operator.

For $\alpha = 1/2$, (15) gives[2]

$$|\rho_o^{1/2} > \equiv |0(\beta)> = exp[-iG_B]|0,0>; -iG_B = \theta(a^\dagger \tilde{a}^\dagger - a\tilde{a}). \tag{16}$$

Thus $|\rho_o^{1/2}>$ is seen to be related to $|0,0>$ by a unitary transformation. Viewed as a two mode state $|\rho_o^{1/2}>$ may easily be recognized as the Caves-Shumaker squeezed state.[20] The notation $|0(\beta)>$ introduced in (16) is to bring out the fact that this state may be viewed as a finite temperature analogue of the vacuum state. Just as the vacuum state $|0,0>$ is annihilated by a and $\tilde{a}$, the thermal vacuum state $|(\beta)>$ is easily seen to be annihilated by a(β) and $\tilde{a}(\beta)$

$$a(\beta)|0(\beta)>= 0, \quad \tilde{a}(\beta)|0(\beta)>= 0, \tag{17}$$

where

$$a(\beta) = e^{-iG_B} a e^{iG_B} = cosh\theta a - sinh\theta \tilde{a}^\dagger, \tag{18a}$$

$$\tilde{a}(\beta) = e^{-iG_B} \tilde{a} e^{iG_B} = cosh\theta \tilde{a} - sinh\theta a^\dagger. \tag{18b}$$

The operators $a(\beta)$, $\tilde{a}(\beta)$, $a^\dagger(\beta)$, $\tilde{a}^\dagger(\beta)$ are thus seen to be related to a, $\tilde{a}$, $a^\dagger$, $\tilde{a}^\dagger$ by a Bogoliubov transformation and as a consequence obey the canonical commutation relations

$$[a(\beta), a^\dagger(\beta)] = [\tilde{a}(\beta), \tilde{a}^\dagger(\beta)] = 1, \tag{19}$$

with all other commutators being zero. In view of (17), we may rightly call $a(\beta)$ ($a^\dagger(\beta)$) and $\tilde{a}(\beta)$ ($\tilde{a}^\dagger(\beta)$) as the thermal annihilation (creation) operators. It should be noted that this very appealing structure emerges only in the symmetric $\alpha = 1/2$ case.

The averages $< A >$ of any operator in this representation are given by

$$< A(a, a^\dagger) >=< 0(\beta)|A(a, a^\dagger)|0(\beta) > . \tag{20}$$

and may be viewed as thermal vacuum expectation values.

[2] The density operator ρ for a displaced harmonic oscillator at a finite temperature is

$$\rho = (1 - e^{-\beta})exp[-\beta(a^\dagger - \mu^*)(a - \mu)] = D(\mu)\rho_o(D^+(\mu)) \tag{21}$$

where $D(\mu) = exp(\mu a^\dagger - \mu^* a)$ is the displacement operator.

The corresponding state in the TFD representation is

$$|\rho^\alpha >= D(\mu)\tilde{D}(\mu)|\rho_o^\alpha > . \tag{22}$$

One may call $|\rho^\alpha >$ given by (22) a thermal coherent state because it is obtained by applying displacement operators on the thermal vacuum state just as an ordinary coherent state is obtained by the action of a displacement operator on the vacuum state. Note that, for $\alpha = 1/2$, the operator $\tilde{D}(\mu)$ containing only the tilde operators cancels out when one calculates averages of functions of a and $a^\dagger$ and may therefore be dropped. This, however, makes the state vector non-tildian. Such a non-tildian representation for the density operator (21) was given by Barnett and Knight[8] and has also been discussed by Fearn and Collett.[9] The tildian representation for $\alpha = \frac{1}{2}$ given by (22) has been obtained by Mann and Revzen.[10]

[3] From the discussion above it follows that the structure of the thermal coherent states bears a close similarity to that of ordinary coherent states:

$$|coherent\ states > \qquad |thermal\ coherent\ state >$$

$$= D(\mu)\ |vacuum > \qquad = D(\mu)\tilde{D}(\mu)|thermal\ vacuum > \tag{23}$$

This provides us with a way of constructing thermal counterparts of known states such as the squeezed coherent states[21] etc. as has been discussed in ref. 11.

For the thermal states discussed above, Mann et al[22] have derived a generalized uncertainty relation

$$< (\Delta P)^2 >< (\Delta Q)^2 >\geq 1/4+ < \Delta P\Delta\tilde{P} >< \Delta Q\Delta\tilde{Q} > \tag{24}$$

The second term on the RHS contains the thermal fluctuations. It is interesting to note that the thermal fluctuations involve correlations between tilde and untilde degrees of freedom.[22]

[4] Finally consider the density operator

$$\rho = \mathcal{N} \sum_{n=o} \frac{\xi^n}{n!} |n >< n| \tag{25}$$

The state $|\rho >$ corresponding to (25), viewed as a two mode state may easily be recognized as a pair coherent state.[23,24]

DYNAMICS

Having seen how to represent a density operator ρ as a state $|\rho^\alpha >$ we turn our attention to the evolution equations for ρ and show how to transcribe them as Schrödinger like equations for the associated states.

Conservative Systems

The evolution equation here is the Liouville von Neumann equation

$$i\frac{\partial \rho}{\partial t} = [H, \rho]. \tag{26}$$

which gives for $\rho^\alpha(t)$

$$i\frac{\partial}{\partial t}\rho^\alpha = [H, \rho^\alpha]. \tag{27}$$

To convert (27) into an equation for $|\rho^\alpha >$, we apply (27) on the state $|I >$

$$i\frac{\partial}{\partial t}\hat{\rho}^\alpha |I >= H\hat{\rho}^\alpha |I > -\hat{\rho}^\alpha H |I > . \tag{28}$$

To obtain the desired equation for $|\rho^\alpha > \equiv \hat{\rho}^\alpha |I >$ we have to move $\hat{\rho}^\alpha$ in the second term on the R.H.S. of (28) next to $|I >$. This is easily done using relation (12) and the hermiticity of H to obtain

$$\frac{\partial}{\partial t}|\rho^\alpha >= -i\hat{H}|\rho^\alpha >, \quad \hat{H} = H - \tilde{H}. \tag{29}$$

Thus we see that starting from the Liouville von-Neumann equation for ρ^α we can derive a Schrödinger like equation for the state $|\rho^\alpha >$ for any value of α and the corresponding Hamiltonian always has the structure (29) independent of α. This happy situation, however is not obtained in the case of dissipative systems.

Dissipative Systems

Dissipative systems are described by master equations which have the following structure

$$i\frac{\partial}{\partial t}\rho = [H, \rho] + L\rho, \tag{30}$$

where the terms $L\rho$ incorporate dissipation. Unlike the conservative case, given (30) there is no general procedure for writing the equation for ρ^α, and hence for the state

$|\rho^\alpha>$. This has the consequence that for dissipative systems the only option available is to work with the state $|\rho>$ which corresponds to $\alpha = 1$.

Consider, for example, the master equation for a linear oscillator

$$\frac{\partial}{\partial t}\rho = -i[H,\rho] + \frac{1}{2}\gamma(\bar{n}+1)(2a\rho a^\dagger - a^\dagger a\rho - \rho a^\dagger a) + \frac{1}{2}\gamma\bar{n}(2a^\dagger\rho a - aa^\dagger\rho - \rho aa^\dagger), \quad (31)$$

where $H = \omega\, a^\dagger a$. Applying $|I>$ on (31) from the right and using (10) the master equation (31), in the TFD notation goes over to

$$\frac{\partial}{\partial t}|\rho> = -i\hat{H}\;|\rho>, \quad (32)$$

where

$$-i\hat{H} = -i\omega(a^\dagger a - \tilde{a}^\dagger\tilde{a}) + \frac{1}{2}\gamma(\bar{n}+1)(2a\tilde{a} - a^\dagger a - \tilde{a}^\dagger\tilde{a}) + \frac{1}{2}\gamma\bar{n}(2a^\dagger\tilde{a}^\dagger - aa^\dagger - \tilde{a}\tilde{a}^\dagger). \quad (33)$$

ALGEBRAIC SOLUTION OF MASTER EQUATIONS

The traditional method for solving master equations in quantum optics proceeds by first transcribing them into c-number Fokker-Planck equations (FPE) using the Glauber-Sudashan P-representation[25–26] or the generalized P-representation of Drummond and Gardiner.[27] By converting master equations into Schrödinger like equations, the TFD notation offers new methods for their exact or approximate solution. It makes master equations amenable to familiar quantum mechanical techniques. This is illustrated below with the help of some examples.

Master Equation for a Linear Oscillator

The master equation (31) for a linear oscillator can be solved exactly in many different ways. The corresponding FPE in the P-representation turns out to have linear drift and constant diffusion and can therefore readily be solved. Here we present an algebraic solution with (32) and (33) as the starting point.

Introducing the operators

$$K_+ = a^\dagger\tilde{a}^\dagger; K_- = a\tilde{a}; \quad K_3 = \frac{1}{2}(a^\dagger a + \tilde{a}^\dagger\tilde{a} + 1); \quad K_o = (a^\dagger a - \tilde{a}^\dagger\tilde{a}), \quad (34)$$

we may rewrite (33) as

$$-i\hat{H} = -i\omega K_o + \gamma(\bar{n}+1)K_- + \gamma\bar{n}K_+ - \gamma(2\bar{n}+1)K_3 + \frac{1}{2}\gamma. \quad (35)$$

The operators K_+, K_- and K_3 generate the SU(1,1) algebra

$$[K_-, K_+] = 2K_3; [K_3, K_\pm] = \pm K_\pm. \quad (36)$$

K_o is a Casimir operator. Solving (32) we get

$$|\rho(t)> = exp(\gamma_o K_o + \frac{1}{2}\gamma t) exp(\gamma_+ K_+ + \gamma_3 K_3 + \gamma_- K_-)|\rho(0)>, \tag{37}$$

where

$$\gamma_+ = \gamma \bar{n} t; \quad \gamma_- = \gamma(\bar{n}+1)t; \gamma_3 = -\gamma(2\bar{n}+1)t; \quad \gamma_o = -i\omega t. \tag{38}$$

Use of the disentangling theorem for SU(1,1)[28]

$$exp(\gamma_+ K_+ + \gamma_3 K_3 + \gamma_- K_-) = exp(\Gamma_+ K_+) exp((log\Gamma_3) K_3) exp(\Gamma_- K_-), \tag{39}$$

where

$$\Gamma_\pm = \frac{2\gamma_\pm sinh\phi}{2\phi cosh\phi - \gamma_3 sinh\phi}, \quad \Gamma_3 = \Big(\frac{2\phi}{2\phi cosh\phi - \gamma_3 sinh\phi}\Big)^2, \tag{40}$$

with

$$\phi^2 = \Big(\frac{\gamma_3^2}{4}\Big) - \gamma_+\gamma_-, \tag{41}$$

and of the fact that K_+, K_- and K_3 have simple actions on $|n, m>$ enable one to solve (32) and hence (31) purely algebraically. Detailed expressions for $\rho_{m,n}(t)$ and the Q-function for an arbitrary initial condition may be found in ref. 12.

Master Equation for a Nonlinear Oscillator

Consider the master equation having the same structure as in (40) but with H given by $H = \omega a^\dagger a + \chi(a^\dagger a)^2$. This master equation has been extensively discussed in the literature[29–34] and its exact solution has been found by solving the corresponding Fokker-Planck equation. Unlike the linear case, the FPE in this case turns out to have non linear drift and non constant diffusion and requires quite some ingenuity for its exact solution. In the TFD approach, however, the exact solution may be found with no more effort than is necessary for the linear case as can be seen from the following considerations.

The Hamiltonian $-i\hat{H}$ in this case differs from (33) only by an extra term $-i[\chi(a^\dagger a)^2 - (\tilde{a}^\dagger \tilde{a})^2]$. This expression may be factorized and expressed in terms of K_3 and K_o as $-2i\chi K_o(K_3 - 1/2)$ giving for $-i\hat{H}$

$$-i\hat{H} = -i(\omega - \chi)K_o + \gamma(\bar{n}+1)K_- + \gamma\bar{n}K_+ - (\gamma(2\bar{n}+1) + 2i\chi K_o)K_3 + \frac{1}{2}\gamma. \tag{42}$$

which is again a linear combination of $K_\mp$ and K_3 with coefficients which depend on K_o. Since K_o is a Casimir operator, the disentangling theorem remains unaffected and the exact solution can be found exactly as in the linear case and is given in ref 12.

Class of Master Equations Describing Coupled Non Linear Oscillators

We may easily extend the considerations given above for the master equation for a single non linear oscillator, to a master equation describing coupled dissipative non linear oscillators. This master equation can again be solved in exactly the same manner as before.[13]

Agarwal and Puri[33] and Tombesi and Mecozzi[34] have considered models describing propagation of elliptically polarized light through a non linear medium. If effects of dissipation are included by adding damping terms as in (31) then, for the case of equal damping parameters for both the modes involved, the corresponding master equations fall in the class discussed above and therefore can be solved exactly.

APPROXIMATE SOLUTIONS OF MASTER EQUATIONS

To illustrate the use of quantum mechanical approximation methods, we consider the master equation for two photon absorption

$$\frac{\partial}{\partial t}\rho = -k(a^{\dagger 2}a^2\rho + \rho a^{\dagger 2}a^2 - 2a^2\rho a^{\dagger 2}), \tag{43}$$

which can be solved exactly using TFD as well as by other methods. The Hamiltonian for this master equation in the TFD notation is given by

$$-i\hat{H} = k(2a^2\tilde{a}^2 - a^{\dagger 2}a^2 - \tilde{a}^{\dagger 2}\tilde{a}^2). \tag{44}$$

Following the standard self consistent linearization techniques, we approximate this by

$$i\hat{H} = k\Delta(t)(a^2 - a^{\dagger 2} + \tilde{a}^2 - \tilde{a}^{\dagger 2}), \tag{45}$$

where $\Delta(t) =< a^2 >=< \tilde{a}^2 >$ is to be determined self consistently. This approximate Hamiltonian evolves an initial coherent state into a squeezed coherent state with a time dependent squeezing parameter. Numerical comparison with the exact results shows fairly good agreement for short times.[14]

TRANSITION TO C-NUMBER EQUATIONS

We have seen that TFD, by transcribing master equations into Schrödinger like equations, enables us to use algebraic techniques for their exact or approximate solution. In some cases, however, we may wish to map $|\rho>$ into a c-number function and the corresponding evolution equation onto a c-number equation. This can be most naturally done by expanding $|\rho>$ in terms of two mode coherent states

$$|\rho>= \int d^2\alpha d^2\beta P(\alpha,\beta)e^{-\alpha\beta}exp(\alpha a^\dagger)exp(\beta\tilde{a}^\dagger)|0,0> . \tag{46}$$

The normalization condition $< I|\rho >= 1$ for $|\rho>$ implies that

$$\int \mathrm{d}^2\alpha \mathrm{d}^2\beta\ P\alpha,\beta) = 1. \tag{47}$$

The expansion (46) may now easily be recognized as the TFD version of the expansion for ρ proposed by Drummond and Gardiner[29] in their work on the positive P-representation and we can make use of their results to associate a positive $P(\alpha, \beta)$ to $|\rho >$ and a genuine FPE to the corresponding Schrödinger like equation. The net result, of course would be the same as applying the positive P-representation directly on ρ and on its evolution equation.

CONCLUSION

We have seen that the TFD formalism, though not specifically developed for that purpose, does prove to be a valuable tool for exact or approximate solution of master equations encountered in quantum optics. By associating a Schrödinger like equation to a given master equation, this formalism permits one to use all the techniques that have been developed for Schrödinger equations. We have demonstrated the usefulness of TFD by obtaining the exact solution of a class of master equations for which the conventional methods lead to rather complicated Fokker-Planck equations. We have also shown that the transition to c-numbers via coherent state expansions naturally leads to the counterpart of the representation for the density operator studied by Drummond and Gardiner in their work on generalized P-representations.

ACKNOWLEDGEMENTS

The author is grateful to Prof. V. Srinivasan, to Prof. G.S. Agarwal and to Ms. P. Shanta for discussions and numerous helpful comments and criticisms.

REFERENCES

1. L. Laplae, F. Mancini and H. Umezawa, *Phys. Rev.* **C10**, 151 (1974).
2. Y. Takahashi and H. Umezawa, *Coll. Phenom.* **2**, 55 (1975).
3. H. Umezawa, H. Matsumoto and M. Tachiki, "Thermofield Dynamics and Condensed States," (1982) (Amsterdam: North Holland).
4. I. Ojima, *Ann. Phys.* **137**, 1(1981).
5. T. Arimitsu "Non-equilibrium thermofield dynamics and thermal processes" Univ. of Tsukuba Preprint (1991) and references therein.
6. T. Tominaga, M. Ban, T. Arimitsu, J. Pradko and H. Umezawa, *Physica* **A149**, 26 (1988).
7. T. Tominaga, T. Arimitsu, J. Pradko and H. Umezawa, *Physica* **A150**, 97(1988).
8. S. M. Barnett and P. L. Knight, *J. Opt. Soc. of Am.* **B2**, 467, (1985).
9. H. Fearn and M. J. Collett, *J. Mod. Opt.* 553 (1988).
10. A. Mann and M. Revzen, *Phys. Lett.* **A134**, 273 (1989).
11. S. Chaturvedi, R. Sandhya, R. Simon and V. Srinivasan, *Phys. Rev.* **A41**, 3969 (1989).
12. S. Chaturvedi and V. Srinivasan, *J. Mod. Opt.* (1991).
13. S. Chaturvedi and V. Srinivasan, *Phys. Rev.* **A** (1991).
14. S. Chaturvedi, P. Shanta and V. Srinivasan, *Opt. Comm.* **78**, 289 (1990).
15. S. Chaturvedi, P. Shanta and V. Srinivasan, *Phys. Rev.* **A43**, 521 (1990).

16. J. A. Crawford, *Nuovo Cimento* **10**, 698 (1958).
17. J. Fiutak and J. Van Kranendonk Can. *J. Phys.* **40**, 1085 (1962).
18. M. Schmutz, *Z. Physik,* **B30**, 97 (1978).
19. S. M. Barnett and B. J. Dalton, *J. Phys.* **A20**, 411 (1987).
20. C. M. Caves and B. L. Schumaker, *Phys. Rev.* **A31**, 3068 (1985).
21. H. P. Yuen, *Phys. Rev.* **A13**, 2226 (1976).
22. A. Mann, M. Revzen, H. Umezawa and Y. Yamanaka *Phys. Lett.* **A140**, 475 (1989).
23. D. Bhaumik, K. Bhaumik and B. Dutta Roy, *J. Phys.* **A9**, 1507 (1976).
24. G. S. Agarwal *J. Opt. Soc. Am.* **15B**, 1940 (1988).
25. E. C. G. Sudarshan, *Phys. Rev. Lett.* **10**, 277 (1963).
26. R. J. Glauber, *Phys. Rev.* **130**, 2529 (1963).
27. P. D. Drummond and C. W. Gardiner, *J. Phys.* **A13**, 2353 (1980).
28. J. H. Eberly and Wodkiéwicz, *J. Opt. Soc. of Am.* **B2**, 458 (1985).
29. G. J. Milburn and C. A. Holmes, *Phys. Rev. Lett,* **56**, 2237 (1986).
30. G. J. Milburn, A. Mecozzi and P. Tombesi, *J. Mod. Opt.* **36**, 1607 (1989).
31. D. J. Daniel and G. J. Milburn, *Phys. Rev.* **A39**, 4628 (1989).
32. V. Peřinova and A. Lukš, *Phys. Rev.*, **A41**, 414 (1990).
33. G. S. Agarwal and R. Puri, *Phys. Rev.* **A40**, 519 (1989).
34. P. Tomebsi and A. Mecozzi, *J. Opt. Soc. Am.* **B4**, 1700 (1987).

SUPERSYMMETRY IN QUANTUM OPTICS

V.A. Andreev*, A.B. Klimov*, P.B. Lerner†

*Lebedev Physical Institute
Moscow, 117924, USSR
†Department of Electrical Engineering
University of Michigan, Ann Arbor, MI 48109, USA

In the terms of Witten's supersymmetric quantum mechanics we will investigate the two quantum optics models. According to the main idea of this theory the supersymmetric hamiltonian H is the anticommutator of two supercharges [1] Q_1, Q_2

$$H = \{Q_1, Q_2\}, \; Q_1^2 = Q_2^2 = 0 \tag{1}$$

In the simplest case such a hamiltonian has the diagonal structure

$$H = \begin{pmatrix} H_1 & 0 \\ 0 & H_2 \end{pmatrix}, \; Q_1 = \begin{pmatrix} 0 & q_1 \\ 0 & 0 \end{pmatrix}, \; Q_2 = \begin{pmatrix} 0 & 0 \\ q_2 & 0 \end{pmatrix}, \; \begin{array}{lcl} H_1 & = & q_1 q_2, \\ H_2 & = & q_2 q_1. \end{array} \tag{2}$$

From this formulae it is easy to see that the spectrums of hamiltonians H_1, H_2 are very similar with each other. Only some lower levels may be different. More exactly, the one of the hamiltonians H_1, H_2 may have more lower levels than the other. This difference between the number of lower levels ΔS we will call the degree of the supersymmetry of the model.

The first model we will consider is the Jaynes-Cummings model with the hamiltonian

$$H_1 = \hbar\omega(a^+ a + \sigma_3) + g(a^+ \sigma_- + a\sigma_+). \tag{3}$$

It's superpartner is [2]

$$H_2 = \hbar\omega(a^+ a + \sigma_3) + g(e^{i\varphi} a^+ \sigma_- + e^{-i\varphi} a\sigma_+). \tag{4}$$

In hamiltonian H_2 the creation-annihilation operators have the phase shift

$$a^+ \rightarrow e^{i\varphi} a^+, \; a \rightarrow e^{-i\varphi} a \tag{5}$$

In the such model the supercharges have the form

Recent Developments in Quantum Optics, Edited
by R. Inguva, Plenum Press, New York, 1993

$$Q_1 = (\hbar\omega)^{-\frac{1}{2}} \begin{pmatrix} 0 & 0 & \frac{1}{2}g & e^{-i\varphi}\hbar\omega a \\ 0 & 0 & \hbar\omega a^+ & \frac{1}{2}ge^{-i\varphi} \\ 0 & 0 & 0 & 0 \\ 0 & 0 & 0 & 0 \end{pmatrix}, \quad Q_2 = (\hbar\omega)^{-\frac{1}{2}} \begin{pmatrix} 0 & 0 & 0 & 0 \\ 0 & 0 & 0 & 0 \\ \frac{1}{2}g & \hbar\omega a & 0 & 0 \\ e^{i\varphi}\hbar\omega a^+ & \frac{1}{2}ge^{i\varphi} & 0 & 0 \end{pmatrix}.$$

The spectrums of the hamiltonians (3), (4) are completely equal so for the Jaynes-Cummings model we have the $\Delta S = 0$ supersymmetry.

Let us discuss the physical meaning of this type of supersymmetry. The hamiltonians H_1 and H_2 are distinguished one from another by the phases of the operators a, a^+. If this phase is constant the hamiltonians H_1, H_2 are absolutely equal and the transformation (5) means the reparametrisation, but if this phase is not constant and has some dynamical sense then these two hamiltonians may have different physical interpretation. Such a situation can arise when some singular lines exist in the space.

It is possible to find the connection between the phase φ in (5) and Berry's phase.

For the definition of the Berry's phase in the Heisenberg representation we will use the following construction. We have the time-dependent hamiltonian $H_1(t)$ and the evolution operator

$$U(t) = Texp(\int^t H_1(\tau)d\tau).$$

With the help of this evolution operator $U(t)$ one can define the new operator $\tilde{H}_1(t)$

$$\tilde{H}_1(t) = U(t)H_1(0)U^{-1}(t)$$

The operators $H(t)$ and $\tilde{H}(t)$ are unitary equivalent to each other

$$\tilde{H}_1(t) = \tilde{U}(t)\tilde{H}_1(t)\tilde{U}^{-1}(t)$$

They constitute the supersymmetrical hamiltonian

$$H = \begin{pmatrix} H_1(t) & 0 \\ 0 & \tilde{H}_1(t) \end{pmatrix}$$

with the supercharges

$$Q_1 = \begin{pmatrix} 0 & \sqrt{H_1}U^{-1} \\ 0 & 0 \end{pmatrix}, \quad Q_2 = \begin{pmatrix} 0 & 0 \\ U\sqrt{H_1} & 0 \end{pmatrix}.$$

In the same manner the pairs $\{H_1(0), H_1(t)\}, \{H_1(0), \tilde{H}_1(t)\}$ constitute the supersymmetrical hamiltonian. The hamiltonian $\tilde{H}_1(T)$ differs from the hamiltonian $H_1(0) = H_1(T)$ by some operator which is the operator of Berry's phase. The Berry's phase for the Jaynes-Cummings model was found in [3]. For the wavefunction

$$\Psi_N = \alpha|N, 0> + \beta|N-1, 1> \tag{6}$$

it has the form

$$\varphi_B = \frac{1}{2}\int_o^T (4\rho^2 N + \Delta^2)^{-\frac{1}{2}}\Delta\dot{\delta}d\tau, \quad \begin{matrix}\Delta = \omega_o - \omega_1, \\ \mu = \rho e^{i\delta}.\end{matrix} \tag{7}$$

The Heisenberg equations of motion give for the operator the operator of Berry's phase

$$\hat{\varphi}_B = \frac{1}{2}\int_o^T (4\rho^2(a^+a + \sigma_3) + \Delta^2)^{-\frac{1}{2}}\Delta\dot{\delta}d\tau. \tag{8}$$

It's eigenvalues for the eigenfunctions (6) coincide with Berry's phase (7). We see that the supersymmetry phase φ (5) can be connected with Berry's phase.

Let us consider the next example,the models of two photo transitions. In this case we will have the degree $\Delta S = 1$.

The hamiltonian H_1 describes the Raman scattering

$$H_1 = \omega_1 a_1^+ a_1 + \omega_2 a_2^+ a_2 + \omega_o \sigma_3 + \lambda(a_1^+ a_2 \sigma_- + a_1 a_2^+ \sigma_+) \tag{9}$$

and the hamiltonian H_2 describes the two photon adsorption

$$H_2 = \omega_1 a_1^+ a_1 + \omega_2 a_2^+ a_2 + \tilde{\omega}_o \sigma_3 + \lambda(a_1^+ a_2^+ \sigma_- + a_1 a_2 \sigma_+),$$

$$\omega_o = \omega_1 - \omega_2, \quad \tilde{\omega}_o = \omega_1 + \omega_2. \tag{10}$$

They are the components of the supersymmetric hamiltonian

$$H = \begin{pmatrix} H_1 & 0 \\ 0 & H_2 \end{pmatrix} + cI, \; c = \frac{1}{2}(\omega_1 + \omega_2), \; I = \begin{pmatrix} 1 & 0 \\ 0 & 1 \end{pmatrix}$$

The supercharges have the form

$$Q_1 = \begin{pmatrix} 0 & 0 & \alpha a_2^+ & \beta a_1 \\ 0 & 0 & \beta a_1^+ & \alpha a_2 \\ 0 & 0 & 0 & 0 \\ 0 & 0 & 0 & 0 \end{pmatrix}, \; Q_2 = \begin{pmatrix} 0 & 0 & 0 & 0 \\ 0 & 0 & 0 & 0 \\ \tilde{\alpha} a_2 & \tilde{\beta} a_1 & 0 & 0 \\ \tilde{\beta} a_1^+ & \tilde{\alpha} a_2^+ & 0 & 0 \end{pmatrix},$$

here $\alpha = \frac{\sqrt{\omega_2}}{f}$, $\tilde{\alpha} = f\sqrt{\omega_2}$, $\beta = \frac{\sqrt{\omega_1}}{fU\pm}$, $\tilde{\beta} = fU \pm \sqrt{\omega_1}$, $f = const$ and $U_\pm$ are the roots of the equation $U + U^{-1} = \lambda(\omega_1\omega_2)^{-\frac{1}{2}}$. The spectrum of the hamiltonian H is

$$E_{n,m}^{ss,\pm} = (n+1)\omega_1 + (m+1) \pm \lambda\sqrt{(n+1)(m+1)}.$$

Each eigenvalue is twofold degenerate with the eigenstate

$$\Psi_{n,m,\pm}^{ss,\pm} = \begin{pmatrix} \begin{pmatrix} |n, m+1> \\ \pm|n+1, m> \end{pmatrix} \\ \pm\begin{pmatrix} |n, m> \\ \pm|n+1, m+1> \end{pmatrix} \end{pmatrix}, \quad n, m = 0, 1, ...$$

Only the groundstate E=0 is nondegenerate. It's eigenstate is

$$\Psi^{ss}_{-1,-1} \equiv \Psi_o = \begin{pmatrix} 0 \\ 0 \\ 0 \\ |0,0> \end{pmatrix}.$$

So we have the supersymmetry degree $\Delta S = 1$.

REFERENCES

1. E. Witten, Nucl. Phys. B, 185 (1981) 513.
2. V. A. Andreev and P. B. Lerner, Phys. Lett. A, 134 (1989) 507.
3. V. A. Andreev, A. B. Klimov and P. B. Lerner, Europhys. Lett. 12 (1990) 101.

POLARIZATION INVARIANCE OF LIGHT FIELDS AND P-SCALAR BIPHOTON LIGHT

V.P. Karassiov

P.N. Lebedev Physical Institute of the USSR
Academy of Sciences, Leninsky prospect 53
Moscow 117924 USSR

ABSTRACT

A classification of polarization states of quantum light fields is given by using their polarization gauge SU(2) invariance and the associated concept of the polarization (P) spin [1]. Specifically, we find a new class of unpolarized light states generated by P-scalar biphotons.

For a last few decades polarization properties of light were widely investigated in both theoretical and experimental aspects (see, e.g., [1-3] and references therein). However, as a rule, the polarization structure of light has been described in terms of the field correlation functions and associated Stokes parameters which are well adapted to classical optics experiments [4] but are inadequate to specific quantum ones (photon counting) [2]. Furthermore, such description ignores, in fact, a specific polarization $SU(2)$ symmetry [1,5,6] of light fields though it has been widely used in the implicit form (through the Stokes parameters) [2-4,7,8]. The aim of this contribution is to give a consequent description of the polarization structure of light within quantum optics using the above polarization $SU(2)$ symmetry.

The starting point of our analysis is the obvious invariance of standard expressions [1,9]

$$H = \sum_{i=1}^{m} \omega_i [\sum_{\sigma=\pm,3} a_\sigma^+(i) a_\sigma(i) - a_o^+(i) a_o(i)], \tag{1a}$$

$$\vec{P} = \sum_{i=1}^{m} \vec{k}_i [\sum_{\sigma=\pm,3} a_\sigma^+(i) a_\sigma(i) - a_o^+(i) a_o(i)] \tag{1b}$$

for the Hamiltonian H and the momentum $\vec{P}$ of the free electromagnetic field with the "m" time-spatial modes

$$A(\vec{r},t) = \sum_{j=1}^{m} (\frac{2\pi}{\omega_i}\nu)^{\frac{1}{2}} \sum_{\sigma=0,\pm,3} \{e_\sigma(j) a_\sigma(j)\ exp[i(\vec{k}_j \vec{r} - \omega_j t)] + h.c.\} \tag{2}$$

Recent Developments in Quantum Optics, Edited
by R. Inguva, Plenum Press, New York, 1993

under the transformations [1,6]

$$a_\sigma(i) \xrightarrow{u} \tilde{a}_\sigma(i) = \sum_{\lambda=\pm} U_{\sigma\lambda} a_\lambda(i), \ \sigma = +,-,$$

$$a_\sigma^+ \xrightarrow{u} \tilde{a}_\sigma^+(i) = (\tilde{a}_\sigma(i))^+, \ u = \|u_{\sigma\lambda}\| \in U(2) \tag{3}$$

We note that eqs. (1) admit, in fact, the more vast group U(3,1) $\supset$ U(2) of polarization transformations, but in quantum optics it reduces to the above U(2) group [1]. It is due to the fact we calculate quantum expectations of physical quantities by averaging on the space $L_{phys} = L_f$ spanned by basis vectors

$$|\{n_i^\sigma\}> = N(\{n_i^\sigma\}) \prod_{i=1}^{m} \prod_{\sigma=\pm} [n_i^\sigma!]^{-\frac{1}{2}} (a_\sigma^+(i))^{n_i^\sigma} |0> \tag{4}$$

which are generated by creation operators $a_\sigma^+(i)$ of photons with transverse (σ = +,-) polarizations (helicities) only [1,2].

The transformations (3) correspond to the U(2) "rotations" of the polarization unit vectors $e_\sigma(i)$ [5] and, therefore, may be interpreted as specific polarization gauge transformations. The generators of the obtained polarization invariance group U(2) are of the form

$$N = \textstyle\sum_{i=1}^m \sum_{\sigma=\pm} a_\sigma^+(i) a_\sigma(i), \quad P_o = \frac{1}{2} \sum_{i=1}^m [a_+^+(i) a_+(i) - a_-^+(i) a - (i)],$$
$$P_\pm = \textstyle\sum_{i=1}^m a_\pm^+(i) a_\mp(i) \tag{5}$$

where N is the total photon number operator and operators P_α are generators of the SU(2) subgroup defining the polarization (P) spin [1,6]. The operators P_α and N satisfy commutation relations

$$[N, P_\alpha] = 0, \quad [P_o, P_\pm] = \pm P_\pm, \quad [P_+, P_-] = 2P_o. \tag{6}$$

We also note that operators P_α do not commute with components $S_\alpha = i \sum_j \sum_{\beta j} A_\beta^{(-)}$ $(j) A_j^{(+)}(j) \varepsilon_{\alpha\beta\gamma}$ of the ordinary spin $\vec{S}$. ($A_\beta^{(\pm)}(j)$) are Fourier transforms of positive/. negative frequency part of (2)) though we have $< \psi|[P_o, S_\alpha]|\psi' >= 0$ for all states $|\psi>, |\psi'> \in L_f$. (A full analysis of this interesting optic (of interrelations between ordinary spin and P-spin) will be reported elsewhere.)

Eqs. (5) imply a physical meaning of different components P_α as quantities measurable in photon count experiments. Specifically, the total helicity $2P_o$ of the field is the difference $(N_+ - N_-)$ of the "right" and "left" photon numbers and Hermitian operators $P_1 = (P_+ + P_-)$ and $P_2 = i(P_+ - P_-)$ determine difference of photon numbers with opposite linear polarizations (cf. [4]). Besides, quantum expectations $< P_\alpha >$ are proportional to the Stokes parameters σ_α in the case of the monochromatic plane waves (cf. [2,5,8]). Therefore one may use P-spin (P_α) as an adequate tool for studying polarization properties of quantum light fields like the usual apparatus of the correlation functions [2]. But unlike the last one we can obtain a more deep insight into the nature of the polarization structure of light using P-spin.

Indeed, as it has been shown in [1,6], we can decompose the Fock space L_f spanned by the vectors (4) into the direct sum

$$L_f = \sum_{JM} \oplus L(JM) \tag{7}$$

of infinite-dimensional subspaces $L(JM)$ which are specified by eigenvalues J, M of the $SU(2)_p$ Casimir operator $P^2 = I/2\,(P_+P_- + P_-P_+) + P_o^2$ and P_o respectively. The spaces $L(JM)$ are generated by basis vectors $|JM, \gamma>$ of the form

$$|JM;\gamma> = \sum C(\{\alpha_i^{\pm}, \beta_{ij}, \sigma_{ij}\}) \prod_{i=1}^{m} (a_{\pm}^{+}(i))^{\alpha_i^{\pm}} \prod_{i<j} (Y_{ij})^{\beta_{ij}} (X_{ij})^{\sigma_{ij}} |0> \tag{8}$$

where operators $Y_{ij} = I/2(a_+^+(i)a_-^+(j) + a_-^+(i)a_+^+(j))$ and $X_{ij} = (a_+^+(i)a_-^+(j) - a_-^+(i)$ $a_+^+(j))$ are solutions of the operator equations

$$[P_o, Y_{ij}] = 0, \quad [P_\alpha, x_{ij}] = 0 \quad \forall \alpha = 0, +, - \tag{9}$$

and may be interpreted as creation operators of P_o-scalar and P-scalar biphoton kinematic clusters respectively. Examples of such states yield generalized coherent states (GCS) associated with Hamiltonians H_{int} of the matter-radiation interaction of the form

$$H_{int} = H_{int}^1 + H_{int}^2,$$

$$H_{int}^1 = \sum_{i<j} (g_{ij} X_{ij} + g_{ij}^* X_{ij}^+), \; H_{int}^2 = \sum_{i<j} (f_{ij} Y_{ij} + f_{ij}^* Y_{ij}^+) \tag{10}$$

Specifically, GCS of the group orbit type $|\xi> = exp[\xi X_{12} - \xi^* X_{12}^+]|0>$ have been discussed together with some related models in [1,6]. It is also of interest to examine GCS of another kind which are eigenfunctions of the operators $P^2, P_o, X_{ij}^+(Y_{ij}^+)$. These GCS are generalizations of the pair CS considered by G. Agarwal in [10].

The decomposition (7) implies some interesting classification of the polarization states of quantum light fields from the physical viewpoint. In particular, for the states $|J0;\gamma>$ specified by the condition $\alpha_i^{\pm} = 0$ and the states $|00;\gamma>$ specified by the condition $\alpha_i^{\pm} = \beta_{ij} = 0$ we have $< P_\alpha > = 0 = < S_\alpha >$ for all spin components that is the characteristic property of unpolarized light (cf. [1-4,7,8]). But, unlike classical (chaotic) unpolarized light, for the states $|00;\gamma>$ and $|J0;\gamma>$ we have some additional characteristics of light depolarization which follow from eq. (9); for example, we have

$$\sigma_{P_o} \equiv < P_o^2 > - (< P_o >)^2 = 0 \; for \; | > = |J0;\gamma>, \tag{11a}$$

$$\sigma_{P_\alpha} \equiv < P_\alpha^2 > - (< P_\alpha >)^2 = 0 \; \forall \alpha = 0, \pm \text{ for } | > = |00;\gamma> \tag{11b}$$

It is also of interest to note that states $|00;\gamma>$ minimize "radial" uncertainty relation for angular momenta operators [11].

Thus, our analysis confirms one H. Lipkin remark about an essentially quantum nature of unpolarized light [8]. Furthermore, it displays mechanisms of the light depolarization at the quantum level and here with the P-spin formalism yields some natural and measurable quantative characteristics of light depolarization: degrees $(I - J/N)$ and $(I - |M|/N)$ of the content of P-scalar and of P_o-scalar biphotons etc.

As a whole above results give a new classification of polarization states of light and open new possibilities in setting optical experiments (cf. [2,3]). Specifically, it is of interest to realize by physical devices some schemes [1] of production and detection of P- and P_o-scalar light. The situation here appears to be similar to that for squeezed light [10,12] which has close relations with P-scalar and especially with P_o-scalar light [1,6]. Specifically, for this aim one may use schemes of two photon lasers [10] or parametric generators in combination with beam splitters.

REFERENCES

1. V. P. Karassiov. Preprint FIAN No 137 1990; *J. Sov. Laser Res.*, v. **12**, No 2 (1990).
2. J. Peřina. *Quantum Statistics of linear and nonlinear optical phenomena* (D. Reidel, Dordrecht, 1984).
3. *Digest of Int. Conf. Quantum Optics* (Jan 5-10 1991, Hyderabad).
4. A. Gerrard and J. M. Burch. *Introduction to Matrix Methods in Optics* (Wiley, London e.a., 1975).
5. V. I. Strazhev, F. I. Fyodorov and P. L. Shkol'nikov. Dokl. *AN Byel. SSR*, v. **26**, 593 (1982).
6. V. P. Karassiov and V. I. Puzyrevsky. *J. Sov. Laser Res.*, v. **10**, 229 (1989).
7. P. Roman. *Nuovo Cimento*, v. **13**, 974 (1959).
8. H.J. Lipkin. *Quantum Mechanics. New Approaches to Selected Topics* (North-Holland, Amsterdam, 1973).
9. C. Itzykson, J.-B. Zuber. *Quantum Field Theory* (McGraw, N.Y., 1980)
10. G.S. Agarwal. *J. Opt. Soc. Am.*, v. **B5**, 1940 (1988)
11. V.I. Manko and V.V. Dodonov. *Trudy FIAN*, v. **183**, 5 (1987).
12. R. Loudon and P.L. Knight. *J. Mod. Optics*, v. **34**, 709 (1987).

GENERALIZED COMMUTATION RELATIONS FOR SINGLE MODE OSCILLATOR

R. Sandhya

School of Physics
University of Hyderabad
Hyderabad - 500 134, India

Recently Greenberg[1] constructed an example of infinite statistics in which all the representations of the symmetric group can occur and the number operator is an infinite series in a and $a^\dagger$. In this paper, we show that we can write down a generalized commutation relation (CR), for a single mode for which the number operators is again an infinite series and contains BE, FD and Greenberg's infinite statistics (IS) as special cases. We discuss the quantization of free fields with infinite statistics following the procedure of Umezawa and Takahashi.[2]

GENERALIZED COMMUTATIONS RELATIONS, NUMBER OPERATOR AND FOCK STATES

Consider the commutation relation

$$aa^\dagger - qa^\dagger a = 1. \tag{1}$$

By taking the hermitian conjugate of this expression it is easily seen that q must be real. The number operator N which satisfies

$$[a, N] = a, \quad [a^\dagger, N] = -a^\dagger \tag{2}$$

is

$$N = \sum_{n=1}^{\infty} \frac{(1-q)^n}{(1-q^n)} a^{\dagger n} a^n, \tag{3}$$

The normalized Fock states for the CR of (1) are

$$|0>, \; a^\dagger|0>, \; \frac{a^{\dagger 2}}{\sqrt{(1+q)}}|0>, \; \frac{a^{\dagger 3}}{\sqrt{(1+q)(1+q+q^2)}}|0>, \; ... \tag{4}$$

For $q = 1$, we recover both the CR's and the number operator of BE statistics. Also the Fock states go over to the well known Fock states of BE statistics, namely

Recent Developments in Quantum Optics, Edited
by R. Inguva, Plenum Press, New York, 1993

$(a^\dagger)^n/\sqrt{n!}|0>$. For $q=-1$, (1) goes over to the anti-commutator of FD statistics. Further, with the additional requirement $a^2 = a^{\dagger 2}, = 0$, the expression in (2) reduces to the known result and for Fock states one obtains the two states $|0>$ and $a^\dagger|0>$. For $q=0$, we obtain from (3)

$$N_{IS} = \sum_{n=1}^{\infty} a^{\dagger n} a^n, \tag{5}$$

which is the result due to Greenberg for a single mode obeying $aa^\dagger = I$. The normalized Fock states in this are $|0>$, $a^\dagger|0>$, $a^{\dagger 2}|0>$ ... We note that the expression for N given here has been obtained purely algebraically by using the commutation relations (1) and has the virtue of reducing to the expression given by Greenberg in the $q=0$ limit.

DISPLACEMENT OPERATOR

For the CR of (1), we now construct a generator $A^\dagger$ for displacements satisfying $[a, A^\dagger] = 1$, given by

$$A^\dagger = a^\dagger\left[1 + \frac{(1-q)}{(1+q)}a^\dagger a + \frac{(1-q)^2}{(1+q+q^2)}a^{\dagger 2}a^2 + \ldots\right], \tag{6}$$

for all q except $q=-1$. From (6), it follows that

$$exp(A^\dagger \alpha)a\ exp(-A^\dagger \alpha) = a - \alpha. \tag{7}$$

For $q=1$, $A^\dagger$ reduces to $a^\dagger$. For $q=0$, we get

$$A^\dagger = a^\dagger(N_{IS}+1) = N_{IS}a^\dagger. \tag{8}$$

For $q=-1$, with $a^2 = a^{\dagger 2} = 0$, $A^\dagger$ reduces to $a^\dagger$. In this case, although $[a, A^\dagger] = 1$, is no longer valid, (7) still holds with α regarded as a Grassmann number anti-commuting with a and $a^\dagger$.

COHERENT STATES

Having constructed $A^\dagger$, the construction of coherent states for arbitrary q, satisfying $a|\alpha >= \alpha|\alpha>$, is immediate. The normalized states satisfying $a|\alpha >= \alpha|\alpha>$, are given by

$$|\alpha >= C\ exp(A^\dagger \alpha)|0> \tag{9}$$

where $C = 1/[<0|exp(\alpha^* A)exp(\alpha A^\dagger)|0>]^{1/2}$. For $q=1$, we get the usual Sudarshan-Glauber coherent states.[3] For $q=-1$, with α regarded as a Grassmann number, we get the Fermionic coherent states of Ohnuki.[4] For $q=0$, the constant C is found to be $\sqrt{1-|\alpha|^2}$ and the normalized coherent states for the Greenberg CR are found to be

$$|\alpha >_{IS}= \sqrt{1-|\alpha|^2}\sum_{n=o}^{\infty}\alpha^n(a^\dagger)^n|0>. \tag{10}$$

In the many mode case of infinite statistics, one finds that the generators of displacement $A_i^\dagger$ satisfying

$$[a_i, A_j^\dagger] = \delta_{ij}, \tag{11}$$

are given by

$$A_i^\dagger = a_i^\dagger + \sum_k a_k^\dagger a_i^\dagger a_k + \sum_{k,l} a_k^\dagger a_l^\dagger a_i a_l a_k + \ldots \tag{12}$$

One, however, has

$$[A_i^\dagger, A_j^\dagger] \neq 0.$$

The construction of the coherent states in many mode case is immediate.

A REALIZATION OF $[x, p] = i$

It is interesting to note that by defining

$$x = \frac{a + A^\dagger}{\sqrt{2}}; \quad p = \frac{a - A^\dagger}{\sqrt{2}i}, \tag{13}$$

we obtain, by virtue of (11), a realization (albeit non hermitian) of the CR's, $[x, p] = i$. It is also easy to see that

$$\frac{1}{2}(x^2 + p^2) = (N + \frac{1}{2}), \tag{14}$$

which is the usual result for a harmonic oscillator. The surprise is that it is valid for all q's (except $q = -1$) of the CR proposed here. Further if the canonical density matrix is taken to be $e^{-\beta N}$ then, $< n >= (e^{\beta} - 1)^{-1}$ for all q except $q = -1$.

SOME SPECIAL STATES

Having constructed the generator $A^\dagger$ of displacement, it is elementary to construct analogues of the bosonic squeezed states[2] for the CR (1).

Using (1) it follows that

$$b = exp\left[\frac{z}{2}(a^2 - A^{\dagger 2})\right] a \; exp\left[-\frac{z}{2}(a^2 - A^{\dagger 2})\right] = (\cos hz)a + (\sin hz)A^\dagger \tag{15}$$

From $a|0 >= 0$ we get $b|z >= 0$, where

$$|z >= exp\left[\frac{z}{2}(a^2 - A^{\dagger 2})\right]|0 >, \tag{16}$$

The analogues of Yuen type squeezed states[2] are given by

$$|z,\alpha>= exp\left[|\frac{z}{2}(a^2 - A^{\dagger 2})\right] exp(A^\dagger \alpha)|0> . \quad (17)$$

They satisfy $b|z,\alpha >= \alpha|z,\alpha >$. It should be noted that for $q = -1$, the Fermi case, since $a^2 = a^{\dagger 2} = 0$, Yuen type states do not exist.

UNCERTAINTY PRODUCTS

If we define $a = X + iP$, one has for the coherent states of infinite statistics, $\Delta X \; \Delta P = \frac{1}{4}(1 - |alpha|^2)$, which can be made as small as possible. For Fock states of I.S., $\Delta X \; \Delta P = \frac{1}{4}$ for $n = 0$ and is $= \frac{1}{2}$ for $n > 0$.

ON QUANTIZATION OF FREE INFINITE STATISTICS FIELDS

Following the procedure of Umezawa and Takahashi, we consider the quantization of the free Schrödinger field, obeying

$$\left[i\frac{\partial}{\partial t} + \frac{1}{2m}\nabla^2\right]\Psi(x,t) = 0 \quad (18)$$

with, $\Psi(x,t) = \sum_k a_k(t) exp[ik.x]$, where $a_k(t) = e^{i\omega_k t} a_k$, $\omega_k = \frac{k^2}{2m}$ and $a_k^\dagger a_l = \delta_{kl}$. We postulate the Hamiltonian, $H = \sum_k \omega_k N_k$ with $N_k = a_k^\dagger a_k + \sum a_l^\dagger a_k^\dagger a_k a_l + . \; . \; .$, one can verify that,

$$i\frac{\partial}{\partial t}\Psi(x,t) = [\Psi(x,t), H] \quad (19)$$

reproduces (18), on using the expressions for H and N. It should be emphasized that $H = \sum_k \omega_k N_k$ cannot be derived from the Lagrangian, which gives (18) and thus this theory cannot be quantized in the canonical procedure.

REFERENCES

1. O. W. Greenberg, *Phys. Rev. Lett.* **64**, 705 (1990); A. B. Govorkov, *Theor. Math. Phys* **54**, 234 (1983).
2. H. Umezawa and Y. Takahashi, *Prog. Theor. Phys.* 9 **14** (1953); Nuclear Phys. 51 193 (1964); Y. Takahashi. An Introduction to Field Quantization (Pergamon Press, 1969).
3. E. C. G. Sudershan, *Phys. Rev. Letts.* **10** 277 (1963); R. J. Glauber, *Phys. Rev.* **131**, 2766 (1963).
4. Y. Ohnuki and T. Kashiwa, *Prog. Theor. Phys* **60**, 548 (1978); S. Chaturvedi, R. Sandhya, V. Srinivasan and R. Simon, *Phys. Rev.* A **41**, 3969 (1990).

QUANTUM JUMPS IN A COHERENTLY SHELVED ATOMIC SYSTEM

B. N. Jagatap* and S. V. Lawande+

*Multidisciplinary Research Section
+Theoretical Physics Division
Bhabha Atomic Research Centre, Bombay, 400 085, India

Intermittency in the resonance fluorescence from a three-level atom has received much attention both theoretically and experimentally in recent years.[1–5] Consider for example a three level atom in V configuration with a ground level coupled to two excited levels by two external coherent fields. One of these excited levels is short lived, having a lifetime in the range of nanoseconds. The transition from ground level to this excited level is strong. The other excited level acts as a metastable level with a lifetime of the order of seconds and is weakly coupled to the ground level. When the atom from the strongly fluorescing level is removed to the forbidden level, the strong fluorescence is interrupted until the atom returns to the ground level by emitting a weak transition photon. The switching on (bright interval) and off (dark interval) of the strong transition fluorescence leads to a generation of "random telegraph" signal as the atom makes random quantum jumps between the forbidden level and strongly fluorescing level.

Interpretation of the two-state atomic telegraph in terms of the "shelved" atom is not in conformity with quantum mechanics. For coherent excitation the atom is expected to be in a superposition of levels. This superposition then must manifest itself in the observed telegraphic fluorescence signal. It has been pointed out by Porrati and Putterman[5] that one of the consequences of such superposition is that a large number of dark intervals must end with emission of strong transition photons as against by emission of weak transition photons alone. The statistics of dark interval is therefore a key issue in the intermittancy of fluorescence.

In case of a three-level atom, the statistics of dark intervals is governed by photon emission which is an incoherent process. Instead if an additional forbidden transition from the metastable level is available, then the "shelved" atom can be put into a long term slowly varying superposition. This is then expected to manifest in the observation of dark intervals. Such a possibility can be studied in a four level system as shown in the inset of Fig. 1. Here the levels $|1>$, $|3>$ and $|4>$ constitute the basic V configuration. The laser Rabi frequencies for the strong and weak transitions are denoted by $2\alpha_1$ and $2\alpha_3$ respectively. The Einstein coefficient for spontaneous emission are given as $2\gamma_{14}$ and $2\gamma_{34}$ respectively. The metastable level $|3>$ is coupled to another forbidden level $|2>$ by external field of Rabi frequency $2\alpha_2$. Averbukh[6]

Recent Developments in Quantum Optics, Edited
by R. Inguva, Plenum Press, New York, 1993

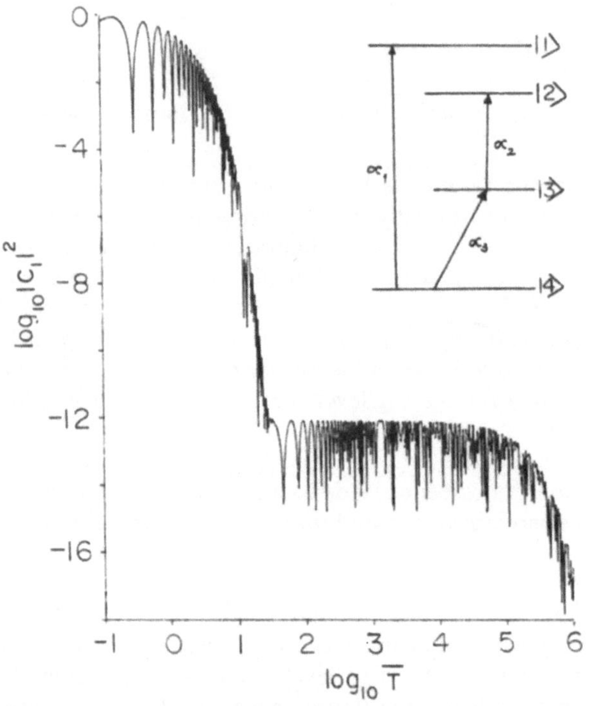

Figure 1. Temporal behaviour of $|C_1(t)|^2 = W_1(t)/\gamma_{14}$ plotted against $\bar{T} = \gamma_{14}t$ for $\alpha_1/\gamma = 10, \alpha_2/\gamma = 10^{-1}$, $\alpha_3/\gamma = 10^{-2}$, $\beta_1/\gamma = 1$, $\beta_2/\gamma = 10^{-6}$, $\Delta_1 = \Delta_2 = 0$, $\Delta_3/\gamma = 5$ and $\gamma = \gamma_{14}$. The inset shows the level scheme.

has studied quantum jumps in a four level system closely related to the scheme of Fig. 1 with $\alpha_3 = 0$. In such a case, quantum jumps occur between two coherently evolving subsystems. It may be mentioned here that the extended V configuration with $\alpha_3 \neq 0$, is quantum mechanically a rich system.

The effective Hamiltonian[5] of the system under consideration is ($\hbar = 1$)

$$H_{eff} = [\omega_{14} - i(\gamma_{14} + \gamma_{12})]A_{11} - [\alpha_1 A_{14} exp(-i\Omega_1 t) + \alpha_1^* A_{41} exp(i\Omega_1 t)]$$

$$+[\omega_{24} - i(\gamma_{24} + \gamma_{23})]A_{22} - [\alpha_2 A_{23} exp(-i\Omega_2 t) + \alpha_2^* A_{32} exp(i\Omega_2 t)]$$

$$+[\omega_{34} - i\gamma_{34}]A_{33} - [\alpha_3 A_{34} exp(-i\Omega_3 t) + \alpha_3^* A_{43} exp(i\Omega_3 t)] \tag{1}$$

Here ω_{ij} is the frequency of transition $|i> \rightarrow |j>$, A_{ij} is the atomic operator $|i><j|$, $2\gamma_{ij}$ is the Einstein coefficient for spontaneous emission for the transition $|i> \rightarrow |j>$ and Ω_i are the laser frequencies. Following Averbukh[6] we have used additional incoherent decay rates apart from the rates γ_{14}, γ_{34} inherent in the quantum jump problem. The wavefunction of the atom plus scattered field is[5]

$$|\psi> = \sum_{i,\{n\}} C_{i,\{n\}} |i, \{n\}> \tag{2}$$

where $\{n\}$ is the number of scattered photons. For null measurement, we have $\{n\} = \{0\}$. The equations of motion for the amplitudes $C_i = C_{i,0}$ may be written in the following form

$$\begin{aligned}
\tfrac{d}{dt}C_1 &= [i\Delta_1 - (\gamma_{12} + \gamma_{14})]C_1 + i\alpha_1 C_4 \\
\tfrac{d}{dt}C_2 &= [i(\Delta_2 + \Delta_3) - (\gamma_{23} + \gamma_{24})]C_2 + i\alpha_2 C_3 \\
\tfrac{d}{dt}C_3 &= (i\Delta_3 - \gamma_{34})C_3 + i\alpha_2 C_2 + i\alpha_3 C_4 \\
\tfrac{d}{dt}C_4 &= i\alpha_1 C_1 + i\alpha_3 C_3
\end{aligned} \tag{3}$$

where Δ_i's are detunings of the lasers from the corresponding atomic transition frequencies. The probability that the next photon of strong transition is recorded by t and $t + dt$ provided that the atom was in $|4>$ at $t = 0$ is

$$W_1(t) = -\gamma_{14}|C_1(t)|^2 \tag{4}$$

where $C_1(t)$ may be obtained from the solutions of Eq. (3). We describe here a particular case which can be analyzed analytically in a somewhat straightforward manner. Without loss of generality we take $\Delta_1 = \Delta_2 = 0$. Rabi frequency α_3 is assumed to be very small as in the normal V configuration. Further we assume $\gamma_{24} = \gamma_{34}$ and $\gamma_{23} \simeq 0$. The expression for $C_1(t)$ then takes the following form

$$\begin{aligned}
C_1(t) = \; & \tfrac{1}{2x} exp(-\beta_1 t/2) sin(\alpha_1 x t) \\
& - \tfrac{2i\alpha_1}{\alpha_3^2} exp[(i\Delta_3 - \beta_2/2)t][A_+ cos\alpha_2 t + iA_- sin\alpha_2 t]
\end{aligned} \tag{5}$$

where

$$A_{\pm} = \frac{\varepsilon_2^2}{i(\Delta_3 + \alpha_2) + \beta_1} \pm \frac{\varepsilon_3^2}{i(\Delta_3 - \alpha_2) + \beta_1}$$

$$\varepsilon_{2,3} = \frac{-\alpha_3^2[i(\Delta_3 \pm \alpha_2) + \beta_1]}{2[\alpha_1^2 - (\Delta_3 \pm \alpha_2)^2 + i\beta_1(\Delta_3 \pm \alpha_2)]}$$

$$\beta_1 = \gamma_{12} + \gamma_{14}, \ \beta_2 = \gamma_{24} + \gamma_{34}, \ x = [1 - \frac{\beta_1^2}{4\alpha_1^2}]^{\frac{1}{2}} \tag{6}$$

Approximate expression for the next photon emission probability $W_1(t)$ can then be obtained from Eq. (4). In Fig. 1, we show the temporal behaviour of $W_1(t)$. It is interesting to note that the long term coherent superposition in levels $|2>$ and $|3>$ as a consequence of "shelving" in the metastable manifold is seen in the dark interval. This is also reflected in the approximate analytical expression (5). Such a behaviour is absent in the normal V configuration. It may be added here that for $\alpha_2 = \pm\Delta_3$, this modulation observed in the dark intervals vanishes and the result is similar to the one obtained for normal V configuration. Lastly we add that it is possible to study such behaviour experimentally using a trapped HgII as suggested by Averbukh.[6]

REFERENCES

1. W. Nagourney, J. Sandburg and H. Dehmelt, *Phys. Rev. Lett.* **56**, 2797 (1986).
2. J. C. Bergquist, R. G. Hulet, W. M. Itano and D. J. Wineland, *Phys. Rev. Let.* **57**, 1699 (1986).
3. H. J. Kimble, R. J. Cook and A. L. Wells, *Phys. Rev.* **A34**, 3190 (1986).
4. G. S. Agarwal, S. V. Lawande and R. D'Souza, *Phys. Rev.* **A37**, 444, (1988).
5. M. Porrati and S. Putterman, *Phys. Rev.* **A39**, 3010 (1989).
6. I. Sh. Averbukh, *Phys. Lett.* **A134**, 298 (1989).

QUANTUM OPTICAL APPROACH TO RECOMBINATION†

S. Ravi

School of Physics
University of Hyderabad
Hyderabad-500 134, India

The study of recombination processes which involve the capture of a free electron by an ion to form a doubly excited state that then radiatively stabilizes has recently attracted much attention. Techniques of scattering theory have been extensively used to study the recombination processes. This formalism becomes inadequate if one is interested in taking into account many events of radiative decay or say if one wants to study photon statistics. To handle these, the density matrix formalism becomes necessary. Also a complete study of laser action can best be handled in the density matrix framework. With these points in view in this paper we present a general density matrix formulation for calculating the recombination probability from a system consisting of many autoionizing (AI) states interacting with many continuum states. We demonstrate how the density matrix framework can be used so as to account for different radiative decay processes. We apply these results to obtain the recombination probability from a system involving two continuum states coupled to an AI state and decaying to a bound state.

We will assume that the diagonalization of the part of the Hamiltonian which accounts for the coupling of the autoionizing states with the continuum has been carried out with a result that we have many orthogonal continuum states denoted by $|\alpha E_\alpha>$ (α represents the continuum index and E_α the energy of the continuum) decaying via spontaneous emission to the bound state $|f>$. The recombination probability to the state $|f>$ can be obtained by calculating the density matrix element ρ_{ff}. The formulation automatically includes all the interference effects between different channels of radiative decay. The total Hamiltonian for our system has the form

$$H = H_A + H_R + H_{AR}, \tag{1}$$

where H_A and H_R are the unperturbed atomic and radiation part of the Hamiltonian, and H_{AR} gives the interaction part which is responsible for recombination. The different parts of H can be written in the form

$$H_A = \sum_\alpha \int E_\alpha |\alpha E_\alpha >< \alpha E_\alpha| dE_\alpha + E_f |f >< f|, \tag{2}$$

† Work done in collaboration with G. S. Agarwal

$$H_R = \sum_{ks} \omega_{ks} a_{ks}^{\dagger} a_{ks}, \tag{3}$$

$$H_{AR} = -\vec{d} \cdot \vec{E} \tag{4}$$

$$\vec{d} = \sum_{\alpha} \int \vec{d}_{\alpha f} |\alpha E_\alpha >< f| dE_\alpha + H.c. \tag{5}$$

Form many situations, particularly those involving many events of the radiative decay, the density matrix formalism is better suited. For example if one were studying recombination in the presence of a laser field, then one has the possibility of the following processes repeatedly occurring: recombination, absorption of laser photon, autoionization, recombination. The complete description of the problem of laser action using autoionization and recombination processes which has attracted considerable attention in the recent years requires density matrix treatment of recombination. Using standard master equation techniques[1] the atomic density operator is found to satisfy the equation

$$\frac{\partial \rho}{\partial t} = -i[H_o, \rho] - \sum \int dE_\alpha dE_\beta s_\beta d_{\alpha f} \cdot d_{\beta f}(|\alpha E_\alpha >< \beta E_\beta| \rho$$

$$+\rho |\alpha E_\alpha >< \beta E_\beta| - 2|f >< f| < \beta E_\beta |\rho| \alpha E_\alpha >, \tag{6}$$

with

$$H_o = \sum_{\alpha} \int E_\alpha |\alpha E_\alpha >< \alpha E_\alpha| dE_\alpha + E_f |f >< f|, \tag{7}$$

$$s_\alpha = 2(E_\alpha - E_\beta)/3c^3. \tag{8}$$

From (6) the matrix element of the density operator ρ which gives the recombination rate to the state $|f>$ is

$$\dot{\rho}_{ff} = 2 \sum_{\alpha\beta} \int s_\alpha \vec{d}_{\alpha f}^{\dagger} \cdot \vec{d}_{\beta f} \rho_{\beta\alpha} dE_\alpha dE_\beta. \tag{9}$$

The recombination probability to the state $|f>$ can be obtained from this as,

$$P_f = \lim_{t \longrightarrow \infty} \rho_{ff}(t). \tag{10}$$

We now introduce quantities ψ's defined by

$$\rho_{\alpha\beta}(t) = \psi_\alpha(t) \psi_\beta^*(t), \tag{11}$$

The wave functions ψ_α are found to satisfy the equations

$$\dot{\psi}_\alpha = -i\Delta_\alpha \psi_\alpha - \sum_{\beta} \int \vec{d}_{\alpha f}^{\dagger} \cdot \vec{d}_{\beta f} \psi_\beta s_\beta dE_\beta, \tag{12}$$

with

$$\Delta_\alpha = (E_\alpha - E_f), \tag{13}$$

For simplicity we ignore in the following, the vectorial nature of $\vec{d}$, though for problems involving magnetic degeneracies the vectorial nature of $\vec{d}$ is important. Defining quantities,

$$K_\alpha(E_\alpha) = \frac{d^\dagger_{\alpha f}}{(z + i\Delta_\alpha)}, \quad L_\beta(E_\beta) = d_{\beta f} s_\beta, \tag{14}$$

$$\hat{\chi}_\beta(z) = \int L_\beta(E_\beta)\hat{\psi}_\beta dE_\beta. \tag{15}$$

and upon taking Laplace transform, (12) can be cast in the form

$$\hat{\chi}_\alpha(z) + m_{\alpha\alpha}\sum_\beta \hat{\chi}_\beta(z) = \hat{f}_\alpha(z), \tag{16}$$

$$m_{\alpha\alpha} = \int K_\alpha(E_\alpha)L_\alpha(E_\alpha)dE_\alpha. \tag{17}$$

$$\hat{f}_\alpha(z) = \int \frac{\psi_\alpha(0)L_\alpha(E_\alpha)dE_\alpha}{(z + i\Delta_\alpha)} \tag{18}$$

The recombination probability (9) can now be rewritten in terms of χ as

$$P_f = \lim_{t\to\infty} \int_o^t \frac{2}{s} |\sum_\beta \chi_\beta(\tau)|^2 d\tau. \tag{19}$$

Explicit determination of the recombination probability to a particular problem of interest would require the form of the diagonalized states $|\alpha E>$ and various dipole matrix elements.

As an application of the above general result we now consider a model system involving two continuum states $|\psi_E)$ and $|\chi_E)$ coupled to the autoionizing (AI) state $|a>$ via configuration interaction and decaying to a responsible configuration interaction[2] produces new continuum states $|E>$ and $|G>$. The recombination probability in this case has the form

$$P_f(t) = \frac{2}{s} - \int_o^\infty (|\chi_1(t) + \chi_2(t)|^2)dt, \tag{20}$$

with the χ's given in Eq. 16. We assume a wave packet structure for the initial state of the system. The details of the calculations are given in Ref 3 and we present here the final result for the recombination probability which has the form

$$P_f = \frac{4\gamma_f}{\Gamma_1 q_{f1}^2}\left[\left(\delta_{E_o} + \frac{\Gamma_1}{\Gamma}q_{f1}\right)^2 + \left(1 - \frac{\Gamma_1}{\Gamma}\frac{q_{f1}}{q_f}\right)^2\right]\frac{1}{\psi^2[\eta^2 + (\delta_{E_o} - \Delta_a)^2]}. \tag{21}$$

where

$$q_{f1} = \frac{V_{fa}}{\pi V_{E_o a} V_{f E_o}}; \; \Gamma_1 = 2\pi|V_{E_o a}|^2; \Delta_a = -\frac{2\gamma_f}{\Gamma q_f \psi}$$

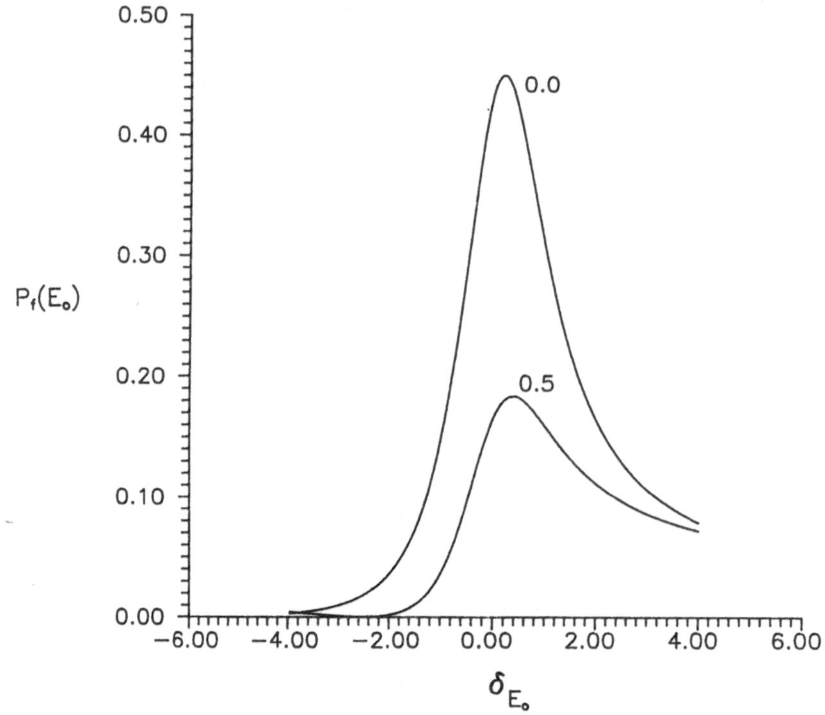

Figure 1. Comparison of recombination probability as a function of incident energy for a single continuum ($\Gamma_2 = 0.5$) system for parameters $q_f = 2, q_{f1} = 5, \gamma_f = 0.1$.

$$\psi = 1 + s\pi(|V_{fE_o}|^2 + |W_{fE_o}|^2);\ \eta = 1 + \frac{\gamma_f}{\Gamma\psi}(1 - \frac{1}{q_f^2}) \tag{22}$$

$$\gamma_f = 2s|V_{af}|^2;\ s = \frac{2}{3}\frac{\omega^3}{c^3}; \delta_{E_o} = \frac{2}{\Gamma}(E_o - E_a)$$

The expression (21) was also recently obtained by Haan and Jacobs[4] using the projection operator approach. The plot of P_f as a function of δ_{E_o} is shown in the figure for a single continuum ($\Gamma_2 = 0$) and for two continuum ($\Gamma_2 = 0.5$). From the figure it is evident that the recombination probability for the latter case is less. This is expected as an increase in Γ_2 leads to an increase in transfer of population into the second continuum and hence a decrease in population to the state $|f>$.

REFERENCES

1. G. S. Agarwal, Quantum optics, Vol. 70 of springer tracts in modern physics, edited by G. Hohler (springer Berlin, 1974).
2. U. Fano, *Phys. Rev.* **A124**, 1866 (1961).
3. S. Ravi and G. S. Agarwal in press.
4. S. L. Haan and V. L. Jacobs, *Phys. Rev.* **A40**, 80 (1989).

GEOMETRIC PHASE ASSOCIATED WITH SPINOR ROTATIONS

Apoorva G. Wagh and Veer Chand Rakhecha

Solid State Physics Division
Bhabha Atomic Research Centre
Bombay 400 085, India

SPINORMAN SPINS A WEB

Ask not my name, just call me a spinor
I have a spin half, yet I am not a minor
I am endowed with a magnetic personality
Precessing on a sphere? That's my speciality
I go round a full circle about any axis one says
And arrive at the starting point opposite in phase
At times I spin around myself eigenically
So the phase keeps accumulating dynamically
Then I take a fancy for parallel transportation
Thus acquiring phase in geometric fashion
And while marvelling at my "Operation SU(2)"
Little wonder your head starts spinning too!

It was thirty five years ago that Pancharatnam[1] encountered an "unexpected geometrical result", viz. the geometric phase, for polarization circuits of light on the Poincare' sphere. Geometric phase was rediscovered recently[2,3] in the context of excursions of a quantal system around closed circuits in parameter space. In this paper, we will confine our attention to the geometric phase acquired by a spinor[4,5] in traversing a closed circuit on the spin sphere, with an emphasis on its observation in neutron interferometric experiments.

The state vectors of a two-state quantum system undergoing special unitary transformations can be represented, to within a phase factor, by points on a sphere. However, changes in the polarization states of a photon[6] or circuits described in NMR interferometry[7] with two-level systems can only be represented on spheres in abstract space. SU(2) operations on the spin state of a spin-$\frac{1}{2}$ particle, in contrast, correspond to real rotations over the sphere of its spin directions.

Aharonov and Anandan[3] separated the dynamical and geometric components of the phase acquired by a spinor during a 2π-precession about an arbitrary axis. We generalize their analysis for an arbitrary closed circuit C (Fig. 1) on the spin sphere,

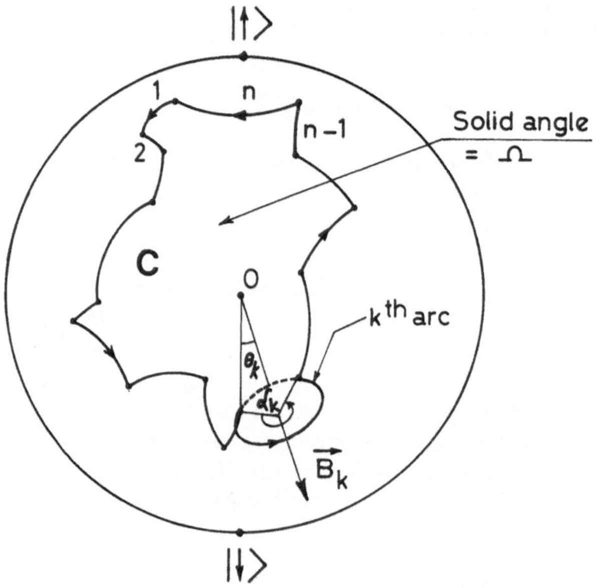

Figure 1. Dynamical and geometric phases acquired by a spinor over an arbitrary closed circuit on the spin sphere.

comprising n arcs representing successive rotations α_k about axes of magnetic inductions $\boldsymbol{B}_k$ over cones of polar angles $\theta_k, k = 1, 2, ..., n$. The phase[5] acquired by the spinor in traversing C is the sum of a dynamical component Φ_D and a geometric component Φ_G given by

$$\Phi_D = -\frac{1}{2}\sum_1^n \alpha_k \cos \theta_k \tag{1a}$$

and

$$\Phi_G = -\frac{1}{2}\Omega, \text{ modulo } 2\pi, \tag{1b}$$

Ω representing the solid angle subtended by C at the centre of the spin sphere. The sign of α_k, determined from its sense relative to $\boldsymbol{B}_k$ is opposite to the sign of the magnetic moment of the particle. The angles α_k are thus always positive for neutron and negative for proton. For a given sequence $\boldsymbol{B}_k$ yielding the circuit C, the dynamical and geometric components and hence the total phase remain the same (cf. eqs. (1)), no matter where one starts (and ends) the excursion.

Neutron is an ideal probe for studying the phases associated with spinor rotations due to its sizeable magnetic moment and zero electric charge. The interaction of a neutron with a uniform magnetic field only causes a precession of its spin vector leaving its space trajectory unaffected. The state-of-the-art of perfect crystal neutron interferometry[8] has also reached a stage where such phase variations can be observed. We have already described a possible experiment[4,5] with two identical π-flippers placed in the two branches of a neutron interferometer. There, a relative rotation of the flippers generates a pure geometric phase and their relative translation results in a pure dynamical phase.

We will address here the case of a neutron passing an rf (radio- frequency) flipper wherein $\boldsymbol{B} = [-B_1 sin(\omega t),\ B_1 cos(\omega t),\ B_o]$, satisfying the resonance conditions, $\omega = 2|\mu|B_o/\hbar$ and $2|\mu|B_1\tau/\hbar = \pi$. Here μ and τ denote the neutron magnetic moment and passage time through the flipper respectively. A spin-up neutron entering the flipper at $t = t_o$ describes a spin trajectory ACA′ (Fig. 2) in a frame rotating about the axis OA with the angular frequency ω. In the lab frame the neutron traverses Trajectory 1, i.e. ABA′, and exits the flipper at $t = t_o + \tau$ in spin-down (A′) state. When compared to a π precession about the y-axis (Trajectory 2), it acquires a geometric phase (cf.eq.1(b)),

$$\Phi_G = -\frac{1}{2}(-2\omega t_o - \omega\tau) = \omega(t_o + \tau/2), \tag{2a}$$

whereas the Φ_D components $-\omega\tau/2\pi$ and $+\omega\tau/2\pi$ accumulated over first and second halves respectively of Trajectory 1 cancel each other. The net phase is thus a pure geometric phase given by eq. (2a). If the length of the rf flipper is increased by a factor $(2n+1)$, it will work as a $(2n+1)\pi$ flipper with

$$\Phi_{1-2} = \Phi_G = \omega[t_o + (n + \frac{1}{2})\tau], \tag{2b}$$

while for a length increase by a factor $2n$, the neutron emerges in the spin-up state with a geometric phase

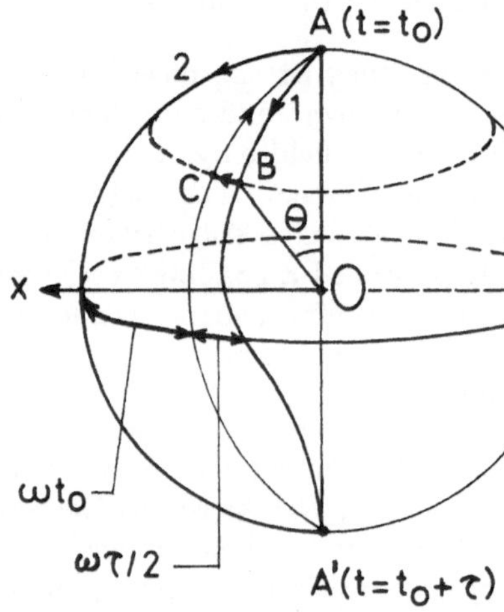

Figure 2. Spin vector trajectories inside an rf flipper.

$$\Phi_{1-2} = -n\omega\tau, \tag{2c}$$

n being a non-negative integer.

We consider a circuit ABCA (Fig. 2). The trajectory AB is described by polar coordinates ($\theta' = \pi u'/\tau$, $\phi' = \omega u'$), where $u' = t - t_o$ and ϕ' is measured from the plane ACA′. The various Φ-components then are:

$$\Phi_G = -\frac{1}{2}[-\int_o^u \omega u' sin(\pi u'/\tau)\pi du'/\tau] = \frac{1}{2}\omega[sin(\pi u/\tau)\tau/\pi - ucos(\pi u/\tau)], \tag{3a}$$

$$\Phi_D^{AB} = -\frac{1}{2}[\int_o^u \omega\ cos(\pi u'/\tau)du'] = -\frac{1}{2}(\omega\tau/\pi)\ sin(\pi u/\tau) \tag{3b}$$

and

$$\Phi_D^{BC} = -\frac{1}{2}(\omega u)(-cos\theta) = \frac{1}{2}\omega u \cdot cos(\pi u/\tau). \tag{3c}$$

The net phase over the loop ABCA is therefore

$$\Phi^{ABCA} = \Phi_G + \Phi_D^{AB} + \Phi_D^{BC} = 0, \tag{3d}$$

irrespective of the polar angle θ of B. Thus all possible loops of the type ABCA are phase holonomic, in the sense that if spinor twins starting at any point on such a loop traverse it in opposite directions, they will be in phase whenever they meet.

Extending the point B to A′, we note that $\Phi^{ABA'CA}$ equals that due to a rotation $\omega\tau$ about OA at A′, i.e.

$$\Phi^{ABA'CA} = -\frac{1}{2}(\omega\tau)(-1) = \frac{1}{2}\omega\tau, \tag{3e}$$

in accordance with eq. (2a).

The experiment of Badurek et al.[9], sketched schematically in Fig. 3, using two rf flippers of angular frequencies ω_1 and ω_2 generates a phase

$$\Phi_{1-2} = (\omega_1 - \omega_2)t + \frac{1}{2}(\omega_1\tau_1 - \omega_2\tau_2). \tag{4}$$

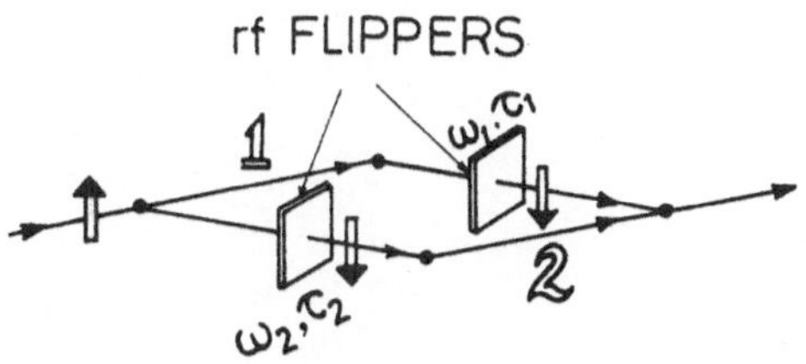

Figure 3. Conceptual depiction of the neutron quantum beats[9] experiment.

The time-proportional term in Φ_{1-2} of eq. (4) manifests itself in macroscopic temporal quantum beats in neutron intensity observed[9] in the experiment.

REFERENCES

1. S. Pancharatnam, *Proc. Ind. Acad. Sci.* **A44**:247 (1956).
2. M. V. Berry, *Proc. R. Soc. London* **A392**:45 (1984).
3. Y. Aharonov and J. Anandan, *Phys. Rev. Lett.* **58**:1593 (1987).
4. A. G. Wagh, *Phys. Lett.* **A146**:369 (1990).
5. A. G. Wagh and V. C. Rakhecha, *Phys. Lett.* **A148**:17 (1990).
6. R. Bhandari and T. Dasgupta, *Phys. Lett.* **A143**:170 (1990).
7. D. Suter, K. T. Mueller and A. Pines, *Phys. Rev. Lett.* **60**:1218 (1988).
8. G. Badurek, H. Rauch and J. Summhammer, *Physica* **B151**:82 (1988).
9. G. Badurek, H. Rauch and D. Tuppinger, *Phys. Rev.* **A34**:2600 (1986).

THE ONE-ATOM MASER AND THE GENERATION OF NONCLASSICAL LIGHT

G. Rempe and H. Walther

Sektion Physik der Universität München and
Max-Planck-Institut für Quantenoptik
8046 Garching, Fed. Rep. of Germany

ABSTRACT

We review our recent work on the nonclassical radiation field of a one-atom maser. The maser cavity is cooled to 0.5 K to exclude thermal photons and has a quality factor of $3\text{x}10^{10}$. Velocity-selected Rydberg atoms pump the maser. The field inside the cavity is measured via the fluctuations in the number of atoms leaving the cavity in the lower maser level which are up to 40% below the Poisson level. This corresponds to photon-number fluctuations 70% below the vacuum-state limit. The one-atom maser is the only maser system which leads to nonclassical radiation even when pumping processes with Poissonian statistics are used. Some application of the one-atom maser to study the quantum measurement process are discussed.

1. INTRODUCTION

The generation of squeezed or nonclassical radiation has been widely discussed recently. There are essentially three phenomena demonstrating the nonclassical character of light: squeezing, photon antibunching, and sub-Poissonian photon statistics (see, for example, Ref. 1). Methods of nonlinear optics are mostly used to generate nonclassical light, but in special cases also single atom fluorescence results in nonclassical radiation. In the following we report on the investigation of sub-Poissonian photon statistics in the one-atom maser.[3] The setup allows to study in detail the generation process in a maser leading to nonclassical radiation even in cases when a Poissonian pumping process is used. In the second part of the paper applications of the one-atom maser to study the quantum measurement will be discussed.

2. REVIEW OF THE ONE-ATOM MASER

The most simple system for a maser is given by a single atom interacting with a single mode of a cavity. At the first glance this seems to be a beautiful toy from the playground of a quantum optics theorist and just another example of a Gedanken

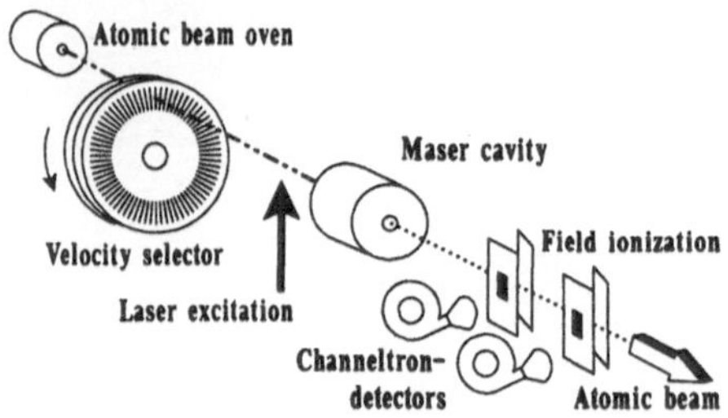

Figure 1. Scheme of the experimental setup. To suppress blackbody-induced transitions to neighboring states, the Rydberg atoms are excited inside the liquid-Helium-cooled environment.

experiment but such a one-atom maser[3] really exists and can be used to study the basic principles of radiation-atom interaction. The advantages of the system are:

(1) it is the first maser which sustains oscillations with less than one atom on the average in the cavity.

(2) this setup allows to study in detail the conditions necessary to obtain nonclassical radiation, especially sub-Poissonian light, and

(3) it is possible to study a variety of phenomena of a quantum field including the quantum measurement process.

It was the enormous progress in constructing superconducting cavities together with the laser preparation of highly excited atoms - Rydberg atoms - that have made the realization of such a one-atom maser possible. Rydberg atoms have interesting properties[4] which make them ideal for such experiments: The probability of induced transitions between neighboring states of a Rydberg atom scales as n^4, where n denotes the principle quantum number. Consequently, a few photons are enough to saturate the transition between adjacent levels. Moreover, the spontaneous lifetime of a highly excited state is very large. We obtain a maser by injecting these Rydberg atoms into a superconducting cavity with a high quality factor. The injection rate is so that on the average there is less than one atom present inside the resonator at any time. A transition between two neighboring Rydberg levels is resonantly coupled to a single mode of the cavity field. Due to the high quality factor of the cavity, the radiation decay time is much larger than the characteristic time of the atom-field interaction, which is given by the inverse of the single-photon Rabi-frequency. Therefore it is possible to observe the dynamics[5] of the energy exchange between atom and field mode leading to collapse and revivals in the Rabi-oscillations.[6,7] Moreover a field is built up inside the cavity when the mean time between the atoms injected into the cavity is shorter than the cavity decay time. The detailed experimental setup of the one-atom maser is shown in Fig. 1. A highly collimated beam of rubidium atoms passes through a Fizeau velocity selector. Before entering the superconducting cavity, the atoms are excited into the upper maser level $63p_{3/2}$ by the frequency-doubled light of a cw ring dye laser. The laser frequency is stabilized onto the atomic transition $5s_{1/2} \rightarrow 63p_{3/2}$ which has a width determined by the laser linewidth and the transit time broadening corresponding to a total of a few

MHz. In this way, it is possible to prepare a very stable beam of excited atoms. The ultraviolet light is linearly polarized parallel to the electric field of the cavity. Therefore only $\Delta m = 0$ transitions are excited by both the laser beam and the microwave field. The superconducting niobium maser cavity is cooled down to a temperature of 0.5 K by means of a ^{3}He cryostat. At such a low temperature the number of thermal photons is reduced to about 0.15 at a frequency of 21.5 GHz. The cryostat is carefully designed to prevent room temperature microwave photons from leaking into the cavity. This would considerably increase the temperature of the radiation field above the temperature of the cavity walls. The quality factor of the cavity is 3×10^{10} corresponding to a photon storage time of about 0.2s. The cavity is carefully shielded against magnetic fields by several layers of cryoperm. In addition, three pairs of Helmholtz-coils are used to compensate the earth's magnetic field to a value of several mG in a volume of 10x4x4 cm^3. This is necessary in order to achieve the high quality factor and to prevent the different magnetic substates of the maser levels from mixing during the atom-field interaction time. Two maser transitions from the $63p_{3/2}$ level to the $61d_{3/2}$ and to the $61d_{5/2}$

The Rydberg atoms in the upper and lower maser levels are detected in two separate field ionization detectors. The field strength is adjusted so as to ensure that in the first detector the atoms in the upper level are ionized, but not those in the lower level.

To demonstrate maser operation, the cavity is tuned over the $63p_{3/2}$ - $61d_{3/2}$ transition and the flux of atoms in the excited state is recorded simultaneously. Transitions from the initially prepared $63p_{3/2}$ state to the $61d_{3/2}$ level (21.50658 GHz) are detected by a reduction of the electron count rate.

In the case of measurements at a cavity temperature of 0.5 K, shown in Fig. 2, a reduction of the $63p_{3/2}$ signal can be clearly seen for atomic fluxes as small as 1750 atoms/s. An increase in flux causes power broadening and a small shift. This shift is attributed to the ac Stark effect, caused predominantly by virtual transitions to neighboring Rydberg levels. Over the range from 1750 to 28000 atoms/s the field ionization signal at resonance is independent of the particle flux which indicates that the transition is saturated. This, and the observed power broadening show that there is a multiple exchange of photons between Rydberg atoms and the cavity field.

For an average transit time of the Rydberg atoms through the cavity of 50 μs and a flux of 1750 atoms/s we obtain that approximately 0.09 Rydberg atoms are in the cavity on the average. According to Poisson statistics this implies that more than 90% of the events are due to single atoms. This clearly demonstrates that single atoms are able to maintain a continuous oscillation of the cavity with a mean number of photons between unity and several hundreds.

3. A NEW SOURCE OF NONCLASSICAL LIGHT

One of the most interesting questions in connection with the one-atom maser is the photon statistics of the electromagnetic field generated in the superconducting cavity. This problem will be discussed in this section.

Electromagnetic radiation can show nonclassical properties,[8,9] that is properties that cannot be explained by classical probability theory. Loosely speaking we need to invoke

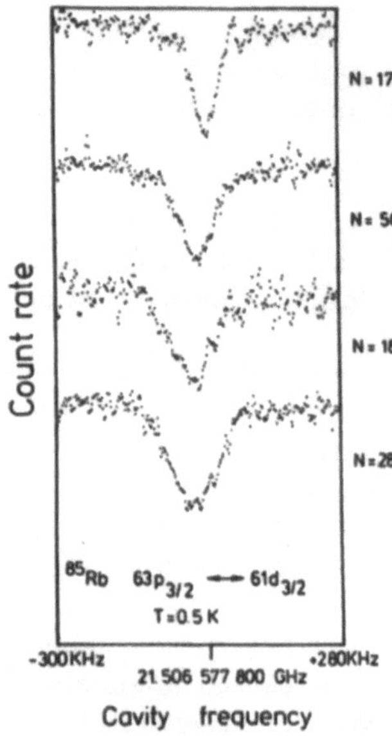

Figure 2. Maser operation of the one-atom maser manifests itself in a decrease of the number of atoms in the excited state. The flux of excited atoms N governs the pump intensity. Power broadening of the resonance line demonstrates the multiple exchange of a photon between the cavity field and the atom passing through the resonator. The measurement was performed with the transition $63p_{3/4} \rightarrow 61d_{3/2}$ having a single photon Rabi-frequency of 10kHz. The Q value of the cavity for this measurement was $4 \cdot 10^9$ which is lower than the one used for the measurements of the $63p_{3/2} \rightarrow 61d_{5/2}$ transition being $3 \cdot 10^{10}$. For the transition $63p_{3/2} \rightarrow 61d_{5/2}$ used also for many of the measurements described in this paper the single photon Rabi-frequency is 40kHz.

"negative probabilities" to get deeper insight into these features. We know of essentially three phenomena which demonstrate the nonclassical character of light: photon antibunching,[10] sub-Poissonian photon statistics[11] and squeezing.[12] Mostly methods of nonlinear optics are employed to generate nonclassical radiation. However, also the fluorescence light from a single atom caught in a trap exhibits nonclassical features.[13,14]

Another nonclassical light generator is the one-atom maser. We recall that the Fizeau velocity selector preselects the velocity of the atoms: Hence the interaction time is well-defined which leads to conditions usually not achievable in standard masers.[15–20] This has a very important consequence when the intensity of the maser field grows as more and more atoms give their excitation energy to the field: Even in the absence of dissipation this increase in photon number is stopped when the increasing Rabi-frequency leads to a situation where the atoms reabsorb the photon and leave the cavity in the upper state. For any photon number, this can be achieved by appropriately adjusting the velocity of the atoms. In this case the maser field is not changed any more and the number distribution of the photons in the cavity is sub-Poissonian,[15,16] that is narrower than a Poisson distribution. Even a number state that is a state of well-defined photon number can be generated[17,18] using a cavity with a high enough quality factor. If there are not thermal photons in the cavity - a condition achievable by cooling the resonator to an extremely low temperature - very interesting features such as trapping states show up.[19] In addition, steady state macroscopic quantum superpositions can be generated in the field of the one-atom maser pumped by two-level atoms injected in a coherent superposition of their upper and lower states.[20]

Unfortunately the measurement of the nonclassical photon statistics in the cavity is not that straightforward. The measurement process of the field invokes the coupling to a measuring device whereby losses lead inevitably to a destruction of the nonclassical properties. The ultimate technique to obtain information about the field employs the Rydberg atoms themselves: Measure the photon statistics via the dynamical behavior of the atoms in the radiation field, that is via the collapse and the revivals of the Rabi-oscillations, that is one possibility. However, since the photon statistics depend on the interaction time which has to be changed when collapse and revivals are measured it is much better to probe the population of the atoms in the upper and lower maser levels when they leave the cavity. In this case, the interaction time is kept constant. Moreover, this measurement is relatively easy since electric fields can be used to perform a selective ionization of the atoms. The detection sensitivity is sufficient so that the atomic statistics can be investigated. This technique maps the photon statistics of the field inside the cavity via the atomic statistics.

In this way, the number of maser photons can be inferred from the number of atoms detected in the lower level.[3] In addition, the variance of the photon number distribution can be deduced from the number fluctuations of the lower-level atoms.[21] In the experiment, we are therefore mainly interested in the atoms in the lower maser level. Experiments carried out along these lines are described in the following section.

4. EXPERIMENTAL RESULTS - A NONCLASSICAL BEAM OF ATOMS

Under steady state conditions, the photon statistics of the field are essentially determined by the dimensionless parameter $\Theta = (N_{ex}+1)^{1/2}\Omega t_{int}$, which can be understood as a pump parameter for the one-atom maser.[15] Here, N_{ex} is the average number of atoms that enter the cavity during the lifetime of the field, t_{int} the time of flight of the atoms through the cavity and Ω the atom-field coupling constant (one-photon Rabi-frequency). The one-atom maser threshold is reached for $\Theta = 1$. At this value and also at $\Theta = 2\pi$ and integer multiples thereof, the photon statistics are super-Poissonian. At these points, the maser field undergoes first-order phase transitions.[15] In the regions between those points sub-Poissonian statistics are expected. The experimental investigation of the photon number fluctuation is the subject of the following discussion.

In the experiments,[22] the number N of atoms in the lower maser level is counted for a fixed time interval T roughly equal to the storage time T_{cav} of the photons. By repeating this measurement many times the probability distribution $P(N)$ of finding N atoms in the lower level is obtained. The normalized variance[23] $Q_a = [< N^2 > - < N >^2 - < N >]/ < N >$ is evaluated and is used to characterize the deviation from Poissonian statistics. A negative (positive) Q_a value indicates sub-Poissonian (super-Poissonian) statistics, while $Q_a = 0$ corresponds to a Poisson distribution with $< N^2 > - < N >^2 = < N >$. The atomic Q_a is related to the normalized variance Q_f of the photon number by the formula

$$Q_a = \epsilon P Q_f (2 + Q_f), \tag{1}$$

which was derived in Ref. 21 with P denoting the probability of finding an atom in the lower maser level. It follows from formula (1) that the nonclassical photon statistics can be observed via sub-Poissonian atomic statistics. The detection frequency ϵ for the Rydberg atoms reduces the sub-Poissonian character of the experimental result. The detection efficiency was 10% in our experiment; this includes the natural decay of the Rydberg states between the cavity and field ionization. It was determined by both monitoring the power-broadened resonance line as a function of flux[3] and observing the Rabi-oscillation for constant flux but different atom-field interaction times.[7] In addition this result is consistent with all other measurements described in the following, especially with those on the second maser phase transition.

We start the discussion of the experimental results by describing measurements in the build-up period of the maser field. For this purpose, the mean number of atoms passing through the cavity during a fixed sampling time interval shorter than the cavity decay photon storage time is varied. For each flux, the normalized variance Q_a of the probability distribution of atoms detected in both the upper and the lower maser level is determined. Experimental results are plotted in Fig. 3. Open circles (full circles) represent the normalized variance of atoms in the lower (upper) maser level leading to a sub-Poissonian (super-Poissonian) atom statistics. About 20,000 experiments are averaged for each data point to keep the uncertainty of Q_a below 1%. For a low temperature of the atomic beam oven, the horizontal error bars are determined by the Poisson statistics of the total flux of atoms. These statistics are measured with the cavity out of resonance so that all atoms leave the cavity in the upper level. The result is given by a Poisson distribution with $Q_a = 0$.

The two solid lines are calculated by a numerical simulation of the maser process which explicitly takes into account the measurement of atoms and the corresponding change of the cavity photon statistics. Details of this procedure are contained in Chapter 5 for the case of a steady-state maser operation. For the comparison with the results of Fig. 3 the method was extended to the transient regime of an increasing maser field starting from the thermal 0.5 K field. The detection efficiency of atoms in the lower and upper maser levels amounts to 10% and 7%, respectively. In the simulation, the Poisson statistics of the flux, the temperature of the cavity field (0.5

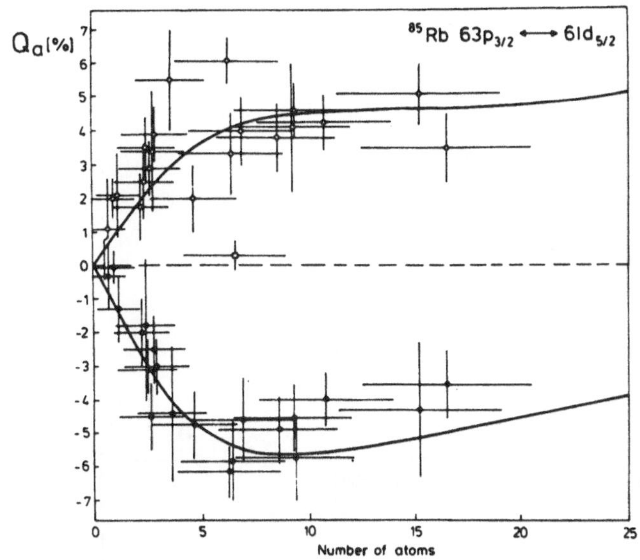

Figure 3. Variance Q_a of the atoms in the lower (open circles) and upper level (full circles) as a function of the tot l number of atoms crossing the cavity.

K), and the damping of the maser field in the time interval between adjacent atoms are also considered. The agreement between the numerical simulation and the experimental results is good.

The sub-Poissonian statistics of atoms in the lower level proves the nonclassical character of the maser field. As is expected from simple statistical arguments, the probability distribution of atoms in the upper maser level is always super-Poissonian. With the single photon Rabi-frequency given by $\Omega = 44$ KHz, $n = 2$ photons are able to induce a 2π-Rabi-nutation of the atom during the atom-field interaction time of $t_{int} = 40\mu s$: $2\Omega(n+1)^{1/2}t_{int} = 6.10$. This number is slightly smaller than 2π to reduce the influence of the velocity distribution on the maser photon statistics. The simulation shows that, averaged over many experiments, the probability distribution of finding n maser photons after about 8 atoms have crossed the cavity has a mean of $< n >= 2$ and a variance of $< n^2 > - < n >^2 = 0.2 < n >$. In the experiment, this value corresponds to $Q_a = -6 \cdot 10^{-2}$. Fig. 3 shows that for a higher flux of atoms (more than 20), the normalized variance of the probability distribution of atoms in the lower level increases slowly indicating that the 2π-trapping condition is not exactly fulfilled.

Now we continue the discussion with out results on the maser in steady state. Experimental results for the transition $63p_{3/2} \leftrightarrow 61d_{3/2}$ are shown in Fig. 4. The measured normalized variance Q_a is plotted as a function of the flux of atoms. The atom-field interaction time is fixed at $t_{int} = 50\mu s$. The atom-field coupling constant Ω is rather small for this transition, $\Omega = 10$kHz. A relatively high flux of atoms $N_{ex} > 10$ is therefore needed to drive the one-atom maser above threshold. The large positive Q_a observed in the experiment proves the large intensity fluctuations at the onset of maser oscillation at $\Theta = 1$. The solid line is plotted according to Eq. (1) using the theoretical predictions for Q_f of the photon statistics.[15,16] The error in the signal follows from the statistics of the counting distribution $P(N)$. About 2×10^4 measurement intervals are needed to keep the error of Q_a below 1%. The statistics of the atomic beam is measured with a detuned cavity. The result is a Poisson distribution. The error bars of the flux follow from this measurement. The agreement between theory and experiment is good.

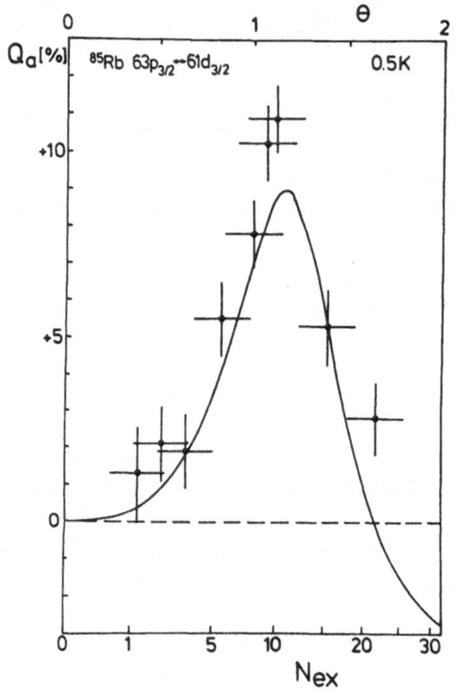

Figure 4. Variance Q_a of the atoms in the lower maser level as a function of flux N_{ex} near the onset of maser oscillation for the $63p_{3/2} \leftrightarrow 61d_{3/2}$ transition (see Ref. 22).

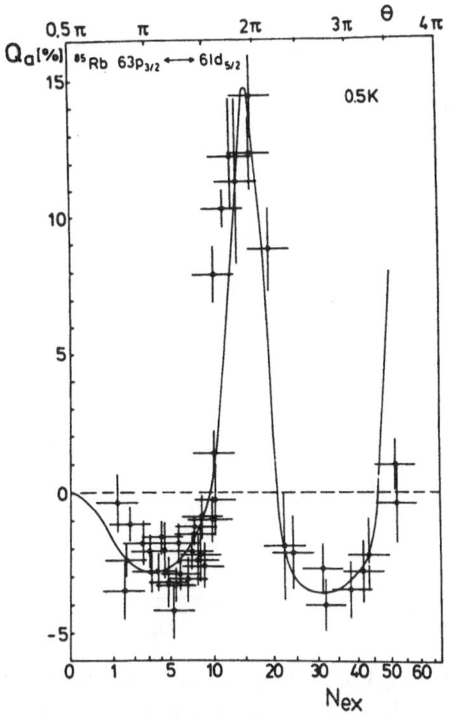

Figure 5. Same as Fig. 3, but above threshold for the $63p_{3/2} \leftrightarrow 61d_{5/2}$ transition (see Ref. 22).

The nonclassical photon statistics of the one-atom maser are observed at a higher flux of atoms or a larger atom-field coupling constant. The $63p_{3/2} \leftrightarrow 61d_{5/2}$ maser transition with Ω =44kHz is therefore studied.

Experimental results are shown in Fig. 5. Fast atoms with an atom-cavity interaction time of $t_{int} = 35\mu s$ are used. A very low flux of atoms of $N_{ex} > 1$ is already sufficient to generate a nonclassical maser field. This is the case since the vacuum field initiates a transition of the atom to the lower maser level, thus driving the maser above threshold. The sub-Poissonian statistics can be understood from Fig. 6, where the probability of finding the atom in the upper level is plotted as a function of the atomic flux. The oscillation observed is closely related to the Rabi-nutation induced by the maser field. The solid curve was calculated according to the one-atom maser theory with a velocity dispersion of 4%. A higher flux generally leads to a higher photon number, but for $N_{ex} < 10$ the probability of finding the atom in the lower level decreases. An increase in the photon number is therefore counterbalanced by the fact that the probability of photon emission in the cavity is reduced. This negative feedback leads to a stabilization of the photon number.[21] The feedback changes sign at a flux $N_{ex} \approx 10$, where the second maser phase transition is observed at $\Theta = 2\pi$. This is again characterized by large fluctuations of the photon number. Here the probability of finding an atom in the lower level increases with increasing flux. For even higher fluxes, the state of the field is again highly nonclassical. The solid line in Fig. 5 represents the result of the one-atom maser theory using Eq. (1) to calculate Q_a. The agreement with the experiment is very good.

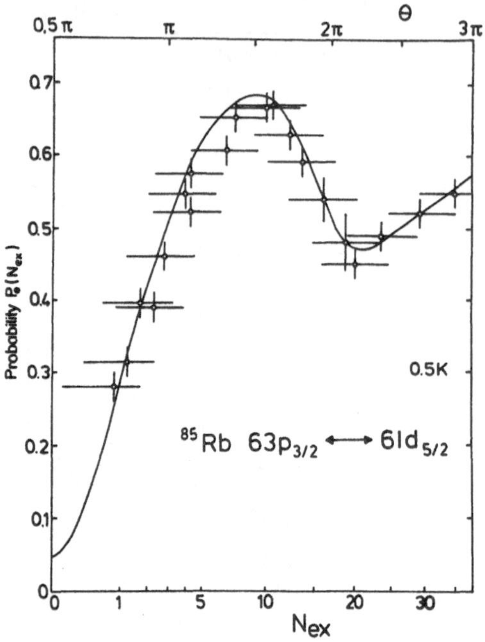

Figure 6. Probability $P_e(N_{ex})$ of finding the atom in the upper maser level $63p_{3/2}$ for the $63p_{3/2} \leftrightarrow 61d_{5/2}$ transition as a function of the atomic flux.

The sub-Poissonian statistics of atoms near $N_{ex} = 30$, $Q_a = -4\%$ and $P = 0.45$ (see Fig. 6) are generated by a photon field with a variance $< n^2 > - < n >^2 = 0.3 \cdot < n >$, which is 70% below the shot noise level. Again, this result agrees with the prediction of the theory.[15,16]

The mean number of photons in the cavity is about 2 and 13 in the regions $N_{ex} \approx 3$ and $N_{ex} \approx 30$, respectively. Near $N_{ex} \approx 15$, the photon number changes abruptly between these two values. The next maser phase transition with a super-Poissonian photon number distribution occurs above $N_{ex} \approx 50$. Sub-Poissonian statistics are closely related to the phenomenon of antibunching for which the probability of detecting a next event shows a minimum immediately after a triggering event. The duration of the time interval with reduced probability is of the order of the coherence time of the radiation field. In our case this time is determined by the storage time of the photons. The Q_a value therefore depends on the measuring interval T. Experimental results for a flux $N_{ex} \approx 30$ are shown in Fig. 7. The measured Q_a value approaches time-independent value for $T > T_{cav}$. For very short sampling intervals, the statistics of atoms in the lower level show a Poisson distribution. This means that the cavity cannot stabilize the flux of atoms in the lower level in a time scale which is short in relation to the intrinsic cavity damping time.

We want to emphasize that the reason for the sub-poissonian atomic statistics is the following: A changing flux of atoms changes the Rabi-frequency via the stored photon number in the cavity. By adjusting the interaction time, the phase of the Rabi-nutation cycle can be chosen so that the probability for the atoms leaving the cavity in the upper maser level increases when the flux and therefore the photon number is enlarged or vice versa. We observe sub-Poissonian atomic statistics in the case where the number

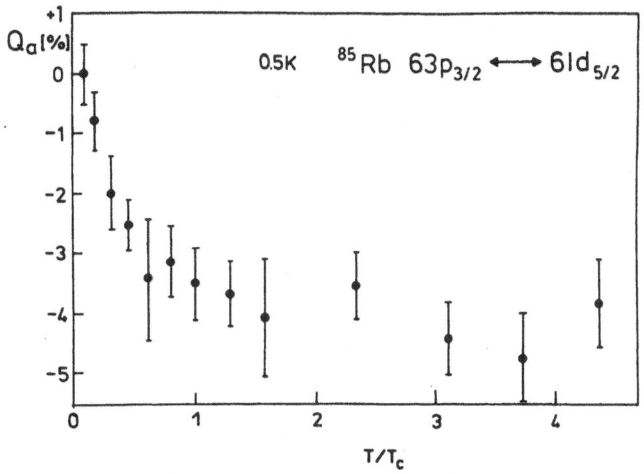

Figure 7. Variance Q_a of the atoms in the lower maser level as a function of the measurement time interval T for a flux $N_{ex} \approx 30$ (see Ref. 22)

of atoms in the lower state is decreasing with increasing flux and photon number in the cavity. The same argument can be applied to understand the nonclassical photon statistics of the maser field: Any deviation of the number of light quanta from its mean value is counterbalanced by a correspondingly changed probability of photon emissions of the atoms. This effect leads to a natural stabilization of the maser intensity by a feedback loop incorporated into the dynamics of the coupled atom-field system.

This feedback mechanism is also demonstrated when the anticorrelation of atoms leaving the cavity in the lower state is investigated. Measurements of these "antibunching" phenomena for atoms is described in the following.

For steady state conditions, experimental results are displayed in Fig. 8. Plotted is the probability $g^{(2)}(t)$ of finding an atom in the lower maser level $61d_{5/2}$ at time t after a first one has been detected at $t = 0$. The probability $g^{(2)}(t)$ was calculated from the actual count rate by normalizing with the average number of atoms determined in a large time interval. Time is given in units of the photon storage time. The error bar of each data point is determined by the number of about 7000 lower level atoms counted. This corresponds to $2 \cdot 10^6$ atoms that have crossed the cavity during the total measurement time. The detection of efficiency is near $\epsilon = 10\%$. The measurements were taken for $N_{ex} = 30$. The time of flight of the atoms through the cavity was $t_{int} = 37\mu s$, leading to $\Theta = 9$.

The fact that anticorrelation is observed shows that the atoms in the lower state are more equally spaced than expected for Poissonian distribution. It means when two atoms enter the cavity close to each other the second one performs a transition to the lower state with a reduced probability.

The experimental results presented here clearly show the sub-Poissonian photon statistics of the one-atom maser field. An increase in the flux of atoms leads to an atomic beam with atoms in the lower maser level showing number fluctuations which are up to 40% below those of a Poissonian distribution found usually in atomic beams. This is interesting, because atoms found in the lower level have emitted a photon to compensate for cavity losses inevitably present in the maser under steady-state condi-

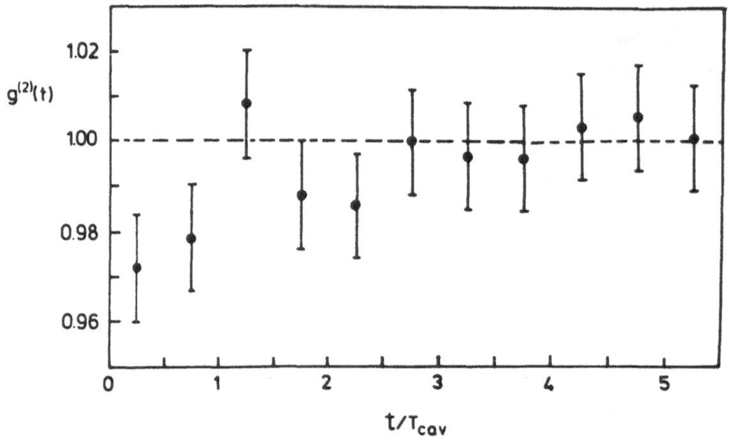

Figure 8. Anticorrelation of atoms in the lower maser state.

tions. This is a purely dissipative phenomenon giving rise to fluctuations, nevertheless the atoms still obey sub-Poissonian statistics.

5. ATOMS IN A NONCLASSICAL LIGHT FIELD

In the experiment it is guaranteed that there is only a single atom in the cavity at a time. But atoms crossing the cavity one after the other are correlated via the common maser field which experiences only a very small damping between two consecutive atoms. Is it possible to see this correlation in the measurement of the atoms? Or in other words: does the improved information we get on the cavity field by measuring it with a first atom correlate with the result we obtain with a second atom? The answer is 'yes'; the details will be discussed in this section.

An important requirement when discussing the measurement process in the one-atom maser is, of course, to have a detection efficiency as high as possible, preferably near unity. Therefore a numerical simulation of the maser process[18] has been performed in which the successive passages of atoms through the cavity and their measurement are explicitly taken into account. In the calculation we are mainly interested in the statistics of the atoms leaving the cavity so that experimental and theoretical results can directly be compared. Details of the calculation can be found in Ref. 21 and will not be repeated here. In the following the main result of the calculation will be presented together with a qualitative discussion of the effect.

Under steady state experimental conditions, all atoms are prepared in the upper maser level. The atom leaves the cavity in a superposition of both maser levels with relative probabilities determined by the Rabi-nutation dynamics. In addition, atom and field are still coupled to form a combined system. Therefore, by detecting an atom measurement of the field is performed. On the average, the atoms emit photons to compensate for cavity losses. But what happens if a particular atom is measured to be in the lower state? We can ask for the conditional probability $g^{(2)}(t) = 1$ of finding the following atom in the lower level as well. Results of a calculation with $N_{ex} = 4$ and $\Theta = 3$ are shown in Fig. 9 for two different detection efficiencies of 10% and

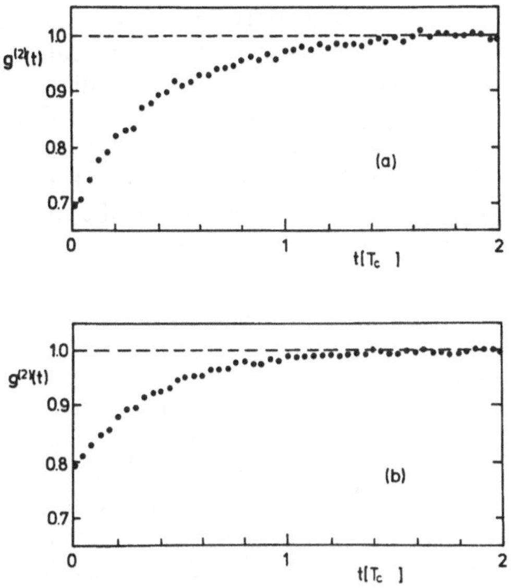

Figure 9: Results of a numerical simulation ($N_{ex} = 4$, $\Theta = 3$) for the normalized probability $g^{(2)}(t)$ of finding an atom in the lower maser level at time t after a first atom was detected in the lower level at time $t = 0$ for two different detection efficiencies: (a) $\epsilon = 10\%$ and (b) $\epsilon = 100\%$. Time is given in units of the cavity storage time T_{cav}.

100%. Both calculations are normalized to the value $g^{(2)}(t) = 1$ valid for a beam of atoms with Poissonian statistics. The main effect displayed in the figure is a reduction of the probability that the next atom is also detected in the lower level. This means that atoms in the lower level show a behavior typical for photon antibunching. It is expected because the parameters of the calculation are chosen in such a way that the nonclassical maser field built up inside the cavity leads to a beam of atoms with number fluctuations below the Poisson level which gives rise to antibunching of atoms. As expected from Fig. 8 and displayed in Fig. 9, this behavior occurs on a time scale comparable with the cavity storage time T_{cav}.

But there is a new more subtle effect showing up in Fig. 9 depending on the detection efficiency. It says that by performing more and more measurements on the atoms the flux of atoms becomes less anticorrelated and therefore more noisy. The higher detection efficiency thus leads to larger fluctuations of the number of lower-level atoms detected.

On the one hand this is not surprising since an atom detected in the lower level has given its energy to the field thus increasing the number of quanta in the cavity. Because of the boson nature this leads to a higher probability for the emission of another photon by the following atom thereby decreasing the amount of anticorrelation. On the other hand an increase of the photon number leads to a larger Rabi-frequency. For nonclassical maser radiation, i.e. when the atoms reabsorb energy from the field before leaving the cavity, a larger Rabi-frequency leads to a higher probability of detecting the next atom in the upper maser level. This clearly contradicts the previous argument. What is the reason? The following discussion addresses this questions.

Even for a detection efficiency near unity, the maser photon number is in general not known exactly. Our knowledge is described by a probability distribution which is broadened by the cavity dissipation and/or thermal radiation emitted by the cavity walls. Only the mean photon number can be calculated within its standard deviation which is given by the width of the probability distribution. As described above and in the previous section, a larger nonclassical field strength leads to a higher probability of detecting an atom in the upper level. Conversely, the measurement of an atom in the excited state increases the mean photon number in the cavity[21]:

$$<n> \rightarrow <n> -Q_f \frac{P}{1-P}$$

which is positive for a nonclassical maser field with $Q_f < 0$. Again, P denotes the probability that the atom makes a transition to the lower maser level inside the cavity. A similar argument can be applied to atoms detected in the lower maser level: If we take into account that the damping of the field in the cavity is small in the time interval between two consecutive atoms, the mean number of maser photons can even be reduced despite the fact that the atom made a transition from the upper to the lower level thereby emitting one photon[21]:

$$<n> \rightarrow <n> +Q_f + 1 - P.$$

The then smaller intensity of the field which causes a lower Rabi-frequency therefore increases the probability of finding the following atom in the lower state as well. Hence this counterintuitive but measurement-induced effect leads to a bunching of atoms in the lower level and decreases the magnitude of the nonclassical fluctuations in the case of a high detection efficiency.

It is emphasized that the discussion only uses mean values which are not subject to the law of energy conservation. Owing to the measurement, the information about the field is improved and the initial photon number distribution is modified to yield the effect described above.

Under steady state conditions, a measurement of an atom which is not state selective does not change the mean number of maser photons in a first approximation. In this case, the photon probability distribution p_n changes according to

$$p_n(t) \rightarrow p_n(t)\cos^2(\Omega t_{int}(n+1)^{1/2}) + p_{n-1}(t)\sin^2(\Omega t_{int} n^{1/2})$$

and the bunching phenomenon disappears for a small detection efficiency because the fluctuations introduced by the measurement process are averaged out. With increasing detection efficiency, the photon probability distribution is changed by the measurement process and, in particular, depend on the state of the atom. After the interaction, the photon number distribution is modified according to

$$p_n(t) \rightarrow N_\downarrow p_{n-1}(t)\sin^2(\Omega t_{int} n^{1/2})$$

for atoms detected in the lower level or by

$$p_n(t) \rightarrow N_\uparrow p_n(t)\cos^2(\Omega t_{int}(n+1)^{1/2})$$

for atoms in the upper level. Here $N_\downarrow$ and $N_\uparrow$ are two normalization constants. Therefore, depending on the outcome of the measurement, the photon probability distribution changes in a different manner and the detection efficiency does not show up as a pure random deletion process. In addition, this effect leads to the result that in case of a high detection efficiency the normalized variance of atoms Q_a is not proportional to this efficiency but is reduced. State selective measurements of the atoms decrease the width of the actual photon number distribution but prevent one-atom maser field from reaching a unique steady state. This effect increases the fluctuations of the mean photon number and the number of atoms detected. It follows that the sub-Poissonian character of the flux of atoms in the lower level is decreased in magnitude in the case of a high detection efficiency near unity. This is not unexpected since the one-atom maser is a quantum system and the measurement process must have an influence.

6. A NEW PROBE OF COMPLEMENTARITY IN QUANTUM MECHANICS

The preceding section discussed how to generate a nonclassical field inside the maser cavity. But this field is extremely fragile because any attenuation causes a considerable broadening of the photon number distribution. Therefore it is difficult to couple the field out of the cavity while preserving its nonclassical character. But what is the use of such a field? In the present section we want to propose a new series of experiments performed inside the maser cavity to test the "wave-particle" duality of nature, or better said "complementarity" in quantum mechanics.

Complementarity[24] lies at the heart of quantum mechanics: Matter sometimes displays wave-like properties manifesting themselves in interference phenomena, and at other times it displays particle-like behavior thus providing "which-path" information. No other experiment illustrates this wave-particle duality in a more striking way than the classic Young's double-slit experiment.[25,26] Here we find it impossible to tell which slit light went through while observing an interference pattern. In other words, any attempt to gain "which-path" information disturbs the light so as to wash out the interference fringes. This point has been emphasized by Bohr in his rebuttal to Einstein's ingenious proposal of using recoiling slits[26] to obtain "which-path" information while still observing interference. The physical positions of the recoiling slits, Bohr argues, are only known to within the uncertainty principle. This error contributes a random phase shift to the light beams which destroys the interference pattern.

Such random-phase arguments illustrating in a vivid way how the "which-path" information destroys the coherent-wave-like interference aspects of a given experimental setup, are appealing. Unfortunately, they are incomplete: In principle, and in practice, it is possible to design experiments which provide "which-path" information via detectors which do not disturb the system in any noticeable way. Such "Welcher Weg"-(German for "which-path") detectors have been recently considered within the context of studies involving spin coherence.[27] In the present section we describe a quantum optical experiment[28] which shows that the loss of coherence occasion by "which-path" information, that is, by the presence of a "Welcher Weg"-detector, is due to the establishing of quantum correlations. It is in no way associated with large random-phase factors as in Einstein's recoiling slits. The two essential ingredients of this novel "Welcher Weg"-detector are the one-atom maser and the quantum-beat concept.[29]

In a quantum-beat experiment atoms are excited by a short laser pulse to a coherent superposition of the two excited states $|a>$ and $|b>$ separated by a frequency difference $\Delta\omega$. They are allowed to decay spontaneously to a lower level $|c>$ as shown in

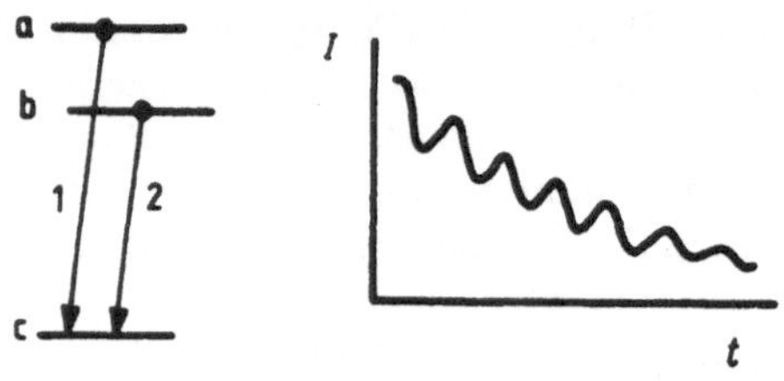

Figure 10. Scheme for a quantum-beat experiment. Coherent superposition of upper levels a and b decay to ground state c. The detector current shows a modulation in addition to the usual exponential decay.

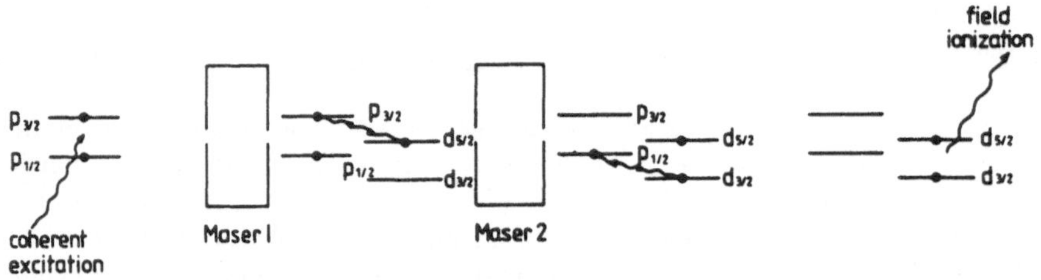

Figure 11. A one-atom maser "Welcher Weg" (which-path) detector. In passing through the first one-atom maser, a Rb atom prepared in a coherent superposition of $63p_{3/2}$ and $63p_{1/2}$ makes a transition from $63p_{3/2}$ to $61d_{5/2}$, and in the second from $63p_{1/2}$ to $61d_{3/2}$. Depending on the maser state - coherent or number state - quantum beats detected via field ionization do or do not occur.

Fig. 10. The spontaneously emitted radiation shows "beats", that is temporal fringes. These beats are a result of the indistinguishability of the two "paths" of excitation $|c> \rightarrow |a> \rightarrow |c>$ and $|c> \rightarrow |b> \rightarrow |c>$ analogous to the interfering paths of the light in the double-slit experiment. Hence the quantum beat experiment represents only the interference side of the coin of complementarity. To recognize its "Welcher Weg"-side we consider the experimental arrangement of Fig. 11. Here we depict the more complicated atomic configuration of a four-level atom and use two consecutive one-atom masers. We excite a coherent superposition of the $63p_{3/2}$ and the $63p_{1/2}$ levels of Rb^{85}. The velocity of the atoms is adjusted such that a transition from $63p_{3/2}$ to $61d_{5/2}$ is guaranteed in the first cavity; moreover the second cavity induces the transition from $63p_{1/2}$ to $61d_{3/2}$. The coherence transferred in this way from the two initially excited states to the two d-states can easily be detected by a field ionization quantum-beat experiment.[30] But why does this experimental setup provide "Welcher Weg"-information? Where is the path-information?

The answers to these questions come to light in the ionization current[28]

$$I(t) = I_0 e^{-\gamma t}[1 + e^{i\Delta\omega t} < \Phi_1^{(f)}, \Phi_2^{(i)} | \Phi_1^{(i)}, \Phi_2^{(f)} > + c.c.]. \qquad (2)$$

Here I_0 denotes the constant component of the current and $\gamma_a = \gamma_b = \gamma$ are the natural linewidth of the two excited $63p_{3/2}$ and $63p_{1/2}$ levels. The initial and final states of the electromagnetic field in the j-th one-atom maser are denoted by $|\Phi_j^{(i)}>$ and $|\Phi_j^{(f)}>$, respectively. Equation (2) spot-lights the crucial role of the field states in this play of quantum beats "to be or not to be".

Let us prepare, for example, both one-atom masers in coherent states[1] $|\Phi_j^{(i)}>=|\alpha_j>$ of large average photon number $<m>=|a_j|^2>>1$. The Poissonian photon number distribution of such a coherent state is very broad, $\Delta m \approx \alpha >> 1$. Hence the two fields are not changed much by the addition of a single photon associated with the two corresponding transitions. We may therefore write

$$|\Phi_j^{(f)}> \cong \alpha_j>$$

which to a very good approximation yields

$$<\Phi_1^{(f)}, \Phi_2^{(i)}|\Phi_1^{9i)}, \Phi_2^{(f)}> \cong <\alpha_1, \alpha_2|\alpha_1, \alpha_2>=1.$$

Thus the interference cross term in Eq. (2) is present so giving rise to quantum beats.

When we, however, prepare both maser fields in number states[17–19] $|n_j>$ the situation is quite different. After the transitions of the atom to the d-states, that is after emitting consecutively the two photons the final states in the two cavities read

$$|\Phi_j^{(f)}> = |n_j+1>$$

and hence

$$<\Phi_1^{(f)}, \Phi_2^{(i)}|\Phi_1^{(i)}, \Phi_2^{(f)}> = <n_1+1, n_2|n_1, n_2+1> = 0$$

that is the coherence cross term vanishes and no quantum beats emerge.

On the first sight this result might seem a bit surprising when we recall that in the case of a coherent state the transitions did not destroy the coherent cross term, that is, it did not affect the temporal interference fringes. However, in the example of number states we can, by simply "looking" at the one-atom maser state, tell which "path" the excitation took. When we find, for example, the first maser cavity in a state of n_1+1 photons we know that the internal atomic "path",

groundstate $\rightarrow$ $63p_{3/2}$ $\rightarrow$ $61d_{5/2}$ $\rightarrow$ ionization

was followed. Likewise the knowledge of having n_2+1 photons in the second cavity allows us to deduce the path of excitation

ground state $\rightarrow$ $61p_{1/2}$ $\rightarrow$ $61d_{3/2}$ $\rightarrow$ ionization.

Hence the states of the electromagnetic field in the cavities play the role of the recoiling plate of the double-slit experiment in extracting which-path information. Moreover it does not matter if we actually look at the states or not. Having the information stored there is enough to destroy the interference!

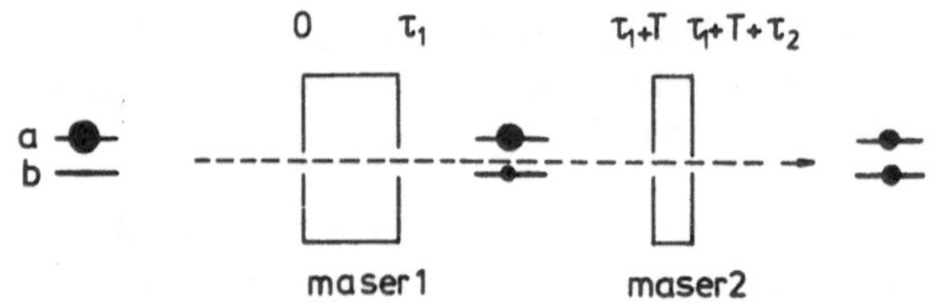

Figure 12. Scheme of the Ramsey experiment.

The use of Rydberg atoms in this experiment delivers us from the necessity of large photon numbers to produce the atomic transitions. In fact the vacuum interaction, that is, spontaneous emission is sufficient to insure the transition. Thus we are relieved of preparing number states $|n_j>$ of large photon numbers and instead prepare our masers in the easiest number state of all, the vacuum. In this case, however, one important condition has to be fulfilled: The beam density must be extremely low to guarantee that the states in the cavities return to the vacuum state before the next atom enters. Otherwise a field different from the vacuum would build up in the cavities and the which-path information would be lost.

It should be pointed out that the beats disappear not only for a number state. For example, a thermal field leads to the same result. In this regard, we note that it is not enough to have an indeterminate photon number to ensure quantum beats. The state $|\Phi_j^{(f)}>$ goes as $a_j^+|\Phi_j^{(i)}>$ where a_j^+ is the creation operator for the j-th maser. Hence the inner product

$$<\Phi_j^{(i)}|\Phi_j^{(f)}> \rightarrow <\Phi_j^{(i)}|a_j^+|\Phi_j^{(i)}> .$$

And in terms of a more general density matrix formalism we have

$$<\Phi^{(i)}|\Phi^{(f)}> \rightarrow \sum_n \sqrt{n+1}\rho_{n,n+1}^{(i)} .$$

Thus we see that an off-diagonal density matrix is needed for the production of beats. For example, a thermal field having indeterminate photon number would not lead to quantum beats since the photon number distribution is diagonal in this case.

The quantum beat experiment in connection with one-atom maser cavities is a rather complicated scheme for a "Welcher Weg"-detector. There is a much simpler possibility which we will discuss briefly in the following. This is based on the logic of the famous "Ramsey fringe" experiment. In this experiment two microwave fields are applied to the atoms one after the other. The interference occurs since the transition from an upper state to a lower state may either occur in the first or in the second interaction region. In order to calculate the transition probability we must sum the two amplitudes and then square, thus leading to an interference term. We will show here only the principle of this new experiment; a more detailed discussion will be the subject of a later paper (see also Ref. 31 which deals with a related application). In the setup we discuss here, the two Ramsey fields are two one-atom maser cavities (see Fig. 12). The atoms enter the first cavity in the upper state and are weakly driven into a lower state $|b>$. That is, each microwave cavity induces a small transition amplitude governed by $m\tau$ where m is the atom-field coupling constant and τ is the time of flight across the cavity.

Now if the quantum state of the initial (final) field in the j-th cavity is given by $\Phi_j^{(i)}(\Phi_j^{(f)})$ then the state of the atom + maser 1 + maser 2 at the various relevant times is given in terms of the coupling constant m_j and interaction times τ_j, initial $|\Phi_j^{(i)}>$ and the final states $|\Phi_j^{(f)}>$ of the j-th maser by:

$$|\psi(0)> = \ |a, \Phi_1^{(i)}, |\Phi_2^{(i)}>$$

$$|\psi(\tau_1)> \cong \ |a, \Phi_1^{(i)}, \Phi_2^{(i)}> -im_1\tau_1|b, \Phi_1^{(f)}, \Phi_2^{(i)}>$$

$$|\psi(\tau_1 + T)> \cong \ |a, \Phi_1^{(i)}, \Phi_2^{(i)}> -im_1\tau_1|b, \Phi_1^{(f)}, \Phi_2^{(i)}> e^{-i\Delta\omega T}$$

$$|\psi(\tau_1 + T + \tau_2)> \cong \ |a, \Phi_1^{(i)}, \Phi_2^{(i)}> -im_1\tau_1|b, \Phi_1^{(f)}, \Phi_2^{(i)}> e^{-i\Delta\omega T}$$

$$-im_2\tau_2|b, \Phi^{(i)}, \Phi_2^{(f)}>$$

where $\Delta\omega$ is the atom-cavity detuning and $T >> \tau_j$ the time flight between the two cavities. If we ask for P_b, the probability that the atom exists cavity 2 in the lower state $|b>$, this is given by

$$\begin{aligned} P &= \left[< \Phi_1^{(f)}, \Phi_2^{(i)}|m_1^*\tau_1 e^{i\Delta\omega T} + < \Phi_1^{(i)}, \Phi_2^{(f)}|m_2^*\tau_2\right] \\ &\cdot\left[|\Phi_1^{(f)}, \Phi_2^{(i)} > m_1\tau_1 e^{-i\Delta\omega T} + |\Phi_1^{(i)}, \Phi_2^{(f)} > m_2\tau_2\right] \\ &= m_1^* m_1 \tau_1^2 + m_2^* m_2 \tau_2^2 + (m_1^* m_2 \tau_1 \tau_2 e^{i\Delta\omega T} < \Phi_1^{(f)}, \Phi_2^{(i)}|\Phi_1^{(i)}, \Phi_2^{(f)} > + \ c.c). \end{aligned}$$

Now in the usual Ramsey experiment $|\Phi_j^{(i)}> = |\Phi_j^{(f)}> = |\alpha_j>$ where $|\alpha_j>$ is the coherent state in the j-th maser which is not changed by the addition of a single photon. Thus the "fringes" appear going as $exp(i\Delta\omega T)$. However, consider the situation in which $|\Phi_j^{(i)}>$ is a number state e.g. the state $|0_j^{(i)}>$ having no photons in the j-th cavity initially; now we have

$$P = m_1^* m_1 \tau_1^2 + m_2^* m_2 \tau_2^2 + (m_1^* m_2 \tau_1 \tau_2 e^{i\Delta\omega T} < 1_1, 0_2|0_1, 1_2 > + \ c.c.).$$

In this case, the one-atom masers are now acting as "Welcher Weg"-detectors, and the interference term vanishes due to the atom-maser quantum correlation.

We note that the more usual Ramsey fringe experiment involves a strong field "$\frac{\pi}{2}$-pulse" interaction in the two regions. This treatment is more involved than necessary for the present purposes. A more detailed analysis of the one-atom maser Ramsey problem will be given elsewhere.

We conclude this section by emphasizing again that this new and potentially experimental example of wave-particle duality and observation in quantum mechanics displays a feature which makes it distinctly different from the Bohr-Einstein recoiling-slit experiment. In the latter the coherence, that is the interference, is lost due to a phase disturbance of the light beams. In the present case, however, the loss of coherence is due to the correlation established between the system and the one-atom maser. Random-phase arguments never enter the discussion. We emphasize that the argument of the number state not having a well-defined phase is not relevant here; the important dynamics is due to the atomic transition. It is the fact that which-path information is made available which washes out the interference cross terms (for a detailed discussion see also Ref. 32).

7. SUMMARY AND CONCLUSION

Nonclassical light as generated in the one-atom maser is characterized by amplitude fluctuations which are below those of the vacuum state. Such a field prepared with a very narrow photon number distribution is nonclassical because of its non-wavelike character. This property can only be observed when classical fluctuations, e.g. fluctuations in the velocity of atoms or dissipation of photons in the cavity walls, are suppressed. The generation of a nonclassical field in the one-atom maser is achieved by controlling the rate of photon emissions by the intensity of the field inside the light source. The intensity is thereby controlled by the intensity itself. This scheme uses a back-coupling mechanism which correlates the probability of photon emission with the instantaneous light amplitude in such a way that any deviation of the intensity from its mean value is counterbalanced by a correspondingly changed rate of photon emissions. Such a quantum nondemolition feedback mechanism is at hand when a single atom with a well-defined velocity interacts with a single mode of the electromagnetic field and the atom field coupling is described by the Rabi-nutation dynamics. This is - in short - the basic principle that allows to generate a nonclassical maser field.

The possibility of tests of the complementarity principle via the one-atom maser is another theme of the present paper. Send an atom prepared in a coherent superposition of two quantum levels through two consecutive one-atom masers. In this way, transfer the population to two neighboring levels. Is the initial coherence between the two atomic states as manifested in an oscillatory ionization signal - temporal interference - preserved? To elucidate this question, imagine starting from coherent cavity fields with a large mean number of photons. In such a coherent state the fluctuations in photon number are large. Hence the two photons emitted in the transfer process, and caught in the cavities, do not change the field state significantly. One cannot deduce information concerning the transfer by studying the cavity fields. Hence the coherence is preserved giving rise to quantum beats. However, if the experiment is started from a state of well-defined photon number it is possible to detect the increase, by one, in photon number induced by the population transfer. This destroys the coherence of the state, just as identifying which slit a photon passes through in the Young's double-slit experiment destroys the coherence and hence the interference fringes in that experiment.

REFERENCES

1. For reviews see Kimble, H. J. and Walls, D., *J. Opt. Soc. Am.* **B4**, 1449 (1987); Loudon, R. and Knight, P. L., *J. Mod. Opt.* **34**, 707 (1987).
2. Diedrich, F. and Walther, H., *Phys. Rev. Lett.* **58**, 203 (1987).
3. Meschede, D., Walther, H. and Müller, G., *Phys. Rev. Lett.* **54**, 551 (1985); for a review see Diedrich, F., Krause, J., Rempe, G., Scully, M. O. and Walther, H., *IEEE J. Quantum Electron.* **24**, 1314 (1988).

4. For a review see the following articles by Haroche, S. and Raimond, J. M., Advances in Atomic and Molecular Physics, Vol. 20, p. 350. Academic Press New York, 1985; Gallas, J. A., Leuches, G., Walther, H. and Figger, H., Advances in Atomic and Molecular Physics, Vol. 20, p. 413. Academic Press New York, 1985.

5. Jaynes, E. T. and Cummings, F. W., Proc IEEE **51**, 89 (1963).

6. See, e.g. Eberly, J. H., Narozhny, N. B. and Sanchez-Mondragon, J. J., *Phys. Rev. Lett.* **44**, 1323 (1980) and references therein.

7. Rempe, G., Walther, H. and Klein, N., *Phys. Rev. Lett.* **58**, 353 (1987).

8. Walls, D. F., *Nature* **280**, 451 (1979); see also articles in: Photons and Quantum Fluctuations (ed. by E. R. Pike and H. Walther), Hilger Bristol, 1988.

9. Walls, D. F., *Nature* **306**, 141 (1983); ibid **324**, 210 (1986); see also the various articles in: Squeezed and Nonclassical Light (ed. by P. Tombesi and E. R. Pike), Plenum Press New York, 1988.

10. First demonstration of photon antibunching: Kimble, H. J., Dagenais, M. and Mandel, L., *Phys. Rev. Lett.* **39**, 691 (1977); *Phys. Rev.* **A18**, 201 (1978); see also Cresser, J. D., Häger, J., Leuchs, G., Rateike, M. and Walther, H.: Dissipative Systems in Quantum Optics, p. 21 Springer Berlin, 1982.

11. First demonstration of sub-Poissonian photon statistics: Short, R. and Mandel, L., *Phys. Rev. Lett.* **51**, 384 (1983).

12. First demonstration of squeezing: Slusher, R. E., Hollberg, L. W., Yurke, B., Mertz, J. C. and Valley, J. F., *Phys. Rev. Lett.* **55**, 2409 (1985); for review, see, e.g., Kimble, H. J. and Walls, D., *J. Opt. Soc. Am.* **B4**, 1449 (1987) and Loudon, R. and Knight, P. L., *J. Mod. Opt.* **34**, 707 (1987).

13. Carmichael, H. J. and Walls, D. F., *J. Phys.* **B9**, 1199 L 43 (1976).

14. Diedrich, F. and Walther, H., *Phys. Rev. Lett.* **58**, 203 (1987).

15. Filipowicz, P., Javanainen, J. and Meystre, P., *Opt. Comm.* **58**, 327 (1986); *Phys. Rev.* **A34**, 3077 (1986); *J. Opt. Soc. Am.* **B3**, 906 (1986).

16. Lugiato, L., Scully, M. O. and Walther, H., *Phys. Rev.* **A36**, 740 (1987).

17. Krause, J., Scully, M. O. and Walther, H., *Phys. Rev.* **A36** (1987) 4547; Krause, J., Scully, M. O., Walther, T. and Walther, H.: *Phys. Rev.* **A39**, 1915 (1989).

18. Meystre, P., *Opt. Lett.* **12**, 699 (1987); Meystre, P., Squeezed and Nonclassical Light (ed. by P. Tombesi and E. R. Pike), p. 115, Plenum New York, 1988.

19. Meystre, P., Rempe, G. and Walther, H., *Opt. Lett.* **13**, 1078 (1988).

20. Sloser, J. J., Meystre, P. and Wright, E. M., *Opt. Lett.* **15**, 233 (1990).

21. Rempe, G. and Walther, H., *Phys. Rev.* **A42**, 1650 (1990).

22. Rempe, G., Schmidt-Kaler, F. and Walther, H. *Phys. Rev. Lett.* **64**, 2783 (1990).

23. Mandel, L., *Opt. Lett.* **4**, 205 (1979).

24. See, e.g. Bohm, D., Quantum Theory, Prentice Hall Englewood Cliffs, 1951, or Jammer, M., The Philosophy of Quantum Mechanics, Wiley New York, 1974.

25. A detailed analysis of Einstein's version of the double-slit experiment is given by Wootters, W. and Zurek, W., *Phys. Rev.* **D19**, 473 (1979); see also Ref. 26.

26. For an excellent presentation of the Bohr-Einstein dialogue see chapter 1 in: Wheeler, J. A. and Zurek, W. H., Quantum Theory and Measurement, Princeton University Press Princeton, 1983, and in particular, the article by Bohr, N., Discussion with Einstein on Epistemological Problems in Atomic Physics.

27. Englert, G.-G., Schwinger, J. and Scully, M. O., *Found. Phys.* **18**, 1045 (1988); Schwinger, J., Scully, M. O. and Englert, B.-G., *Z. Physics* **D10**, 135 (1988); Scully, M. O., Englert, B.-G. and Schwinger, J., *Phys. Rev.* **A40**, 1775 (1989).

28. Scully, M. O. and Walther, H., *Phys. Rev.* **A39**, 5229 (1989).

29. Haroche, S., High Resolution Laser Spectroscopy, p. 253, Springer Berlin, 1976; see also Chow, W. W., Scully, M. O. and Stoner, J. O., *Phys. Rev.* **A11**, 1380 (1975).

30. Leuchs, G., Smith, S. and Walther, H., Laser Spectroscopy IV (ed. by H. Walther and K. W. Rothe), p. 255, Springer Verlag Berlin, 1979; Leuchs, G. and Walther, H., *Z. Phys.* **A293**, 93 (1979).

31. Krause, J., Scully, M. O., Walther, H., *Phys. Rev.* **A34**, 2032 (1986).

32. Scully, M. O., Englert, G.-G. and Walther, H., *Nature* **351**, 111 (1991).

OBSERVATION OF A NONCLASSICAL BERRY'S PHASE IN QUANTUM OPTICS†

R.Y. Chiao, P. G. Kwiat, I.H. Deutsch, and A. M. Steinberg

Department of Physics
University of California
Berekeley, CA 94720

ABSTRACT

The question of the classical vs. quantum nature of Berry's phases in optics is discussed. We present the results of a quantum optics experiment demonstrating a nonclassical Berry's phase. This experiment involved coincidence detection of photon pairs produced in parametric fluorescence, in conjunction with a Michelson interferometer in which one member of each pair acquired a geometrical phase due to a cycle in polarization states. The experiment constitutes the first observation of Berry's phase at the single photon level. We have verified that each interfering photon was essentially in an n=1 Fock state, by means of a beamsplitter following the interferometer combined with a triple coincidence technique. The results can be interpreted in terms of a nonlocal collapse of the wave function.

We shall discuss optical manifestations of Berry's phase, with an eye towards addressing the following question: When are these manifestations classical, and when are they quantum in nature? (Berry's phase[1] is a geometrical phase which can be acquired by a quantum system in addition to the usual dynamical phase when the system evolves cyclically back to its initial state.) A closely related question has arisen recently:[2] Should one always view the geometrical phase in optics as a *classical* Hannay angle, rather than a *quantum* Berry's phase? (The Hannay angle[3] is the difference between the Berry's phases of two adjacent quantum states, obtained in the correspondence principle limit when the quantum number, here the photon number, is large.) In the optical experiments on Berry's phases which have been done up to now,[4–5] what has actually been measured is the phase acquired by a classical electromagnetic wave after a cycle of transformations. A case has been made that the resulting phase must be interpreted as a Hannay angle rather than as a Berry's phase.[2]

However, we note that in the case of optics, the quantum system can in principle consist of an individual photon. This is true when one prepares the system in a one-photon Fock state, passes it through an interferometer, and detects it in such a

†Manuscript of an invited talk by R.Y. Chiao for the International Conference on Quantum Optics in Hyderabad, India, January 5-10, 1991.

Recent Developments in Quantum Optics, Edited
by R. Inguva, Plenum Press, New York, 1993

manner that only this one photon is detected. Then the observed phase, geometrical or dynamical, is that acquired by the ket of the one-photon system. In such a case, we shall argue that the observed phase must be interpreted as a nonclassical, quantum phase, even though it may coincide in its numerical value with the phase of a classical electromagnetic wave that has gone through the same interferometer.

For example, consider a Mach-Zehnder interferometer (see Fig. 1 (a)), into one of whose input ports is injected a one-photon Fock state, and two detectors D1 and D2 placed at its two output ports. For simplicity, we first consider only a dynamical phase which results from the translational motion of one of the mirrors. A generalization to include a geometrical phase will then be obvious. We postpone for the moment the discussion of how such a one-photon state is prepared. For now, we simply note that we can infer that a one-photon Fock state has traversed the interferometer from the fact that the two detectors D1 and D2 never give simultaneous clicks. This is true since a single photon is indivisible, and thus must decide at the final beamsplitter to go *either* to D1 *or* to D2. Therefore for a one-photon state, coincidences between D1 and D2 do not occur. In contrast, for a two-photon state it is possible at the final beam splitter for one photon to go to D1 and the other to go to D2, so that coincidences between D1 and D2 do occur. Similarly, this conclusion also follows for an n-photon state ($n \geq 2$) and for a coherent state. In the classical limit, the beam splitter divides the amplitude of a classical electromagnetic wave, so that coincidences will always occur. In fact, there exists an inequality giving a lower bound to the coincidence rate for classical light, which we will discuss below. We have observed violations of this inequality which mean that both the light and the phase are nonclassical.

For comparison, in Fig. 1(b) we sketch a schematic Mach-Zehnder interferometer for neutrons. (In actual experiments, a Bonse-Hart interferometer was used.) Again let us inject a one-neutron Fock state into one of its input ports, and observe the interference signal by means of two detectors D1 and D2 placed at its output ports. The observed phase in this case is that acquired by the ket of the neutron, and is clearly quantum mechanical in nature. Again, the two detectors D1 and D2 can never give simultaneous clicks, since a single neutron is indivisible. However, since the neutron is a fermion we are guaranteed by the Pauli exclusion principle not to have an occupation number greater than one, so that there exists no classical wave limit. Coherent states do not exist for neutrons. Hence, the phase in this case is always exclusively quantum mechanical in nature.

Since both the one-photon and the one-neutron experiments depicted in Fig. 1 defy classical explanation, it is necessary to ascribe a quantum mechanical character to both phases. *There is no fundamental difference between one-photon and one-neutron interference: In both cases, a single particle interferes with itself.*

Comparisons between various levels of descriptions of optics and mechanics may be helpful in this connection:

OPTICS	MECHANICS
I. Rays in geometrical optics	I. Trajectories in classical mechanics
Upon first quantization:	*Upon first quantization:*
II. One-photon wave optics:	II. One-particle quantum mechanics:
One-photon Maxwell's equations	Schrödinger's equation
III. Many-photon wave optics:	III. Many-body quantum mechanics:
Coherent states; Maxwell's eqs.	No classical limit for fermions
Upon second quantization:	*Upon second quantization:*
IV. Quantum field theory	IV. Quantum field theory

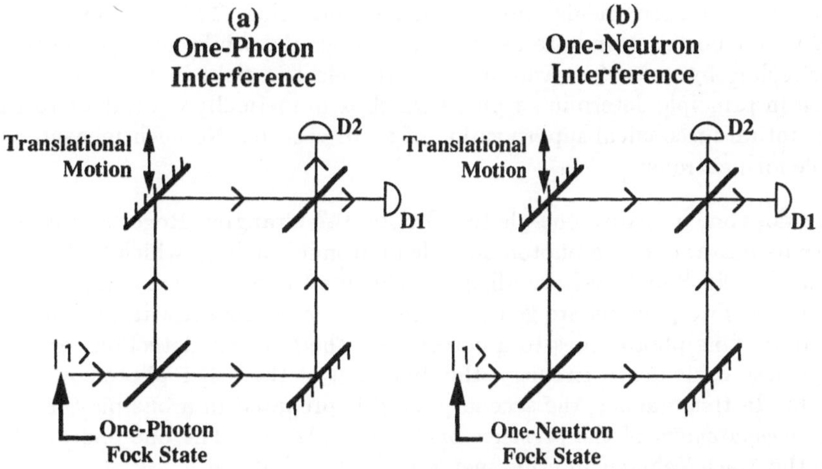

Figure 1. Comparison of one-photon interference and one-neutron interference.

At the most fundamental level, interference in both optics and mechanics arises from superposition of kets in Hilbert space. In particular, the phase which is observed in one-photon interference experiments is the phase acquired by kets, not the phase acquired by classical electromagnetic waves. This phase happens to coincide in numerical value with that calculated by means of classical Maxwell's equations, because the one-photon quantum Maxwell's equations are formally identical to the classical equations. Planck's constant does not explicitly appear in the quantum equations, because of a fortuitous cancellation due to the masslessness of the photon. Nevertheless, level II of optics is just as quantum in nature as level II of mechanics.

The above table should be viewed with some caveats. Although one cannot write down a single-particle relativistic quantum mechanical theory for photons without the introduction of negative energy solutions, the same is true for massive particles. The one-particle quantum theories are *approximate* theories, valid for low energies, when pair creation and annihilation are neglected. An effective nonrelativistic theory for the photon can be constructed for quasimonochromatic fields, with the carrier frequency playing the role of the mass.[6] There is another caveat. A fundamental distinction between bosons and fermions exists, in that a *local* measurement of field strength can be made for boson fields, but not for fermion fields. This difference stems from the fact that bosons commute at space-like intervals, while fermions anticommute. In particular, by a *local* measurement of the electric field in the case of photons, one can in principle determine a phase which is numerically equivalent to the phase of a quantum-mechanical superposition of state vectors. No such measurements are possible for neutrons.

How does one prepare a one-photon Fock state? Grangier, Roger and Aspect[7] have done so by means of a two-photon cascade in atomic calcium, which had already been used in Einstein-Podolsky-Rosen-like experiments to demonstrate violations of Bell's inequalities. Two photons are generated in an atomic cascade within nanoseconds of each other. One photon flies to a detector on the left. Its detection serves to gate, through fast coincidence circuitry, the detection of the other photon, which flies to the right. In this manner, the second photon is prepared in a one-photon Fock state by the *measurement* of the presence of the first photon. The one-photon state then enters the Mach-Zehnder interferometer as discussed above.

In our experiment, we replaced the calcium source by a more convenient two-photon light source, which utilizes spontaneous parametric down-conversion of uv laser photons in a $\chi^{(2)}$ nonlinear crystal.[8,9] In this fluorescence process, a single uv photon with a sharp spectrum is spontaneously converted inside the crystal into two photons ("signal" and "idler") with broad, conjugate spectra centered at twice the uv wavelength, conserving energy and momentum. The light was prepared in an entangled state consisting of a pair of photons whose energies, although individually broad in spectrum, sum up to a sharp quantity E because they were produced from a single uv photon whose energy E was sharp. This entangled state is given by

$$|\Psi >_{in} = \int dE' A(E')\ |1>_{E'}\ |1>_{E-E'}, \qquad (1)$$

where $A(E') = A(E-E')$ is the complex probability amplitude for finding one photon with an energy E', i.e., in the n=1 Fock state $|1>_{E}$, and one photon with an energy $E-E'$, i.e. in the $n=1$ Fock state $|1>_{E-E'}$. According to the standard Copenhagen interpretation, the meaning of this entangled state is that when a measurement of the energy of one photon results in a sharp value E', there is a sudden collapse of the wavefunction such that instantly at a distance, the other photon, no matter how remote, also possesses a sharp value of energy $E - E'$. Thus energy is conserved. Entangled states, i.e., coherent sums of product states, such as the one given by Eq. (1), result in Einstein-Podolsky-Rosen-like effects which are nonclassical and nonlocal.

We prepared the entangled state of energy, Eq. (1), by means of parametric fluorescence in the $\chi^{(2)}$ nonlinear optical crystal potassium dihydrogen phosphate (KDP), excited by a single mode ultraviolet (uv) argon ion laser operating at λ = 351.1 nm.[8,9] The uv laser beam was incident normally on the KDP input face. We employed type I phase matching, so that both signal and idler beams were horizontally polarized. Coincidences in the detection of conjugate photons were then observed. The KDP crystal was 10 cm long and cut such that the c-axis was 50.3° to the normal of its input face. We selected for study signal and idler beams both centered at λ=702.2 nm which emerged at -1.5° and +1.5°, respectively, with respect to the uv beam.

In Fig. 2, we show a schematic of the experiment. The idler photon (upper beam) was transmitted through the "remote" filter F1 to the detector D1, which was a cooled RCA C31034A-02 photomultiplier. The signal photon (lower beam) entered a Michelson interferometer, inside one arm of which were sequentially placed two zero-order quarter waveplates Q1 and Q2. The fast axis of the first waveplate Q1 was fixed at 45° to the horizontal, while the fast axis of the second waveplate Q2 was slowly rotated by a computer-controlled stepping motor. After leaving the Michelson the signal beam impinged on a second beamsplitter B2, where it was either transmitted to detector D2 through filter F2, or reflected to detector D3 through filter F3. Filters F2 and F3 were identical: They both had a broad bandwidth of 10 nm centered at 702 nm. Detectors D2 and D3 were essentially identical to D1. Coincidences between D1 and D2 and between D1 and D3 were detected by feeding their outputs into constant fraction discriminators and coincidence detectors after appropriate delay lines. We used EGG C102B coincidence detectors with coincidence window resolutions of 1.0 ns and 2.5 ns, respectively. Also, triple coincidences between D1, D2 and D3 were detected by feeding the outputs of the two coincidence counters into a third coincidence detector (a Tektronix 11302 oscilloscope used in a counter mode). The various count rates were stored on computer every second.

Our particular arrangement of quarter waveplates in the Michelson interferometer has been shown previously to generate Pancharatnam's phase on the *classical* level.[5] Upon appropriate detection after the interferometer, we observed at the *quantum* level the interference fringes resulting from this phase. To calculate the phase, we use the generalized Poincaré sphere[9] shown in Fig. 3, where polarization states are referred to space-fixed axes, and not to the direction of light propagation, as in the ordinary Poincaré sphere. (We do so in order to avoid extraneous discontinuities upon reflection from mirrors.) The first quarter waveplate Q1 converts horizontal linear polarization, represented by point A on the equator of the sphere, into circular polarization, represented by B at the north pole. This transformation of polarizations is represented by a geodesic arc AB. Then Q2 converts the circular polarization back to linear polarization (C on the equator), but with an axis rotated from the horizontal by θ, the angle between the fast axes of Q2 and Q1, in real space. On the sphere, the azimuthal angle from A to C is 2θ. After reflection from the mirror, the linear polarization is unchanged with respect to space-fixed axes, and is again represented on the generalized Poincaré sphere by the same point C. After reentering Q2, this is converted to circular polarization represented by D (the south pole), completing

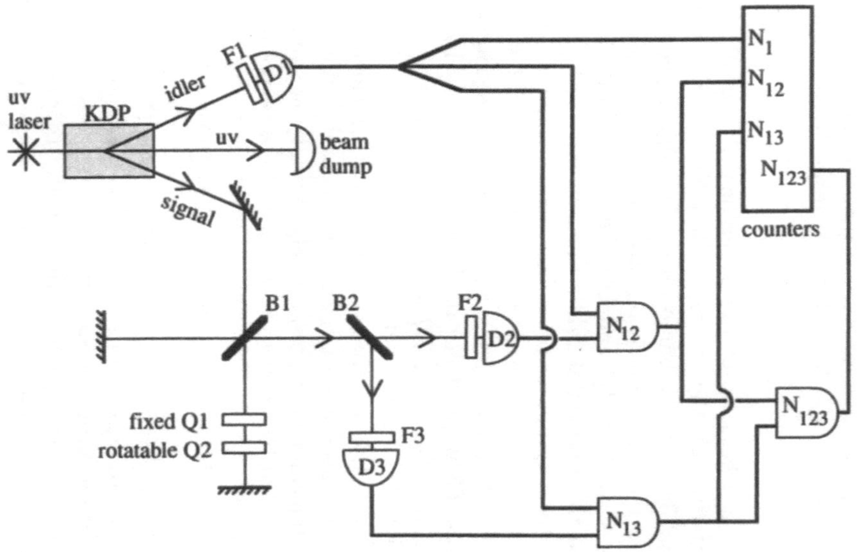

Figure 2. Schematic of experiment to observe an optical Berry's phase on the quantum level. D1, D2, and D3 are photomultipliers, Q1 and Q2 quarter waveplates, and B1 and B2 beamsplitters. Logical "AND" symbols denote coincidence detectors.

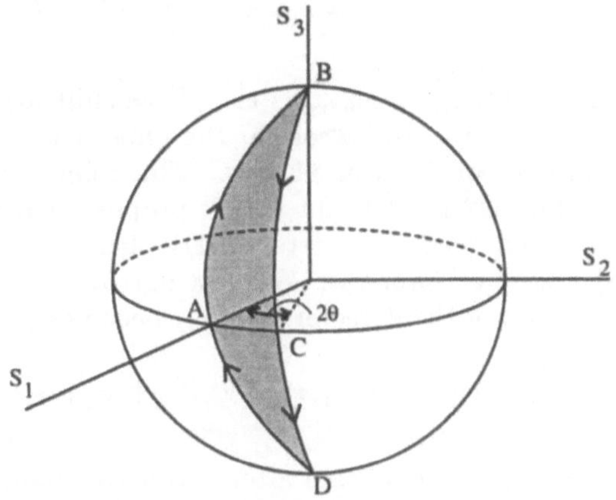

Figure 3. Generalized Poincaré sphere, where S_1, S_2 and S_3 are Stokes parameters. Circuit ABCDA represents a round trip through waveplates Q1 and Q2, where θ is the angle between their fast axes. Berry's phase, here Pancharatnam's phase, is 2θ for this circuit.

geodesic arc BCD. Then Q1 reconverts the circular polarization back to horizontal linear polarization, generating geodesic arc DA. Thus a cycle in polarization states is completed, represented by the circuit ABCDA. Pancharatnam's phase is minus one-half the solid angle subtended by the circuit with respect to the center of the sphere. For this circuit, the phase is equal to 2θ.

We took data both outside and inside the white light fringe region where the usual interference in singles detection occurs. We report here only on data taken outside this region, where the optical path length difference was at a fixed value much greater than the coherence length of the signal photons determined by the filters F2 and F3. Hence the fringe visibility seen by detectors D2 and D3 in singles detection was essentially zero.

Here we present a simplified quantum analysis of this experiment. Elsewhere we have presented a more comprehensive analysis based on Glauber's correlation functions.[10] The state of the light after the Michelson interferometer is given by

$$|\Psi>_{out} = \int dE' A(E')|1>_{E'}|1>_{E-E'} \{1+\exp(i\phi(E-E'))\}/2^{\frac{1}{2}}, \quad (2)$$

where $\phi(E-E') = 2\pi\Delta L/\lambda_{E-E'} + \phi_{Berry}$ is the phase shift arising from the optical path length difference ΔL of the Michelson for the photon with energy $E-E'$, plus the Berry's phase contribution for this photon. The coincidence rate N_{12}(or N_{13}) between detectors D1 and D2 (or D1 and D3) is proportional to the probability of finding at the same time one photon at detector D1 placed at r_1, and one photon at detector D2 (or D3) placed at r_2(or r_3). When a narrowband filter F1 centered at energy E' is placed in front of the detector D1, N_{12} becomes proportional to

$$|\Psi'_{out}(r_1,r_2,t)|^2 = |<r_1,r_2,t|\Psi>'_{out}|^2 \propto 1+cos\phi, \quad (3)$$

where the prime denotes the output state after a von Neumann projection onto the eigenstate associated with the sharp energy E' upon measurement. Therefore, the phase ϕ is determined at the sharp energy $E-E'$. In practice, the energy width depends on the bandwidth of the filter F1 in front of D1, so that the visibility of the fringes seen in coincidences should depend on the width of this *remote* filter. This fringe visibility will be high, provided that the optical path length difference of the Michelson does not exceed the coherence length of the *collapsed* signal photon wavepacket, determined by F1. If a sufficiently broadband remote filter F1 is used instead, such that the optical path length difference is much greater than the coherence length of the collapsed wavepacket, then the coincidence fringes should disappear.

In Fig. 4, we show data which confirm these predictions. In the lower trace (squares) we display the coincidence count rate between detectors D1 and D3, as a function of the angle θ between the fast axes of waveplates Q1 and Q2, when the remote filter F1 was narrow, i.e., with a bandwidth of 0.86 nm. The calculated coherence length of the collapsed signal photon wavepacket (570 μm) was greater than the optical path length difference at which the Michelson was set (220 μm). The visibility of the coincidence fringes was quite high, viz., 60% ± 5%.[11] This is in contrast to the low visibility, viz., less than 2%, of the singles fringes detected by D3 alone (not shown). For comparison, in the upper trace (triangles) we display the coincidence count rate versus θ when a broad remote filter F1, i.e., one with a bandwidth of 10 nm, was substituted for the narrow one. The coherence length of the

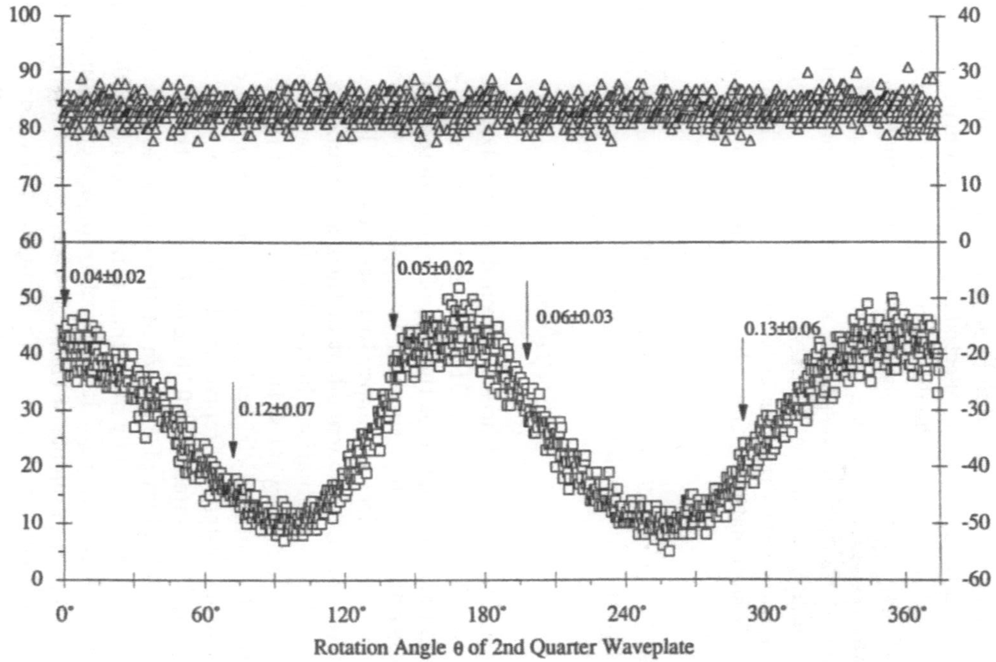

Figure 4. Interference fringes (lower trace, squares) for an unbalanced Michelson with a slowly varying Berry's phase, observed in coincidences between D3 and D1 with a narrow remote filter F1. With a broad F1, these fringes disappear (upper trace, triangles). They also disappear when detected by D3 alone. Vertical arrows indicate where the anti-correlation parameter a was measured (see text).

collapsed signal photon wavepacket was thus only 50 μm. The coincidence fringes in this case have indeed disappeared, as predicted.

In light of the observed violations of Bell's inequalities,[12] it is incorrect to interpret these results in terms of an ensemble of conjugate signal and idler photons which possess definite, but unknown, conjugate energies before filtering and detection. Any observable, e.g., energy or momentum, should not be viewed as a local, realistic property carried by the photon *before it is actually measured.*

The function of the second beamsplitter B2 was to verify that the signal beam was composed of photons in an n=1 Fock state. In such a state, the photon, due to its indivisibility, will be either transmitted or reflected at the beamsplitter, but not both. Thus coincidences between D2 and D3 should never occur, except for rare accidental occurrences of two pairs of conjugate photons within the coincidence window. However, if the signal beam were a classical wave, then one would expect an equal division of the wave amplitude at the 50% beamsplitter, and hence frequent occurrences of coincidences. An inequality, which was strongly violated in our experiment, places a lower bound on this coincidence rate for classical light (see below). This verifies the essentially n=1 Fock state nature of the light, and confirms the previous result of Hong and Mandel.[13]

The vertical arrows in Fig. 4 indicate the points at which triple coincidences were measured. Let us define the anticorrelation parameter[7]

$$a \equiv N_{123}N_1/N_{12}N_{13}, \tag{4}$$

where N_{123} is the rate of triple coincidences between detectors D1, D2 and D3, N_{12} is the rate of double coincidences between D1 and D2, and N_{13} is the rate of double coincidences between D1 and D3, and N, is the rate of singles detections by D1 alone. From the Cauchy-Schwartz inequality, we show in Appendix A that the inequality $a \geq 1$ holds in general for classical (perhaps stochastic) light. The equality $a = 1$ holds for coherent states $|\alpha_s, \alpha_i >$, independent of their amplitudes α_s, α_i. Since in our experiment the amplitude fluctuations in the double coincidence pulses led to a triple coincidence detection efficiency η less than unity, we should reduce the expected value of a accordingly. The modified classical inequality is $a \geq \eta$. We calibrated our triple coincidence counting system by replacing the two-photon light source by an attenuated light bulb, and measured $\eta = 0.70 \pm 0.07$. During the data run of Fig. 4 (lower trace), we measured values of a shown at the vertical arrows. The average value of a is 0.08 ± 0.04, which violates by more than thirteen standard deviations the predictions of any classical wave theory. It is therefore incorrect to interpret these results in terms of a stochastic ensemble of classical waves, in a semiclassical theory of photoelectric detection.[14] Classical waves with conjugate, but random, frequencies could conceivably yield the observed interference pattern, but they would also yield many more triple coincidences than were observed.

In conclusion, we have observed Berry's phase for photons in essentially n=1 Fock states, in a way which excludes with very high probability any possible classical explanation. These results can be explained in terms of the nonlocal collapse of the wavefunction. We conclude that this and other optical phases originate fundamentally at the quantum level, but under special circumstances, can survive the correspondence principle limit onto the classical level.[4a]

ACKNOWLEDGEMENTS

This work was supported by ONR under grant N00014-90-J-1259. We thank Y. Aharonov for helpful discussions, J. F. Clauser, E. D. Commins and H. Nathel for the loan of electronics, and R. Tyroler for the loan of the Michelson.

APPENDIX A: Derivation of an inequality for classical light

In this appendix we show that the anticorrelation parameter defined in Eq. (4), satisfies the inequality $a \geq 1$ for any classical field (i.e., any field generated by a classical current source, perhaps stochastic). From Eq. (4) this inequality can be written,

$$N_{123}N_1 \geq N_{12}N_{13}. \qquad (A1)$$

For simplicity, consider here a two-mode field representing the signal and idler photons, although all the arguments to be presented can easily be extended to the more realistic multimode case. In terms of creation/annihilation operators for these modes, the counting rates are given by the normally ordered expectation values

$$N_1 = \eta_1 < a_i^\dagger a_i >, \qquad (A2a)$$

$$N_{12} = \eta_1\eta_2 < a_s^\dagger a_i^\dagger a_s a_i >, \qquad (A2b)$$

$$N_{13} = \eta_1\eta_3 < a_s^\dagger a_i^\dagger a_s a_i >, \qquad (A2c)$$

$$N_{123} = \eta_1\eta_2\eta_3 < (a_s^\dagger)^2 a_i^\dagger (a_s)^2 a_i >, \qquad (A2d)$$

where η_1, η_2, η_3 are the net detection efficiencies of detectors D1, D2, and D3 respectively (including reflectivity of beam splitter B2), and angle brackets denote a trace with the density operator, $< O > \equiv Tr(\rho O)$. Since detectors D1 and D2 are placed symmetrically at the output ports of beam splitter B2, and one input port is unused, counting rates in these detectors can be described by the same number operator $a_s^\dagger a_s$. The density operator for a classical stochastic field can always be written in terms of the Glauber-Sudarshan diagonal P-representation,

$$\rho = \int d\alpha_s d\alpha_i \; P(\alpha_s, \alpha_i)|\alpha_s, \alpha_i >< \alpha_s, \alpha_i|, \qquad (A3)$$

where P is a positive, nonsingular function. Given such a state the counting rates of Eq. (A2) can be expressed as

$$N_1 = \eta_1 \overline{|\alpha_i|^2}, \qquad (A4a)$$

$$N_{12} = \eta_1\eta_2 \overline{|\alpha_s|^2|\alpha_i|^2}, \qquad (A4b)$$

$$N_{13} = \eta_1\eta_3 \overline{|\alpha_s|^2|\alpha_i|^2}, \qquad (A4c)$$

$$N_{123} = \eta_1\eta_2\eta_3 \overline{|\alpha_s|^4|\alpha_i|^2}, \qquad (A4d)$$

with the bar denotes the c-number average,

$$\overline{f(\alpha_s, \alpha_i)} \equiv \int d\alpha_s d\alpha_i P(\alpha_s, \alpha_i) f(\alpha_s, \alpha_i). \qquad (A5)$$

with f and P positive functions.

In order to prove Eq. (A1), we make use of the following Cauchy-Schwartz inequality. Given any two real functions f and g, and a *positive* real probability function P, these must satisfy,

$$(\bar{f})^2(\bar{g})^2 \geq (\overline{fg})^2. \qquad (A6)$$

The equality will hold whenever the probability function P is proportional to a Dirac delta-function, or whenever the function f and g are proportional to one another (given an arbitrary P). The former will be true when the field is a pure coherent state. Note that Eq. (A6) is valid only for positive functions P. Therefore in application to problems where P represents the Glauber-Sudarshan P-function, Eq. (A6) pertains only to classical fields.

Applying this to the problem at hand, choose

$$f(\alpha_s, \alpha_i) = \sqrt{\eta_1\eta_2\eta_3}|\alpha_s|^2|\alpha_i|, \qquad (A7a)$$

$$g(\alpha_s, \alpha_i) = \sqrt{\eta_1}|\alpha_i|. \qquad (A7b)$$

Since these real positive functions application of Eq. (A6) is valid and yields,

$$(\overline{|\alpha_s|^4|\alpha_i|^2})(\overline{|\alpha_i|^2}) \geq (\overline{|\alpha_s|^2|\alpha_i|^2})^2. \qquad (A8)$$

From Eqs. (A4), this is equivalent to Eq. (A1), the desired result.

REFERENCES

1. M. V. Berry, Proc. Roy. Soc. (London) **A392**, 45 (1984). For a review of optical manifestations of Berry's phases, see R. Y. Chiao, *Proc 3rd International Symposium on the Foundations of Quantum Mechanics*, Tokyo, 1989, p. 80.
2. G. S. Agarwal and R. Simon, "Berry Phase, Interference of Light Beams, and Hannay Angle" (Preprint).
3. J. H. Hannay, *J. Phys.* **A18**, 221 (1985).
4. R. Y. Chiao and Y. S. Wu, *Phys. Rev. Lett.* **57**, 933 (1986); A. Tomita and R. Y. Chiao, *ibid.* **57**, 937 (1986) and **59**, 1789 (1987); R. Y. Chiao, A. Antaramian, K. M. Ganga, H. Jiao, S. R. Wilkinson, and H. Nathel, *Phys. Rev. Lett.* **60**, 1214 (1988); W. R. Tompkin, M. S. Malcuit, R. W. Boyd, and R. Y. Chiao, *J. Opt. Soc. Am.* **B7**, 230 (1990); M. Kitano and T. Yabuzaki, *Phys. Lett.* **A142**, 321 (1989).
5. S. Pancharatnam, *Proc. Ind. Acad. Sci.* **A44**, 247 (1956); R. Bhandari and J. Samuel, *Phys. Rev. Lett.* **60**, 1210 (1988); T. H. Chyba, L J. Wang, L. Mandel, and R. Simon, *Opt. Lett.* **13**, 562 (1988); R. Simon, H. J. Kimble, and E. C. G. Sudarshan, *Phys. Rev. Lett.* **61**, 19 (1988); R. Bhandari, *Phys. Lett.* **A133**, 1 (1988).

6. I. H. Deutsch and J. C. Garrison, *Phys. Rev.* **A43**, 2498 (1991).
7. P. Grangier, G. Roger, and A. Aspect, *Europhys. Lett.* **1**, 173 (1986).
8. P. G. Dwiat, W. A. Vareka, C. K. Hong, H. Nathel, and R. Y. Chiao, *Phys. Rev.* **A41**, 2910 (1990), and references therein.
9. P. G. Kwiat and R. Y. Chiao, *Phys. Rev. Lett.* **66**, 88 (1991), and references therein.
10. R. Y. Chiao, P. G. Kwiat, and A. M. Steinberg, "The Energy-Time Uncertainty Principle and the EPR Paradox: Experiments Involving Correlated Two-Photon Emission in Parametric Down-Conversion" (to be published in *the Proceedings of the Workshop on Squeezed States and Uncertainty Relations*, University of Maryland, College Park, MD, March 29, 1991).
11. The slightly nonsinusoidal component in Fig. 4 (lower trace) can be explained by a slight wedge in Q2, in conjunction with the fact the signal beam was incident on Q2 off center.
12. J. G. Rarity and P. R. Tapster, *Phys. Rev. Lett.* **64**, 2495 (1990), and references therein.
13. C. K. Hong and L. Mandel, *Phys. Rev. Lett.* **56**, 58 (1986).
14. J. F. Clauser, *Phys. Rev.* **D9**, 853 (1974).

INVARIANTS AND SCHRÖDINGER UNCERTAINTY RELATION FOR NONCLASSICAL LIGHT

V. I. Man'ko

Levedev Physics Institute
117333 Moscow, Leninsky pr., 53

INTRODUCTION

The aim of this work is to discuss integrals of the motion and uncertainty relations and to obtain the distribution function of photons in squeezed and correlated light for the multimode case. The distribution function of photons in squeezed light for one mode field was discussed by Schleich and Wheeler,[1] by Agarwal and Adam,[2] and by Chaturvedi and Srinivasan.[3] The photon distribution function for squeezed and correlated light[4,5] was discussed by Dodonov, Klimov and Man'ko.[6] This distribution function depends not only on squeezing parameters, but also on correlation parameters connected with the Schrödinger uncertainty relation[7]

$$\delta q \delta p \geq \frac{\hbar}{2} \frac{1}{\sqrt{1-r^2}},$$

where the parameter r is the correlation coefficient of the position and momentum

$$r = (\delta q \delta p)^{-1} \left\{ \frac{1}{2} \langle \hat{q}\hat{p} + \hat{p}\hat{q} \rangle - \langle \hat{q} \rangle \langle \hat{p} \rangle \right\}.$$

The states with nonzero parameter r we call the correlated states. In the next section we will consider the problem of how to find the states which minimize the Schrödinger uncertainty relation. For such states, instead of the Schrödinger inequality, we have the equality

$$\delta q \delta p = \frac{\hbar}{2} \frac{1}{\sqrt{1-r^2}}.$$

These states describe squeezed and correlated light. We will demonstrate in the next section how these states are naturally created for quantum parametric oscillators.

The case of the photon distribution function for the two-mode squeezed light was considered by Caves, Zhu, Milburn and Schleich.[8] The multidimensional generalization of the expression for the distribution of photons in squeezed light in terms of Hermite polynomials of several variables may be formulated to derive this expression on the basis of the results[5,9,10] for a nonstationary parametric multidimensional oscillator.

Recent Developments in Quantum Optics, Edited
by R. Inguva, Plenum Press, New York, 1993

To obtain the photon distribution function we will consider the nonstationary multidimensional oscillator using the integral of motion method. The linear integral of motion for one-dimensional parametric oscillator was found in Ref. 11. We will apply the linear integrals of motion for multidimensional parametric oscillator[10] to construct wave functions of squeezed and correlated states and to expand these functions in terms of the Fock states. The coefficients of these expansions give the expression determining the photon distribution function under consideration. In fact, these coefficients have been calculated in terms of Hermite polynomials of several variables.[5,10] Our aim is to apply these expressions for describing the photon distribution function and to discuss the wavy behavior of this distribution function.

ONE MODE PHOTON DISTRIBUTION FUNCTION

Let us discuss the nonstationary oscillator system with the Hamiltonian

$$\hat{H} = \frac{\Omega^2(t)+1}{4}(\hat{a}\hat{a}^\dagger + \hat{a}^\dagger\hat{a}) + \frac{\Omega^2(t)-1}{4}(\hat{a}^2 + \hat{a}^{\dagger 2}). \tag{1}$$

Here the annihilation and creation operators $\hat{a}$ and $\hat{a}^\dagger$ obey the boson commutation relations

$$[\hat{a}, \hat{a}^\dagger] = 1. \tag{2}$$

The dimensionless frequency $\Omega(t)$ has arbitrary dependence on time and is chosen to satisfy the initial condition $\Omega(0) = 1$. In terms of coordinate $\hat{Q}$ and momentum $\hat{P}$ the annihilation operator $\hat{a}$ is expressed as follows

$$\hat{a} = \frac{1}{\sqrt{2}}(\hat{Q} + i\hat{P}). \tag{3}$$

The Hamiltonian (1) may be rewritten in the matrix form if we will introduce the two-vectors

$$\hat{\vec{q}} = \begin{pmatrix} a \\ a^\dagger \end{pmatrix}, \quad \hat{\vec{q}}^{\,T} = (\hat{a}, \hat{a}^\dagger), \tag{4}$$

and the 2×2-matrix

$$B(t) = \frac{1}{2}\left\| \begin{matrix} \Omega^2(t)-1, & \Omega^2(t)+1 \\ \Omega^2(t)+1, & \Omega^2(t)-1 \end{matrix} \right\|. \tag{5}$$

Then we have

$$\hat{H} = \frac{1}{2}\hat{\vec{q}}^{\,T} B(t)\hat{\vec{q}}. \tag{6}$$

In coordinate representation the Hamiltonian (1) takes the form of the Hamiltonian for the oscillator with time-dependent frequency

$$\hat{H} = \frac{\hat{p}^2}{2} + \frac{\Omega^2(t)\hat{Q}^2}{2}. \tag{7}$$

We will demonstrate now how the evolution of the parametric oscillator with varying frequency produces the squeezed and correlated photon mode from the initially coherent state. To show this let us use the linear integral of the motion constructed in Ref. 11

$$\hat{A}(t) = \frac{i}{\sqrt{2}}[\epsilon(t)\hat{P} - \dot{\epsilon}(t)\hat{Q}]. \tag{8}$$

This operator is the integral of the motion if the complex c-number function $\epsilon(t)$ satisfies the classical equation of motion

$$\ddot{\epsilon}(t) + \Omega^2(t)\epsilon = 0. \tag{9}$$

If the initial conditions for this function are chosen to be

$$\epsilon(0) = 1, \quad \dot{\epsilon}(0) = i1, \tag{10}$$

the integrals of motion $\hat{A}(t)$, $\hat{A}^\dagger(t)$ obey the boson commutation relation

$$[\hat{A}(t), \hat{A}^\dagger(t)] = 1. \tag{11}$$

Due to this the complete set of solutions of the Schrödinger equation

$$i\frac{\partial\psi(Q,t)}{\partial t} = -\frac{1}{2}\frac{\partial^2\psi(Q,t)}{\partial Q^2} + \frac{1}{2}\Omega^2(t)Q^2\psi(Q,t), \tag{12}$$

may be constructed using the *ground* state wave function $\psi_0(Q,t)$ which obeys to the Schrödinger equation and to the equation

$$\hat{A}(t)\psi_0(Q,t) = 0. \tag{13}$$

This normalized function has the form

$$\psi_0(Q,t) = \pi^{-1/4}\epsilon^{-1/2}\exp\left\{\frac{i\dot{\epsilon}Q^2}{2\epsilon(t)}\right\}. \tag{14}$$

Acting on this function by displacement operator $D(\alpha) = \exp[\alpha\hat{A}^\dagger(t) - \alpha^*\hat{A}(t)]$ which is integral of motion we obtain the coherent state wave function

$$\psi_\alpha(Q,t) = \psi_0(Q,t)\exp\left\{-\frac{|\alpha|^2}{2} + \frac{\sqrt{2}\alpha Q}{\epsilon(t)} - \frac{\alpha^2}{2}\frac{\epsilon^*(t)}{\epsilon(t)}\right\}, \tag{15}$$

which satisfies the Schrödinger equation (12) and the condition

$$\hat{A}(t)\psi_\alpha(Q,t) = \alpha\psi_\alpha(Q,t). \tag{16}$$

Expanding the analytic function $\exp(|\alpha|^2/2)\psi_\alpha(Q,t)$ into series with respect to complex number α we obtain the Fock state wave functions

$$\Psi_n(Q,t) = \psi_0(Q,t)\left(\frac{\epsilon^*}{2\epsilon}\right)^{n/2}(n!)^{-1/2}H_n\left(\frac{Q}{|\epsilon|}\right). \tag{17}$$

The phase ϕ_ϵ of the classical trajectory $\epsilon(t)$ determines the geometrical phase (Berry phase). This phase is $\phi = -\left(n+\frac{1}{2}\right)\phi_\epsilon$. If we calculate for all the coherent states $\psi_\alpha(Q,t)$ (15) the variances $(\delta Q)^2$ and $(\delta P)^2$ we obtain

$$(\delta Q)^2 = \frac{|\epsilon(t)|^2}{2}, \tag{18}$$

$$(\delta P)^2 = \frac{|\dot{\epsilon}(t)|^2}{2}. \tag{19}$$

The correlation coefficient is determined by the complex classical trajectory of the oscillator $\epsilon(t)$ and is given by the formula

$$r = |\epsilon\dot{\epsilon}|^{-1}(|\epsilon\dot{\epsilon}|^2 - 1)^{1/2}. \tag{20}$$

Thus, the constructed set of coherent states (15) satisfies the condition of minimizing the Schrödinger uncertainty relation. It is worthy to point out that the second momenta (18)-(20) do not depend on the quantum number α labeling coherent states and are the same as for the *ground* state $\psi_0(Q,t)$ (14).

From the point of view of the eigenvalue equation (16), the constructed states are the coherent states. So, the scalar product of these states is given by the formula

$$\langle\alpha|\beta\rangle = \exp\left\{-\frac{|\alpha|^2}{2} - \frac{|\beta|^2}{2} + \alpha^*\beta\right\}, \tag{21}$$

and we have the usual completeness condition

$$\frac{1}{\pi}\int d^2\alpha|\alpha\rangle\langle\alpha| = \hat{1}. \tag{22}$$

These properties are connected only with commutation relations of Heisenberg-Weyl group generators realized by the integrals of motion $\hat{A}(t)$ and $\hat{A}^\dagger(t)$ (11). At the same time from the point of view of physical properties the states (15) are the squeezed and correlated states. Squeezing is characterized by the ratio of the variances of the coordinate and momentum and the variances in the usual ground state

$$k_q = \frac{(\delta Q)^2}{(1/2)} = |\epsilon|^2, \quad k_p = (\delta P)^2/(1/2) = |\dot{\epsilon}|^2. \tag{23}$$

Thus we conclude that if at the initial moment $t = 0$ we have the usual coherent state of the oscillator with $r = 0$, $\delta Q^2 = \delta P^2 = \frac{1}{2}$ due to varying frequency this state evolves and becomes the squeezed and correlated state with the wave function (15).

In order to calculate the photon distribution function in the squeezed and correlated light for one-mode case which is described by the function (15) we have to evaluate the overlap integral of the initial Fock state $|n, t = 0\rangle$ (17) with the squeezed state $|\beta, t\rangle$ (15)

$$c_{n\beta}(t) = \langle n, t = 0|\beta, t\rangle. \tag{24}$$

We can calculate this integral considering the propagator

$$G(\alpha^*, \beta, t) = \langle\alpha, t = 0|\beta, t\rangle \exp\left(\frac{|\alpha|^2}{2} + \frac{|\beta|^2}{2}\right)$$

since the propagator (Green function in the coherent state representation) determines the generating function for the coefficients (24). The propagator satisfies the system of equations (see Ref. 5)

$$\begin{aligned}\hat{A}_{\alpha^*}(t)G(\alpha^*, \beta, t) &= \beta G(\alpha^*, \beta, t),\\ \hat{A}^\dagger_{\alpha^*}(t)G(\alpha^*, \beta, t) &= \frac{\partial}{\partial\beta}G(\alpha^*, \beta, t).\end{aligned} \tag{25}$$

Solving this system of equation we have for the propagator $G(\alpha^*, \beta, t)$ the expression

$$G(\alpha^*, \beta, t) = \frac{1}{\xi^{1/2}} \exp\left\{-\frac{1}{2}\frac{\alpha^{*2}\eta}{\xi} + \frac{\alpha^*\beta}{\xi} + \frac{\eta^*}{2\xi}\beta^2\right\}. \tag{26}$$

The complex parameters $\xi(t)$ and $\eta(t)$ are defined as follows

$$\xi(t) = \frac{1}{2}[\epsilon(t) - i\dot{\epsilon}(t)], \tag{27}$$

$$\eta(t) = -\frac{1}{2}[\epsilon(t) + i\dot{\epsilon}(t)]. \tag{28}$$

Comparing this function with the generating function for Hermite polynomials

$$\exp[2tx - x^2] = \sum_{n=0}^{\infty} \frac{H_n(t)}{n!} x^n, \tag{29}$$

we have

$$C_{n\beta}(t) = \frac{(n!)^{-1/2}}{\xi^{1/2}} \exp\left[\frac{\eta^*\beta^2}{2\xi} - \frac{|\beta|^2}{2}\right]\left(\frac{\eta}{\xi}\right)^n \frac{1}{2^{n/2}} \cdot H_n\left(\frac{\beta}{\sqrt{2\eta}}\right). \tag{30}$$

The photon distribution function $W_n(\beta)$ for the squeezed and correlated light described by the wavefunction $\psi_\beta(Q,t)$ (15) has the form

$$W_n(\beta) = |C_{n\beta}(t)|^2 = \frac{2^{-n}}{n!}\left|\frac{\eta}{\xi}\right|^{2n} \frac{1}{|\xi|} \exp(-|\beta|^2)$$
$$\times \exp\left[\mathrm{Re}\left(\frac{\eta^*\beta^2}{\xi}\right)\right]\left|H_n\left(\frac{\beta}{\sqrt{2\eta}}\right)\right|^2. \tag{31}$$

In the case $\eta = 0$, $\xi = 1$ or for $t = 0$ this function is equivalent to Poisson distribution function

$$W_n^0(\beta) = \{\exp(-|\beta|^2)\}\frac{|\beta|^{2n}}{n!}. \tag{32}$$

Since the Hermite polynomial has the wavy behavior, the photon distribution function (31) for squeezed and correlated light has also the wavy behavior.[6] The parameters $\xi(t)$ and $\eta(t)$ determined by the classical trajectory according to formula (27) and (28) coincide with the parameters for the symplectic group $Sp(2,R)$, since the formula for the integrals of motion $\hat{A}(t)(\hat{A}^\dagger(t))$ (8) may be rewritten in the form

$$\begin{pmatrix}\hat{A}(t)\\ \hat{A}^\dagger(t)\end{pmatrix} = \begin{pmatrix}\xi(t) & \eta(t)\\ \eta^*(t) & \xi^*(t)\end{pmatrix}\begin{pmatrix}\hat{a}\\ \hat{a}^\dagger\end{pmatrix}. \tag{33}$$

The linear transform (33) is the canonical transform conserving the boson commutation relation. It means that

$$|\xi(t)|^2 - |\eta(t)|^2 = 1. \tag{34}$$

The matrix of the transform (33) belongs to the group $SU(1,1)$. The real matrix

$$\Lambda(t) = \begin{pmatrix}\lambda_1(t) & \lambda_2(t)\\ \lambda_3(t) & \lambda_4(t)\end{pmatrix}, \tag{35}$$

where

$$\lambda_1 = Re\epsilon(t),\ \lambda_2(t) = -\mathrm{Re}\dot{\epsilon},\ \lambda_3 = -\mathrm{Im}\epsilon(t),\ \lambda_4(t) = \mathrm{Im}\dot{\epsilon}(t), \tag{36}$$

belongs to the group $Sp(2,R)$ or $SL(2,R)$. This matrix determines linear integrals of motion $\hat{P}_0(t)$ and $\hat{Q}_0(t)$ which coincide at the initial moment $t = 0$ with the position $\hat{P}$ and momentum $\hat{Q}$ operators

$$\begin{pmatrix} \hat{P}_0(t) \\ \hat{Q}_0(t) \end{pmatrix} = \Lambda(t) \begin{pmatrix} \hat{P} \\ \hat{Q} \end{pmatrix}. \tag{37}$$

Thus, we see that the correlated and squeezed state (15) may be expressed as the Gaussian function and the dependence of this function on time is completely given in terms of the dependence of this function on the symplectic group $Sp(2,R)$ parameters $\xi(t)$, $\eta(t)$ or $\lambda_i(t)$, $(i = 1,2,3,4)$ which themselves depend on the time. Thus the squeezed state photon distribution function (31) is also dependent on the symplectic group parameters.

We will write down now the propagator of the considered model in Wigner representation. It is given by the following formula[5]

$$G(p,q,t) = 2\{\det[\Lambda(t) + 1]\}^{-1/2} \exp\{i\vec{q}^{T} A(t)\vec{q}\}, \tag{38}$$

where $\vec{q}^{T} = (p,q)$ and the 2×2-matrix $A(t)$ is defined as follows

$$A(t) = \frac{1}{2 + \mathrm{Re}\epsilon(t) + \mathrm{Im}\dot{\epsilon}(t)} \begin{pmatrix} -2\mathrm{Im}\epsilon(t) & \mathrm{Im}\dot{\epsilon}(t) - \mathrm{Re}\epsilon(t) \\ \mathrm{Re}\epsilon(t) - \mathrm{Im}\dot{\epsilon}(t) & 2\mathrm{Re}\dot{\epsilon}(t) \end{pmatrix}. \tag{39}$$

The matrix $A(t)$ may be expressed in terms of symplectic matrix $\Lambda(t)$ as

$$A(t) = \Sigma(\Lambda(t) + 1)^{-1}(\Lambda(t) - 1), \tag{40}$$

where

$$\Sigma = \begin{pmatrix} 0 & 1 \\ -1 & 0 \end{pmatrix}. \tag{41}$$

Parametric excitation of the light mode produces the squeezed and correlated light from the initially coherent light. This phenomenon is connected with another effect. If, at the initial moment, $t = 0$ we had n-photon state $|n,0\rangle$, then at the moment t there exists the probability to excite m photons. This probability may be expressed as follows[5]

$$W_n^m(t) = \left(E + \frac{1}{2}\right)^{-1/2} \frac{n!}{m!} \left(P_{(m+n)/2}^{(m-n)/2}\left(E + \frac{1}{2}\right)\right)^2, m \geq n, \tag{42}$$

where

$$E = \frac{|\epsilon|^2 + |\dot{\epsilon}|^2}{4} = \frac{\lambda_1^2 + \lambda_2^2 + \lambda_3^2 + \lambda_4^2}{4}. \tag{43}$$

Thus for a given symplectic matrix $\Lambda(t)$ the squeezing phenomenon is connected with the excitation of photons which is determined by the same matrix $\Lambda(t)$ as the squeezing and correlation parameters. The variance $(\delta Q)^2$ may be rewritten in terms of symplectic matrix $\Lambda(t)$ as follows

$$(\delta Q)^2 = \frac{1}{2}[\lambda_1^2(t) + \lambda_3^2(t)]. \tag{44}$$

The variance $(\delta P)^2$ is expressed as follows

$$(\delta P)^2 = \frac{1}{2}[\lambda_2^2(t) + \lambda_4^2(t)]. \tag{45}$$

Thus we have obtained very simple connection of symplectic matrix $\Lambda(t)$ parameters with squeezing parameters k_q and k_p. We have

$$k_q = \lambda_1^2 + \lambda_3^2, \quad k_p^2 = \lambda_2^2 + \lambda_4^2.$$

The correlation coefficient is also expressed in terms of $\Lambda(t)$-matrix parameters

$$r = [(\lambda_1^2 + \lambda_3^2)(\lambda_2^2 + \lambda_4^2)]^{-1/2}[(\lambda_1^2 + \lambda_3^2)(\lambda_2^2 + \lambda_4^2) - 1]^{1/2}. \tag{46}$$

Thus we have $r = (k_p k_q)^{-1/2}(k_p k_q - 1)^{1/2}$.

Let us now discuss briefly what will be changed in our results if the mode is described by the Hamiltonian (1) with extra linear terms

$$\hat{H} = \frac{1}{2}\hat{\vec{q}}^T B(t)\hat{\vec{q}} + f(t)\hat{a} + f^*(t)\hat{a}^\dagger. \tag{47}$$

The linear integral of motion for this system will have extra c-number term, but all the parameters like variances and correlation coefficient do not depend on the time-dependent function $f(t)$. They are given by the same formula (44)-(46) with the same matrix $\Lambda(t)$ as in the case of parametric oscillators without any external force.

POLYMODE PHOTON DISTRIBUTION FUNCTION

We have studied in the previous section the one-mode parametric oscillator. The scheme of the consideration had the following points. First we have found linear integrals of motion with boson commutation relations. Then the ground-like state has been constructed as the solution to the equation which expresses the action of the annihilation operator onto the wave function of this state. The *ground* state wave function turned out to be a Gaussian wave packet with squeezing and correlation of position and momentum. The squeezing and correlation depend on time and this dependence is completely determined by the complex classical trajectory of the parametric oscillator. Acting onto ground state wave function by the displacement operator we have obtained the coherent state wave function. This state being coherent state from the point of view that it is eigenstate of the boson annihilation operator – linear integral of motion turned out to be squeezed and correlated state as well as the ground state. The squeezing and correlation parameters do not depend on the label of the coherent state. Since the coherent state wave function is the Gaussian wave packet, the Fock states are easily calculated in terms of Hermite polynomials of one variable by expanding this Gaussian function into a series and comparing this expansion with the generating function for Hermite polynomials. The important point is the calculation of the propagator in the coherent state representation. This propagator satisfies the system of equation determined by the linear integrals of motion.

The solution to this system is the Gaussian function dependent on the parameters of the initial and final coherent states. Expanding this function with respect to one of this parameters, we obtain the transition amplitude from the Fock state to the coherent state. Modulus squared of this amplitude gives the transition probability from the Fock to the coherent state. At the same time this transition probability coincides with the photon distribution function for the one-mode squeezed and correlated light. Thus, the known result[5,10] for the transition probabilities of the parametric oscillator may be reinterpreted as the calculation of the photon distribution function for the squeezed and correlated light.

We will follow this scheme for the case of the polymode light. The generalization of the Hamiltonians (1) and (6) in the polymode case looks as follows

$$\hat{H} = \frac{1}{2}\hat{\vec{q}}^{T} B(t)\hat{\vec{q}}, \quad (\hbar = 1), \tag{48}$$

where

$$\hat{\vec{q}}^{T} = (\hat{a}_1, \hat{a}_2, \ldots a_N, \hat{a}_1^{\dagger}, \hat{a}_2^{\dagger}, \ldots, \hat{a}_N^{\dagger}), \tag{49}$$

and $2N \times 2N$ matrix $B(t)$ has the block structure

$$B(t) = \begin{pmatrix} b_1(t) & b_2(t) \\ b_3(t) & b_4(t) \end{pmatrix}, \tag{50}$$

with four time-dependent $N \times N$ matrices $b_i(t)$, $i = 1,2,3,4$. Following to the introduced scheme we will find the $2N$ linear integrals of motion – $2N$ nonhermitian operators which are the components of $2N$-vector

$$\hat{\vec{I}}^{T}(t) = (\hat{A}_1(t), \hat{A}_2(t), \ldots \hat{A}_N(t), \hat{A}_1^{\dagger}(t), \hat{A}_2^{\dagger}(t), \ldots \hat{A}_N^{\dagger}(t)). \tag{51}$$

This vector is connected with $2N$-vector $\hat{\vec{q}}$ by the relation

$$\hat{\vec{I}}(t) = M(t)\hat{\vec{q}}. \tag{52}$$

The $2N \times 2N$-matrix $M(t)$ obeys the equation

$$-i\dot{M}(t) = M(t)\Sigma B(t), \tag{53}$$

where the $2N \times 2N$-matrix Σ has the form

$$\Sigma = \begin{pmatrix} 0 & 1_N \\ -1_N & 0 \end{pmatrix}. \tag{54}$$

The $N \times N$-matrix 1_N is the unity matrix. The matrix $M(t)$ consists of four time-dependent $N \times N$-matrices

$$M(t) = \begin{pmatrix} \xi(t) & \eta(t) \\ \eta^*(t) & \xi^*(t) \end{pmatrix}. \tag{55}$$

The initial conditions for the matrix $M(t)$ are chosen as follows

$$\xi(0) = 1_N, \quad \eta(0) = 0. \tag{56}$$

In that case the commutation relations for the components of the integral of motion vector $\vec{I}(t)$ are the boson commutation relations

$$[\hat{A}_i(t), \ \hat{A}_k^{\dagger}(t)] = \delta_{ik}, \tag{57}$$

$$[\hat{A}_i(t),\ \hat{A}_k(t)] = 0, \tag{58}$$

Now we can find the *ground* state wave function as the solution to the equation (in the coordinate representation)

$$\hat{\vec{A}}(t)\psi_0(\vec{Q},t) = 0. \tag{59}$$

We have[5] the normalized function

$$\psi_0(\vec{Q},t) = (2\pi)^{-N/4}(\det\lambda_p)^{-1/2}\exp\left\{-\frac{1}{2}\vec{Q}^T\lambda_p^{-1}\lambda_q\vec{Q}\right\}, \tag{60}$$

where the $N\times N$-matrices λ_p and λ_q are expressed in terms of matrices $\xi(t)$ and $\eta(t)$

$$\lambda_p = \frac{i}{\sqrt{2}}[\xi(t)-\eta(t)],\quad \lambda_q = \frac{1}{\sqrt{2}}[\xi(t)+\eta(t)]. \tag{61}$$

The coherent state wave function satisfying the eigenvalue equation

$$\hat{\vec{A}}(t)\psi_{\vec{\alpha}}(\vec{Q},t) = \vec{\alpha}\psi_{\vec{\alpha}}(\vec{Q},t), \tag{62}$$

has the form

$$\psi_{\vec{\alpha}}(\vec{Q},t) = \psi_0(\vec{Q},t)\exp\left\{i\vec{Q}^T\lambda_p^{-1}\vec{\alpha} - \frac{1}{2}|\vec{\alpha}|^2 + \frac{1}{2}\vec{\alpha}^T\lambda_P^*\lambda_P^{-1}\vec{\alpha}\right\}. \tag{63}$$

If we calculate for the state $\psi_{\vec{\alpha}}$ (63) the variance matrix for the observables $\hat{\vec{q}}$ it may be expressed in terms of the initial variance matrix $\sigma(0)$ and all the time-dependence is determined by the classical trajectory of the parametric oscillator $M(t)$ (55)

$$\sigma(t) = \Sigma M^T(t)\Sigma\sigma(0)\Sigma M(t)\Sigma. \tag{64}$$

The matrix Σ is given by the formula (54). For initially coherent state the matrix $\sigma_{\vec{\alpha}}(0)$ has the form

$$\sigma_{\vec{\alpha}}(0) = \frac{1}{2}\begin{pmatrix} 0 & 1_N \\ 1_N & 0 \end{pmatrix}. \tag{65}$$

Thus for the polymode case the variance matrix does not depend on the parameter $\vec{\alpha}$ and is the same as for *ground* state $\Psi_0(Q,t)$. The extra linear terms in Hamiltonian of the type

$$\hat{H}_e = \vec{f}(t)\hat{\vec{a}}^\dagger + \vec{f}^*(t)\hat{\vec{a}}, \tag{66}$$

change the form of the linear integrals of motion (51) but do not influence the squeezing and correlation parameters which are given by the same variance matrix $\sigma(t)$ (64). The matrix $M(t)$ is conjugate to the matrix which belongs to the symplectic group $Sp(2N,R)$. Thus for the polymode squeezed light the statistical properties may be parameterized by the symplectic group matrix elements. As well as in one-mode case the state $\psi_{\vec{\alpha}}(Q,t)$ is the coherent state from the point of view that it is an eigenstate of the boson annihilation operator $\hat{\vec{A}}(t)$. But this state is squeezed and correlated state from the point of view of the usual coordinates and momenta (or their linear combinations $\hat{\vec{a}}$ and $\hat{\vec{a}}^\dagger$).

Now we will calculate the transition amplitude between these squeezed and correlated states. According to Ref. 5 and 9, this amplitude $G(\vec{\alpha}^*, \vec{\beta}, t)$ (the propagator in the coherent states representation) obeys the system of $2N$ equations

$$\left[\xi(t)\frac{\partial}{\partial\vec{\alpha}^*} + \eta(t)\vec{\alpha}^*\right]G(\vec{\alpha}^*, \vec{\beta}, t) = \vec{\beta}G(\vec{\alpha}^*, \vec{\beta}, t),$$
$$\left[\eta^*(t)\frac{\partial}{\partial\vec{\alpha}^*} + \xi^*(t)\vec{\alpha}^*\right]G(\vec{\alpha}^*, \vec{\beta}, t) = \frac{\partial}{\partial\vec{\beta}}G(\vec{\alpha}^*, \vec{\beta}, t). \tag{67}$$

The solution to the system of equation gives the propagator in the coherent states representation for the polymode case

$$G(\vec{\alpha}^*, \vec{\beta}, t) = [\det \xi(t)]^{-1/2} \exp\left[-\frac{1}{2}\vec{\alpha}^{*T}\xi^{-1}\eta\vec{\alpha}^* + \vec{\alpha}^{*T}\xi^{-1}\vec{\beta} + \frac{1}{2}\vec{\beta}^T\eta^*\xi^{-1}\vec{\beta}\right]. \tag{68}$$

Since this propagator is the generating function for the transition amplitude between Fock and coherent state we have for the transition probability $W_{\vec{n}}(\vec{\alpha})$ between the Fock and squeezed state the expression

$$W_{n_1 n_2,\ldots n_N}(\alpha_1, \alpha_2, \ldots \alpha_N) = |\det \xi(t)|^{-1}\frac{e^{-|\vec{\alpha}|^2}}{n_1!n_2!\ldots n_N!}\left|H^{\{\xi^{-1}\eta\}}_{n_1,n_2,\ldots,n_N}(\eta^{-1}\vec{\alpha})\right|^2 \tag{69}$$

The transition probability is identical to the photon distribution function for the polymode squeezed and correlated light. It is expressed in terms of Hermite polynomials of several variables. The generating function for these polynomials has the Gaussian form

$$\exp\left\{-\frac{1}{2}\vec{a}^T R\vec{a} + \vec{a}^T R\vec{x}\right\} = \sum_{\vec{m}=0}^{\infty}\frac{a_1^{m_1}a_2^{m_2}\ldots a_N^{m_N}}{m_1!m_2!\ldots m_N!}H^{\{R\}}_{m_1 m_2,\ldots m_N}(\vec{x}) \tag{70}$$

For the case $\eta = 0$, $\xi(t) = 1_N$ the distribution function (67) is the product of one-dimensional Poisson distributions. Since the Hermite polynomials have wavy behavior the photon distribution function (69) has also the wavy behavior as well as for the one-mode light. For two-mode squeezed light the Hermite polynomials may be expressed in terms of other special functions. Thus for the case $N = 2$ and

$$R = \begin{pmatrix} 0 & 1 \\ 1 & 0 \end{pmatrix}, \tag{71}$$

we have[5]

$$H^{\{R\}}_{n_1 n_2}(x_1, x_2) = (-1)^{\mu_{n_1 n_2}}\mu_{n_1 n_2}!$$
$$\times\, x_1^{(n_1 - n_2 + |n_1 - n_2|)/2}x_2^{(n_2 - n_1 + |n_2 - n_1|)/2}$$
$$\times\, L^{|n_1 - n_2|}_{\mu_{n_1 n_2}}(x_1 x_2) \tag{72}$$
$$\mu_{n_1 n_1} = \min(n_1, n_2).$$

In this case the photon distribution function of two-mode squeezed light is described by Laguerre polynomials. This relation gives the photon distribution function for two-mode light (compare with Ref. 8). The existence of nontrivial geometrical phase and the phenomenon of squeezing and correlation are closely related. If there are squeezing and correlation there are nontrivial geometrical phase and vise-versa. This may be seen even in the one-mode case for which squeezing and correlation parameters (18) - (20) are determined by the modulus of classical trajectory complex functions $\epsilon(t)$ and $\dot{\epsilon}(t)$,, and the geometrical phase is determined by the phase of this function. It is worthy to point out that all the quantum properties as second momenta of photon distribution function, as well as the distribution function itself, are completely determined only by the property of the classical trajectory (in the one-mode case and in the polymode case, too). Thus, studying the macroscopic classical parametric oscillator system we can obtain all the quantum properties of the corresponding quantum-mechanical microscopic system. In the WKB-approximation, this property may be easily generalized for more complicated nonlinear parametric quantum systems because the correction to the propagator for these systems is reduced in the approximation to the propagator of nonstationary quadratic quantum systems.

ACKNOWLEDGMENTS

The author thanks Professors R. Glauber, W. Schleich and E. C. G. Sudarshan for their fruitful discussions. The author is deeply thankful to Miss J. Williams for help in preparing this manuscript.

REFERENCES

1. W. Schleich and J. Wheeler, Nature **326**, 574 (1987).
2. G. S. Agarwal and G. Adam, Phys. Rev. A **39**, 6259 (1989).
3. S. Chaturvedi and V. Srinivasan, Phys. Rev. A **40**, 6095 (1989).
4. V. V. Dodonov, E. V. Kurmyshev and V. I. Man'ko, Phys. lett. A **79**, 150 (1980).
5. V. V. Dodonov and V. I. Man'ko, in *Proceedings of Lebedev Physics Institute*, v. 183, 1989, Nauka Publ. Moscow (in Russian) ["Invariants and Evolution of Nonstationary Quantum Systems," ed. M. A. Markov, Nova Science Publ. N.Y.].
6. V. V. Dodonov, A. B. Klimov, and V. I. Man'ko, Phys. lett. A **149**, 225 (1990).
7. E. Schrödinger, Ber. Kgl. Akad. Wiss., **296**, (1930).
8. C. M. Caves, C. Zhu, G. J. Milburn, and W. Schleich, Phys. Rev. A **43**, 3854 (1991). After this talk about calculating the photon distribution function for squeezed polymode light Professor Schleich kindly informed me about the photon distribution function for squeezed two-mode light obtained in the quoted paper and expressed in terms of Luagerre polynomials. I am very thankful to Prof. Schleich for informing me about this result before the publication. The general polymode case f distribution function for squeezed photons, expressed in terms of Hermite polynomials of several variables may be reduced for two-mode light to the Luagerre polynomials but even for this case of two modes there are the special parameters of squeezing and correlation when the distribution function for photons is not expressed in terms of Luagerre polynomial only. The complete classification of the extra cases is determined by the classification of different types of Hermite polynomials of two variables given in Ref. 5.
9. V. V. Dodonov, I. A. Malkin and V. I. Man'ko, Intern. J. Theor. Phys. **14**, 37 (1975).
10. I. A. Malkin, V. I. Man'ko and D. A. Trifonov. J. Math. Phys. **14**, 576 (1973).
11. I. A. Malkin and V. I. Man'ko, Phys. Lett. A **32**, 243 (1970).

NOISE IN TWO MODE SQUEEZED STATES AND THERMO FIELD DYNAMICS

H. Umezawa

The Theoretical Physics Institute
University of Alberta
Edmonton, Alberta T6G 2J1 Canada

ABSTRACT

This report starts with a short resume of thermal squeezed coherent states formulated in terms of thermo field dynamics (TFD). This is followed by a study of two mode squeezed states which shows how the TFD mechanism creates noise in pure states. It is then discussed how this concept of noise in pure states applied to quantum field theory leads to a TFD formalism for thermal physics for closed systems.

1. INTRODUCTION

There are many kinds of noise in physical systems, e.g. quantum noise, thermal noise, etc. It has been thought that all of these noises except the quantum noise appear only in mixed states in the sense of quantum statistical mechanics. This situation changed when the thermo field dynamics (TFD), which is a real-time operator formalism for thermal effects in quantum field systems, showed that the thermal noise can be treated as a noise in pure states [1,2,3]. This is the TFD mechanism for creation of noise in pure states. This mechanism has been applied also to the Hawking radiation around the black hole [4]. This line of study provided an interesting idea about origin of heat in our world [5]. However, a very straightforward way of demonstrating noise in pure states created by the TFD mechanism has been provided by the squeezed states in quantum optics [6,7,8,9,10,11]. This development together with studies of thermal coherent states [12,8,13] naturally opened application of TFD to quantum optics. The noise in squeezed states can be interpreted in both ways. Noise in one part of the system can be obtained from coarse graining applied to the other part of the system. However the TFD mechanism provides a simple view of noise in pure states without any need of statistical average. This point will be touched in the next section.

This analysis of noise can be applied to both the quantum mechanics and the quantum field theory. However, when we consider the very particular and significant case, that is, the thermal effects in our world, the quantum field system is demanded. It has been shown that with infinite number of degrees of freedom a system of quantum fields exhibits a variety of phases depending on various forms of particle condensation

Recent Developments in Quantum Optics, Edited by R. Inguva, Plenum Press, New York, 1993

in vacuum. In this way, in a system of quantum fields both the microscopic and macroscopic objects can coexist; the properties associated with the vacuum with particle condensation are said to be macroscopic, while the quantum excited states are the states of microscopic particles. This line of thought can be extended to thermal effects in our world which can also be understood as the degrees of freedom associated with the vacuum with particle condensation. This will be discussed in section 3.

2. NOISE IN PURE STATE

Thermal Squeezed Coherent States

Consider two commuting sets of oscillator operators a and $\tilde{a}$:

$$[a, a^\dagger] = [\tilde{a}, \tilde{a}^\dagger] = 1. \tag{2.1}$$

When we consider the generators bilinear in these oscillator operators, we find the following transformation operators:

$$S_\lambda = exp[(\lambda a^{\dagger 2} - \lambda^* a^2)/2], \tag{2.2}$$

$$\tilde{S}_\eta = exp[(\eta^* \tilde{a}^{\dagger 2} - \eta \tilde{a}^2)/2], \tag{2.3}$$

$$U_\theta = exp[-(\theta a^\dagger \tilde{a}^\dagger - \theta^* \tilde{a} a)]. \tag{2.4}$$

The generators linear in oscillator operators give

$$D_\alpha = exp(\alpha a^\dagger - \alpha^* a), \tag{2.5}$$

$$\tilde{D}_\gamma = exp(\gamma^* \tilde{a}^\dagger - \gamma \tilde{a}). \tag{2.6}$$

Let $|0>>$ denote the vacuum for oscillator operators:

$$a|0>>= \tilde{a}|0>>= 0. \tag{2.7}$$

Then action of D_α or $\tilde{D}_\gamma$ on $|0>>$ creates the coherent states for $a-$ or $\tilde{a}$-operators, while S_λ or $\tilde{S}_\eta$ creates the squeezed states. Action of U_θ on $|0>>$ results in the two mode squeezed state. Thus U_θ is called the two mode squeezing operator. On the other hand, according to TFD, when the parameter θ is real, $|0(\beta)>>= U_\theta|0>>$ is the thermal state with the inverse temperature β, because $< 0(\beta)|A|0(\beta) >$ for any operator A consisting of a and $a^\dagger$ only is equal to the thermal average $Tr\rho A/Tr\rho$ with the dendity matrix operator $\rho = exp(-\beta H)$, where H is the dynamical Hamiltonian. Then, the parameter θ is related to the temperature as

$$cosh\theta = [1 - e^{-\beta\omega}]^{-\frac{1}{2}} \tag{2.8}$$

with quantum energy ω. Thus, U_θ is called also the thermal operator. *This indicates a close relationship between two mode squeezed states and thermal states.*

The state $U_\theta S_\lambda \tilde{S}_\eta D_\alpha \tilde{D}_\gamma|0>>$ represents a combination of thermal, squeezed and coherent effects. This is the basic instrument in the application of TFD to the subject of thermal squeezed coherent states.

Since there is an excellent review on TFD formalism for quantum optics which is prepared by S. Chaturvedi and V. Srinivasan and is reported in this conference by Chaturvedi [14], I do not enter this subject here. In the following I will discuss the subject of noises formulated in TFD.

Noise and Bogoliubuv Transformation

Let us focus our attention to the two mode squeezing operator U_θ with real parameter θ and write it as $U_\theta = exp(-i\theta G)$ with the generator

$$G = i(a\tilde{a} - \tilde{a}^\dagger a^\dagger). \tag{2.9}$$

This changes the vacuum $|0>>$ into

$$|0(\theta)>= U_\theta|0>>. \tag{2.10}$$

The annihilation operators $\xi(\theta)$ and $\tilde{\xi}(\theta)$ are defined by

$$\xi(\theta)|0(\theta)>= \tilde{\xi}(\theta)|0(\theta)>= 0, \tag{2.11}$$

which gives $a = U_\theta^{-1}\xi(\theta)U_\theta$ and $\tilde{a} = U_\theta^{-1}\tilde{\xi}(\theta)U_\theta$ because we require

$$[\xi(\theta), \xi^\dagger(\theta)] = [\tilde{\xi}(\theta), \tilde{\xi}^\dagger(\theta)] = 1. \tag{2.12}$$

The structure of G shows that the new vacuum $|0(\theta)>$ contains $a\tilde{a}$-pair condensations. Calculation shows that the above transformation is the particular case of the Bogoliubov transformation

$$\begin{bmatrix} a \\ \tilde{a}^\dagger \end{bmatrix}^\mu = B^{-1}(\theta)^{\mu\nu}\begin{bmatrix} \xi(\theta) \\ \tilde{\xi}^\dagger(\theta) \end{bmatrix}^\nu, \tag{2.13}$$

$$[a^\dagger, -\tilde{a}]^\mu = [\xi^\dagger(\theta), -\tilde{\xi}(\theta)]^\nu B(\theta)^{\nu\mu} \tag{2.14}$$

The general form of the Bogoliubov transformation matrix $B(\theta)$ can be put in the form

$$B(\theta) = e^{-\tau_3 \ln b_L}\begin{bmatrix} u & v \\ v & u \end{bmatrix}e^{\tau_3 \ln b_R}, \tag{2.15}$$

where $\tau_i (i = 1 \sim 3)$ are the Pauli matrices. The commutation relation (2.12) requires the condition $u^2 - v^2 = 1$ for real parameters u and v. The matrix $B(\theta)$ depends on three parameters, v, b_L, b_R. However the above operator U_θ with real θ gives the particular choice; $b_L = b_R = 0$ and $u = cosh\theta$. The consideration in this section makes use of a particular choice of these parameters, it can easily be extended to general choices of the parameters. We introduce the number parameters by $n \equiv< 0|a^\dagger a|0>$ and $\tilde{n} \equiv< 0|\tilde{a}^\dagger \tilde{a}|0>$. We then have $n = \tilde{n}$. Then the magnitude of the parameters u

and v is given in terms of the number parameter n as $n = v^2$. It is common to use the three parameters (n, α, s) instead of (v, b_L, b_R) by rewriting (2.15) as [15,16,17]

$$B(\theta) = (1+n)^{\frac{1}{2}} e^{s\tau_3} \begin{bmatrix} 1 & -f^{\alpha} \\ -f^{1-\alpha} & 1 \end{bmatrix}, \tag{2.16}$$

The f is defined by $f = \frac{n}{1+n}$. In the particular case under consideration we have $s = 0$ and $\alpha = 1/2$.

We have pointed out previously that the two mode squeezing operator acts like a thermal operator, suggesting that a two mode squeezed state contains thermal-like noise induced by the TFD mechanism, although it is a pure state. To see this we analyze the uncertainty relation associated with the canonical variables. We now move from the oscillator variables to the canonical variables, defining the bare coordinates, (Q, P) and $(\tilde{Q}, \tilde{P})$, by

$$Q_a = \sqrt{\frac{1}{2}}(a + a^{\dagger}), \ P_a = i\sqrt{\frac{1}{2}}(a^{\dagger} - a), \tag{2.17}$$

$$\tilde{Q}_a = \sqrt{\frac{1}{2}}(\tilde{a} + \tilde{a}^{\dagger}), \ \tilde{P}_a = -i\sqrt{\frac{1}{2}}(\tilde{a}^{\dagger} - \tilde{a}) \tag{2.18}$$

and similarly for the physical coordinates $(Q_\xi(\theta), P_\xi(\theta))$ and $(\tilde{Q}_\xi(\theta), \tilde{P}_\xi(\theta))$ replacing a and $\tilde{a}$ with $\xi(\theta)$ and $\tilde{\xi}(\theta)$ above. From the above definitions, one can prove the canonical commutation relations:

$$[Q_a, P_a] = -[\tilde{Q}_a, \tilde{P}_a] = i \tag{2.19}$$

$$[Q_\xi(\theta), P_\xi(\theta)] = -[\tilde{Q}_\xi(\theta), \tilde{P}_\xi(\theta)] = i. \tag{2.20}$$

We then again find the Bogoliubov transformation

$$\begin{bmatrix} Q_a \\ \tilde{Q}_a \end{bmatrix}^{\mu} = B^{-1}(\theta)^{\mu\nu} \begin{bmatrix} Q_\xi(\theta) \\ \tilde{Q}_\xi(\theta) \end{bmatrix}^{\nu}, \begin{bmatrix} P_a \\ \tilde{P}_a \end{bmatrix}^{\mu} = B^{-1}(\theta)^{\mu\nu} \begin{bmatrix} P_\xi(\theta) \\ \tilde{P}_\xi(\theta) \end{bmatrix}^{\nu}. \tag{2.21}$$

We can then derive the following uncertainty relation for the bare coordinates [11]:

$$< (\Delta P_a)^2 >< (\Delta Q_a)^2 > \geq \frac{1}{4} + < \Delta P_a \Delta \tilde{P}_a >< \Delta Q_a \Delta \tilde{Q}_a > . \tag{2.22}$$

Here ΔA is defined by $\Delta A = A - < A >$ and the expectation value of A means $< A > = < T|A|T >$ for any state vector $|T>$ with the following properties: $|T>$ is such a state that (i) it is factorizable as

$$|T> = |T_\xi> \otimes |T_{\tilde{\xi}}> \tag{2.23}$$

where $|T_\xi>$ and $|T_{\tilde{\xi}}>$ are Fock states of $\xi^{\dagger}$ and $\tilde{\xi}^{\dagger}$, respectively, and that (ii) it is symmetric under exchange of a and b,

$$|T> = |T>^{\sim}, \ < T| = < T|^{\sim}. \tag{2.24}$$

Here tilde means the exchange between a and $\tilde{a}$. The physical vacuum $|0>$ is an example of $|T>$ because it is a state with pair condensation.

The equation (2.22) shows that the bare coordinates contain, not only the quantum fluctuations, but also thermal-like fluctuations which contribute to the uncertainty relation through the term $<\Delta P_a \Delta \tilde{P}_a><\Delta Q_a \Delta \tilde{Q}_a>$. This consideration shows an interesting implication that the same bosonic Bogoliubov transformation creates the same amount of thermal-like fluctuation or thermal-like noise *whatever the cause for the transformation is.* Thus we may define "thermal-like noise" to be that induced by any Bogoliubov transformation [18]. It has been shown [19,9] also that the Bogoliubov transformation induces a thermal-like expansion of the Wigner function, which is consistent with the above uncertainty relation.

To see where this fluctuation comes from, it is instructive to recall that the Bogoliubov transformation induces the $a\tilde{a}$-pair condensation in the vacuum. In condensing in the vacuum, every a-quantum is accompanied by a $\tilde{a}$-quantum, making the $a-$ and $\tilde{a}-$quanta *correlated* with each other. As a result, a measurement of observables associated only with a inevitably contains an effect which cannot be controlled by the a-system only, manifesting the thermal-like noise.

Suppose the following free Hamiltonians for a- and $\tilde{a}$-oscillators:

$$H_o^1 = \omega_1 a^\dagger a, \; H_o^2 = \omega_2 \tilde{a}^\dagger \tilde{a}. \tag{2.25}$$

The a- and $\tilde{a}$-systems with empty vacuum carry zero energy as $<<0|H_o^i|0>>=0$ with $i=1,2$, while those systems with the vacuum $|0(\theta)>$ have $\omega_1 n$ and $\omega_2 n$, where n is the number parameter introduced previously. These energy changes, $<0(\theta)|H_o^1|0(\theta)> - <<0|H_o^1|0>>$, are the thermal-like (or heat-like) energies associated with the noise.

To make a point of this argument clear we consider the following experimental situation. Suppose that a and $\tilde{a}$ represent two incoming radiations which are spatially separated. Assume that their initial vacuum is the empty one $|0>>$. The experimental set up is such that these two radiation waves enter a crystal and interact with each other through the crystal medium, and finally come out as two spatially separated waves with the new vacuum $|0(\theta)>$. In this case the interaction through the crystal medium has the particular structure which induces the Bogoliubov transformation. Therefore, the noise and heat-like energy in the last state is induced by the crystal medium interaction; the origin of noise is dynamical. However, when we confine our attention to the last state only, we find two spatially separated systems sharing the vacuum and carrying the noise, *although they have no mutual interactions.* This is just the right situation for TFD mechanism to cause noise in pure states [18,20].

3. THERMAL PHYSICS FOR A CLOSED QUANTUM FIELD SYSTEM

Thermal Degrees of Freedom and Quantum Fields

The noise in pure states discussed in the last section leads us to a question asking if we could formulate a thermal physics on this concept. However this is impossible due to a couple of reasons. First, the consideration in the last section is based on a quantum mechanical model so that the Fock space built on the empty vacuum $|0>>$ is equivalent to the one made on $|0(\theta)>$. Here we used the terminology that, when any state vector in one Fock space is a superposition of state vectors of another, these two Fock spaces are said to be equivalent. It is impossible to have inequivalent Fock spaces for a quantum system as far as we consider a system with finite number of degrees of freedom (i.e. quantum mechanical system). Therefore, we should consider a system of quantum fields, because quantum fields carry an infinite number of degrees of freedom. We can extend the consideration in the last section to quantum fields by introducing the momentum suffices as a_k and $\tilde{a}_k$. The entire argument in the last section can be extended to this quantum field system. Thus the Bogoliubov transformation connecting $(a_k, \tilde{a}_k)$ to $(\xi_k, \tilde{\xi}_k)$ creates noise in pure states.

Need of quantum field theory for thermal physics for closed systems is intuitively understood for the following reasons. First, the thermal physics in the past mostly dealt with open systems, whose thermal behavior is controlled by a heat bath. Since a heat bath carries an infinite number of degrees of freedom, the number of degrees of freedom associated with the whole system including the heat bath should be infinite. Since quantum fields carry infinite degrees of freedom, it can be hoped that quantum field systems might exhibit some features of open systems (such as dissipative nature), because their degrees of freedom are inexhaustible. There have been some arguments suggesting that, since no measurements in quantum field systems can include the whole degrees of freedom; the system might behave like an open system. However, since the dynamical Hamiltonian H obtained from the Lagrangian is the operator of the dynamical energy only, we cannot expect that the dynamical Hamiltonian can take care of thermal phenomena.

On the other hand, we know a merit of infinite degrees of freedom associated with quantum field theory in describing many different phases for a many body system. Presence of infinitely many inequivalent Fock spaces is the simple reason for appearance of many phases. A well known example is a metal which can appear either in a normal conducting state or in superconducting state. All the inequivalent Fock spaces are distinguished by a form of particle condensation in vacuum. When particles with spin up only are condensed in vacuum, the vacuum exhibits a macroscopic spin order, breaking the spin-rotational symmetry of the Hamiltonian. This is called spontaneous breakdown of spin-rotational symmetry. This illustrates how particle condensation creates some macroscopic behavior of vacuum. When a Hamiltonian is invariant under spin rotation, even when spin-rotation is spontaneously broken, all the vacuums with a different direction of macroscopic vacuum spin have the same dynamical energy; these vacua are degenerate in dynamical energy. Then a mode moving through these vacua carries zero dynamical energy. It is due to this vacuum degeneracy that the so called Goldstone mode of gapless energy appears with spontaneous breakdown of symmetries. This shows that the quantum field theory has a capacity for describing many phases of a system. However, this is also a weakness of the quantum field theory, because it shows the lack of uniqueness of solutions. In reality, different phases appear at different temperature. This suggests that the usual quantum field theory is missing some significant degrees of freedom in order to be a theory for thermal physics for closed systems. Since we are searching for a theory for a closed system, we expect that it is a conserving system controlled by a certain Hamiltonian. However, we have seen that the dynamical Hamiltonian cannot describe the thermal effects. Furthermore, the fact that thermal effects make any excited states dissipative seems to suggest that the new Hamiltonian including the thermal degrees of freedom may be lower unbounded. Indeed, the Hamiltonian in TFD is known to have such features. This will be explained in the next subsection. However, this lower unboundedness may be likely to make the system entirely unstable.

We need a very sophisticated mechanism to avoid such an instability. The fact that any bosonic Bogoliubov transformation induces noise in pure states with pair condensation suggests that this sophisticated mechanism which is supposed to create the thermal noise may be related to the Bogoliubov transformation. In the next section we will see how TFD takes care of thermal phenomena in closed quantum field systems through a very sophisticated mechanism based on particle condensation.

Basics of TFD

In this section we try to build a quantum field formalism for thermal physics for closed systems on the basis of the concept of noise in pure states developed in section 2. The thermal degree of freedom does arise from two sets of fields, i.e. nontilde and tilde fields. The relation between nontilde and tilde fields can be arbitrary if the Bogoliubov transformation is dynamically induced; then, the relation should be controlled by the nature of dynamics which causes the transformation. However we are now considering closed systems and therefore the transformation cannot be of dynamical origin but should be induced by a self consistent mechanism. Therefore, the relations between the two sets of fields are controlled by some self consistency conditions called tilde conjugation rules.

First we try to extract from the above consideration some basic properties of the tilde conjugation rules. Let ω_k denote the energy of a-particle. Then, the wave function of the a-particle behaves as $exp[-i\omega_k t]$. In order for the Bogoliubov transformation to be able to mix a_k with $\tilde{a}_k^\dagger$ at any time, the wave function of $\tilde{a}$-particle should be $exp[i\omega_k t]$. This implies that the imaginary number i changes its sign, and therefore that a c-number becomes its complex conjugate under the tilde conjugation. We recall that the number parameter is given by the vacuum expectation value of the operator $a_k^\dagger a_k$ which is equal to the vacuum expectation value of $\tilde{a}_k^\dagger \tilde{a}_k$. Thus we consider $\tilde{a}_k^\dagger \tilde{a}_k$ as the tilde conjugate of $a_k^\dagger a_k$. The ordering in this operator is important because a_k and $a_k^\dagger$ do not commute with each other.

We thus assume that the tilde conjugation does not change the ordering among operators. We recall also that the thermal vacua are states with $a\tilde{a}$-pair condensate. Therefore, they are invariant under the tilde conjugation. We introduced the parameter α in the general definition of the Bogoliubov matrix B in (2.16). Unless $\alpha = 1/2$ the ket-vacuum is not the Hermitian conjugate of bra-vacuum. Therefore, the transformations are not necessarily unitary. Summarizing these arguments, in the following we set up the basic relations in TFD.

In TFD, to any operator A is associated its tilde conjugate $\tilde{A}$. The purely dynamical behavior is described by the nontilde operators. However, thermal phenomena require the tilde operators. The tilde conjugation rules are summarized as follows:

$$(AB)^{\sim} = \tilde{A}\tilde{B}, \tag{3.1}$$

$$(c_1 A + c_2 B)^{\sim} = c_1^* \tilde{A} + c_2^* \tilde{B}, \tag{3.2}$$

$$(A^\dagger)^{\sim} = \tilde{A}^\dagger, \tag{3.3}$$

$$(\tilde{A})^{\sim} = A, \tag{3.4}$$

$$|0(\theta)>^{\sim} = |0(\theta)>, \tag{3.5}$$

$$< 0(\theta)|^{\sim} = < (\theta)|. \tag{3.6}$$

Here c_1 and c_2 are any two c-numbers and A and B stand for any two operators. This tilde conjugation rules uniquely determines $\tilde{A}$ when A is given.

Let us now study the Hamiltonian $\hat{H}$ in TFD. Since in the Schroedinger representation the time dependence of the thermal vacua is given by $exp[-i\hat{H}t]$, invariance of vacua under tilde conjugation rules, (3.5) and (3.6), is preserved in time when and only when

$$(\hat{H})^{\sim} = -\hat{H}. \tag{3.7}$$

When a Lagrangian is given, it leads to the dynamical energy operator H. However, this cannot be the Hamiltonian $\hat{H}$, because it does not generate the time change in tilde operators. Furthermore, $\hat{H}$ is conditioned by the requirement that the Heisenberg equation for nontilde operators should not change by thermal degree of freedom, implying that $\hat{H}$ should not contain any product of nontilde and tilde operators. Then, the relation (3.7) uniquely determines $\hat{H}$ as

$$\hat{H} = H - \tilde{H}. \tag{3.8}$$

It is essential to make a distinction between $\hat{H}$ and H, whose eigenvalues or expectation values are denoted by $\hat{E}$ (*the hat-energy*) and E (*the dynamical energy*), respectively. The thermal vacua are required to be eigenstates of $\hat{H}$. Furthermore, according to the tilde conjugation rules, thermal vacua should be invariant under tilde conjugation. Thus, we have

$$\hat{H}|0(\theta)> = \hat{E}|0(\theta)>, \tag{3.9}$$

$$|0>^{\sim} = |0>. \tag{3.10}$$

Then the real eigenvalue $\hat{E}$ turns our to be zero,

$$\hat{E} = 0, \tag{3.11}$$

from

$$(\hat{H})^{\sim} = -\hat{H}, \tag{3.12}$$

$$(\hat{E})^{\sim} = \hat{E}. \tag{3.13}$$

Thus the thermal vacua are the zero hat-energy states which are invariant under tilde conjugation. It is due to the negative sign in front of $\tilde{H}$ in (3.8) that there is a continuous set of such thermal vacua. The continuous parameter θ in the thermal vacuum of (3.9), which may have multi-components, is intended to label each member in this set of continuously degenerate thermal vacua, $\{|0(\theta)>\}$. This continuous degeneracy of thermal vacua in hat-energies is a crucial feature of thermal field theory.

To connect a thermal vacuum with another, we introduce the operator $\hat{G}$ through the following operation:

$$|0(\theta+d\theta)>= e^{id\theta\hat{G}}|0(\theta)>, \tag{3.14}$$

$$<0(\theta+d\theta)| =<0(\theta)|e^{-id\theta\hat{G}}. \tag{3.15}$$

Then $\hat{G}$ can be chosen to satisfy

$$[\hat{H},\hat{G}] = 0 \tag{3.16}$$

This dictates two statements. The first is that $\hat{G}$ is an operator of zero hat-energy mode. The second is that there is the invariance of Hamiltonian under the $\hat{G}$-transformation, (3.16). Furthermore, this $\hat{G}$-symmetry is *spontaneously broken*, according to (3.14) and (3.15) which shows that

$$\hat{G}|0(\theta)>\neq 0. \tag{3.17}$$

The degeneracy of the thermal vacua in hat-energies is associated with breakdown of this symmetry and $\hat{G}$ acts as the Goldstone mode operator. As it is well known in quantum field theory, the relation (3.17) leads us to conclude that the norm of the state $\hat{G}|0(\theta)>$ diverges, implying that $\hat{G}$ is pathological. The common cure for this is first to write $\hat{G}$ as spatial integration of a local operator and then to smear out the local operator with a square integrable function $f(\vec{x})$. At the end of calculation the limit $f(\vec{x}) \to 1$ is performed. Since this mathematical trick is a familiar one in theory of spontaneous breakdown of symmetries, here we do not enter this problem any further.

The continuous set of vacua $\{|0(\theta)>\}$ implies the continuous set of Fock spaces denoted by $\{\mathcal{H}(\theta)\}$, each $\mathcal{H}(\theta)$ being constructed on each $|0(\theta)>$. In other words, according to the previous paragraph, the set $\{\mathcal{H}(\theta)\}$ consists of all possible Fock spaces, in each of which the algebraic relation (3.16) is realized. This situation is same as those cases of spontaneous breakdown of symmetries such as ferromagnetism and superconductivity. Note that any finite number of operation of $\hat{G}$ does not change the Fock space; in order for two Fock spaces to be inequivalent to each other, their vacuum condensation should differ by an infinite number of particles, requiring infinite power of $\hat{G}$.

Now we clearly know what is the thermal degree of freedom. The thermal degree of freedom is the freedom of moving through the continuous set of thermal vacua.

The above consideration can be applied to time (τ)-dependent nonequilibrium systems of quantum fields by making θ dependent on time $\tau, \theta(\tau)$. (The simpler notations, $|0(\tau)>$ for $|0(\theta(\tau))>$ and so on, will be used from now on.) Here we use the notation τ for time associated with thermal change. According to this picture the thermal vacuum has to move through many Fock spaces inequivalent to each other. We describe this move as the move through the set $\{\mathcal{H}(\theta)\}$ with time (τ)-dependent θ.

Then we introduce the generator of τ-translation in $|0(\tau)>, \hat{Q}(\tau)$, called *the thermal generator* through

$$|0(\tau + d\tau) >= e^{id\tau\hat{Q}(\tau)}|0(\tau) >, \tag{3.18}$$

$$< 0(\tau + d\tau)| =< 0(\tau)|e^{-id\tau\hat{Q}(\tau)}. \tag{3.19}$$

Clearly from (3.14) and (3.15), this operator is proportional to $\hat{G}$:

$$\hat{Q}(\tau) = \frac{d\theta}{d\tau}\hat{G}. \tag{3.20}$$

The thermal average of a dynamical (i.e. nontilde) operator $A(t)$ is given by

$$\bar{A}(\tau) =< 0(\tau)|A(t)|0(\tau) > . \tag{3.21}$$

Note that the quantity $\bar{A}(\tau)$ does not depend on t because of t-translational invariance of the vacua (3.11). This illustrates the usefulness of distinguishing τ-time from t-time. The τ-time is the measurement time generated by $\hat{Q}(\tau)$. Thus the drastic effect appearing in thermal field theory is that $\hat{Q}$ adds to the Hamiltonian $\hat{H}$ to generate the total time-translation of the system. An important point to be noted is that *we never need to consider inner products of state vectors belonging to different Fock spaces, because the bra- and ket state have always a common time τ.* This $\hat{Q}$-effect has not appeared in equilibrium TFD, because $\dot{n} = 0$ makes $\hat{Q}$ vanishing in stationary states.

We now apply the above argument to the free field with the dynamical operator:

$$H_o = \int d^3k\omega_k a_k^\dagger a_k. \tag{3.22}$$

Then (3.8) indicates that the free Hamiltonian is

$$\hat{H}_o = \int d^3k\omega_k[a_k^\dagger a_k - \tilde{a}_k^\dagger\tilde{a}_k]. \tag{3.23}$$

Then, the relation (3.16) reads as

$$[\hat{H}_o, \hat{G}] = 0. \tag{3.24}$$

Furthermore, the commutation relation (3.24) enables us to find [21] that $\hat{G}$ has three components:

$$\theta\hat{G} = \int d^3k\sum_{i=1}^{3}\theta_{i,k}\hat{G}_{i,k} \tag{3.25}$$

with

$$\hat{G}_{1,k} = i(a_k\tilde{a}_k - \tilde{a}_k^\dagger a_k^\dagger), \tag{3.26}$$

$$\hat{G}_{2,k} = i(a_k\tilde{a}_k + \tilde{a}_k^\dagger a_k^\dagger), \tag{3.27}$$

$$\hat{G}_{3,k} = i(a_k^\dagger a_k + \tilde{a}_k^\dagger \tilde{a}_k). \tag{3.28}$$

Thus the parameter θ has three components, which are the three parameters appearing in the general form of Bogoliubov transformation discussed in section 2. These generators form the $SU(1,1)$ algebra which is the familiar one in quantum optics. When a thermal situation changes, it was shown in [21] that $\hat{Q}$ takes the form

$$\hat{Q}(t) = i \int d^3k \bar{a}_k^\mu P_k(t)^{\mu\nu} a_k^\nu \tag{3.29}$$

with

$$P_k(t) = B_k^{-1}(t)\Big(\frac{d}{dt}B_k(t)\Big). \tag{3.30}$$

When we have a system of interacting fields with $H = H_o + H_{int}$, we can formulate the perturbation calculation with the unperturbed Hamiltonian given by $\hat{H}_o - \hat{Q}$. The unperturbed propagator including the $\hat{Q}$-effect has been calculated [22]. Since $\hat{G}_1$ and $\hat{G}_2$ has the pair structure familiar in the BCS Hamiltonian for superconductivity, *we know that $\hat{Q}$ cannot be treated as a perturbation interaction.* This is the reason why *we include $\hat{Q}$ in the unperturbed Hamiltonian.*

Short-Time Thermal Effects in Particle Reactions

Being equipped with the time dependent TFD presented in the last subsection, an immediate application is the particle reactions, because particle reactions take place in a closed system. The TFD formalism with $\hat{Q}$-effect above tells us that there should be a short time thermal effect even when the initial and final temperatures are very low. This is what has been expected by many high energy physicists working on quark-gluon plasmas. A merit of our method is that, using the free propagator with $\hat{Q}$-effect as the unperturbed propagator and H_{int} for interaction vertices, we can calculate the reaction matrix. Such a calculation is in progress.

ACKNOWLEDGEMENT

The author would like to thank Dr. Y. Yamanaka for illuminating discussions and careful reading of this paper. This work was supported by NSERC, Canada and the Dean of Science, the University of Alberta.

REFERENCES

1. L. Leplae, H. Umezawa, and F. Mancini. Derivation and application of the boson method in superconductivity. Physics Reports, 10C, 153 (1974).
2. Y. Takahashi and H. Umezawa. Collect. Phenom., 2, 55 (1975).
3. H. Umezawa, H. Matsumoto, and M. Tachiki. Thermo Field Dynamics and Condensed States. North-Holland, 1982.
4. W. Israel. Phys Lett., 10, 1018 (1980).
5. R. Laflamme. Physica, 158A, 58 (1989).
6. B. Yurke and M. Potasek, Phys. Rev., A36, 3464 (1987).

7. S. M. Barnett and P. L. Knight. J. Opt. Am., B2, 467 (1985).
8. A. Mann and M. Revzen, Phys. Lett., A134, 273 (1989).
9. Y. S. Kim and Ming Li. Phys. Lett., A139, 443 (1989).
10. A. Mann, M. Revzen, and H. Umezawa, Phys. Lett., 139, 197 (1989).
11. A. Mann, M. Revzen, H. Umezawa, and Y. Yamanaka. Phys. Lett., 140, 475 (1989).
12. T. Garavaglia, Phys. Rev., A38, 4365 (1988).
13. A. Mann, M. Revzen, K. Nakamura, H. Umezawa, and Y. Yamanaka. J. math. Phys., 30, 2883 (1989).
14. Chaturvedi. Thermofield Dynamics and its Applications to Quantum Optics, In G. S. Agarwal, editor, Proceedings for International Conference on Quantum Optics at Hydrabad, 1991.
15. H. Umezawa and Y. Yamanaka. Advances in Physics, 37, 531 (1988).
16. T. Arimitsu, H. Umezawa, and Y. Yamanaka. J. Math. Phys., 28, 2741 (1987).
17. I. Hardman, H. Umezawa, and Y. Yamanaka. J. Math. Phys., 28, 2925 (1987).
18. H. Umezawa and Y. Yamanaka. Physica, 170A, 291 (1991).
19. M. Berman. Phs. Rev., A40, 2057 (1989).
20. H. Umezawa. Origin of non-quantum noise and time dependent thermo field dynamics. In T. Arimitsu, editor, Third International Workshops on Thermal Quantum Field Theories. North- Holland, 1990.
21. H. Umezawa and Y. Yamanaka. Thermal degree of freedom in thermo field dynamics. Univ. of Alberta preprint, 1990.
22. H. Umezawa and Y. Yamanaka. Short-time high temperature effects in particle reactions. Univ. of Alberta preprint, 1991.

INTERACTION OF ATOMS WITH SQUEEZED RADIATION

R. R. Puri

Theoretical Physics Division
Bhabha Atomic Research Center
Bombay 400 085, India

INTRODUCTION

We will briefly review some aspects of the interaction of two-level atoms and harmonic oscillators with the squeezed radiation from a degenerate parametric amplifier (DPA) in an optical cavity. We discuss the effects of interaction of atoms with the multimode field transmitted outside the cavity as well as those arising due to the interaction with the intracavity field.

The dynamics of the field inside an optical cavity containing a non-linear optical crystal which creates two photons of frequency $\omega_p/2$ on absorbing a photon of frequency ω_p by the parametric process is described by the master equation ($\hbar = 1$)

$$\frac{\partial \varrho}{\partial t} = -(i/2)[\varepsilon a^{+2} + \varepsilon^* a^2, \varrho] + \kappa(n_{th}+1)(2a\varrho a^+ - a^+a\varrho - \varrho a^+ a)$$

$$+ \kappa n_{th}(2a^+a\varrho - aa^+\varrho - \varrho aa^+), \qquad (1)$$

for the density matrix ϱ of the field in a frame rotating with the frequency $\omega_p/2$. In Eq. (1), $\varepsilon = |\varepsilon| exp(2i\phi)$ is the parametric driving rate, n_{th} is the average number of thermal photons in the cavity and 2κ is the rate of decay of the cavity field. The system approaches the steady state if $|\varepsilon| < \kappa$. It can be shown that, in a frame rotating with frequency $\omega_p/2$ the two-time correlation functions of the output field, described by the annihilation and creation operators a_{out} and a^+_{out}, in the steady state are given by

$$< a_{out}(t+\tau)a_{out}(t) >$$

$$= -i\varepsilon\kappa[\lambda(2\kappa n_{th} + |\varepsilon|)exp(-\eta|\tau|) - \eta(2\kappa n_{th} - |\varepsilon|)exp(-\lambda|\tau|)]/2\lambda\eta|\varepsilon|,$$

$$< a^+_{out}(t+\tau)a_{out}(t) >$$

$$= \kappa[\lambda(2\kappa n_{th} + |\varepsilon|)exp(-\eta|\tau|) + \eta(2\kappa n_{th} - |\varepsilon|)exp(-\lambda|\tau|)]/2\lambda\eta. \qquad (2)$$

Recent Developments in Quantum Optics, Edited
by R. Inguva, Plenum Press, New York, 1993

where

$$\lambda = \kappa + |\varepsilon|, \; \eta = \kappa - |\varepsilon|, \tag{3}$$

The output has a vanishing mean amplitude and, it can be shown that, it is squeezed. The environment outside the cavity is, therefore, said to constitute a "squeezed vacuum." We investigate the dynamical behaviour of atoms interacting with such a squeezed vacuum.

The Master Equation for Interaction with Squeezed Vacuum

We assume that a system of N identical atoms interacts collectively with the squeezed vacuum whose statistical properties are characterized by Eqs. (2) and is also driven by a coherent field of frequency ω_l. The atomic dynamics, in electric dipole and the Rotating Wave Approximations is then described by the equation[1]

$$\frac{\partial \varrho}{\partial t} = -i[H_o, \varrho] - i[\alpha^* c exp(i\omega_l t) + h.c., \varrho] - i\sqrt{2\gamma}[a^+_{out}(t) c exp(i\omega_p t/2) + h.c., \varrho] \tag{4}$$

for the atomic density matrix ϱ. H_o is the free atomic hamiltonian and 2γ is the rate of decay of the atomic oscillations. c and c^+ are, respectively, the atomic lowering and raising operators. For a system of harmonic oscillators, c and c^+ obey the commutation relation $[c, c^+]$=1 whereas for a system of N two-level atoms, $c = S_-$ and $c^+ = S_+$, where $S_\pm$ along with S_z obey the angular momentum commutation relations $[S_+, S_-] = 2S_z$, $[S_z, S_\pm] = \pm S_\pm$. For a system of harmonic oscillators, $H_o = \omega_o c^+ c$ whereas for a system of two-level atoms, $H_o = \omega_o S_z$. The fluctuating variables $a_{out}(t)$ and $a^+_{out}(t)$ in the stochastic equation (4) are operators. It is, however, more convenient to rewrite it in the following form[2] in terms of c-number fluctuating variables $\beta_\pm(t)$,

$$\frac{\partial \varrho}{\partial t} = -i[H_o, \varrho] - i[\alpha^* c exp(i\omega_l t) + h.c., \varrho]$$

$$- i\sqrt{2\gamma}[\beta_+(t) c exp(i\omega_p t/2) + h.c., \varrho] + \gamma[2c\varrho c^+ - c^+ c\varrho - \varrho c^+ c], \tag{5}$$

where $\beta_\pm(t)$ are such that,

$$< \beta_+(t+\tau)\beta_-(t) >= \kappa[\lambda(2\kappa n_{th} + |\varepsilon|) exp(-\eta|\tau|) + (2\kappa n_{th} - |\varepsilon|) exp(-\lambda|\tau|)/2\lambda\eta,$$

$$< \beta_-(t+\tau)\beta_-(t) >= -i\kappa\varepsilon[\lambda(2\kappa n_{th} + |\varepsilon|) exp(-\eta|\tau|)$$

$$- \eta(2\kappa n_{th} - |\varepsilon|) exp(-\lambda|\tau|)/2\lambda\eta|\varepsilon|, \tag{6}$$

Finding an exact analytic solution of Eq. (5) for arbitrary λ, η and γ is, in general, a formidable task. However, exact solutions can be found for a variety of atomic systems in the "white noise" limit $\lambda, \mu >> \gamma$. In this limit, the two-time correlation functions (6) reduce to[1]

$$< \beta_+(t+\tau)\beta_-(t) >= \bar{N}\delta(\tau),$$

$$< \beta_+(t+\tau)\beta_+(t) >= m^*\delta(\tau),$$

$$< \beta_-(t+\tau)\beta_-(t) >= m\delta(\tau). \tag{7}$$

where

$$\bar{N} = \bar{n} + \bar{n}_{th},$$

$$\bar{n} = (\lambda^2 - \eta^2)^2/4\lambda^2\eta^2, \; \bar{n}_{th} = (\lambda+\eta)^2 n_{th}(\lambda^2+\eta^2)/2\lambda^2\eta^2,$$

$$m = (\lambda^2 - \eta^2)(\lambda^2+\eta^2)exp(i\phi)/4\lambda^2\eta^2 + (\lambda+\eta)^2 n_{th}(\lambda^2-\eta^2)/2\lambda^2\eta^2 exp(i\phi). \tag{8}$$

Note that

$$|m|^2 = [\bar{N}(\bar{N}+1)^2 - \bar{n}_{th}(\bar{N}+1)^2]/[\bar{N} - \bar{n}_{th} + 1] \tag{9}$$

Thus, $|m|^2 = \bar{n}(\bar{n}+1)$ for $n_{th} = 0$ whereas $|m|^2 < \bar{N}(\bar{N}+1)$ for $n_{th} \neq 0$. In a frame rotating with $\omega_p/2$ Eq. (5) averaged over the δ-correlated fluctuations reduces to

$$\frac{\partial \varrho}{\partial t} = -i\delta[H_o, \varrho] - i[\alpha c^+ exp(-i\delta_o t) + c.c., \varrho] + L_{sq}\varrho \tag{10}$$

where $\delta_o = \omega_l - \omega_p/2, \delta = \omega_o - \omega_p/2$ and L_{sq}, describing the interaction with the squeezed vacuum bath, is given by

$$L_{sq}\varrho = \gamma(\bar{N}+1)(2c\varrho c^+ - c^+c\varrho - \varrho c^+c) + \bar{N}\gamma(2c^+\varrho c - cc^+\varrho - \varrho cc^+)$$

$$+ m^*\gamma(2c\varrho c - cc\varrho - \varrho cc) + m\gamma(2c^+\varrho c^+ - c^+c^+\varrho - \varrho c^+c^+) \tag{11}$$

Eq. (11) can be written in the following compact and often useful form

$$L_{sq}\varrho = \gamma(1+\beta)(2R\varrho R^+ - R^+R\varrho - \varrho R^+R) + \gamma\beta(2R^+\varrho R - RR^+\varrho - \varrho RR^+) \tag{12}$$

where

$$R = \mu exp(-i\phi)c + \nu exp(i\phi)c^+, R^+ = \nu c exp(-i\phi) + \mu c^+ exp(i\phi), m = |m| exp(2i\phi) \tag{13}$$

with

$$\mu^2 - \nu^2 = 1, \; \mu = [\bar{n} + \bar{n}_{th} + \beta + 1]/[1+2\beta], \tag{14}$$

and β is the positive solution of the equation

$$\beta^2 + \beta + |m|^2 - (\bar{n}_{th} + \bar{n} + 1)(\bar{n}_{th} + \bar{n}) = 0. \tag{15}$$

Clearly, $\beta = 0$ for n_{th}=0. Also, since $\mu^2 - \nu^2 = 1$, R and R^+ obey the same commutation relations as c and c^+. In the following we present solution of Eq. (10) for some specific atomic models.

Harmonic Oscillators in Squeezed Vacuum

The solution of Eq. (10) for $\alpha = \delta_o = 0$ in the case of harmonic oscillators can be shown to be given by

$$\varrho(t) = (1/\pi\chi) \int d^2\xi exp(-i\xi^* R^+)exp(-i\xi R)exp(-\chi(t)|\xi|^2)G(\xi, \xi^*) \tag{16}$$

where

$$G(\xi, \xi^*, t) = Tr[exp(i\xi f(t)R)exp(i\xi^* f^*(t)R^+)\varrho(0)], \tag{17}$$

$f(t) = exp(-\gamma t)$, $\chi(t) = [1 - exp(-2\gamma t)](1 + \beta)$. It is found that the evolution of the expectation values of c and c^+ in the case of harmonic oscillators do not depend on $\bar{N}$ or m. Consequently, the presence of squeezed bath does not affect the position and the width of the spectral lines. The intensity of the spectral lines is, however, influenced by squeezing because the quadratic functions of c and c^+ depend on $\bar{n}$ and m.

Some very interesting and unusual effects of squeezed bath are, however, shown by the radiation from a system of two-level atoms.

Two-Level Atoms in Squeezed Bath

For a system of two-level atoms, an exact solution of Eq. (10) can be found for $N = 1$ and $\delta_o = 0$. It is found that[1], in the absence of an external drive (α=0) the two quadratures, $< S_x >= [exp\ (i\phi)\ <\ S_+ >\ + exp\ (-i\phi)\ <\ S_- >]/2$ and $< S_y >= [exp\ (i\phi) < S_+ > - exp\ (-i\phi) < S_- >]/2i$ of the atomic dipole decay at different rates, $\gamma_x = \gamma(2N + 1 - 2m)$ and $\gamma_y = \gamma(2\bar{N} + 1 + 2m)$ respectively whereas the population inversion $< S_z >$ decays at the rate $\gamma_z = 2\gamma(2\bar{n} + 1)$. For $\bar{n} >> 1$, and in the case of an ideal DPA ($|m|^2 = \bar{n}(\bar{n} + 1))\gamma_x \sim \gamma/4\bar{n}, \gamma_y \sim 2\gamma(2\bar{n} + 1)$ i.e. by increasing $\bar{n}$, $< S_x >$ can be made to decay at an arbitrarily slow rate but the rate of decay of $< S_y >$ is then enhanced. This is reflected in the narrowing of the absorption spectrum of a weak probe by the atom for large $\bar{n}$.[3]

In the presence of an external field, an exact solution of Eq. (10) has been derived in Ref. 4 for $\delta = \delta_o = 0$. However, the most interesting effects of squeezing are predicted when the Rabi frequency is much stronger than other rates in the system. In this limit, the master equation (10) for $\delta_o = 0$ can be reduced to a simple form by making the transformations

$$R_z = [|\alpha|\{S_+ exp(i\theta) + S_- exp(-i\theta)\} + 2\delta S_z]/\Gamma, R_x = [S_+ exp(i\theta) - S_- exp(-i\theta)]/2i,$$

$$R_y = [2|\alpha|S_z - \delta\{S_+ exp(i\theta) + S_- exp(-i\theta)\}]/\Gamma, \tag{18}$$

where $\alpha = exp(i\theta)|\alpha|$, $\Gamma^2 = 4|\alpha|^2 + \delta^2$. The R's also obey the angular momentum commutation relations. In terms of the transformed operators, Eq. (10) with $\delta_o = 0$ reads

$$\frac{\partial\varrho}{\partial t} = -i\Gamma[R_z,\varrho] + A_-[2R_-\varrho R_+ - R_+R_-\varrho - \varrho R_+R_-] + A_+[2R_+\varrho R_- - R_-R_+\varrho - \varrho R_-R_+]$$

$$+ A_z[2R_z\varrho R_z - R_zR_z\varrho - \varrho R_zR_z] + \text{other terms}, \tag{19}$$

where

$$A_\pm = \gamma|\mu r_\pm - \nu exp(\pm 2i\psi)r|^2/4, r_\pm = 1 \pm \delta/\Gamma, A_z = 4\gamma|\alpha|^2|\mu + \nu exp(2i\psi)|^2, \tag{20}$$

with $\psi = \theta - \phi$ and the "other terms" in Eq. (19) contribute factors like $exp(\pm i\Gamma t)$ and $exp(\pm 2i\Gamma t)$ in a frame rotating with the modified Rabi frequency Γ. These terms can be ignored in the secular approximation, $N\gamma/2\Gamma << 1$. The fluorescent spectrum

$$S(v) \sim Re\int exp[i(\nu - \omega_l)\tau] < S_+(\tau)S_- > \tag{21}$$

may then be evaluated by using the expression for the two-time correlation function $< S_+(\tau)S_- >$ which, for N=1 is found to be given by

$$< S_+(\tau)S_- >= exp(-\tilde{\gamma}\tau)[A_-r_+^2 exp(i\Gamma\tau) + A_+r_-^2 exp(-i\Gamma\tau)]/(A_+ + A_-)$$

$$+ 4|\alpha|^2[exp(-\gamma_z\tau)(1/4 + (A_+ - A_-)/\gamma_z) - (A_+ - A_-)/\gamma_z] \tag{22}$$

where

$$\gamma_z = (A_+ + A_-)/4, \tilde{\gamma} = A_+ + A_- + A_z. \tag{23}$$

For N=2, on the other hand

$$< S_+(\tau)S_- >= [\{A_-r_+^2 exp(i\Gamma t) + A_+r_2 exp(-i\Gamma t)\}\{f_1(t)(A_+ + A_-) + f_2(t)A_-\}$$

$$+8|\alpha|^2\{\phi_1 A_-^2 + A_+^2(\phi_1 + 2\phi_2)\}]/2[A_+^2 + A_-^2 + A_+A_-]$$

$$+ 4|\alpha|^2(A_+^2 - A_-^2)^2/[\{A_-^2(2A_+ + A_-^2)\}\Gamma^2], \tag{24}$$

where

$$f_1(t) = exp(-xt)[(D + Z)exp(Zt/2) - (D - Z)exp(-Zt/2)]/2Z,$$

$$f_2(t) = -(A_+ - A_-)exp(-xt)[exp(Zt/2) - exp(-Zt/2)]/2Z,$$

$$\phi_1(t) = exp\{-(A_+ + A_-)t\}[(\sqrt{A}_+ + \sqrt{A}_-)exp\{(\sqrt{A_+A_-})t\}$$

$$+(\sqrt{A_+}-\sqrt{A_-})exp\{-(\sqrt{A_+A_-})t\}]/2\sqrt{A_+},$$

$$\begin{aligned}\phi_2(t) &= (A_+-A_-)exp\{-(A_++A_-)t\}[exp\{(\sqrt{A_+A_-})t\}\\ &-exp\{-(\sqrt{A_+A_-})t\}]/4\sqrt{A_+A_-},\end{aligned}$$

$$\begin{aligned} x &= \{A_z+3(A_++A_-)/4\},\\ Z &= (1/2)[A_+^2+A_-^2+14A_+A_-]^{\frac{1}{2}},\\ D &= -i\Gamma+3(A_++A_-)/2.\end{aligned} \qquad (25)$$

These expressions show that for $N=1$ as well as for $N=2$ the incoherent part of the spectrum consists of three peaks one centered at the driving field frequency ω_l and the other two at $\omega_l \pm \Gamma$. The width of the side peaks which is determined by $\tilde{\gamma}$ for $N=1$ is always greater in a squeezed bath compared with that in an unsqueezed reservoir. For $N=2$, each of the side peaks is a sum of two Lorentzians of width $x\pm$Z.

The central peak for $N=1$ is a Lorentzian of width γ_z whereas for $N=2$ it is a sum of two Lorentzian of width $A_++A_-\pm\sqrt{A_+A_-}$. For δ=0, $A_+=A_-$ and hence the central peak for $N=1$ as well as for $N=2$ consists of only one Lorentzian of width A_+. In that case $A_+=2\gamma_z$ and

$$\begin{aligned}\gamma_z=\gamma[2n+1-2\sqrt{n}(n+1)cos(2\psi)] &\approx \gamma[\sqrt{(n+1)}-\sqrt{n}]^2,\ for\ \psi=0\\ &\approx \gamma/4n,\ for\ n>>1.\end{aligned} \qquad (26)$$

Thus, the central peak may acquire a subnatural width in the presence of squeezing.

Further, note that the height of the side peaks for $N=1$ as well as for $N=2$ is determined by $A_\pm$. Thus, if A_+=0 i.e. if

$$\psi=0 \text{ and } \delta/\Gamma=-[\sqrt{(\bar{n}+1)}-\sqrt{\bar{n}}]/[\sqrt{(\bar{n}+1)}+\sqrt{\bar{n}}] \qquad (27)$$

then the peak at $\omega_l-\Gamma$ would disappear whereas if

$$\psi=\pi \text{ and } \delta/\Gamma=[\sqrt{(\bar{n}+1)}-\sqrt{\bar{n}}]/[\sqrt{(\bar{n}+1)}+\sqrt{\bar{n}}] \qquad (28)$$

then the spectral component at $\omega_l+\Gamma$ would vanish. Note also that for $A_+\to 0$

$$\phi_1(t)\to exp(-A_-t)[1+tA_-],\ \phi_2(t)\to -tA_-exp(A_-t)/2 \qquad (29)$$

and for $A_-\to 0$

$$\phi_1(t)\to exp(-A_+t),\ \phi_2(t)\to -tA_-exp(A_+t)/2. \qquad (30)$$

These results, in turn, lead to a non-exponential decay of the atomic population and to a non-Lorentzian spectral profile at the centre for $N=2$. For $N=1$, the decay is always exponential and the spectrum Lorentzian.

The Steady State Solution

Some very interesting unusual properties of the atomic interaction with a squeezed bath are exhibited in the steady state.[5,6] The steady state solution of Eq. (10) for $\delta_o = 0$ for a system of N two-level atoms in an ideal bath ($|m|^2 = \bar{n}(\bar{n}+1)$) is given by[5]

$$\varrho_{ss} = \sum_{m,n} c_m c_n^* < \phi_m|\phi_n > |\psi_m >< \psi_n|, \tag{31}$$

where $|\psi_m >$ and $|\phi_m >$ are respectively the eigenstates of R and R^+ with eigenvalues $2(N/2-m)\sqrt{\bar{n}(\bar{n}+1)}$, $m = 0, 1, 2....N$ and

$$c_m = \Gamma(m + i\alpha_o^* - i\delta/2\gamma)/\Gamma(m + i\alpha_o + i\delta/2\gamma), \tag{32}$$

where $\alpha_o = (\alpha\mu - \alpha^*\nu)/\gamma$ and $\Gamma(z)$ is the Gamma function. For $\alpha = \delta = 0$, the steady state solution (31) assumes a simpler form:

$$\varrho_{ss} = |\psi_o >< \psi_o|, \tag{33}$$

for N even and

$$\varrho_{ss} = (R)^{-1}(R^+)^{-1} \tag{34}$$

for N odd. Thus, for even N, the steady state is a pure state which is in contrast with the steady state reached on interaction with an ordinary thermal bath which is always a mixed state (except in the trivial case when the steady state is a ground state). In the state $|\psi_o >$ the occupation probability of the atomic states is such that starting with the occupation of the ground state, the alternating states are occupied leaving the states in between them empty. This is shown in Fig. 1 where we have plotted the probability $p_m =< m|\varrho|m >$ of the occupation of the collective atomic state $|m >$ as a function of m for ϱ of Eq. (33) for N=40 and compared it with the behaviour of p_m for atoms interacting with an unsqueezed bath. A pure state results also when $\alpha \neq 0$ but $\delta = 0$ and α is such that for some positive integer $p \leq N$,

$$\alpha_o = -2i\sqrt{\bar{n}(\bar{n}+1)}(N/2 - p). \tag{35}$$

The steady state in this case is $|\psi_p >< \psi_p|$. The occupation probability of the atomic states predicted for $\varrho_{ss} = |\psi_p >< \psi_p|$ exhibits oscillatory behaviour which is a signature of pure quantum effects. Also, for these states, $\Delta S_x \Delta S_y = | < S_z > |/2$. Such states, for which the equality holds in the Heisenberg uncertainty relation, are called the intelligent spin states.

It has not been possible to obtain an exact steady state solution of Eq. (10) even for α=0 for a system of N two-level atoms in case of a non ideal squeezed bath i.e. if $|m|^2 \neq n(n+1)$ except for $N = 1$ and 2. The exact steady state solution for $N = 1$ in the absence of an external drive in a non-ideal bath is given by

$$\varrho_{ss} = [1/(\bar{N}+1)]exp[-1n\{(\bar{N}+1)/\bar{n}\}S_z]. \tag{36}$$

For $N = 2$ we find that ϱ_{ss} for $\alpha = 0$ may be written in the form

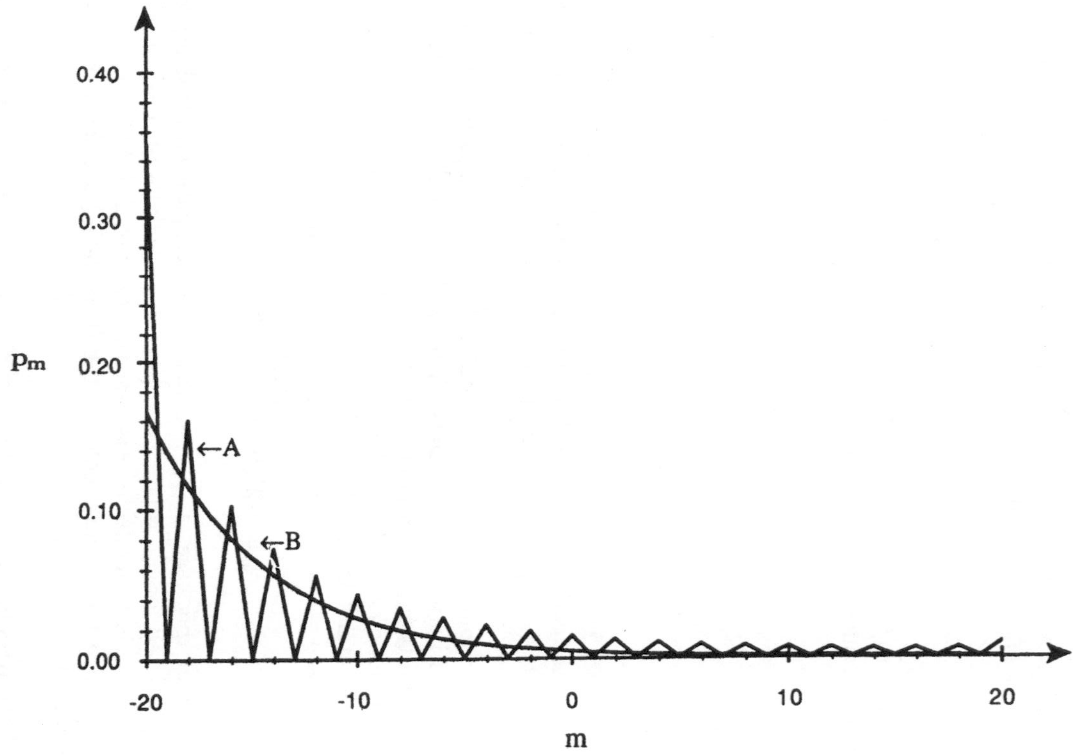

Figure 1. The probability of occupation of collective atomic states $|m>$ as a function of m for $\alpha = \delta = 0$ and for $N = 40$. Curve A is for the atoms interacting with a squeezed bath with $\bar{n} = 5$ and B is for the atoms interacting with an ordinary thermal reservoir having a mean of 5 photons.

$$\varrho_{ss} = \alpha_3 exp[a_1 S_z^2 + a_2 S_z + a_3(S_+^2 + S_-^2)], \quad (37)$$

where

$$a_1 = (1/2)ln(\lambda_1\lambda_2/\lambda_3^2), a_2 = -(cos(\theta))/2)ln(\lambda_1/\lambda_2),$$

$$a_3 = (sin(2\theta)/4)ln(\lambda_1/\lambda_2),$$

$$\lambda_{1,2} = [\alpha_1 + \alpha_3\{(\alpha_1 - \alpha_3)^2 + 4\alpha_4^2\}^{\frac{1}{2}}]/2, tan(\theta) = -\alpha_4/(\lambda_1 - \alpha_1),$$

$$\alpha_1 = [(\bar{n} + \bar{n}_{th})\{1 + 2\beta(1+\beta)\} + \{1 + 3\beta(1+\beta)\}]/[\{1 + 2(\bar{n} + \bar{n}_{th})\}\{1 + 3\beta(1+\beta)\}],$$

$$\alpha_2 = \beta(1+\beta)/[1 + 3\beta(1+\beta)],$$

$$\alpha_3 = [(\bar{n} + \bar{n}_{th})\{1 + 2\beta(1+\beta)\} - \beta(1+\beta)]/[\{1 + 2(\bar{n} + \bar{n}_{th})\}\{1 + 3\beta(1+\beta)\}],$$

$$\alpha_4 = -|m|/[1 + 2(\bar{n} + \bar{n}_{th})]. \quad (38)$$

The exact steady state solution of Eq. (10) for $\alpha = 0$ for a system of harmonic oscillators in a non-ideal bath is given by

$$\varrho_{ss} = [1/(1+\beta)]exp[-ln\{(1+\beta)/\beta\}R^+R] \quad (39)$$

where $R = \mu c + \nu c^+$.

It is interesting to see the relation of these solutions with the form of the density matrix[7]

$$\varrho = exp[-\sum_i b_i A_i], \quad (40)$$

where b's are constants, obtained by requiring that it maximizes the entropy

$$S = -Tr[\varrho ln(\varrho)] \quad (41)$$

subject to the constraints that the expectation value of the operators A_i's be constant i.e.

$$Tr[\varrho A_i] = a_i \quad (42)$$

The value of b_i's can be determined in terms of a_i's from (40) and (42). In the case of a system interacting with an ordinary uncorrelated bath such that its average energy remains constant, there is only one constraint so that $A_o = H$ where H is the system hamiltonian. Thus the equilibrium density matrix, $exp(-\lambda_o H)$, of a system interacting with an ordinary thermal bath maximizes the entropy (41) subject to the constraint that its average energy i.e. the expectation value of its hamiltonian H

remains constant. In the case of interaction with a squeezed bath, a comparison of the steady state solutions (36), (37) and (39) with (40) shows that these solutions maximize the entropy (41) such that for a single two-level atom $< S_z >$ and $< S_\pm >$; for two two-level atoms $< S_z >, < S_z^2 >$, and $< S_\pm^2 >$; and for harmonic oscillators $< a^+a >, < a^2 >$ and $< a^{+2} >$ remain constant. Thus, apart from the average energy (which is proportional to $< S_z >$ for two-level atoms and $< a^+a >$ for harmonic oscillators) the squeezed bath preserves certain correlations in the atomic observables. The kind of correlations preserved, however, depend on the kind of the atomic system interacting with the squeezed bath. It will be interesting to know the general nature of the constraints under which the maximization of the entropy (40) can determine the steady state solution of (10) with $\alpha = 0$ in a non-ideal bath.

Interaction of Atoms with Coloured Bath

Now we discuss the characteristic features of the dynamics of an atomic system interacting with a squeezed bath having finite correlation time. First, we discuss the case of N two-level atoms driven strongly by a resonant external field[8] by rewriting Eq. (10) with $\delta = 0$ in terms of the R operators defined in Eq. (18) to obtain

$$\begin{aligned}\frac{\partial \varrho}{\partial t} &= -2i|\alpha|[R_z, \varrho] - i[\beta_y(\theta, t)R_z, \varrho] + (1/4)[2R_-\varrho R_+ - R_+R_-\varrho - \varrho R_+R_-],\\ &+ (1/4)[2R_+\varrho R_- - R_-R_+\varrho - \varrho R_-R_+]\\ &+ (1/2)[2R_z\varrho R_z - R_zR_z\varrho - \varrho R_zR_z] + \text{other terms},\end{aligned} \tag{43}$$

where

$$\beta_y(\theta, t) = \beta_+ exp(i\theta) + \beta_- exp(-i\theta). \tag{44}$$

The "other terms" in Eq. (43) have the factors of the kind $exp(\pm 2i|\alpha|t)$ and exp $(\pm 4i|\alpha|t)$ in a frame rotating at the frequency $2|\alpha|$. These, in turn, make contributions of the order of $\gamma/|\alpha|$, $\mu/|\alpha|$ and $\lambda/|\alpha|$ to the solution of (43) and hence can be neglected if $|\alpha| >> \lambda, \mu, \gamma$. In this limit, Eq. (43) results in the following equations for the expectation values of the atomic operators

$$< \dot{R}_z >= -\gamma < R_z >,$$

$$< \dot{R}_+ >= [-2i|\alpha| - i\beta_y(\theta, t) - 3\gamma/2] < R_+ > . \tag{45}$$

Since $\beta(\theta, t)$ is a Gaussian process, Eq. (45) can be averaged over it by the standard methods. By ignoring the terms like $exp(-\gamma t)$ in the resulting equation for $< R_\pm >$ averaged over the distribution of β, we find that the spectrum consists of three peaks centered at ω_o and at $\omega_o \pm 2|\alpha|$. It is found that (a) the central component has the natural line width. Note that in the white noise limit, the central peak could be wider or narrower than the natural width depending on the relative phase between the driving field and that of the squeezed bath, (b) the width of the side peaks is given by

$$\gamma[3/2 + 2\{\bar{n} - \sqrt{\bar{n}(\bar{n}+1)}sin(2\psi)\}] \rightarrow \gamma[1 + 1/4\bar{n}]/2 \text{ for } \psi = \pi/4 \text{ and } \bar{n} >> 1 \tag{46}$$

i.e. the width may become subnatural. Recall that in the case of a broadband field the side peaks are always broader in the squeezed vacuum compared with the unsqueezed vacuum.

The effects of the finite correlation time of the bath modes on the emission and absorption spectrum from a single two-level atom in the absence of an external drive have also been studied.[9,10] In this case, the equations can only be solved numerically (see, however, Ref. (9) for some approximate analytic results). It is found that[10] the line width of the absorption spectrum decreases with an increase in the width of the squeezed bath spectrum at a fixed intensity $\bar{n}$. Thus, the prediction of line narrowing in the ideal case of a broadband squeezed vacuum may not hold for the realistic fields having finite band-width.

Dynamics of Atoms Inside a Squeezed Cavity

Next, we discuss the characteristics of the radiation from a system of atoms inside a DPA cavity. The cavity now contains, besides the non-linear optical crystal, N atoms interacting collectively with the field. The dynamical evolution of the cavity field and the atoms is described by the master equation.

$$\frac{\partial \varrho}{\partial t} = -i(\omega_p/2)[a^+a, \varrho] - i\omega_o[S_z, \varrho] - i[H, \varrho] + L\varrho \tag{47}$$

where

$$H = [g^*a^+c + gac^+] + [\varepsilon exp(-i\omega_p t/2)a^{+2} + \varepsilon^* exp(i\omega_p t/2)a^2]/2$$
$$+ [\alpha exp(-i\omega_p t)c^+ + \alpha^* exp(i\omega_p t)c], \tag{48}$$

and

$$L\varrho = \kappa(2a\varrho a^+ - a^+a\varrho - \varrho a^+a) + \Gamma(2c\varrho c^+ - c^+c\varrho - \varrho c^+c) \tag{49}$$

The first term in Eq. (48) describes the interaction between the cavity field and the atoms, represented by the operators c, c^+, the second describes the parametric generation of two photons from the pump at the rate ε and the third describes the action of the external drive on the atoms. Eq. (49) describes the dissipation of the cavity field at the rate 2κ and that of the atomic oscillations at the rate 2Γ.

Eq. (47) can be solved exactly in the case of atoms modeled as harmonic oscillators[11] i.e. when $[c, c^+] = 1$. In this case, with α=0, it is found that the system acts as an oscillator if the coupling g between the atoms and the cavity field is such that $g > |\kappa - \gamma - \varepsilon|/2$. Thus the threshold for oscillations in a squeezed cavity is lower than that in an unsqueezed cavity which corresponds to ε=0.

Apart from the case of harmonic oscillators, finding an exact analytic solution of Eq. (47) is a formidable task. However, exact analytic solutions can be found in the limiting cases of (i) strong field damping, and (ii) strong atomic damping. The field variables in case (i) and the atomic variables in case (ii) can then be eliminated adiabatically. In case of strong field damping, the adiabatic elimination of the field from Eq. (47) results in the following equation for the atomic density matrix ϱ_a in a frame rotating at $\omega_p/2$

$$\frac{\partial \varrho_a}{\partial t} = -i\delta[c^+c, \varrho_a] - i|\varepsilon|\gamma[g^2c^{+2} + g^*c^2]/\kappa - i[\alpha exp(-i\omega_p t)c^+ + \alpha^* exp(i\omega_p t)c] + L\varrho_a, \tag{50}$$

where $L\varrho$ is given by Eq. (10) with $\delta = \omega_o - \omega_p/2$ and

$$\bar{n} = |\varepsilon|^2/(\kappa^2 - |\varepsilon|^2),\ \gamma = |g|^2\kappa/(\kappa^2 - |\varepsilon|^2),\ |m| = \sqrt{\bar{n}(\bar{n}+1)}. \tag{51}$$

Except for the parametric term, $c^{+2} + c^2$, this equation has the form of the equation for the atoms interacting with the field transmitted out of the cavity. In the case of harmonic oscillators, the parametric term results in the lowering of the oscillation threshold whereas in the case of a single two- level atom it does not contribute because then $S_+^2 = S_-^2 = 0$. The studies of the consequences of the parametric term in the case of N two-level atoms for $N > 1$ are underway.

In case when the atomic damping is very strong, the adiabatic elimination of the atomic variables results in the following equation for the density matrix ϱ_f of the cavity field[12]

$$\frac{\partial \varrho_f}{\partial t} = -i\delta_o[a^+a, \varrho_f] - i[\varepsilon a^{+2} + \varepsilon^* a^2, \varrho_f]/2 + \kappa(2a\varrho_f a^+ - a^+a\varrho_f - \varrho_f a^+ a)$$

$$+ [A_1(a\varrho_f a^+ - \varrho_f a^+ a) + A_2(a\varrho_f a^+ - aa^+\varrho_f) + \text{h.c.}], \tag{52}$$

where $\delta_o = \omega - \omega_p/2$ and A's are given in terms of the two-time steady-state correlation functions of the c-operators. It is found that the system reaches a steady state if

$$\kappa + Re(A_1) - Re(A_2) > 0, |\kappa + A_1 - A_2 - i\delta_o| > |\varepsilon|^2, \tag{53}$$

where $Re(z)$ denotes the real part of z. $2(Re(A_1) - Re(A_2))$ is the rate of absorption of the cavity photons by the atoms. The first of the equations in (53) therefore is a condition on the optical gain processes whereas the second is a restriction on the squeezing for the system to reach a steady state. In case of an unsqueezed cavity i.e. ε=0, the second condition is, of course, redundant. In a squeezed cavity, however, one may encounter a situation in which the first condition is satisfied but not the second. Then it is not the optical gain processes but the cavity mode squeezing that would be responsible for instability.

REFERENCES

1. C. W. Gardiner, *Phys. Rev. Lett.* **56**, 1917 (1986).
2. H. Ritsch and P. Zoller, *Phys. Rev.* **A38**, 4657 (1988).
3. H. Ritsch and P. Zoller, *Optics Comm.* **64**, 523 (1987).
4. H. J. Carmichael, A. S. Lane and D. F. Walls, *Phys. Rev. Lett.* **58**, 2539 (1987).
5. G. S. Agarwal and R. R. Puri, *Opt. Comm* **69**, 267 (1989); *Phys. Rev.* **A41**, 3782 (1990).
6. G. M. Palma, P. Knight, *Phys. Rev.*, **A39**, 1962 (1962).

7. R. R. Tucci, *J. Mod. Phys.*, **B5**, 545 (1991).
8. A. S. Parkins, *Phys. Rev.* **A42**, 4352 (1990).
9. A. S. Parkins and C. W. Gardiner, *Phys. Rev.* **A40**, 3796 (1989).
10. H. Ritsch and P. Zoller, *Phys. Rev. Lett.*, **61**, 1097 (1988).
11. G. S. Agarwal and S. DattaGupta, *Phys. Rev.* **A39**, 2961 (1989).
12. G. S. Agarwal, *Phys. Rev.* **A40**, 4138 (1990).

WIGNER FUNCTION FOR COHERENT SUPERPOSITION OF TWO SQUEEZED STATES

J.K. Sharma

Physics Department
University of Jodhpur
Jodhpur (Rajasthan) India

INTRODUCTION

Squeezed state of radiation field is a minimum uncertainty product state, but in this state the uncertainty in quadrature component of the electric field is reduced at the cost of increased uncertainty in other component. The product of these uncertainties is still equal to that corresponding to a vacuum state or coherent state. A non-linear oscillator can generate superposition of two coherent states[1] as well as superposition of two squeezed states.[2] In this paper, we obtain the Wigner function for coherent superposition of two squeezed states. We show that it may be negative in some range, and it displays oscillations. Expectation values of some quantum operators has been calculated as a classical average of corresponding classical functions in Weyl's rule over Wigner function.

WIGNER FUNCTION FOR COHERENT SUPERPOSITION OF TWO SQUEEZED STATES

The squeezed state $|z;\mu,\nu>$ is defined by eigenvalue equation

$$(\mu\hat{a}+\nu\hat{a}^{+})|z;\mu,\nu>=(\mu z+\nu z^{*})|z;\mu,\nu> \tag{1}$$

Here $\hat{a}$ and $\hat{a}^{+}$ are annihilation and creation operators of single mode radiation field, μ and ν have been restricted to real numbers satisfying relation $\mu^2-\nu^2=1$, z is any complex number $x+iy$. In the limit $\mu=1,\ \nu=0$, the $|z;\mu,\nu>$ state reduces to the coherent state $|z>$. One may also work with operators $\hat{x}$ and $\hat{p}$ defined as

$$\hat{x}=2^{-\frac{1}{2}}(\hat{a}+\hat{a}^{+});\quad \hat{p}=\frac{2^{-\frac{1}{2}}}{i}(\hat{a}-\hat{a}^{+}). \tag{2}$$

The operators $\hat{x}$ and $\hat{p}$ may be identified as amplitudes of two quadrature phases of the electric field

$$\hat{E}(\vec{r},t)=\chi(\vec{r})(\hat{x}cos\omega t+\hat{p}\ sin\omega t). \tag{3}$$

Recent Developments in Quantum Optics, Edited
by R. Inguva, Plenum Press, New York, 1993

The wavefunction for a squeezed state is

$$< x|z;\mu,\nu >= \pi^{-\frac{1}{4}}(\frac{c+1}{c-1})^{\frac{1}{4}} exp[\frac{c+1}{c-1}(Rez)^2] \text{x}$$

$$exp[-\frac{1}{2}\frac{c+1}{c-1}x^2 + \sqrt{2}\frac{(cz+z^*)}{c-1}x]. \tag{4}$$

Here $c = \mu/\nu$. Using this, the wavefunction for coherent superposition of two states

$$|\chi >= N[|z_1;\mu,\nu > + e^{i\phi}|z_2;\mu,\nu >] \tag{5}$$

may be obtained as

$$\chi(x) = N[\chi_1(x) + e^{i\phi}\chi_2(x)] \tag{6}$$

where N is a normalization factor.

The Wigner function $W(q,p)$ for the state $|\chi >$ is given by the relation[3]

$$W(q,p) = \frac{1}{\pi}\int_{-\infty}^{\infty} e^{-i2px}\chi^*(q-x)\chi(q+x)dx \tag{7}$$

From (6) and (7), we obtain

$$W(q,p) = N^2\{W_1 + W_2 + 2\sqrt{W_1W_2}exp\left[\frac{1}{2}\frac{c+1}{c-1}(x_2-x_1)^2 + \frac{1}{2}\frac{c-1}{c+1}(y_2-y_1)^2\right] \text{x}$$

$$cos[\sqrt{2}q(y_2-y_1) - \sqrt{2}p(x_2-x_1) + (x_2-x_1)(y_1+y_2) + \phi]\}. \tag{8}$$

Here $W_1(q,p)$ and $W_2(q,p)$ are Wigner functions for squeezed states $|z_1;\mu,\nu >$ and $|z_2;\mu,\nu >$ respectively and $z_1 = x_1 + iy_1, z_2 = x_2 + iy_2$. Since $W(q,p)$ is a quasi-probability satisfying relation

$$\int_{-\infty}^{\infty}\int_{-\infty}^{\infty} W(q,p)dqdp = 1, \tag{9}$$

we obtain from (8) and (9),

$$N = 2^{-\frac{1}{2}}\left[1 + cos\Phi exp[-\frac{c+1}{c-1}\frac{(x_2-x_1)^2}{2} - \frac{c-1}{c+1}\frac{(y_2-y_1)^2}{2}]\right]^{-\frac{1}{2}}, \tag{10}$$

where $\Phi = (x_1+x_2)(y_2-y_1) + \phi$.

It is clear from relation (8) that $W(q,p)$ may take negative values in some range and that it displays oscillation. Using (8), we can easily calculate averages of quantum operators in the state $|\chi >$. We obtain

$$< \hat{x} >= \frac{x_1 + x_2}{\sqrt{2}}, \quad < \hat{p} >= \frac{y_1 + y_2}{\sqrt{2}},$$

$$< (\Delta\hat{x})^2 >=< (\hat{x}- < \hat{x} >)^2 >= \frac{1}{2}\frac{c-1}{c+1} + N^2(x_1 - x_2)^2,$$

$$< (\Delta\hat{p})^2 >=< (\hat{p}- < \hat{p} >)^2 >= \frac{1}{2}\frac{c+1}{c-1} + N^2(y_1 - y_2)^2,$$

$$< \frac{(\Delta\hat{x})(\Delta\hat{p}) + (\Delta\hat{p})(\Delta\hat{x})}{2} >= N^2(x_1 - x_2)(y_1 - y_2). \tag{11}$$

RESULTS

The Wigner function for coherent superposition of two squeezed states is not positive definite and it displays oscillations.

REFERENCES

1. P. Tombesi and A. Mecozzi, "Generation of . . . technique," *J. Opt. Soc. Am.* **4**:1700 (1987).
2. B. C. Sanders, "Superposition of two squeezed vacuum states and interference effects," *Phys. Rev. A.* **39**:4284.
3. W. H. Louisell, "Operator Algebra" in: "Quantum Statistical Properties of Radiation," John Wiley (1973).

EXPONENTIAL OF A QUADRATIC IN BOSON OPERATORS AND SQUEEZED STATES

Anil K. Roy and C. L. Mehta

Physics Department
Indian Institute of Technology
New Delhi 110 016, India

1. INTRODUCTION

We consider harmonic oscillator states with $\hat{a}$, $\hat{a}^\dagger$ as annihilation and creation operators which satisfy the commutation relation $[\hat{a},\ \hat{a}^\dagger] = 1$. The vacuum state $|0>$ is the eigenstate of boson annihilation operator a with eigenvalue 0. Any unitarily transformed vacuum $U|0>$ must, therefore, be the eigenstate of the unitarily transformed annihilation operator, $U\hat{a}U^\dagger$, with eigenvalue 0, where U is a unitary operator. Examples of these states are:

Coherent states, where the unitary operator U is the displacement operator

$$D(\alpha) = exp(\alpha\hat{a}^\dagger - \alpha^*\hat{a}). \tag{1.1}$$

with

$$D(\alpha)\hat{a}D^\dagger(\alpha) = \hat{a} - \alpha. \tag{1.2}$$

Squeezed vacuum states, where the unitary operator U is the squeeze operator

$$S(\sigma) = exp\{\frac{1}{2}(\sigma^*\hat{a}^2 - \sigma\hat{a}^{\dagger 2})\}, \tag{1.3}$$

and

$$S\hat{a}S^\dagger = \hat{a}\ \cosh\ r + \hat{a}^\dagger e^{i\phi} \sinh\ r, \tag{1.4}$$

with

$$\sigma = re^{i\phi}. \tag{1.5}$$

Squeezed coherent states, where the unitary operator is the product $S(\sigma)D(\alpha)$. The eigenvalue equation than becomes:

Recent Developments in Quantum Optics, Edited
by R. Inguva, Plenum Press, New York, 1993

$$(\hat{a}\ \cosh\ r + \hat{a}^{\dagger}e^{i\phi}\sinh\ r)|\sigma,\alpha> = \alpha\ |\sigma,\alpha>, \tag{1.6}$$

with

$$|\sigma,\alpha> = S(\sigma)D(\alpha)\ |0> . \tag{1.7}$$

All these states have been studied extensively [1-3]. In the present paper we consider a unitary operator $U(\sigma,\beta)$ in which the exponent is a general quadratic in boson operators:

$$U(\sigma,\beta) = exp\{\frac{1}{2}(\sigma^{*}\hat{a}^{2} - \sigma\hat{a}^{\dagger 2}) - i\beta\ (\hat{a}^{\dagger}\hat{a} + \hat{a}\hat{a}^{\dagger})\} \tag{1.8}$$

The unitarity of the operator requires that β must be real and σ may be any complex parameter. We then define a state $|\alpha,\sigma,\beta>$ as:

$$|\alpha,\sigma,\beta> = D(\alpha)\ U(\sigma,\beta)\ |0> \tag{1.9}$$

and consider its properties.

2. UNITARY TRANSFORMATION AND THE EIGENVALUE EQUATION

A straight forward application of the result:

$$e^{\hat{A}}\hat{B}e^{-\hat{A}} = \hat{B} + [\hat{A},\hat{B}] + \frac{1}{2!}[\hat{A},[\hat{A},\hat{B}]]\ +\ ... \tag{2.1}$$

gives the following expression for the unitarily transformed operator $\hat{a}$:

$$U\hat{a}U^{\dagger} = k\hat{a} + m\hat{a}^{\dagger}. \tag{2.2}$$

Where

$$k = C\ -\ 2i\beta S, \tag{2.3}$$

and

$$m = \sigma S, \tag{2.4}$$

with the notation

$$C = \cosh\ \lambda, \tag{2.5}$$

$$S = (\sinh\ \lambda)/\lambda, \tag{2.6}$$

$$\lambda^{2} = r^{2}\ -\ 4\beta^{2}. \tag{2.7}$$

and

$$\sigma = re^{i\phi}. \tag{2.8}$$

The transformations of the operators such as $\hat{a}^\dagger$ or the coordinate $\hat{q}$ and its canonical conjugate momentum $\hat{p}$ can be written down using Eq. (2.2). From the eigenvalue equation

$$D(\alpha)U(\sigma,\beta)\hat{a}U^\dagger(\sigma,\beta)D^\dagger(\alpha)D(\alpha)U(\sigma,\beta)\;|0> = 0,$$

we readily find that:

$$(k\hat{a} + m\hat{a}^\dagger)\;|\alpha,\sigma,\beta> = (k\alpha + m\alpha^*)\;|\alpha,\sigma,\beta>. \tag{2.9}$$

Note that the case $\beta = 0$ corresponds to the well studied Squeezed states [4]. It may also be noted from Eq. (2.9) and the properties of the displacement operator that we may write:

$$D(\alpha)U(\sigma,\beta) = U(\sigma,\beta)D(k\alpha + m\alpha^*). \tag{2.10}$$

3. EXPECTATION VALUES AND UNCERTAINTIES

We next consider the expectation values of various operators in the state $|\alpha,\sigma,\beta>$. Using standard definition of the expectation value:

$$<\hat{a}> = <\alpha,\sigma,\beta|\hat{a}|\alpha,\sigma,\beta>, \tag{3.1}$$

and of the uncertainties in $\hat{q}$ and $\hat{p}$:

$$(\Delta q)^2 =<\hat{q}^2> - <\hat{q}>^2, \text{ and } (\Delta p)^2 = <p^2> - <\hat{p}>^2, \tag{3.2}$$

we readily obtain the uncertainties in $\hat{q}$ and $\hat{p}$:

$$(\Delta q)^2 = \frac{1}{2}[1 + 2rS(rS - 2\beta S\;\sin\;\phi - C\;\cos\;\phi)], \tag{3.3}$$

$$(\Delta p)^2 = \frac{1}{2}[1 + 2rS(rS + 2\beta S\;\sin\;\phi + C\;\cos\;\phi)]. \tag{3.4}$$

Note that the uncertainties in $\hat{q}$ and $\hat{p}$ are independent of the parameter α. We also find that the case $\beta = 0$ is consistent with the result obtained by Stoler [2].

4. CONDITION FOR MUS AND SQUEEZING

The uncertainty product is given by:

$$(\Delta q)^2(\Delta p)^2 = \frac{1}{4}[1 + 4r^2S^2(C\;\sin\;\phi - 2\beta S\;\cos\;\phi)^2]. \tag{4.1}$$

The state $|\alpha,\sigma,\beta>$ is a MUS (minimum uncertainty product state) iff:

$$\tan \phi = 2\beta \tanh \lambda / \lambda \tag{4.2}$$

If we require squeezing in q−parameter we obtain from Eq. (3.3) a relation

$$r < \lambda \coth \lambda \cos \phi + 2\beta \sin \phi. \tag{4.3}$$

Similarly squeezing in p−parameter requires:

$$r < -(\lambda \coth \lambda \cos \phi + 2\beta \sin \phi). \tag{4.4}$$

Since r is positive, only one of the two possibilities, i.e., Eq. (4.3) or (4.4) can be satisfied at a time. If $r > |\lambda \coth \lambda \cos \phi + 2\beta \sin \phi|$ squeezing in neither q nor in p is possible. We thus find that the condition for squeezing and/or MUS can be obtained by adjusting the parameters β and ϕ. This is the generalization of the condition for squeezing which we wish to draw attention to.

5. BUNCHING AND ANTI-BUNCHING

In this section we calculate the fluctuation in the number of photons. For this we are interested in calculating:

$$(\Delta n)^2 - <\hat{n}> = <\hat{a}^{\dagger 2}\hat{a}^2> - <\hat{n}>^2 . \tag{5.1}$$

We readily obtain:

$$(\Delta n)^2 - <\hat{n}> = r^2 S^2 [2|\alpha|^2 + 2r^2 S^2 + 1] - 2Re(\alpha^2 k m^*). \tag{5.2}$$

Eq. (5.2) shows that we may get bunching or anti-bunching by adjusting the parameters α, r, β. For the case $\alpha = 0$, we get:

$$(\Delta n)^2 - <\hat{n}> = <\hat{n}> [2 <\hat{n}> + 1]. \tag{5.3}$$

which is always positive giving bunching of photons. On the other hand, if we study the case of squeezed states in the dominance of coherent contribution, i.e. if $|\alpha| \neq 0$ and $rS << 1$, we find that:

$$(\Delta n)^2 - <\hat{n}> = r^2 S^2 (2|\alpha|^2 + 1) - 2|\alpha|^2 r S \cos \psi. \tag{5.4}$$

Here ψ is the phase of the complex quantity $(\alpha^2 k m^*)$ and $|k| \simeq 1$. Hence $< \Delta n >^2 - < n >$ may be negative in this case, leading to antibunching.

REFERENCES

1. R. J. Glauber, Phys. Rev, 131, 2766 (1963).
2. D. Stoler, Phys. Rev. D, 1, 3217 (1970).
3. H. P. Yuen, Phys. Rev. A, 13, 2226 (1976).
4. R. Loudon and P.L. Knight, J. Mod Opt., 34, 709 (1987).

CORRELATED AND SQUEEZED COHERENT STATES OF NONSTATIONARY QUANTUM SYSTEMS

V. V. Dodonov*, A. V. Klimov†, V. I. Man'ko†

*Moscow Physics Technical Institute
Moscow, USSR

†PN Lebedev Physical Institute
Moscow, USSR

Correlated quantum states were introduced in ref. (1) as the states minimizing the left-hand side of the generalized uncertainty relation by Schrödinger and Robertson (2-4)

$$\sigma_p \sigma_q (1 - r^2) \geq \hbar^2/4 \tag{1}$$

σ_p and σ_q are variances of the canonically conjugated momentum and coordinate operations: $\sigma_q = < q^2 > - < q >^2$; r is their correlation coefficient:

$$r = \frac{\sigma_{pq}}{\sqrt{\sigma_p \sigma_q}}, \quad \sigma_{pq} = \frac{1}{2} < pq + qp > - < p >< q > \tag{2}$$

The wave functions of the correlated states in the coordinate representation are Gaussian exponentials with the leading term like

$$\psi(x) \sim exp\Big[- \frac{x^2}{4\sigma_q}\Big(1 - \frac{ir}{\sqrt{1 - r^2}}\Big) + ...\Big] \tag{3}$$

In a special case of zero correlation coefficient we have well-known squeezed states with a real squeezing parameter (a general state (3) can be also treated as the squeezed state with a complex squeezing parameter; however, we think that the name "correlated state" is more adequate).

The simplest way to create the state (3) is to use some time-parametric process. For example, if we consider a quantum harmonic oscillator with time-dependent frequency ω(t), then the solutions of Schrodinger equation can be chosen (5) in the form of the Gaussian exponentials with the leading term

$$\psi(x, t) \sim exp\Big(\frac{i\dot{\beta}}{2\beta} x^2 + ...\Big) \tag{4}$$

Recent Developments in Quantum Optics, Edited
by R. Inguva, Plenum Press, New York, 1993

where $\beta(t)$ is a solution of a classical equation of motion

$$\ddot{\beta} + \omega^2(t)\beta = 0 \tag{5}$$

satisfying the subsidiary condition

$$\dot{\beta}\beta^* - \dot{\beta}^*\beta = 2i \tag{6}$$

(hereafter we use dimensionless variables, assuming $\hbar = 1$, etc). Comparing (3) and (4) one express the variances and the correlation coefficient in terms of β-function as follows (6) from

$$\sigma_q = \frac{1}{2}|\beta|^2, \ \sigma_p = \frac{1}{2}|\dot{\beta}|^2, \ r^2 = 1 - |\beta\dot{\beta}|^{-2} \tag{7}$$

If the frequency depends on time according to inverse linear law:

$$\omega(t) = \omega_o/(1 + \lambda t) \tag{8}$$

then Eq. (5) admits exact solutions for which the correlation coefficient is conserved in time: $r = \lambda/2\omega_o$. Moreover, the squeezing coefficient normalized by instant value of the frequency equals unity: $\omega^2(t)\ \sigma_q/\sigma_p \equiv 1$. Thus, we have an example of correlated but nonsqueezed state.

Eq. (5) itself looks like Schrödinger equation with "effective potential" $\omega^2(t)$. Introducing the energy reflection coefficient from this "barrier" R we can get the following inequalities for the maximum values of correlation and squeezing coefficients (7,8):

$$|r| \leq \frac{2\sqrt{R}}{1+R}, \ \left(\frac{1-\sqrt{R}}{1+\sqrt{R}}\right)^2 \leq \omega_f^2 \frac{\sigma_q}{\sigma_q} \leq \left(\frac{1+\sqrt{R}}{1-\sqrt{R}}\right)^2 \tag{9}$$

In the case of the parametric resonance, when

$$\omega^2(t) = \omega_o^2(1 + \mathcal{H} cos 2\omega_o t), \ \mathcal{H} << 1, \tag{10}$$

one can obtain an approximate solution of Eq. (5)

$$\beta(t) \sim \frac{1}{\sqrt{\omega_o}}\left[cosh(\mathcal{H}\omega_o t)e^{-i\omega_o t} - i\ sinh(\mathcal{H}\omega_o t)e^{-i\omega_o t}\right] \tag{11}$$

(satisfying Eq. (6)). Then the squeezing coefficient oscillates with the frequency $2\omega_o$, but its maximal and minimal values at each time instant are confined between the limits

$$exp(-4\mathcal{H}\omega_o t) \leq \omega_o^2 \frac{\sigma_q}{\sigma_p} \leq exp(4\mathcal{H}\omega_o t) \tag{12}$$

Thus, the only problem is to realize the model of an oscillator with time-dependent frequency in a physical experiment. Consider a resonator occupied with some space uniform linear medium with time-dependent dielectric and magnetic permeabilities

$\varepsilon(t)$ and $\mu(t)$. Introducing the vector potential as usual: $B = rotA$, and choosing the gauge ϕ=0, $divA = 0$ one obtains the following consequence of Maxwell's equation

$$\frac{c}{\mu t}\Delta A = \frac{1}{c}\frac{\partial}{\partial t}\left[\varepsilon(t)\dot{A}\right] = -\dot{D} \tag{13}$$

For simplicity we confine ourselves to the case of "one- dimensional electrodynamics," when $A(t,x,y,z) = (0,0,A(t,x))$. Then Eq. (13) can be derived from the Lagrangian density

$$\mathcal{L} = \frac{1}{2}\left\{\frac{\varepsilon(t)}{c^2}\left(\frac{\partial A}{\partial t}\right)^2 - \frac{1}{\mu(t)}\left(\frac{\partial A}{\partial x}\right)^2\right\} \tag{14}$$

Then canonical momentum density equals

$$\prod = \partial\mathcal{L}/\partial A_t = \frac{\varepsilon(t)}{c^2}\dot{A} = -\frac{1}{c}D \tag{15}$$

and the hamiltonian density is equal to

$$\mathcal{Y} = \prod A - \mathcal{L} = \frac{1}{2}\left\{\frac{1}{\varepsilon(t)}D^2 + \frac{1}{\mu(t)}\left(\frac{\partial A}{\partial x}\right)^2\right\} \tag{16}$$

Formally, it is the same as in the case of a medium with constant parameters, although in the present case $\varepsilon(t)$ and $\mu(t)$ can be arbitrary functions of time. Introducing the mode expansion of fields $D(t,x)$ and $A(t,x)$:

$$D(t,x) = \sqrt{\frac{2c}{L}}\sum_n q_n(t)sin\left(\frac{n\pi}{L}x\right) \tag{17}$$

$$A(t,x) = \frac{\sqrt{2c}}{L}\sum_n p_n(t)sin\left(\frac{n\pi}{L}x\right) \tag{18}$$

(L is the length of the resonator) we can transform the field Hamiltonian to the sum of independent oscillator mode Hamiltonians:

$$H = \int_0^L \mathcal{Y}(t,x)dx = \frac{1}{2}\sum_n\left\{\frac{ck_n^2}{\mu(t)}p_n^2 + \frac{c}{\varepsilon(t)}q_n^2\right\},\ k_n = \frac{n\pi}{L} \tag{19}$$

Time-dependent frequency of each mode is given by the relation

$$\omega_n^2(t) = \frac{c^2k_n^2}{\varepsilon(t)\mu(t)} \tag{20}$$

Eq. (19) elucidates the physical significance of quadrature components of electromagnetic field: the generalized coordinate is related to the electric displacement vector D, while the generalized momentum is related to the magnetic induction vector B or to the vector potential A (see also in this connection ref. (9)).

Another method of generating squeezed states of the field inside the empty resonator consists in exciting field modes though their interaction with moving walls of the resonator. This problem was investigated in refs. (10,11). It cannot be reduced to the corresponding mathematics is much more complicated. Therefore, we give here only qualitative results obtained in refs. [10,11]. If the wall vibrates with twice frequency according to Eq. (10) (with the replacement of $\omega^2(t)$ by $L(t)$), then the maximum degree of squeezing of the resonance field mode is given by the same formula (12) provided $\omega_o t << 1$.

Some other interesting related problems, such as the nonstationary Casimir effect or spontaneous radiation of a charged oscillator from an arbitrary quantum state (including squeezed or correlated ones) were considered in refs. (12,13).

REFERENCES

1. V. V. Dodonov, E. V. Kurmyshev, V. I. Man'ko: *Phys. Lett.,* 1980, v. **79A**, N 2,3 p. 150-152.
2. E. Schrodinger: Ber. Kgl. Akad. Wiss., Berlin, 1930, p. 296-303.
3. H. P. Robertson: Phys. Rev., 1930, v. 35, N 5, p. 667.
4. V. V. Dodonov, V. I. Man'ko: In "Invariants and the evolution of nonstationary quantum systems," Proceedings of Lebedev Physics Institute, vol. 183, p. 3-101 (Nova Science, Commack, 1989).
5. I. A. Malkin, V. I. Man'ko: Phys. Lett., 1970, v. 31A, N4, p. 243-244.
6. V. V. Dodonov, V. I. Man'ko: Proceedings of Lebedev Physics Institute, vol. 183, p. 103-261.
7. V. V. Dodonov, A. B. Klimov, V. I. Man'ko: Phys. Lett., 1989, v. 134A, N4, p. 211-216.
8. V. V. Dodonov, A. B. Klimov, V. I. Man'ko: Proceedings of Lebedev Physics Institute, vol. 200 (1991).
9. M. Born, L. Infeld: Proc. Roy. Soc. London, ser. A, 1934, v. 147, p. 522; 1935, v. 150, p. 141.
10. V. V. Dodov, A. B. Klimov, V. I. Man'ko: Phys. Lett., 1990, v. 149A, N4, p. 225-228.
11. V. V. Dodonov, A. B. Klimov, V. I. Man'ko: Proceedings of the Workshop "Squeezing and Correlated States," Moscow, 3-7 December 1990, Soviet Journal of Laser Research, 1991.
12. V.V. Dodonov, A. B. Klimov, V. I. Man'ko: Phys. Lett., 1989, v. 142A, N8, 9, p. 511-513.
13. V. V. Dodonov, O.V. Man'ko, V. I. Man'ko: Proceedings of Lebedev Physics Institute, v. 192, p. 204-220 (1988).

NONCLASSICAL PROPERTIES OF STATES GENERATED BY THE EXCITATIONS ON A COHERENT STATE†

K. Tara

School of Physics
University of Hyderabad
Hyderabad 500134, India

1. INTRODUCTION

In the past few years there has been considerable interests in attempts to produce non-classical states of light such as squeezed states and photon number states. The squeezed states have reduced fluctuations in one field quadrature when compared with the coherent states.[1] In this paper we consider the state obtained by repeated application of the photon creation operator on the coherent state. Such a state has a nonzero field amplitude and is shown to exhibit non-classical properties like the squeezing in one of the quadratures of the field, and sub- Poissonian photon statistics. We calculate different quasiprobability functions for fields in such states and also the distribution function for one of the field quadratures. In the last section we discuss how such states can be generated in nonlinear processes in cavities.

2. THE STATE $|\alpha, m>$ AND ITS NONCLASSICAL PROPERTIES

We introduce the state $|\alpha, m>$ defined by

$$|\alpha, m> = \frac{a^{\dagger m}|\alpha>}{\sqrt{<\alpha|a^m a^{\dagger m}|\alpha>}} = \frac{a^{\dagger m}|\alpha>}{\sqrt{m! L_m(-|\alpha|^2)}}, \tag{2.1}$$

where $|\alpha>$ is a coherent state, m is an integer and $L_m(x)$ is the Laguerre polynomial of order m^2. In the limit $\alpha \to 0$ $(m \to 0)$ the state $|\alpha, m>$ reduces to Fock state (coherent state). Thus it is a state intermediate between Fock state and coherent state.

We first examine fluctuation characteristics of the radiation field which is in the state (2.1). Let us consider the field quadrature x defined by

$$x = \frac{a\, exp(i\phi) + a^{\dagger} exp(-i\phi)}{2}. \tag{2.2}$$

†Work done in collaboration with G.S. Agarwal

Recent Developments in Quantum Optics, Edited by R. Inguva, Plenum Press, New York, 1993

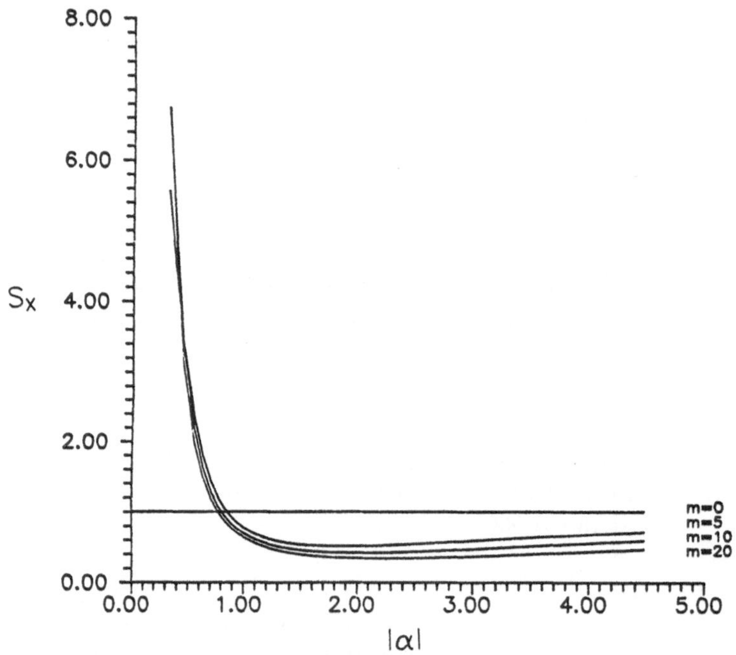

Figure 1. Uncertainty in field quadrature x, S_x as a function of $|\alpha|$ for different values of m.

For the state $|\alpha, m>$ the fluctuations in x will be

$$(\Delta x)^2 =< x^2 > - < x >^2$$

$$= \Big\{(L_m^{(2)}(-|\alpha|^2)L_m(-|\alpha|^2) - (L_m^{(1)}(-|\alpha|^2))^2)2|\alpha|^2 cos(2(\theta+\phi))$$

$$-2(L_m^{(1)}(-|\alpha|^2))^2|\alpha|^2 - (L_m(-|\alpha|^2))^2$$

$$+2(m+1)L_{m+1}(-|\alpha|^2)L_m(-|\alpha|^2)\Big\} \Big/ 4(L_m(-|\alpha|^2))^2. \tag{2.3}$$

where $L_m^{(k)}(x)$ is the associated Laguerre polynomial.[2]

We show in Fig. 1 the quantity $S_x = 4(\Delta x)^2$ as a function of the parameter $|\alpha|$ for different values of m. For $m \neq 0,\ \alpha \neq 0$ the Fig. 1 shows values of S_x less than one implying that the quadrature x of the field is squeezed. We get almost 50% squeezing over a wide range of parameters.

We next study the characteristics of the number distribution of the field in the state $|\alpha, m>$. We calculate the parameter Q defined by

$$Q(\alpha, m) = \frac{< (a^\dagger a)^2 > - < a^\dagger a >^2}{< a^\dagger a >} \tag{2.4}$$

which measures the deviation from the Poisson distribution. We find that the mean number of photons and the parameter $Q(\alpha, m)$ are given by

$$\bar{n} = \frac{(m+1)L_{m+1}(-|\alpha|^2)}{L_m(-|\alpha|^2)} - 1 \tag{2.5}$$

$$Q(\alpha, m) = \frac{\left\{((m+2)L_{m+2}(-|\alpha|^2) - L_{m+1}(-|\alpha|^2))(m+1)L_m(-|\alpha|^2) - ((m+1)L_{m+1}(-|\alpha|^2))^2\right\}}{L_m(-|\alpha|^2)((m+1)L_{m+1}(-|\alpha|^2) - L_m(-|\alpha|^2))} \tag{2.6}$$

In Fig. 2 we show the mean number of photons (Eq. (2.5)) for different values of $|\alpha|^2$ and m. In Fig. 3 we display the parameter $Q(\alpha, m)$ as a function of $|\alpha|$ for different values of m. The values of $Q(\alpha, m)$ less than one signify the sub-Poissonian statistics of the field. For $\alpha \neq 0,\ m \neq 0$ we see that the field in the state $|\alpha, m>$ exhibits significant amount of sub-Poissonion statistics.

We next calculate different quasiprobability distributions for the state $|\alpha, m>$. These distributions provide a convenient way of studying the non-classical properties of fields. The Glauber-Sudarshan P function associated with the state $|\alpha, m>$ is found to be

$$P(z) = \frac{exp(|z|^2 - |\alpha|^2)}{m! L_m(-|\alpha|^2)} \frac{\partial^{2m}}{\partial z^{*m} \partial z^m} \delta^{(2)}(z - \alpha). \tag{2.7}$$

Thus the quasiprobability distribution P is highly singular. This is quite typical of states exhibiting non-classical character. The properties of other quasiprobability distributions like Q function and the Wigner function can be found in the paper which is now published.[3]

3. PRODUCTION OF THE STATE $|\alpha, m>$

In this section we discuss how the states $|\alpha, m>$ can be produced in principle. Consider the passage of the excited two level atoms through a cavity. The atom makes transition from the excited state to the ground state by emitting a photon.[4]

Let the initial state of the atom-field system be $|\alpha>|e>$ where $|\alpha>$ is the coherent state of the field. The interaction Hamiltonian has the form

$$H = \hbar(gS^+a + g^*S^-a^\dagger). \tag{3.1}$$

For interaction times such that $gt << 1$, the state at time t can be approximated by

$$|\psi(t)> \cong |\alpha>|e> - ig^*a^\dagger|\alpha>|g> . \tag{3.2}$$

From Eq. (3.2) we observe that if the atom is detected to be in the ground state $|g>$, then the state of the field is reduced to $a^\dagger|\alpha>$ i.e. to $|\alpha, 1>$. Thus we can in principle produce the state $|\alpha, 1>$.

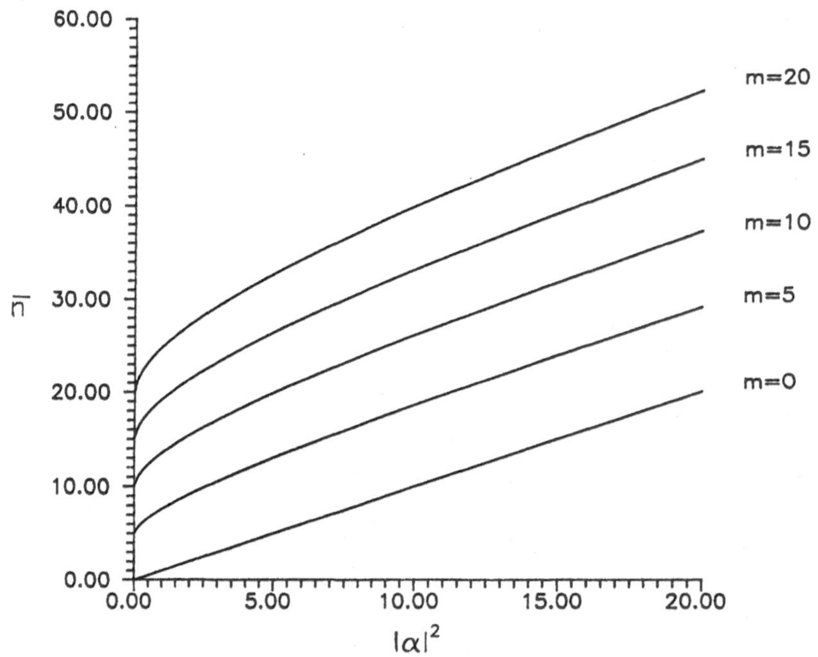

Figure 2. Mean number of photons n as a function of $|\alpha|^2$ for different values of m.

Note that if we send a ground state atom through the cavity, then the detection of the atom in the excited state leaves the field in the coherent state $|\alpha>$ itself indicating that the field in such a situation has no non-classical character. The above discussion also points out the very fundamental distinction between absorption and emission processes. We create non-classical character in emission and not in absorption.

An extension of the above arguments to multiphoton processes would imply that the states $|\alpha, m>$ can be produced in multiphoton emission process. The state $|\alpha, m>$ may also be produced by other methods such as those based on special state reduction and feedback methods.[5] For example consider the process of parametric amplification in which the signal ('a'- mode) and idler ('b'-mode) are generated. These two modes are strongly correlated. Let us assume that initially the strong field is in the state $|\alpha>$. One can show that if b mode is measured in the Fock state $|m>$, the state of the a-mode is reduced to $|\alpha, m>$.

In conclusion we have introduced a new class of states which are generated by the action of photon creation operator on a coherent state and shown the important non-classical properties such states possess.

ACKNOWLEDGEMENTS

The work of K. Tara was supported by the Council of Scientific and Industrial Research, Government of India.

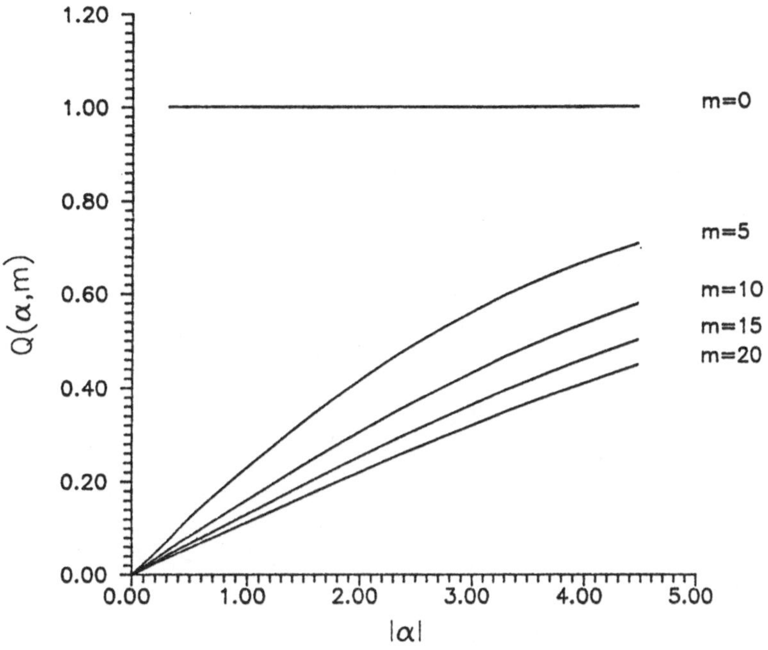

Figure 3. $Q(\alpha, m)$ as a function of $|\alpha|$ for different values of m.

REFERENCES

1. R. J. Glauber, *Phys. Rev.* **130**, 2529 (1963); *Ibid.*, 131, 2766 (1963).
2. I. S. Gradshteyn and I. M. Ryzhik, Table of Integrals, Series and Products (Academic Press, New York, 1965).
3. G. S. Agarwal and K. Tara, *Phys. Rev.* **A43**, 492 (1991).
4. J. Krause, M. O. Scully, T. Walther, and H. Walther, *Phys. Rev.* **A39**, 1915 (1989).
5. H. P. Yuen, *Phys. Rev. Lett.* **56**, 2176 (1986); G. Bjork and Y. Yamamoto, *Phys. Rev.* **A37**, 4229 (1988); K. Watanabe and Y. Yamamoto, *Phys. Rev.* **A38**, 3556 (1988); G. S. Agarwal, *Quant. Opt.* **2**, 1 (1990).

PHASE PROPERTIES OF SQUEEZED NUMBER STATES

R. Nath and Pradumn Kumar*

Department of Physics and Astrophysics
University of Delhi
Delhi (India)

ABSTRACT

We have studied the phase properties of the squeezed number states by evaluating the expression for the phase probability distribution and the phase variance. In addition, the expression for the photon-number distribution of the squeezed phase states has been evaluated.

1. INTRODUCTION

There has been a renewed interest in the study of problems associated with the quantum optical phase [1] with the advent of theoretical and experimental work on squeezed states of Light [2]. Squeezed states have phase sensitive noise and it is important therefore to examine the phase properties of these states. Recently Vaccaro and Pegg [3] have studied the phase properties of squeezed states with particular reference to squeezed vacuum by using the recent formalism of Pegg and Barnett [4] in which the existence of a hermitian optical phase operator, $\hat{\phi}_\theta$, has been shown. The exponential phase operators, $Exp(\pm i\hat{\phi}_\theta)$, are thus unitary and the relation $\cos^2\hat{\phi}_\theta + \sin^2\hat{\phi}_\theta = 1$ is satisfied. This formalism, thus overcomes several difficulties associated with the earlier approach of Susskind and Glogower based on non-unitary character of exponential phase operators, $\widehat{Exp}(\pm i\phi)$ [5].

In this manuscript, we have studied the phase properties of the quasi-photon number states so called squeezed number states [6] by evaluating the expression for the phase probability distribution $P_n(\theta)$ in the weak squeezing limit. This expression has been used to calculate the variance of the optical phase operator in these states. We have further utilized this analysis to evaluate the expression for the photon-number distribution for the squeezed phase states introduced by the authors in an earlier paper [7].

*research fellow

Recent Developments in Quantum Optics, Edited
by R. Inguva, Plenum Press, New York, 1993

2. PHASE PROBABILITY DISTRIBUTION FOR SQUEEZED NUMBER STATE

A complete set of (s+1) orthonormal phase states is defined as [4]

$$|\theta_p> = (s+1)^{-\frac{1}{2}} \sum_{n=o}^{s} e^{in\theta_p}|n> \tag{2.1}$$

where $|n>$ are the $(s+1)$ number states which span the $(s+1)$-dimensional state space ψ and

$$\theta_p = \theta_o + \frac{2\pi p}{s+1} \ with \ p = 0, 1, 2, ...s.$$

The value of θ_o is arbitrary; it defines a particular basis set of $(s+1)$ mutually orthogonal phase states. The phase states $|\theta_p>$ are the eigenstates of hermitian phase operator, $\hat{\phi}_\theta$, defined as

$$\hat{\phi}_\theta = \sum_{p=o}^{s} \theta_p|\theta_p><\theta_p| \tag{2.2}$$

Evidently the phase probability distribution $P_n(\theta_p)$ for the squeezed number state $\hat{S}(r)|n>$ is given by the expression

$$P_n(\theta_p) = |<\theta_p|\hat{S}(r)|n>|^2 \tag{2.3}$$

where $\hat{S}(r)$ is the squeezing operator and it provides a Bogolibubov transformation of the annihilation and creation operators as [6,7].

$$\hat{S}\hat{a}\hat{S}^+ = \hat{a}\cosh r + \hat{a}^+ \sinh r = \hat{b}$$

$$\hat{S}\hat{a}^+\hat{S}^+ = \hat{a}^+ \cosh r + \hat{a}\sinh r = \hat{b}^+ \tag{2.4}$$

where r is the squeezing parameter, assumed to be real. The matrix element $<\theta_p|\hat{S}(r)|n>$ can now be written as:

$$<\theta_p|\hat{S}(r)|n> = (s+1)^{-\frac{1}{2}} \sum_{\ell=o}^{s} e^{-i\ell\theta_p} <\ell|\hat{S}(r)|n> \tag{2.5}$$

From the expression of the photon-number distribution of a squeezed number state [6], we have

$$<\ell|\hat{S}(r)|n> = \frac{(n!\ell!)^{\frac{1}{2}}(\frac{1}{2}\tanh r)^{\frac{n-\ell}{2}}}{(\cosh r)^{(\ell+\frac{1}{2})}} \sum_m \frac{(-1)^m(\frac{1}{2}\tanh r)^{2m}(\cosh r)^{2m}}{m!(\ell-2m)![\frac{1}{2}(n-\ell)+m]!} \tag{2.6}$$

where

$$\frac{1}{2}(\ell-n) \leq m \leq \ell/2$$

In the weak squeezing limit [8] i.e. for small r, we put

$$\tanh r \simeq r \textit{ and } \cosh r \simeq 1 + \frac{1}{2}r^2$$

in eq. (2.6) to obtain

$$< \ell|\hat{S}(r)|n >= \frac{(n!\ell!)^{\frac{1}{2}}(\frac{1}{2}r)^{\frac{n-\ell}{2}}}{(1+\frac{1}{2}r^2)^{(\ell+\frac{1}{2})}} \sum_m \frac{(-1)^m(\frac{1}{2}r)^{2m}}{m!(\ell-2m)![\frac{1}{2}(n-\ell)+m]!} \tag{2.7}$$

Using the above expression in eq. (2.5), we get

$$< \theta_p|\hat{S}(r)|n >= (s+1)^{-\frac{1}{2}} \sum_{\ell=o}^{s} \frac{e^{-i\ell\theta_p}(n!\ell!)^{\frac{1}{2}}}{(1+\frac{1}{2}r^2)^{(\ell+\frac{1}{2})}} \sum_m \frac{(-1)^m(\frac{1}{2}r)^{2m+\frac{n-\ell}{2}}}{m!(\ell-2m)![\frac{1}{2}(n-\ell)+m]!} \tag{2.8}$$

The factorials in eq. (2.8) are valid for non-negative integers, so to order r^2 in the squeezing parameter, the matrix element $< \theta_p|\hat{S}(r)|n >$ gets contribution only from the following values i.e.

$$\ell = n, \quad n \pm 2, \quad n \pm 4$$

Hence eq. (2.8) simplifies to:

$$< \theta_p|\hat{S}(r)|n >= \frac{(s+1)^{-\frac{1}{2}}e^{-in\theta_p}}{(1+\frac{1}{2}r^2)^{(n+\frac{1}{2})}}\Big[\Big\{1 - \frac{n(n-1)r^2}{4}\Big\}$$
$$+\frac{r}{2}\Big\{\sqrt{n(n-1)}e^{2i\theta_p} - \sqrt{(n+2)(n+1)}e^{-2i\theta_p}\Big\}$$

$$+\frac{r^2}{8}\Big\{\sqrt{n(n-1)(n-2)(n-3)}e^{4i\theta_p} + \sqrt{(n+1)(n+2)(n+3)(n+4)}e^{-4i\theta_p}\Big\}\Big]. \tag{2.9}$$

Thus the expression for the phase probability distribution can be written as

$$P_n(\theta_p) = | < \theta_p|\hat{S}(r)|n > |^2 = (s+1)^{-1}[1 + A_n r\cos 2\theta_p + B_n r^2 \cos 4\theta_p] \tag{2.10}$$

where

$$A_n = \{\sqrt{n(n-1)} - \sqrt{(n+2)(n+1)}\}$$

$$\begin{aligned} B_n = \ & \tfrac{1}{4}\{\sqrt{n(n-1)(n-2)(n-3)} \\ & +\sqrt{(n+1)(n+2)(n+3)(n+4)} \\ & -2\sqrt{n(n-1)(n+2)(n+1)}\} \end{aligned} \tag{2.11}$$

The density of phase states is $\frac{(s+1)}{2\pi}$, so in the continuum limit as $s \to \infty$, the phase probability distribution becomes

$$P_n(\theta) = \frac{1}{2\pi}[1 + A_n r\cos 2\theta + B_n r^2 \cos 4\theta] \tag{2.12}$$

where we have dropped the subscript p on the phase angle θ. We note that $P_n(\theta)$ is normalized i.e.,

$$\int_{\theta_o}^{\theta_o+2\pi} P_n(\theta)d\theta = 1$$

For $n = 0$, the expression (2.12) reduces to the one obtained by Vaccaro and Pegg [3] for a weakly squeezed vacuum while for $r = 0$, it reduces to a constant value $(2\pi)^{-1}$, which we get for an ordinary number state, $|n>$.

Now the expectation value of any function $f(\hat{\phi}_\theta)$ of the phase operator $\hat{\phi}_\theta$ for the squeezed number state is given by

$$< f(\hat{\phi}_\theta) >= \int_{\theta_o}^{\theta_o+2\pi} P_n(\theta)f(\theta)d\theta \tag{2.13}$$

We can now easily show that

$$< \hat{\phi}_\theta >= (\theta_o + \pi) + (\frac{1}{2}rA_n \sin 2\theta_o + \frac{1}{4}r^2 B_n \sin 4\theta_o)$$

and

$$< \hat{\phi}_\theta^2 >= (\frac{4\pi^2}{3} + 2\pi\theta_o + \theta_o^2) + \frac{1}{2}rA_n[2(\theta_o + \pi)\sin 2\theta_o + \cos 2\theta_o]$$
$$+\frac{1}{8}r^2 B_n[4(\theta_o + \pi)\sin 4\theta_o + \cos 4\theta_o]$$

Thus the phase variance is

$$< \Delta\hat{\phi}_\theta^2 >= \frac{\pi^2}{3} + \frac{1}{2}rA_n \cos 2\theta_o + \frac{1}{4}r^2[(\frac{B_n}{2} + \frac{A_n^2}{2})\cos 4\theta_o - \frac{A_n^2}{2}] \tag{2.14}$$

We note that the above result is dependent on θ_o, that is, on the choice of particular 2π phase window. For $r = 0$, the above result reduces to the one obtained for an ordinary number state $|n>$ while for $n = 0$ it agrees, as expected, with the result obtained by Vaccaro and Pegg [3] for the weakly squeezed vacuum. The above analysis can also be extended to study the photon-number distribution for the squeezed phase states $\hat{S}(r)|\theta_p>$ [7]. Evidently the photon-number distribution for these states is given by

$$P(n) = | < n|\hat{S}(r)|\theta_p > |^2.$$

In the weak squeezing limit, we find

$$P(n) = (s+1)^{-1}[1 - A_n r \mathrm{Cos}2\theta + B_n r^2 \mathrm{Cos}4\theta]$$

$$< n >= (s/2)[1 + 2r\mathrm{Cos}4\theta + 8r^2\mathrm{Cos}4\theta]$$

and

$$< n^2 >= [(s^2/3 + s/6)] + [(2s^2/3 - 2s/3)r\mathrm{Cos}2\theta] + [(8s^2/3 + 28s/3)r^2\mathrm{Cos}4\theta]$$

The second-order correlation function is given as:

$$g^{(2)}(0) = 1 + (\sigma^2 - < n >)/ < n >^2 = (4/3) - 8r^2\mathrm{Cos}^2 2\theta$$

We thus notice that in the limit $r \rightarrow 0$, the above results are the same as that of an ordinary phase state.

3. CONCLUSION

In this paper, we have obtained the expression for the phase probability distribution $P_n(\theta)$ for the squeezed number state in the weak squeezing limit. This distribution has been used to obtained the variance, $< \Delta\hat{\phi}_\theta^2 >$, which depends on the parameters θ_o and r. So, by a suitable choice of the 2π phase window, we can minimize the variance for a given value of squeezing parameters. Finally, we have also studied the statistical properties of the squeezed phase radiation.

REFERENCES

1. R. E. Slusher, L. W. Hollberge, B. Yurke, J. C. Mertz and J. F. Valley; Phys. Rev. Lett. 55, 2409 (1985). D. F. Walls; Nature 306, 141 (1983). R. Loudon and P. L. Knight; J. Mod. Opt. 34, 709 (1987).
2. S. M. Barnett and P. L. Knight; Opt. Commun. 58, 290 (1986). S. M. Barnett and D. T. Pegg; J. Mod. Opt. 36, 7 (1989). R. Lynch; J. Opt. Soc. Am. B4, 1723 (1987).
3. Vaccaro and D. T. Pegg; Opt. Commun. 70, 529 (1989).
4. S. M. Barnett and D. T. Pegg; J. Mod. Opt. 36, 7 (1989). D. T. Pegg and S. M. Barnett; Phys. Rev. A39, 1665 (1989). R. Nath and P. Kumar; Opt. Commun. 76, 51 (1990).
5. L. Susskind and J. Glogower; Physics 1, 49 (1969).
6. M. S. Kim, F. A. M. de Oliveira and P. L. Knight; Opt. Commun. 72, 99 (1989). H. P. Yuen; Phys. Rev. A. 13, 2226 (1976).
7. R. Nath and P. Kumar; J. Mod. Opt. (To be published).
8. Hong-Yi Fan and H. R. Zaidi; Opt. Commun. 68, 143 (1988).

OPTICAL DOUBLE RESONANCE FROM A REGULAR ARRAY OF N THREE-LEVEL ATOMS: A SOURCE OF MACROSOPIC SQUEEZED LIGHT

Richard D'Souza+, Arundhati S. Jayarao*, and S.V. Lawande*

+Spectroscopy Division and *Theoretical Physics Division
Bhabha Atomic Research Centre
Trombay, Bombay 400 085

One of the theoretical proposals to generate macroscopic squeezed light is the resonance fluorescence from N two-level atoms distributed at regular positions.[1] This represents a macroscopic light source producing squeezed coherent states and squeezed vacuum states simultaneously. We present here the squeezed light produced in optical double resonance from N identical three-level atoms.

A multimode light field may be written as,

$$E(\vec{r},t), = E^{(+)}(\vec{r},t) + E^{(-)}(\vec{r},t) \tag{1}$$

where $E^{(+)}$ and $E^{(-)} = (E^{(+)})^{\dagger}$ respectively are the positive and negative frequency parts of the electric field. The condition for squeezing can be written in terms of the normally ordered variance,[2]

$$<: (\Delta E)^2 :> < 0 \tag{2}$$

We consider N identical three-level atoms with unequally spaced levels in the V configuration interacting resonantly with two linearly polarized, monochromatic applied fields of frequencies ω_1 and ω_2 (see inset of Fig. 1). If the measurements are carried out in a manner such that the incident laser fields do not contribute, the electric field operators $E_i^{(+)}$ (i=1,2) are related to the atomic operators by[3,4]

$$E_1^{(+)}(\vec{r},t) \longrightarrow \vec{g}_1 A_{31}(t), \quad E_2^{(+)}(\vec{r},t) \longrightarrow \vec{g}_2 A_{32}(t) \tag{3}$$

where $E_1^{(+)}$ and $E_2^{(+)}$ correspond to the positive frequency part of the electric field emitted from $|1> \longrightarrow |3>$ and $|2> \longrightarrow |3>$ transitions respectively. $\vec{g}_1$ *and* $\vec{g}_2$ are factors which depend on the spatial dimension $\vec{r}$ and the dipole moments of the atom. Cooperative effects are ignored since the atoms are assumed to be sufficiently far apart from each other, i.e. the wavelengths of the driving fields are much smaller than the distance between the neighbouring atoms. Under this assumption, we have the following factorization ansatz,

Recent Developments in Quantum Optics, Edited
by R. Inguva, Plenum Press, New York, 1993

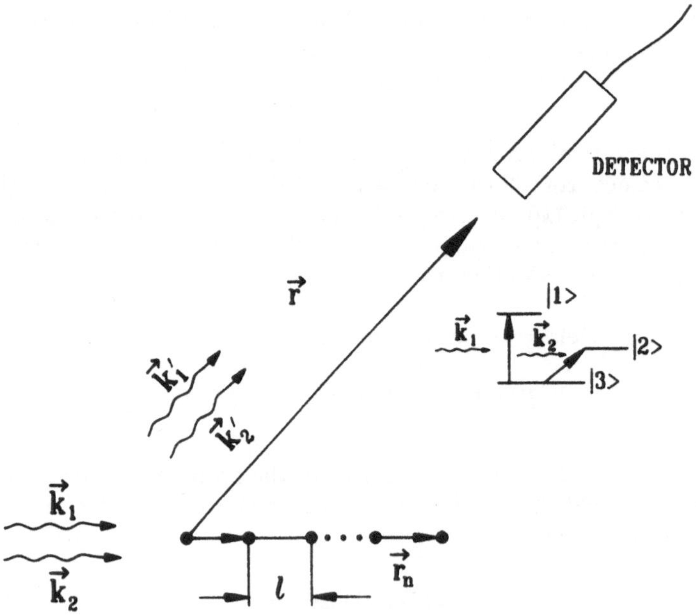

Figure 1. Schematic diagram of an array of three-level atoms interacting with two external fields with wave vectors $\vec{k}_1$ and $\vec{k}_2$. The inset shows a three-level atom in a V-configuration.

$$< E_{in}^{(+)}(\vec{r},t)E_{im}^{(\pm)}(\vec{r},t) >=< E_{in}^{(+)}(\vec{r},t) >< E_{im}^{(\pm)}(\vec{r},t) > \quad (n \neq m; i = 1,2) \tag{4}$$

Incorporating the above conditions, we obtain, after some straight forward algebra, the following expression for the normally ordered variance:

$$<: (\Delta E_i)^2 :>= 2\sum_{i=1}^{N} \Big[< E_{in}^{(-)}(\vec{r},t)E_{in}^{(+)}(\vec{r},t) >$$

$$- \frac{1}{2}\Big(< E_{in}^{(+)}(\vec{r},t) > + < E_{in}^{(-)}(\vec{r},t) > \Big)^2\Big] \quad (i = 1,2) \tag{5}$$

where,

$$< E_{1n}^{(+)}(\vec{r},t) >= \vec{g}_1 < A_{31}(t - |\vec{r} - \vec{r}_n|/c) > \tag{6a}$$

$$< E_{2n}^{(+)}(\vec{r},t) >= \vec{g}_2 < A_{32}(t - |\vec{r} - \vec{r}_n|/c) > \tag{6b}$$

$$< E_{1n}^{(-)}(\vec{r},t)E_{1n}^{(+)}(\vec{r},t) >= |g_1|^2 < A_{13}(t - |\vec{r} - \vec{r}_n|/c)A_{31}(t - |\vec{r} - \vec{r}_n|/c) >$$

$$= |g_1|^2 < A_{11}(t - |\vec{r} - \vec{r}_n|/c) > \tag{6c}$$

$$< E_{2n}^{(-)}(\vec{r},t)E_{2n}^{(+)}(\vec{r},t) >= |g_2|^2 < A_{23}(t - |\vec{r} - \vec{r}_n|/c)A_{32}(t - |\vec{r} - \vec{r}_n|/c)$$

$$= |g_2|^2 < A_{22}(t - |\vec{r} - \vec{r}_n|/c) > \tag{6d}$$

We now assume that the atoms are arranged on a one dimensional lattice defined by a fundamental translation vector $\vec{l}$. For convenience, the reference frame is centered on the first atom of this linear chain. $\vec{r}_n$ represents the position vector of the nth atom from the origin. The point of observation, $\vec{r}$, is assumed to be much larger than the linear dimensions of the lattice as shown in Fig. 1.

We now solve the master equation for N three-level atoms numerically to obtain the steady-state values of the polarizations, $< A_{ij} >_{ss}$, and the atomic level populations, $< A_{ii} >_{ss}$. After the initial transients have died out the expectation values of the atomic operators evolve as follows,

$$< A_{31}(t - |\vec{r} - \vec{r}_n|/c) >=< A_{31} >_{ss} \, exp[i(\vec{k}_1 \cdot \vec{r}_n - \omega_1 t - \phi_1 + \omega_1|\vec{r} - \vec{r}_n|/c)] \tag{7a}$$

$$< A_{32}(t - |\vec{r} - \vec{r}_n|/c) >=< A_{32} >_{ss} \, exp[i(\vec{k}_2 \cdot \vec{r}_n - \omega_2 t - \phi_2 + \omega_2|\vec{r} - \vec{r}_n|/c)] \tag{7b}$$

$$< A_{11}(t - |\vec{r} - \vec{r}_n|/c) >=< A_{11} >_{ss}, < A_{22}(t - |\vec{r} - \vec{r}_n|/c) >=< A_{22} >_{ss} \tag{7c}$$

Using eqns. 5, 6 and 7 we obtain

$$<: (\Delta E_i)^2 :>= -N|g_i|^2\{|X_i|^2[2|F_i|cos(2\psi_i) + 1] - 2Y_i\} \quad (8)$$

where,

$$F_i = \frac{1}{N}\sum_{n=1}^{N} exp[2i(\vec{k}_i - \vec{k}_i') \cdot \vec{r}_n] \quad (9a)$$

$$X_1 =< A_{13} >_{ss}, Y_1 =< A_{11} >_{ss}, X_2 =< A_{23} >_{ss}, Y_2 =< A_{22} >_{ss} \quad (9b)$$

$$\psi_i = \vec{k}_i' \cdot \vec{r} - \omega t + argX_i + \frac{1}{2}argF_i - \phi_i, k_i' = (\omega_i/c)(\vec{r}/|\vec{r}|) \quad (9c)$$

For the case of atoms in a one dimensional lattice, F_i simplifies to,

$$|F_i| = \frac{1}{N}\left|\frac{sin[(\vec{k}_i - \vec{k}_i') \cdot \vec{l}N]}{sin[(\vec{k}_i - \vec{k}_i') \cdot \vec{l}]}\right| \quad (10)$$

From eqn. 8 it is evident that to obtain squeezing we require,

$$|X_i|^2(2|F_i|cos(2\psi_i) + 1) > 2Y_i \quad (11)$$

Choosing the maximum values for $|F_i|$ and $cos(\psi_i)$, i.e., $\psi_i = 0$ and $|F_i| = 1$, the inequality (11) reduces to,

$$3|X_i|^2 > 2Y_i \quad (12)$$

For some values of the Rabi frequencies α_1 and α_2, it is possible to satisfy inequality (12), thus obtaining squeezed light in the directions of observation which satisfy the condition,

$$(\vec{k}_i - \vec{k}_i') \cdot \vec{l} = n\pi \quad n = 0, \pm 1, \pm 2, \cdots; i = 1, 2 \quad (13)$$

Figs. 2-3 show the behaviour of the variance of the electric field as a function of the Rabi frequencies. It is evident that the amount of squeezing has a sensitive dependence on the Rabi frequencies. Squeezing disappears for large or very small driving fields. Further, it is possible to achieve squeezed radiation from both the transitions simultaneously.

Similary, the expectation values of the electric field can be calculated,

$$< E_i >= 2N|X_i|^2|G_i|cos(\psi_i') \quad (i = 1, 2) \quad (14)$$

where

$$\psi_i' = \omega_i t + \phi_i - \vec{k}_i' \cdot \vec{r} - argX_i - argG_i \quad (15)$$

$$|G_i| = \frac{1}{N}\left|\frac{sin[(\vec{k}_i - \vec{k}_i') \cdot \vec{l}N/2]}{sin[(\vec{k}_i - \vec{k}_i') \cdot \vec{l}/2]}\right| \quad (16)$$

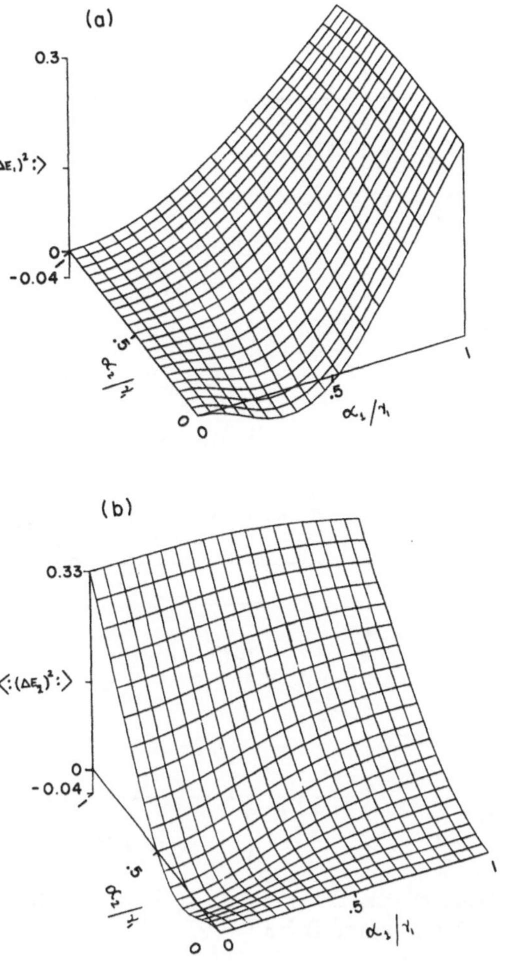

Figure 2. Plot of variances $<: (\Delta E_i)^2 :>$ $(i = 1, 2)$ as a function of Rabi frequencies α_1 and α_2.

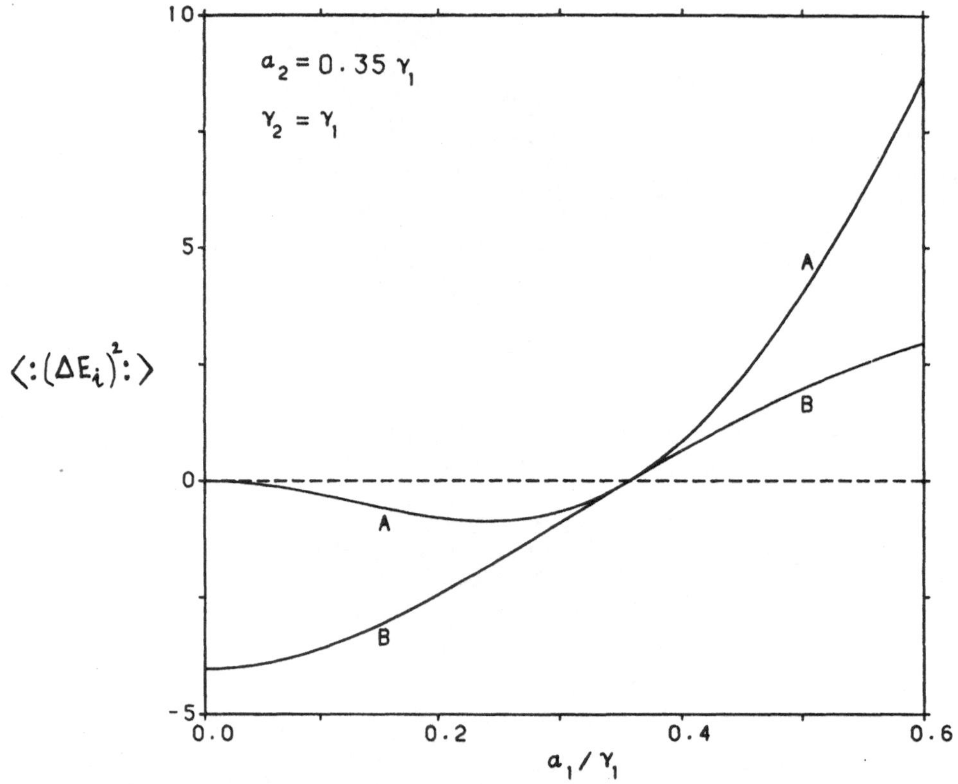

Figure 3. Plot of variances $<:(\Delta E_i)^2:>$ $(i = 1,2)$ as a function of α_1. Curve A denotes $i = 1$ and curve B for $i = 2$.

We observe from eqn. 16 that in addition to squeezing there is also an ordinary intereference pattern formed by the scattering of coherent light. The maxima and minima of the scattered intensity satisfy respectively the conditions.

$$(\vec{k}_i - \vec{k}'_i) \cdot \vec{l} = \begin{cases} 2n\pi \\ (2n+1)\pi \end{cases} \quad n = 0, \pm 1, \pm 2, \ldots \ (i = 1, 2) \tag{17}$$

We may conclude that in the directions given by eqn. 13, the resonance fluorescence from N regularly arrayed three-level atoms exhibit macroscopic squeezing. Further, eqn. 17 shows that the squeezed light from the bright fringes of the interference pattern will correspond to squeezed coherent states and that from the dark fringes to squeezed vacuum states. Mathematically we have,

$$(\vec{k}_i - \vec{k}'_i) \cdot \vec{\ell} = \begin{cases} 2n\pi & \text{for squeezed coherent states} \\ (2n+1)\pi & \text{for squeezed vacuum states} \end{cases}$$

We may add here that the V-configuration alone gives squeezed radiation from both the transitions. Preliminary work on the cascade and the Raman configuration shows that squeezed light is emitted only from the lower and Rayleigh transitions respectively. Details will be reported elsewhere.[5]

REFERENCES

1. W. Vogel and D. G. Welsch, *Phys. Rev. Lett.*, **54** 1802 (1985).
2. D. F. Walls, *Nature* **306**, 141 (1983); R. Loudon and P. L. Knight, *J. Mod. Opt.* **34**, 709 (1987).
3. G. S. Agarwal, in *Quantum Optics*, Vol. 70 of Springer Tracts in Modern Physics, edited by G. Hohler (Springer-Verlag, Heidelberg, 1974).
4. H. J. Kimble and L. Mandel, *Phys. Rev.* **A13**, 2123 (1976); L. mandel, *Phys. Rev.* **A28**, 929 (1983).
5. R. D'Souza, A. S. Jayarao, and S. V. Lawande, *Mod. Phys. Lett.*, 1992 (in press).

PHOTON CORRELATION AND SUCCESSIVE EMISSION FROM A TWO LEVEL ATOM IN A BROADBAND SQUEEZED VACUUM

B.N. Jagatap*, S.V. Lawande+ and Q. V. Lawande+

*Multidisciplinary Research Section, +Theoretical Physics Division
Bhabha Atomic Research Centre
Bombay-400 085, India

Radiative decay and spectroscopic properties of an atom that is damped by a squeezed vacuum has received considerable attention in recent years.[1,2] Interaction of a two-level atom with a broadband squeezed vacuum leads to inhibition of the phase decay of the atom when compared with the normal vacuum. Specifically, the component of the atomic polarization that is in phase with the low noise quadrature phase of the field experiences reduced fluctuations and hence decays slower than the other component which experiences increased fluctuations. Consequently, the fluorescence spectrum of a coherently driven two-level atom in a broadband squeezed vacuum exhibits dependence on the relative phase between the driving field and squeezed vacuum and departs considerably from the usual Mollow triplet in the normal unsqueezed vacuum. The relative phase and other parameters of the squeezed vacuum can now be specified to obtain subnatural widths in the fluorescence and absorption spectra.[1,2]

The phase dependent decay of the atomic dipole in squeezed vacuum is also manifested in the statistics of the photon emission.[3] In particular the second order intensity correlation function $g^{(2)}(\tau)$ in a broadband squeezed vacuum exhibits a quantum jump like behaviour. In this paper we report some results on the photon statistics of resonance fluorescence in a broadband squeezed vacuum in terms of the waiting time distribution $w(\tau)$. The essential distinction between $g^{(2)}(\tau)$ and $w(\tau)$ is that they respectively define the probability[4] of emitting any photon and the very next photon at time $t+\tau$ following emission of a photon at time t. We show that the distribution function $w(\tau)$ exhibits the essential manifestation of the inhibition of fluorescence in the squeezed vacuum. We further examine the temporal correlations between the emissions of photons belonging to the two sidebands [5,6] of the phase sensitive Mollow triplet in the squeezed vacuum.

Consider a two-level atom of transition frequency ω_a between an excited level $|1>$ and ground level $|2>$ interacting with a classical driving field of frequency ω_L. The master equation for the reduced atomic density operator $\rho(t)$ describing the interaction of this atom with a broadband squeezed vacuum has the form[1,2]

$$\frac{d\rho}{dt} = -i[H_o, \rho] - \gamma(N+1)(A_{11}\rho + \rho A_{11} - 2A_{21}\rho A_{12})$$

$$-\gamma N(A_{22}\rho + \rho A_{22} - 2A_{12}\rho A_{21})$$

$$-2\gamma|M|exp(-i\phi)A_{12}\rho A_{21} - 2\gamma|M|exp(i\phi)A_{21}\rho A_{21} \quad (1)$$

where

$$H_o = \alpha(A_{12} + A_{21}) + \Delta A_{11}$$

$$\Delta = \omega_a - \omega_L,\ \phi = 2\phi_L - \phi_s \quad (2)$$

Here A_{ij} is the atomic operator $|i><j|$, 2α is the laser Rabi frequency, 2γ is the Einstein coefficient for spontaneous emission in the normal vacuum and ϕ_L, ϕ_s are respectively the phases for the laser field and the squeezed vacuum. The parameters N and $|M|$ characterize the squeezing. Further $|M| \leq (N^2+N)^{\frac{1}{2}}$, where the equality sign holds for minimum uncertainty squeezed state.

We are interested in the statistical properties of fluorescence with well separated sidebands.[5,6] In that case it is convenient to introduce dressed operators B_{ij} related to the atomic operators A_{ij} through a canonical transformation

$$A_{11,22} = \frac{1}{2}(1 \mp r)B_{11} + \frac{1}{2}(1 \pm r)B_{22} \pm \frac{1}{2}(1-r^2)^{\frac{1}{2}}(B_{12} + B_{21})$$

$$A_{12} = (A_{21})^{\dagger} = \frac{1}{2}(1-r^2)^{\frac{1}{2}}(B_{11} - B_{22}) - \frac{1}{2}(1-r)B_{12} + \frac{1}{2}(1+r)B_{21} \quad (3)$$

where

$$r = \frac{\Delta}{2\Gamma},\ \Gamma = (\alpha^2 + \frac{1}{4}\Delta^2)^{\frac{1}{2}} \quad (4)$$

When the generalized Rabi frequency Γ is much greater than $N\gamma$, we may make the secular approximation on Eq. (1) and obtain approximate evolution equations of one time expectation values of the dressed operator $< B_{ij} >$, whose solutions are,

$$< B_{11} > = \frac{\gamma_o + r\gamma}{2\gamma_o}(1 - e^{-2\gamma_o t}) + < B_{11}(0) > e^{-2\gamma_o t}$$

$$< B_{12} > = < B_{12}(0) > e^{-(2i\Gamma + \gamma_1)t} \quad (5)$$

where

$$\gamma_o = \gamma[(N + \frac{1}{2})(1+r^2) + |M|(1-r^2)cos\phi]$$

$$\gamma_1 = \gamma[(N + \frac{1}{2})(3-r^2) - |M|(1-r^2)cos\phi] \quad (6)$$

The steady state intensity and the second order intensity correlation function $G^{(2)}(\tau)$ are then obtained as

$$< I >_{ss} = 2\gamma < A_{11} >_{ss} = (\gamma/\gamma_o)\xi, \quad \xi = \gamma_o - r^2\gamma \tag{7}$$

$$G^{(2)}(\tau) = \lim_{t\to\infty} 4\gamma^2 < A_{12}(t)A_{12}(t+\tau)A_{21}(t+\tau)A_{21}(t) >$$

$$= (\gamma/\gamma_o)^2[\xi^2 + r^2\xi(\gamma_o - \gamma)e^{-2\gamma_o t} - (1-r^2)\gamma_o\xi e^{-\gamma_1 t}cos(2\Gamma t)] \tag{8}$$

Following Kim and Knight,[4] we introduce conditional probability $Q(\tau)$ such that

$$\tilde{w}(Z) = \frac{\tilde{Q}(Z)}{1+\tilde{Q}(Z)}, \quad \tilde{Q}(Z) = \frac{\tilde{G}^{(2)}(Z)}{< I >_{ss}} \tag{9}$$

where in general $\tilde{A}(z)$ is Laplace transform of $A(t)$. The delay function $w(\tau)$ can then be obtained by taking inverse Laplace transform of $\tilde{w}(Z)$. In Fig. 1 we have shown the temporal behaviour of $w(\tau)$ in a squeezed vacuum for $\phi = \pi$ (curve 2) and $\phi = 0$ (curve 3) compared with the behaviour in the normal unsqueezed vacuum. In the initial times $\gamma\tau << 1$, $w(\tau)$ is higher in squeezed vacuum in comparison to the normal vacuum. Further the oscillatory behaviour of $w(\tau)$ in normal vacuum is damped rapidly in squeezed vacuum. The most important feature is however exhibited for $\gamma\tau >> 1$. Here the squeezed vacuum with $\phi = \pi$ shows significantly higher probability of counting the very next photon as compared to the normal vacuum or squeezed vacuum with $\phi = 0$. The plateau for next photon emission for counting times $\tau > \gamma^{-1}$ in squeezed vacuum with $\phi = \pi$ is manifestation of inhibition of atomic decay.

We now examine the quantum correlation between the photons of two sidebands of the Mollow triplet in squeezed vacuum. These correlations in the normal vacuum have been measured experimentally by Aspects et al.[5] It is clear from Eq. (5) that the dressed operators B_{12} and B_{21} generate spectral components at frequencies $\omega_L \pm 2\Gamma$ respectively. We may therefore introduce the photon detection operators [6]

$$D_+^L = -\frac{1}{2}(1-r)B_{12} = (D_-^L)^\dagger$$

$$D_+^R = -\frac{1}{2}(1+r)B_{21} = (D_-^R)^\dagger \tag{10}$$

corresponding to the detection of radiation from left (L) and right (R) sidebands. The conditional probability $Q_{ab}(\tau), a, b = R, L$ then can be defined as

$$Q_{ab}(\tau) = \frac{G_{ab}^{(2)}(\tau)}{< I_a >_{ss}} \tag{11}$$

where

$$G_{ab}^{(2)}(\tau) = \lim_{t\to\infty} 4\gamma^2 < D_+^a(t)D_+^b(t+\tau)D_-^b(t+\tau)D_-^a(t) >$$

$$< I_a >_{ss} = \lim_{t\to\infty} 2\gamma < D_+^a(t)D_-^a(t) > \tag{12}$$

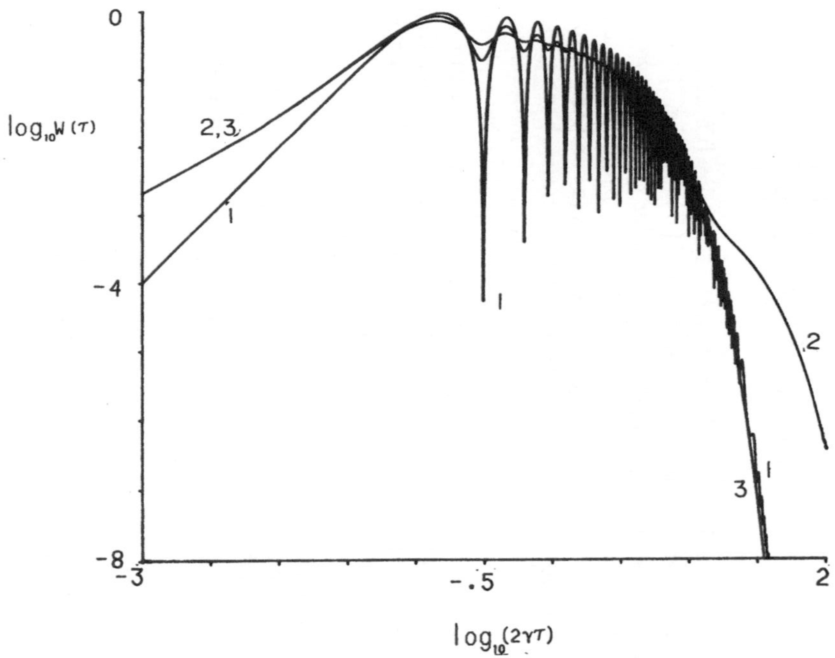

Figure 1: Delay function $w(\tau)$ for total emission with $\Gamma/2\gamma = 10$, $\Delta/2\gamma = 2$ and $M = (N^2+N)^{\frac{1}{2}}$. Curves 1-3 correspond to $(N,\phi) = (0,0)$, $(2,\pi)$ and $(2,0)$ respectively.

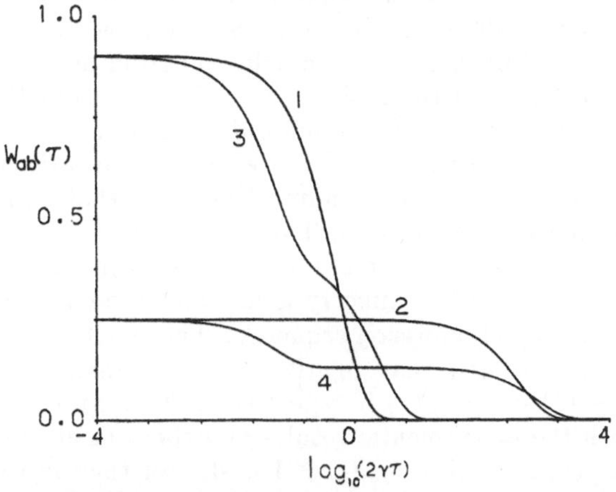

Figure 2. Sideband delay functions $w_{aa}(\tau)$, $a = L, R$ for $r = 0.9$ and $2\gamma = 1$. Curves 1-4 correspond to $(a,N,\phi) = (R,0,0)$, $(L,0,0)$, $(R,10,\pi)$ and $(L,10,\pi)$ respectively. Curves 1,2,4 are scaled up by $10.^2$

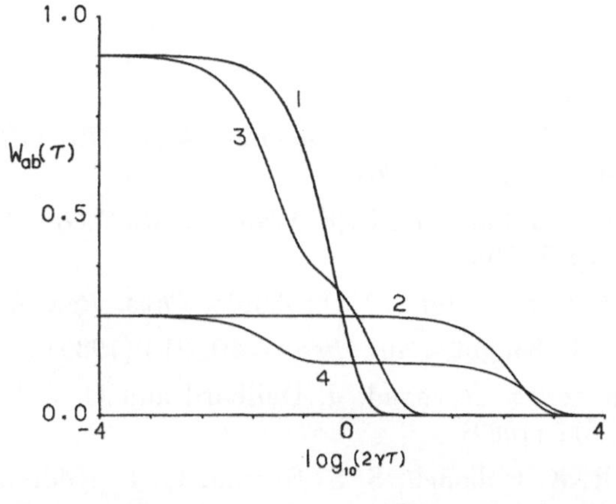

Figure 3. Sideband delay functions $w_{ab}(\tau), a,b = L,R (a \neq b)$. Curves 1-4 correspond to $(a,b,N,\phi) = (L,R,0,0)$, $(R,L,0,0)$, $(L,R,10,\pi)$ and $(R,L,10,\pi)$ respectively. Other data is same as in Fig. 2. Curves 2,4 are scaled up by $10.^2$

The corresponding delay functions $w_{ab}(\tau)$ obtained from Eq. (9) is simply the probability density that the atom emits the very next photon of type b at time $t+\tau$ following the emission of one photon of type a at time t. The delay function $w_{ab}(\tau), a, b = L, R$ can then be used to characterize the sideband emission.

In the study of the correlated emission of the photon of the sidebands the interesting situation is one where the detuning is very large ($r \rightarrow 1$). We therefore discuss here the temporal behaviour of the delay function $w_{ab}(\tau)$ for $r \rightarrow 1$. It may be noted form Eq. (6) that in the limit $r \rightarrow 1$, the delay functions exhibit only a weak dependence on the phase ϕ. The temporal behaviour for $\phi = \pi$ and $\phi = 0$ are therefore only marginally different. In Fig. 2 we have shown $w_{LL}(\tau)$ and $w_{RR}(\tau)$ in a squeezed ($\phi = \pi$) and the normal vacuum. Note here that $w_{RR}(0) = w_{LL}(0) = 0$ in normal as well as in squeezed vacuum. This is the antibunching property. Further $w_{LL}(\tau) = w_{RR}(\tau)$ holds only when the atom interacts with a normal vacuum. In the case of a squeezed vacuum, this symmetry is lost and the emission sequence R-R is greatly favoured over the L-L emission sequence. The functions $w_{LR}(\tau)$ and $w_{RL}(\tau)$ are shown in Fig. 3. Note here that $w_{LR}(0) \cong 1$ and remains constant over a time interval whereas $w_{RL}(\tau) << w_{LR}(\tau)$ irrespective of the nature of the vacuum. This is in agreement with the experimental results of Aspect et al. In squeezed vacuum this initial time domain for which $w_{LR}(t) \cong 1$ is shorter than in the normal vacuum. Moreover $w_{LR}(\tau)$ and $w_{RL}(\tau)$ exhibit evolution on two widely separated time scales in squeezed vacuum. This results in the formation of two distinct emission plateaux in the squeezed vacuum.

In conclusion we have used the delay function $w(\tau)$ to characterize the phase dependent decay of a two-level atom in a broadband squeezed vacuum. The inhibition of fluorescence in squeezed vacuum is reflected in the long time behaviour of $w(\tau)$. Further the sideband delay functions are employed for studying the quantum correlations of photons belonging to sidebands in squeezed vacuum.

REFERENCES

1. C. W. Gardiner and M. J. Collett, *Phys. Rev.* **A31**, 3761 (1985); C. W. Gardiner Phys. Rev. Lett. 56, 1917 (1986).
2. H. J. Carmichael, A. S. Lane and D. F. Walls, *J. Mod. Opt.* **34**, 821 (1987); Phys. Rev. Lett. 58, 2539 (1987).
3. R. D'Souza, A. S. Jayarao and S. V. Lawande, *Phys. Rev.* **A41**, 4083 (1990).
4. M. S. Kim and P. L. Knight, *Phys. Rev.* **A40**, 215 (1989).
5. A. Aspect, G. Roger, S. Reynaud, J. Dalibard and C. Cohen Tannoudji, *Phys. Rev. Lett.* **45**, 617 (1980).
6. G. S. Agarwal, R. K. Bullough, S. S. Hassan, G. P. Hildred and R. R. Puri, in Coherence, Cooperation and Fluctuations ed. F. Haake, L M. Narducci and D. F. Walls (Cambridge Univ. Press 1986) pp. 115-131.

TIME DEPENDENT SPECTRA OF A THREE-LEVEL ATOM IN SQUEEZED VACUUM

Arundhati S. Jayarao*, Richard D'Souza+, and S.V. Lawande*

*Theoretical Physics Division and +Spectroscopy Division
Bhabha Atomic Research Centre
Trombay, Bombay 400 085 India

The recent successful generation[1] of squeezed light has opened new avenues in the studies of atom-field interaction. In recent works the radiative decay[2] and the spectroscopic properties[3–7] of an atom interacting with a broadband squeezed vacuum have been considered. Quite recently, cooperative effects arising from the interaction of many atoms with the squeezed vacuum field have been studied.[8]

Further understanding of the fluorescence dynamics can be achieved in the framework of time-dependent physical spectra (TDS).[9] TDS of two[10] and three-level atom[11,12] interacting with the unsqueezed vacuum have been studied. More recently, TDS of a two level atom interacting with a broadband squeezed vacuum has been carried out.[13]

We present here the transient spectra of a three-level atom in a cascade configuration interacting with a broadband squeezed vacuum and resonantly driven by two strong coherent fields as shown in Fig. 1. The bandwidth of the squeezing is assumed to be sufficiently broad that the squeezed vacuum appears as a δ-correlated squeezed white noise to the atom. The parameters characterizing the effect of the squeezed vacuum are N_i, $M_i = |M_i| exp(i\phi_{is})$, with $|M_i|^2 \leq N_i(N_i + 1)$, $i = 1, 2$. The equality holds for a minimum uncertainty squeezed state. $\phi_i = 2\phi_{iL} - \phi_{is}$ is the phase difference of the driving field and the squeezed vacuum. Using analytic expressions for the auto-correlation functions in the Eberly-Wodkiewicz counting rate definition of TDS[9] we obtain analytic expressions for the upper $I_1(t, D, \Gamma)$ and lower $I_2(t, D, \Gamma)$ spectra. $I_1(t, D, \Gamma)$ and $I_2(t, D, \Gamma)$ are normalized by the steady state values of $< A_{12}A_{21} >=< A_{11} >$ and $< A_{23}A_{32} >=< A_{22} >$ respectively. The upper spectrum is displayed in Fig. 2 for various squeezing parameters. A general feature of the fluorescent spectra is the phase sensitive nature of squeezing (see Fig. 2). This is evident from the line-narrowing for $\phi_1 = \phi_2 = \pi$ [see Fig. 2(b)] and line-width broadening for $\phi_1 = \phi_2 = 0$ [see Fig. 2(c)] of the central peak as compared to the line-width of the spectra for the unsqueezed vacuum [see Fig. 2(a)]. However, the sidebands are always broadened. This is consistent with the results obtained for the steady-state spectra derived using the Wiener-Khintchine theorem.[7]

Recent Developments in Quantum Optics, Edited
by R. Inguva, Plenum Press, New York, 1993

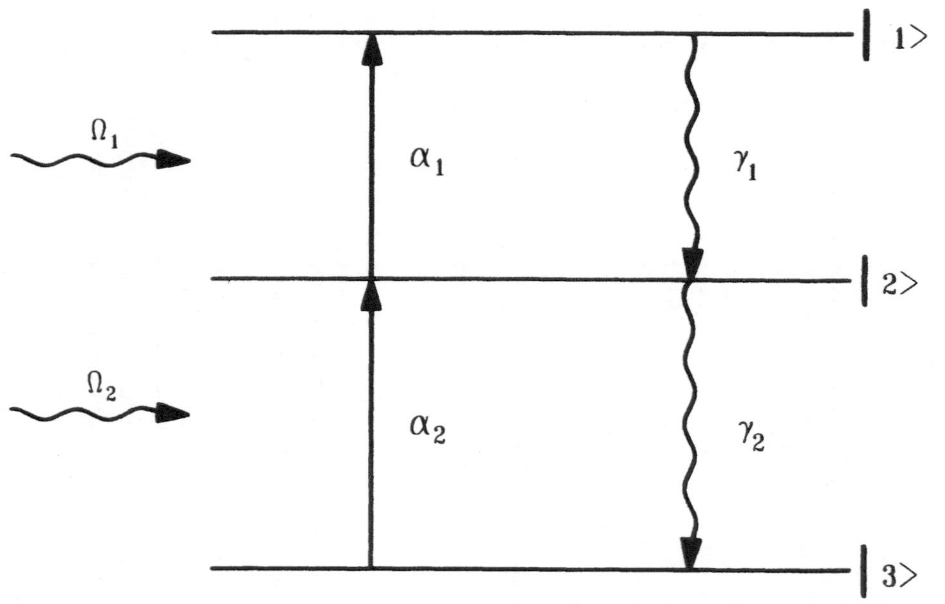

Figure 1. Schematic diagram of a three-level atom in a cascade configuration. Ω_1, Ω_2 are the external driving fields, $2\alpha_1$, $2\alpha_2$ are the Rabi frequencies and $2\gamma_1$, $2\gamma_2$ are the Einstein-A coefficients.

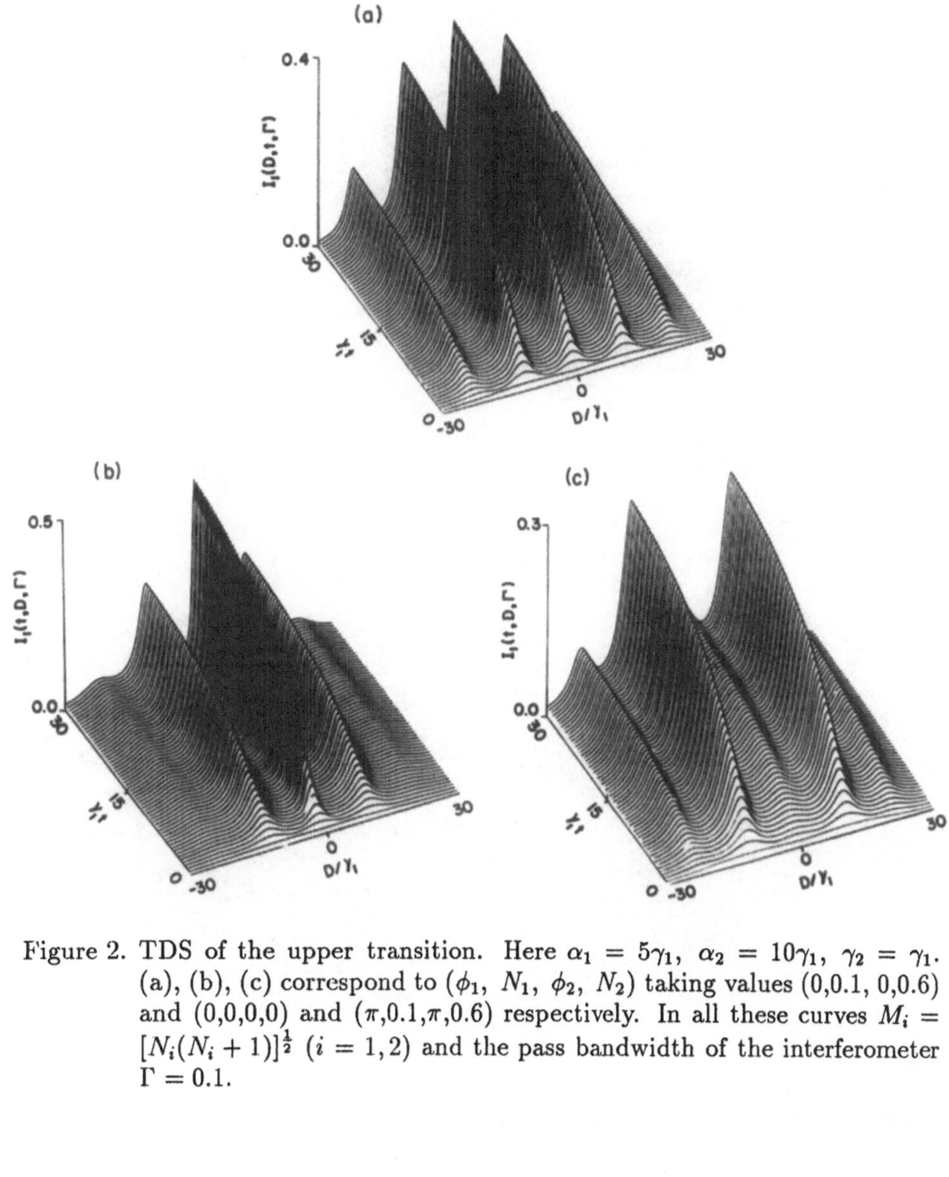

Figure 2. TDS of the upper transition. Here $\alpha_1 = 5\gamma_1$, $\alpha_2 = 10\gamma_1$, $\gamma_2 = \gamma_1$. (a), (b), (c) correspond to (ϕ_1, N_1, ϕ_2, N_2) taking values (0,0.1, 0,0.6) and (0,0,0,0) and (π,0.1,π,0.6) respectively. In all these curves $M_i = [N_i(N_i+1)]^{\frac{1}{2}}$ ($i = 1, 2$) and the pass bandwidth of the interferometer $\Gamma = 0.1$.

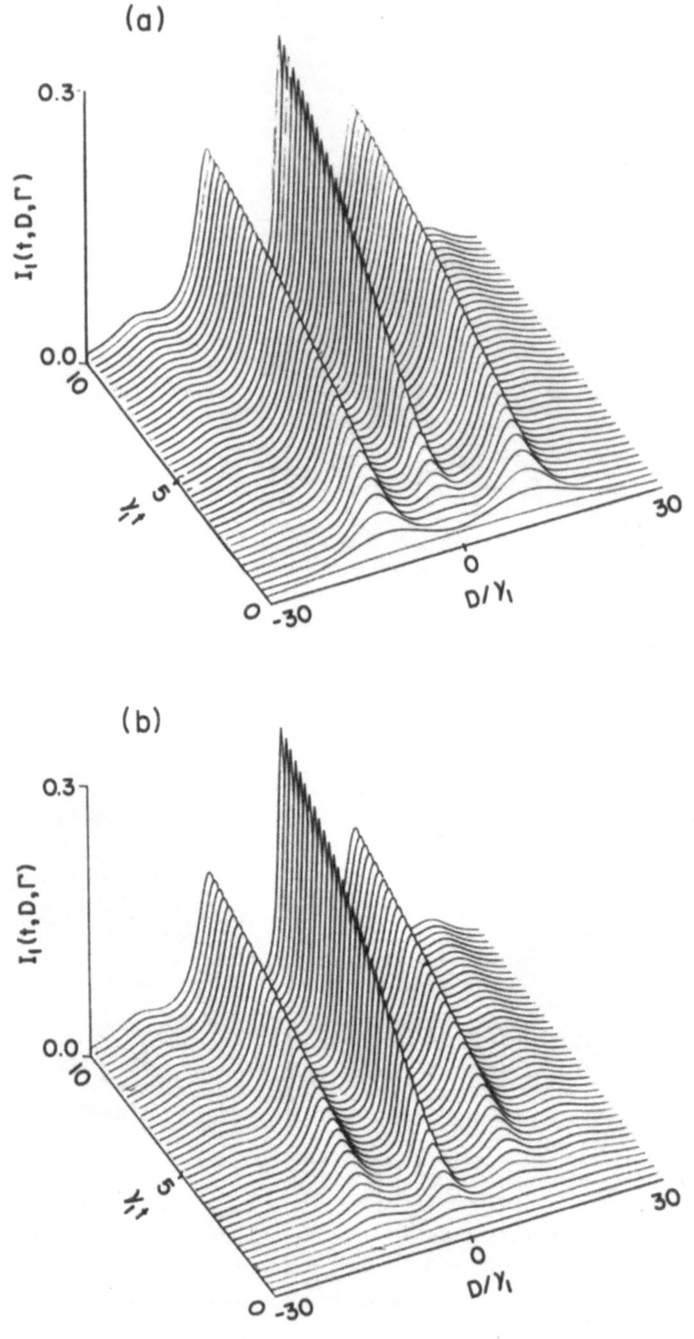

Figure 3. Upper spectrum from an atom prepared initially in dressed states. (a) corresponds to the dressed state $|\psi_1>$ and (b) to $|\psi_2>$. Here $\phi_1 = \phi_2 = \pi$, $N_1 = 0.1$ and $N_2 = 0.6$. Other data as in Figure 2.

Further, it is of interest to study the effects of the initial state preparation of the atom on the development of the transient spectrum. In Fig. 3(a) and 3(b) we display the temporal development of the upper spectra when the atom was initially in the dressed state $|\psi_1>$ and $|\psi_2>$ respectively. $|\psi_i>$ corresponds to the eigenstates of the interaction Hamiltonian of the atom and radiation field.[14] When the atom is prepared in the dressed state $|\psi_1>$, the central peak is initially absent and only the nearest sidebands develop prominently. The development of the farthest sideband is also delayed as seen in Fig. 3(a). In contrast, when the atom is prepared in the dressed state $|\psi_2>$ all the peaks develop simultaneously as seen in Fig. 3(b). However the spectrum for the earlier case develops more rapidly into the steady state spectra than the latter.

The temporal behavior of the spectra when the atom is prepared in atomic state $|1>$ is similar to the case shown in Fig. 3(a). The transient spectra when the atom is prepared in atomic states $|2>$ or $|3>$ or dressed state $|\psi_3>$ is similar to the case shown in Fig. 3(b).

The more interesting case of the temporal behavior of the spectra for the general off-resonant case, along with the detailed derivations will be reported elsewhere.

REFERENCES

1. R. E. Slusher, L. W. Hollberg, B. Yurke, J. C. Mertz, and J. F. Valley, Phys. Rev. Lett., 55, 2409 (1985); M. W. Maede, P. Kumar, and J. Shapiro, Opt. Lett. 12, 161 (1988); L. Wu, H. J. Kimble, J. L. Hall, and Wu, Phys. Rev. Lett., 57, 2520 (1988).
2. C. W. Gardiner, Phys. Rev. Lett., 56, 1917 (1986).
3. H. J. Carmichael, A. S. Lane, and D. F. Walls, Phys. Rev. Lett., 58, 2539 (1987); J. Mod. Opt. 34, 821 (1987).
4. A. S. Parkins and C. W. Gardiner, Phys. Rev. A37, 3867 (1988).
5. A. S. Shumovsky and T. Quang, J. Phys. B22, 131 (1989).
6. R. D'Souza, A. S. Jayarao and S. V. Lawande, Phys. Rev. A41, 4083 (1990).
7. B. N. Jagatap, Q. V. Lawande, and S. V. Lawande, Phys. Rev. A43, 535 (1991).
8. Z. Ficek, Phys. Rev. A42, 611 (1990) and references therein.
9. J. H. Eberly and K. Wodkiewicz, J. Opt. Soc. Am. 67, 1252 (1977).
10. J. H. Eberly, C. V. Kunasz, and K. Wodkiewicz, J. Phys. B13, 217 (1980); T. Ho and H. Rabitz, Phys. Rev. A37, 1576 (1988).
11. N. Lu, P. R. Berman, A. G. Yokh, Y. S. Bai, T. W. Mossberg, Phys. Rev. A33, 3956 (1986); Y. S. Bai, T. W. Mossberg, N. Lu, and P. R. Berman, Phys. Rev. Lett., 57, 1692 (1986).
12. A. S. Jayarao, S. V. Lawande, Phys. Rev. A39, 3464 (1989).
13. A. Joshi and S. V. Lawande, Phys. Rev. A41, 2822 (1990).
14. S. V. Lawande, R. D'Souza, and R. R. Puri, Phys. Rev. A36, 3228 (1987).

EFFECT OF SQUEEZED RADIATION FIELD ON A TWO LEVEL ATOM

Swapan Mandal and Pradip N. Ghosh

Department of Physics
University College of Science
92, A.P.C. Road, Calcutta - 700009, India

ABSTRACT

A perturbative approach has been used to consider the interaction of a two-photon coherent state with a two level atom. In addition to the usual emission and absorption terms the transition probability contains a few interference terms. The absorption coefficient is given by the quasiphoton number and the quantum noise energy. A method of calculating the quasiphoton number in the two-photon coherent states has been discussed.

The coherent state of the radiation field is the closest approach towards a classical state from a quantum mechanical one. The fluctuations of the two quadrature components in a coherent state become equal and eventually reduce to the minimum values allowed by the uncertainty principle. The concept of attainable uncertainty has changed dramatically after the proposal of two photon coherent states (TCS)[1]. The TCS are also known as squeezed coherent states. It is to be remembered that the TCS is actually a special type of squeezed states with minimum uncertainty product.

The electromagnetic field interacting with a two level atom is one of the fundamental problems in quantum optics, electrodynamics and spectroscopy. The population difference in a two level atom irradiated by a single mode of the electromagnetic field prepared in a Fock state $|n>$, oscillates with the frequency $(n+1)^{\frac{1}{2}}$, resulting in the collapse and revival of the oscillation if the field was prepared in a coherent state. The question does naturally arise about what change would be expected if the field was prepared in a two-photon coherent state. This problem was considered by Milburn[2] and very recently by Satyanarayana et. al[3] to obtain the ringing revivals of the atomic inversion. Both of these papers[2-3] were based on the assumption that the direction of squeezing was parallel to the direction of the coherent excitation of the field. This situation added an extra advantage for calculating the photon number distribution and hence the inversion.

In the present report we calculate the transition probability of a two level atom interacting with a TCS. We assume that the atomic levels are very sharp and hence the broadening effects are excluded. Moreover, the field mode which is excited in a TCS is treated as a quantum reservoir to which the atom is coupled. In the present

Recent Developments in Quantum Optics, Edited
by R. Inguva, Plenum Press, New York, 1993

derivation we shall work in the dipole approximation.[4] The perturbative method which we shall follow is particularly useful in calculating the atomic level shifts [5-6] with the participation of squeezed radiation.

TRANSITION PROBABILITY

We consider a single mode of the electromagnetic field with frequency ω interacting with a two level atom. The mode is excited in a TCS and all other modes are in the ground state. The resonance frequency ω_{12} of the two level atom is close to ω and the transition is dipole allowed. If the atom is in state 1 at $t = 0$, then the probability of finding the atom in the state 2 at time t under the effect of the TCS mode is give by[4]

$$|C(2,t|1,0)|^2 = \frac{\omega_{12}^2|\mu_{12}|^2(1-cos^2\theta)}{2\hbar\varepsilon_o L^3}\int_o^t dt_1 \int_o^t dt_2 exp[i\omega_{12}(t_1-t_2)]$$

$$Tr[\rho(0)\{(1+a^+a)exp[i\omega(t_1-t_2)] + a^+a exp[-i\omega(t_1-t_2)]$$

$$+ a^2 exp[-i\omega(t_1+t_2)] + a^{+2}exp[i\omega(t_1+t_2)]\}] \tag{1}$$

where L is the length of the cavity. The angle between the propagation vector and μ_{12} has been denoted by θ. The radiation field density operator is simply given by

$$\rho(0) = |\beta><\beta|\{|0><0|\} \tag{2}$$

with

$$Tr\rho(0) = 1 \tag{3}$$

The two-photon coherent state (TCS) obeys the following eigenvalue equation[1]

$$b|\beta>=\beta|\beta> \tag{4}$$

where

$$b = \mu a + \nu a^+ \tag{5}$$

subject to the condition

$$|\mu|^2 - |\nu|^2 = 1 \tag{6}$$

The parameters μ, ν and β are complex numbers. It is easy to establish that the operators b and b^+ obey the following commutation relation

$$[b, b^+] = 1 \tag{7}$$

The Yuen operators b and b^+ are linearly related with the usual photon annihilation and creation operators (5). The inverse relations, for which a and a^+ are expressed as linear combinations of b and b^+ are

$$a = \mu^* b - \nu b^+ \tag{8a}$$

$$a^+ = \mu b^+ - \nu^* b \tag{8b}$$

The following relations may easily be obtained

$$a^2 = \mu^{*2} b^2 + \nu^2 b^{+2} - 2\mu^* \nu b^+ b - \mu^* \nu \tag{9a}$$

$$a^{+2} = \mu^2 b^{+2} + \nu^{*2} b^2 + 2\mu\nu^* b^+ b - \mu\nu^* \tag{9b}$$

$$a^+ a = (|\mu|^2 + |\nu|^2) b^+ b - \mu\nu b^{+2} - \mu^* \nu^* b^2 + |\nu|^2 \tag{9c}$$

The equation (1) is simplified by using the equations (2-8,9) to obtain the following result

$$|C(2,t|1,0)|^2 = \frac{\omega_{12}^2}{2\hbar\varepsilon_o L^3} \frac{|\mu_{12}|^2(1-cos^2\theta)}{\omega}[\{1 + |\nu|^2 + |\beta|^2(|\mu|^2$$

$$-|\nu|^2)\}\frac{4sin^2(\omega_{12}+\omega)\frac{t}{2}}{(\omega_{12}+\omega)^2} + \{|\nu|^2 + |\beta|^2(|\mu| - |\nu|)^2\}\frac{4sin^2(\omega_{12}-\omega)\frac{t}{2}}{(\omega_{12}-\omega)^2}$$

$$+\{(|\mu| - |\nu|)^2|\beta|^2 - |\mu||\nu|\}\frac{4cos\omega t(cos\omega t - cos\omega_{12}t)}{\omega_{12}^2 - \omega^2}] \tag{10}$$

where we have considered μ, ν and β as real. The transition rate is obtained by differentiating the equation (10).

$$W = \frac{d}{dt}|C(2,t|1,0)|^2 = \frac{\omega_{12}^2}{\hbar\varepsilon_o L^3}\frac{|\mu_{12}|^2(1-cos^2\theta)}{\omega}$$

$$\text{x}[\{1 + |\nu|^2 + (|\mu| - |\nu|)^2|\beta|^2\}\frac{sin(\omega_{12}+\omega)t}{\omega_{12}+\omega}$$

$$+\{|\nu|^2 + (|\mu| - |\nu|)^2|\beta|^2\}\frac{sin(\omega_{12}-\omega)t}{\omega_{12}-\omega} + \{(|\mu| - |\nu|)^2 - |\mu||\nu|\}$$

$$\text{x}\{-\frac{2\omega sin2\omega t}{\omega_{12}^2 - \omega^2} + \frac{sin(\omega+\omega_{12})t}{\omega_{12}-\omega} + \frac{sin(\omega_{12}-\omega)t}{\omega_{12}+\omega}\}] \tag{11}$$

For near resonance (i.e. $\omega \to \omega_{12}$) the absorption rate assumes the form,

$$W_{abs} = \frac{\omega_{12}^2}{\hbar\varepsilon_o L^3}\frac{|\mu_{12}|^2(1-cos^2\theta)}{\omega}\{(|\mu| - |\nu|)^2|\beta|^2 + |\nu|^2\}\frac{sin(\omega_{12}-\omega)t}{\omega_{12}-\omega} \tag{12}$$

where we have neglected the terms in equation (11) which do not contribute significantly. For large t and $\omega \to \omega_{12}$ the equation (12) reduces to

$$W_{abs} = \frac{\pi\omega_{12}}{\hbar\varepsilon_o L^3}|\mu_{12}|^2(1-cos^2\theta)\{(|\mu| - |\nu|)^2|\beta|^2 + |\nu|^2\} \tag{13}$$

Therefore, it is clear that for large values of squeezing for which $|\mu| \sim |\nu|$, the transition probability (eqn. 13) is dominated by the squeezing parameter $|\nu|$ whatever be the value of $|\beta|^2$. An evaluation of $|\beta|^2$ will lead to the absorption coefficient as a function of the photon number n. Thus the transition is possible even when $|\beta|^2$ is zero. When no quasiphoton is present, the energy of the TCS is nonzero and is termed as quantum noise energy.[1] The quantum noise energy is one of the fundamental characteristic of two photon coherent states. For very large squeezing the atom is exposed to the quantum noise energy which makes the transition possible. For $\nu \sim 0, |\beta|^2 \sim |\alpha|^2$, hence the result coincides exactly with that of a coherent state.[4] It is also to be noted that for very small values of $|\nu|$ (i.e. small squeezing) for which $(|\mu| - |\nu|)^2 \neq 0$, the transition probability would be simply proportional to the quasiphoton number $|\beta|^2$ of the radiation field.

CALCULATION OF $|\beta|^2$

The projection of the TCS on the Fock state $|n>$ is given by[1]

$$< n|\beta >= (\mu n!)^{-\frac{1}{2}}(\frac{|\nu|}{2|\mu|})^{\frac{n}{2}} exp(-\frac{1}{2}|\beta|^2 + \frac{\nu^*}{2\mu}\beta^2)H_n(\frac{\beta}{[2\mu\nu]^{\frac{1}{2}}}) \tag{14}$$

where H_n is the Hermite polynomial of order n. The photon number distribution is simply obtained as

$$p_n = | < n|\beta > |^2 = \frac{1}{|\mu|n!}\frac{1}{2^n}(\frac{|\nu|}{|\mu|})^n exp(-|\beta|^2 + \frac{\nu^*}{2\mu}\beta^2 + \frac{\nu}{2\mu^*}\beta^{*2})$$

$$H_n(\frac{\beta}{[2\mu\nu]^{\frac{1}{2}}})H_n(\frac{\beta^*}{[2\mu^*\nu^*]^{\frac{1}{2}}}) \tag{15}$$

β, μ and ν are real and the corresponding equation (15) reduces to

$$p_n = \frac{1}{|\mu|n!}(\frac{|\mu|}{|\nu|})^n exp[-|\beta|^2(1 - \frac{|\mu|}{|\nu|})|H_n[\frac{|\beta|}{(2|\mu||\nu|)^{\frac{1}{2}}}]|^2 \tag{16}$$

It is easy to show that the equation (16) coincides exactly with the photon number distribution in a coherent state if $|\nu| \to 0$ for which $|\beta|^2 \to |\alpha|^2$. From the definition of $|\beta|^2$

$$|\beta|^2 =< \beta|b^+b|\beta >= \sum_n[(|\mu|^2 + |\nu|^2)n + |\mu||\nu|(\{(n+1)(n+2)\}^{\frac{1}{2}}$$

$$+\{n(n-1)\}^{\frac{1}{2}}) + |\nu|^2]p_n \tag{17}$$

Therefore the evaluation of the quasiphoton number $|\beta|^2$ for a particular value of μ is possible either graphically or numerically from the equations (16) and (17).

In conclusion, we have studied the transition probability and hence the transition rate of a two level atom interacting with a single mode of the radiation field excited in a TCS. We have assumed that the field is unchanged during the interaction so that a perturbative calculation is possible.

ACKNOWLEDGEMENTS

One of us (S.M.) is thankful to the University Grants Commission for award of a Senior Research Fellowship.

REFERENCES

1. H. P. Yuen, Phys. Rev. A13, 2226 (1976).
2. G. Milburn, Optica Acta 31, 671 (1984).
3. M. V. Satyanarayana, P. Rice, R. Vyas and H. J. Carmichael, JOSA B6, 228 (1989).
4. W. H. Louisell, Quantum Statistical Properties of Radiation (Willey, New York, 1973) P 273.
5. G. J. Milburn, Phys. Rev. A34, 4882 (1986).
6. G. W. Ford and R. F. O'Connell, JOSA B4, 1710 (1987).

SQUEEZING OF A WEAK LIGHT BEAM USING INTERACTION WITH AN INTENSE LIGHT BEAM IN A KERR MEDIUM

Hari Prakash, Naresh Chandra and Ranjana Prakash

Department of Physics
University of Allahabad
Allahabad, India

By squeezing[1] of light, we mean variance of a quadrature amplitude taking a value which is less than that for vacuum. Like antibunching[2], this is also a quantum effect. Quantum effects [9-7] including squeezing[5-7] were known to exist in principle since long and squeezed states have been used[7] under the name of minimum uncertainty states. Perhaps two-photon absorption is the first studied nonlinear phenomenon (in 1970) which converts classical light into a quantum state exhibiting both antibunching[4,6] and squeezing.[6] Squeezing is now known to be generated in a number of interactions and some experimental observations have also been reported.[8]

Recently, Shirasaki and Hauz[9] showed that a weak light beam can be squeezed using an intense light beam of identical frequency and polarization by making them incident on a Mach-Zehnder interferometer containing identical Kerr media in its two arms. An intense light beam is known to self-squeeze[10,11] in propagation through Kerr medium. Very large squeezing occurs[11] when the dimensionless amplitude is of the order of the reciprocal of the dimensionless nonlinear susceptibility. Shirasaki and Hauz[9] consider a different approximation in which the dimensionless intensity is of the order of the reciprocal of dimensionless nonlinear susceptibility.

Here we show that weak beam can be squeezed with the help of an intense light beam in a Kerr medium without requiring a Mach- Zehnder interferometer if the two have orthogonal linear polarization. For the same total length of Kerr medium, this scheme produces a larger squeezing. The hamiltonian for light in a Kerr medium in Schrodinger representation is (with c = $\hbar$ = 1)

$$H_s = H_o + H',\ H_o = \omega N,\ H' = g_1 : N^2 : + g_2(a_x^{\dagger 2} + a_y^{\dagger 2})(a_x^2 + a_y^2), \tag{1}$$

where z-axis has been taken along $\vec{k}$ and $a_{x,y}$ are annihilation operator for modes polarized linearly along the x and y directions, $N = N_x + N_y$, $N_{x,y} = a_{x,y}^\dagger a_{x,y}$, and :: denotes the normal form of the included operator. In terms of the basis of right and left handed circular polarizations (natural definition; not traditional) R and L, i.e., with $a_{R,L} = (a_x \mp i a_y)/\sqrt{2}$, and $N_{R,L} = a_{R,L}^\dagger a_{R,L}$, eq. (1) takes the form

$$H_s = H_o + H',\ H_o = \omega N,\ H' = g_1 : N^2 : + 4 g_2 N_R N_L,\ N = N_R + N_L, \tag{2}$$

Recent Developments in Quantum Optics, Edited
by R. Inguva, Plenum Press, New York, 1993

Since $[H_o, H'] = 0$, the interaction picture Hamiltonian is $H_I = H'$ and the time-evolution operator is $exp[-iH_I t]$. In time interval t, $a_{R,L}$ evolves to $a_{R,L}(t) = exp[-2iGN]exp[\pm 2iG_2 M]a_{R,L}$, where $G = G_1 + G_2, G_{1,2} = g_{1,2}t$, $N = a_x^\dagger a_x + a_y^\dagger a_y$, $M = N_R - N_L = i(a_y^\dagger a_x - a_x^\dagger a_y)$, and therefore $a_x = (a_R + a_L)/\sqrt{2}$ evolves to $a_x(t) = exp[-2iGN][cos(2G_2M)a_x + sin(2G_2M)a_y]$.

For light in a single mode, $H' = g : N^2 :$, $N = a^\dagger a$, $g = g_1 + g_2$. For this case, as is seen to evolve to $a(t) = exp[-2iGN]a$. In the Shirasaki-Hauz[9] scheme, weak signal and strong pump modes (operators a and b) are incident on a Mach-Zehnder interferometer. In the two arms, the modes have operators $a_\pm \equiv (a \pm b)/\sqrt{2}$. In length l of the Kerr medium, they evolve to $a_\pm(t) = exp[-2iGN_\pm]$, $N_\pm \equiv a_\pm^\dagger a_\pm$, where $G = gt = gl$ and the output signal operator is $a(t) = (a_+(t) + a_-(t))/\sqrt{2}$ given by $a(t) = exp(iGN)[cos(GM)a + isin(GM)b]$, with mutually commuting operators N and M defined by $N = N_+ + N_- = a^\dagger a + b^\dagger b$, $M = N_+ - N_- = a^\dagger b + b^\dagger a$.

The two schemes can be studied jointly by considering

$$a(t) = exp(iK_1N)[cos(K_2M)a + isin(K_2M)b], \tag{3}$$

with $K_1 = K_2 = -2G$, $a = a_x$, $b = ia_y$ for our scheme and $K_1 = K_2 = -G$ for Shirasaki-Hauz scheme. Here, $N = a^\dagger a + b^\dagger b$, $M = a^\dagger b + b^\dagger a$.

Consider the approximations that (i) nonlinear susceptibility is small($|K_{1,2}| << 1$), (ii) b mode is intense ($< b^\dagger b > >> 1$), and (iii) a mode is weak in such a way that $< b^\dagger b > \sim |K_{1,2}|^{-1} >> < a^\dagger a > \sim 1$. Here $< 0 >$ stands for $Tr(\rho 0)$ for any operator 0. If the intense b mode is initially in the coherent state $|\beta>$, direct calculations give

$$< a(t) > = e^{-i(\omega t - \phi)}[\mu < a > + \nu < a^\dagger >], \tag{4}$$

$$< a^2(t) > = e^{-2i(\omega t - \phi)}[\mu^2 < a^2 > + 2\mu\nu < a^\dagger a > + \nu^2 < a^{\dagger 2} > + \mu\nu], \tag{5}$$

$$< a^\dagger(t)a(t) > = (|\mu|^2 + |\nu|^2) < a^\dagger a > + \mu^*\nu < a^{\dagger 2} > + \mu\nu^* < a^2 > + |\nu|^2, \tag{6}$$

with $\mu = 1 + iK_2|\beta|^2, \nu = iK_2\beta^2$. If we define quadrature component

$$Q(t) = (a(t)exp[i(\omega t - \phi + \theta) + h.c.)/\sqrt{2}, \tag{7}$$

where θ is an arbitrary constant and h.c. means hermitian conjugate, and put $\beta = |\beta|exp(i\theta_\beta)$, its variance at time t is

$$< \Delta Q^2 >_t - \frac{1}{2} = 2X^2 sin^2\psi - X sin2\psi + (|\mu|^2 + |\nu|^2)[< \Delta\bar{Q}^2 >)0 - \frac{1}{2}], \tag{8}$$

where $X \equiv |\nu| = |K_2\beta^2|$, $\psi = \theta_\beta + \theta$, $\bar{Q} = (aexp(i\bar{\theta}) + h.c.)/\sqrt{2}$, $exp(i\bar{\theta}) = \sigma/|\sigma|$, $\sigma = \mu exp(i\theta) + \nu^* exp(-i\theta)$ and $< \Delta\bar{Q}^2 >_0$ is variance of $\bar{Q}$ at time $t = 0$. In equation (9), the terms involving X may be enough negative to result in squeezing of the signal. If it is coherent initially $< \Delta Q^2 >_t - \frac{1}{2} = 2X^2 sin^2\psi - X sin2\psi$. This is maximum when $sin2\psi = 1/(1 + X^2)^{\frac{1}{2}}$, $tan2\psi = 1/X$, i.e., with ψ in the first or third quadrant and this maximum value is

$$< \Delta Q^2 >_t - \frac{1}{2} = X^2 - X(1+X^2)^{\frac{1}{2}}. \qquad (9)$$

Note that this is zero for $X = 0$, decreases monotonically on increasing X and $\rightarrow -\frac{1}{2}$ as $X \rightarrow \infty$.

Thus, squeezing can be observed by coupling with a quadrature component whose phase $\omega t - \phi_1 + \theta$ is such that $\psi = \theta_\beta + \theta$ lies in 1st or 3rd quadrant and has $tan 2\psi - X^{-1} < 0$. By choosing suitably large X any large squeezing can be obtained but, for large X, the phase adjustment becomes more demanding.

For the Shirasaki-Hauz scheme[9] using lengths l of the Kerr medium, $X = |K_2\beta^2| = |G\beta^2| = |gl\beta^2|$, while for our scheme with a single length 2ℓ of the same Kerr medium, $X = |K_2\beta^2| = |2G_2\beta^2| = |4g_2 l\beta^2|$. The ratio of the value of X for our scheme to that for the Shirasaki-Hauz scheme is $4g_2/g_1 = 4g_2/(g_1+g_2)$, which is always greater than one (for liquids and for crystals, $g_2/g_1 = 3$ and $1/2$ respectively, giving $4g_2/g_1 = 3$ and $4/3$ respectively). Hence our scheme produces a larger squeezing using the same total length of the Kerr medium besides not requiring a Mach-Zehnder interferometer.

For signal polarization $\hat{\varepsilon} = \varepsilon_x \hat{e}_x + \varepsilon_x \hat{e}_x$, and othogonal pump polarization, exactly similar calculations give $X = |4g_2 l\beta^2(\varepsilon_x^2 + \varepsilon_y^2)|$, which is maximum (=1) for plane polarization and zero for circular polarization. One should thus use orthogonally linearly polarized signal and pump modes.

REFERENCES

1. See, e.g., review articles, D. F. Walls, *Nature* **306**, 141 (1983), R. Loudon and P. L. Knight, *J. Mod. Optics* **34** (1987).
2. See, e.g., review articles, H. Paul and P. Mohr, *Rev. Mod. Phys.* **54**, 1061 (1982), M. C. Teich and B. E. A. Saleh, *Progress in Optics, Vol. XXI,* p. 1 (North Holland Pub. Co., Amsterdam, 1988).
3. R. J. Glauber, *Quantum Optics and Electronics* (lectures at Les Houches, 1964), Gordon and Greach, 1965. T. F. Jordan, *Phys. Lett.* **11**, 289 (1964), M. M. Miller and E. A. Mishkin, *Phys. Lett* **24A**, 188 (1967), P. P. Bertrand and E. A. Mishkin, *Phys. Lett.* **25A**, 204 (1967). N. Chandra and H. Prakash, *Lett. Nuovo Cim.* **49**, 1196 (1970); *Indian J. pure appl. Phys.* **9**, 677 and 767 (1971).
4. N. Chandra and H. Prakash, *Phys. Rev.* **A1**, 1696 (1970).
5. B. R. Mollow and R. J. Glauber, *Phys. Rev.* **160**, 1076 (1967), D. Stoler, *Phys. Rev.* **D1**, 3257 (1970); D4, 1925 (1971).
6. N. Chandra and H. Prakash, *Indian J. pure appl. Phys.* **9**, 677 (1971).
7. E. Y. C. Lu, *Lett. Nuovo Cim.* **2**, 1241 (1971); **3**, 585 (1972), H. Prakash, N. Chandra and Vachaspati, *Ann. Phys.* **85**, 1 (1974); *J. Phys.* **A**, L161 (1974). Some other quantum states have also been used for representation of density operators. See, e.g., K. E. Kahill, *Phys. Rev.* **180**, 1244 (1969), N. Chandra and H. Prakash, *Phys. Rev.* **D4**, 1936 (1973), H. Prakash, N. Chandra and Vachaspati, *Indian J. pure appl. Phys.* **14**, 48 (197x).
8. See references in M. C. Teich and B. E. A. Saleh, *Phys. Today* **28**, 26 (1990).
9. M. Shirasaki and H. A. Hauz, *J. Opt. Soc. Am.* **B7**, 30 (1990).
10. R. Tanas and S. Kielich, *Opt. Comm.* **30**, 443 (1979).
11. H. Prakash, N. Chandra and R. Prakash, to be published; presented in *Workshop on Quantum Optics*, Goa 1989.

SQUEEZING AND QUANTUM NOISE QUENCHING IN OPTICALLY PUMPED ATOMIC CLOCKS

G. M. Saxena and B. S. Mathur

National Physical Laboratory
Dr. K.S. Krishnan Road
New Delhi - 110012, India

INTRODUCTION

The application of diode laser in optically pumped Rubidium atomic clock has resulted in improved performance (1) due to their narrow linewidth and less fluctuation in intensity. However, the intensity and frequency fluctuations due to quantum noise in the diode laser limit the frequency stability of the standard. The squeezing of the diode laser light may reduce its frequency fluctuations to a large extent and in turn improve the frequency stability of the atomic clock. In this paper we shall discuss the effect of squeezing of diode laser light on the performance of Rubidium atomic clock.

PRINCIPLE

Rubidium (Rb) and Cesium (Cs) atomic clocks work on the principle of phase locking a voltage controlled oscillator (VCXO) to the frequency of the field independent hyperfine (hf) transitions between ground state (gs) h.f. sublevels. In the case of Rb atomic clock these h.f. sublevels are F=2, m_F=0 and F=1, m_F=0 of ^{87}Rb isotope as shown in Fig. 1. To obtain the necessary correction signal to phase lock VCXO to the h.f. transition frequency; at first a state of population inversion is created between the above mentioned hyperfine sublevels using tuned diode laser.

On applying a r-f field to the population inverted assembly of atoms a net transition from F=2 to F=1 level results. To restore the steady state of the population inversion more light is absorbed in ^{87}Rb absorption cell. Thus the emerging light from the absorption cell shows a dip in its intensity at the resonance. If the resonant r-f field is phase modulated at a very low frequency, the photodetector output yields a composite signal containing the modulating frequency and its harmonics riding on d.c. signal. As the resonant r-f field is derived from a 5 MHz VCXO, application of the correction signal after phase sensitive detection to EFC (Electronic Frequency Control) of VCXO transfers the frequency stability of the ^{87}Rb h.f. transition to VCXO and a Rb atomic clock is realized.

The diode lasers have good intensity stability with narrow line width and this is reflected in better frequency stability of the atomic clocks. However, due to quantum

Recent Developments in Quantum Optics, Edited
by R. Inguva, Plenum Press, New York, 1993

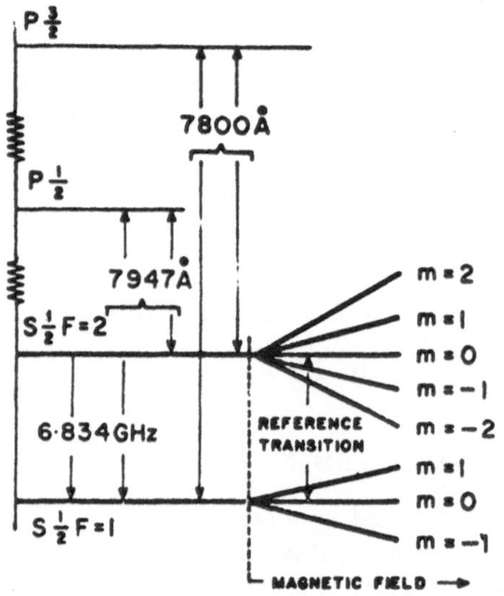

Figure 1. Energy level diagram of ^{87}Rb isotope.

noise the frequency and intensity stability of the diode laser is limited and consequently the improvement in the diode laser pumped Rb atomic clock over the conventional r.f. discharge Rb lamp and filter cell combination is also restricted. To reduce the effect of quantum noise on the diode laser pumped Rb atomic clock, squeezing of laser light is one of the possibilities. We shall now discuss the application of squeezed light to diode laser pumped Rb clock. It is well known semi-classically that for a laser operating well above threshold the amplitude fluctuations are small and occur in the vicinity of a given value $\sqrt{\bar{n}}$ and $\bar{n}$ is the steady state number of the emitted photons. The uncertainties in laser phase and amplitude fluctuations are related to the variances in canonical variable X_1 and X_2 respectively by the relation (2).

$$<(\delta a_1)^2>=\Delta X_1^2(\phi_o) \quad (1)$$

and

$$<(\delta\phi)^2>=\frac{1}{\bar{n}}\Delta X_2^2(\phi_o), \quad (2)$$

where ϕ_o is the instantaneous phase of the semi-classical electric field amplitude, δa_1 and $\delta\phi$ are the amplitude and phase fluctuation respectively.

The fractional frequency fluctuations $(\delta f/f_o)$ in the Rb frequency and time standard due to quantum noise can be expressed as (3)

$$\frac{\delta f}{f_o}=\frac{1}{f_o}(\frac{\delta^2 f}{\delta\nu\delta I})I\delta\nu. \quad (3)$$

Where f, ν, I and f_o denote, respectively, the hyperfine transition frequency, the laser frequency, the laser intensity and unperturbed ^{87}Rb 0-0 hyperfine frequency. $\delta\nu$ is the frequency fluctuations in the laser light. With squeezed laser light let $\Delta\nu$ be the laser frequency fluctuation ($\Delta\nu < \delta\nu$). The fractional frequency fluctuations in the clock frequency may be written as:

$$\frac{\Delta f}{f_o} = \frac{1}{f_o}(\frac{\delta^2 f}{\delta\nu\delta I})I_s\Delta\nu. \tag{4}$$

Here $I_s = \eta I$ with η being the efficiency of the squeezing device interposed between the laser and the absorption cell.

To evaluate $\Delta\nu$ consider ^{87}Rb atoms as two level system. Using Jaynes-Cummings (JC) model (4,5) for describing the interaction between single mode squeezed light and Rb atoms, JC hamiltonian in rotating wave approximation (RWA) is given by

$$H = \hbar\omega_o S_3 + \hbar w a^+ a + \hbar\lambda(a^+ S_- + aS_+); \tag{5}$$

where ω_o and ω are the frequencies of two level atom and the field respectively and λ is the coupling constant. S_3, S_+ and S_- define atomic inversion, excitation and relaxation operators respectively. The time development operator corresponding to hamiltonian (5) is defined as

$$U(t) = esp(-i\frac{Ht}{h}). \tag{6}$$

The density matrix operator $\rho(t)$ of the system at t is time evolved through the relation.

$$\rho(t) = U(t)\rho(o)U^+(t). \tag{7}$$

$\rho(o)$ defines the system initially and is the direct product of density matrices of initially squeezed field and the atom

$$\rho(o) = \rho_f(o) \otimes \rho_a(o). \tag{8}$$

Here $\rho_f(o)$ and $\rho_a(o)$ define the initial state of the field and atom in the ground state with the following forms,

$$\rho_f(o) = |\xi, z >< z, \xi| \tag{9}$$

and

$$\rho_a(o) = |g >< g|, \tag{10}$$

where $|g>$ is the ground state of Rb atoms. Now the phase fluctuations (cf. Eq. 2) can be expressed as

$$\begin{aligned} < (\delta\phi)^2 > \; &= \tfrac{1}{\bar{n}}\Delta X_2^2(\phi_o) \\ &= \tfrac{1}{4\bar{n}}[2\textstyle\sum_m m < m|\rho_f(t)|m > -2Re\sum_m \{(m+1)(m+2)\}^{\frac{1}{2}} \\ &< m|\rho_f(t)|m+2> +1 - (2Im\textstyle\sum_m (m+1)^{\frac{1}{2}} < m|\rho_f(t)|m+1>)^2]. \end{aligned} \tag{11}$$

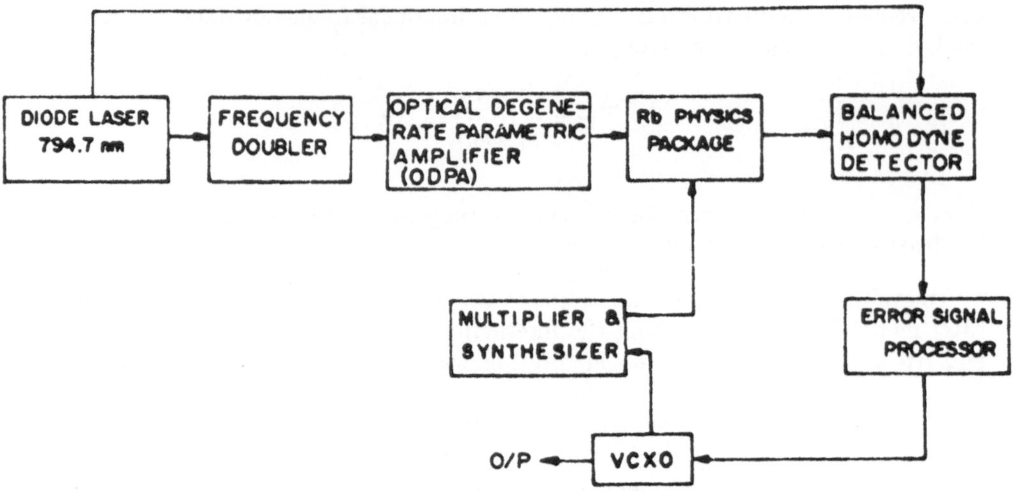

Figure 2. Block diagram of the proposed squeezed diode laser pumped Rubidium Atomic Frequency Standard.

The frequency fluctuation is time derivative of the phase fluctuation $[<(\delta\phi^2>]^{\frac{1}{2}}$ i.e.

$$\Delta\nu = \frac{d}{dt}[<(\delta\phi)^2>]^{\frac{1}{2}} = \frac{1}{\sqrt{\bar{n}}}\frac{d}{dt}[\Delta X_2^2(\phi_o)]^{\frac{1}{2}}. \tag{12}$$

In the above derivation the interaction of two level Rb atom with coherent squeezed light is considered. To have a realistic picture we should consider the system consisting of N identical two-level atoms interacting with single mode radiation field including atomic relaxation. If the life time has $P(t)$ distribution then the state of the field, after it has interacted with large number of non-interacting atoms is given by the density matrix

$$\tilde{\rho}_f(t) = \int_o P(t)\rho_f(t)dt. \tag{13}$$

If γ^{-1} is the average life time and $P(t) = \gamma e^{-\gamma t}$ then

$$\begin{aligned} <(\Delta\nu)> &= \tfrac{d}{dt}[<(\overline{\delta\phi}^2>]^{\frac{1}{2}} \\ &= \tfrac{1}{\sqrt{\bar{n}}}\tfrac{d}{dt}[\overline{\Delta X}_2^2(\phi_o)]^{\frac{1}{2}}, \end{aligned} \tag{14}$$

where $<(\overline{\delta\phi})^2>$ and $<\overline{\Delta X}_2^2(\phi_o)>$ correspond to the case with atomic relaxation. On putting the value of $<(\Delta\nu)>$ from eqn. 14 into eqn. 4 the effect of squeezing on the performance of atomic clocks is obtained. The computation and analysis of our results will be reported in a separate paper. The experimental set-up of the proposed squeezed diode laser pumped Rubidium atomic clock is shown in Fig. 2.

ACKNOWLEDGEMENTS

Authors are grateful to Prof. C.L. Mehta, IIT Delhi for the fruitful discussions with him on squeezed states of light. They are thankful to Prof. S.K. Joshi, DCSIR for his encouragement for this work.

REFERENCES

1. J. Venier and C. Audoin, *The Quantum Physics of Atomic Frequency Standard* Hilder, Bristol, U.K. (1989).

2. J. Bergou, M. Orszag, M. O. Scully and K. Wodkiewicaz, *Phys. Rev.* **A39** 5136 (1989).

3. J. C. Camparo and R. P. Furehold, *J. Appl. Phys.* **59**, 3313 (1986).

4. Zou Ming - Liang and Guo - Guang - can, *J. Phys. B. At. Mol. Opt. Phys.* **22** 2205 (1989).

5. J. R. Kuklinski and J. L. Modajczyk, *Phys. Rev.* **A37** 3175 (1988).

PROPERTIES OF SQUEEZED BINOMIAL STATES AND SQUEEZED NEGATIVE BINOMIAL STATES

Amitabh Joshi and S. V. Lawande

Bhabha Atomic Research Center
Bombay, 400 085
India

The effect of squeezing on binomial and negative binomial state has been studied in terms of quasiprobability of Wigner function and their photon number distributions. The results presented for squeezed binomial (negative binomial) states may be useful when one makes transients from squeezed coherent states to squeezed number (quasi-thermal) states.

The recent developments in the generation of number states [1], localized one-photon states [2] and proposals for generating binomial states [3] etc. have inspired theorists to have a possibility of generation of squeezed number states etc. In other words the input to a squeezing device can be considered not only in terms of a vacuum state or a coherent state but also as a superposition of number states.

Here, we consider the squeezing of input fields which are described in terms of the superposition of number states i.e. either binomial or negative binomial weighted sum of number states. The binomial state $|p, N>$ is intermediate between a number state and a coherent state [4]. This state reduces to either a coherent state ($p \to 0, N \to \infty$) or a number state (p=1) under two extreme conditions of parameters. Thus it is useful in studying systematic changes in some observables as one moves from a coherent state to a number state or vice versa.

The negative binomial state $|q, w>$ may be useful in carrying out a transient study from a coherent state ($q \to 0, w \to \infty, \bar{n} = wq/(1-q)$) to a state ($w = 1, \bar{n} = q/(1-q)$, a pure state) whose photon number distribution is identical to thermal state (mixed state). We designate this pure state as a quasi-thermal state. Thus negative binomial states can provide transient studies from coherent to thermal fields for those variables which depend only on the diagonal elements of the field density matrix. The negative binomial field distribution reduces to a coherent field ($q \to 0, w \to \infty, \bar{n} = \omega q/(1-q)$) or a thermal field ($w = 1, \bar{n} = q/(1-q)$) distribution under two extreme condition of parameters. By using squeezed binomial (neg-binomial) states we study the systematic changes in quasiprobability of the Wigner function and photon number distribution function as we move down gradually from a squeezed coherent state to a squeezed number (quasi-thermal) state.

The density matrix for squeezed number state can be defined as follows

$$\hat{\rho} = \hat{S}(r)|n><n|\hat{S}^{+}(r) \tag{1}$$

where $\hat{S}(r)$ is the squeeze operator and has the following form [5]

$$\hat{S}(r) = (cosh(r))^{-\frac{1}{2}}exp[-\frac{1}{2}(tanh(r)a^{+2})](cosh(r)^{-a^{+}a}exp[\frac{1}{2}(tanh(r))a^{2}] \tag{2}$$

The photon number distribution $P(\ell) =< \ell|\varrho|\ell >$ is a measure of probability of photon in the field. For the squeezed number state $S(r)|n>$ the photon number distribution has the following form

$$P_{sn}(\ell) = \frac{n!\ell!}{(cosh(r))^{2\ell+1}}(\frac{1}{2}tanh(r))^{n-\ell}\text{x}\left|\sum_{m}\frac{(-1)^{m}(\frac{1}{2}tanh(r))^{2m}(cosh(r))^{2m}}{m!(\ell-2m)![\frac{1}{2}(n-\ell)+m]!}\right|^{2}$$

$$(\frac{1}{2}(\ell-n) \leq m \leq \frac{1}{2}\ell) \quad \begin{matrix}\text{for } (\ell-n) \text{ even}\\ \text{for } (\ell-n) \text{ odd}\end{matrix}$$

$$= 0 \tag{3}$$

The photon number distribution $p_{sb}(l)$ for the squeezed binomial state is

$$P_{sb}(\ell) = \sum_{n=0}^{N} P_{sn}(\ell) \cdot \frac{N!}{n!(N-n)!}p^{n}(1-p)^{N-n} \quad (0 \leq p \leq 1) \tag{4}$$

The expression of Eq. (4) reduces to Eq. (3) in the limit $p \to 1$, and it represents the number distribution for a squeezed coherent state under other limit $p \to 0,\ N \to \infty$. Thus by varying p and keeping $\bar{n} = Np = constant$ we can study the transients in the photon number distribution as one moves from a squeezed coherent state to a squeezed number state.

The photon number distribution for a squeezed negative binomial state can be written as

$$P_{snb}(\ell) = \sum_{n=0}^{\infty} P_{sn}(\ell) \cdot \frac{(n+w-1)!}{(w-1)!n!}q^{n}(1-q)^{w} \quad (0 \leq q < 1) \tag{5}$$

This expression represents the photon number distribution for a squeezed coherent state under the condition $q \to 0,\ w \to \infty$, and it reduces to photon number distribution of a squeezed thermal state when $w = 1,\ \bar{n} = q/(1-q)$. Hence, with Eq. (5) we can study the transients in the photon number distribution as one moves from a squeezed coherent distribution to a squeezed thermal distribution. In Figs. 1 and 2 we have plotted the photon number distribution of squeezed binomial and squeezed negative binomial states respectively.

In all these distributions there are pairwise oscillations resulting from the quadratic or the two photon nature of the squeeze operator $S(r)$. The interpretation of these oscillations can be obtained using the appropriate Wigner function for the various field states.

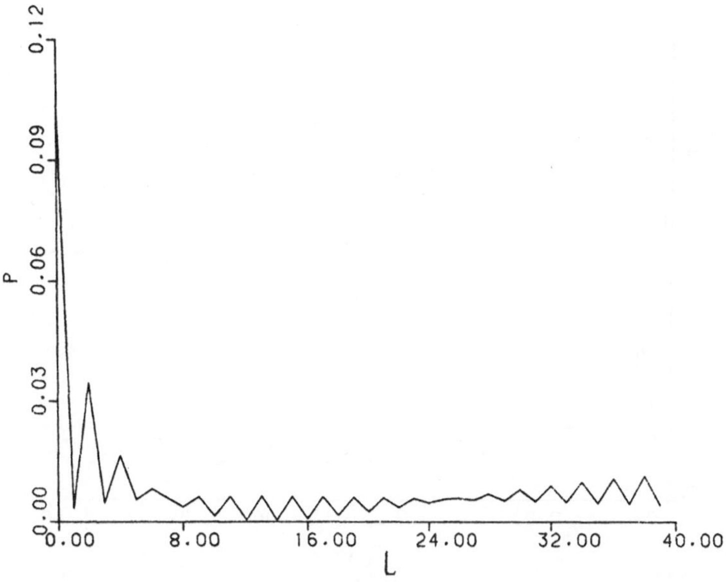

Figure 1. Photon number distribution $P_{sb}(\ell)$ of a squeezed binomial state when squeeze parameter r=2 and initial average photon number $\bar{n}$ = 2, with N=3,p=0.666.

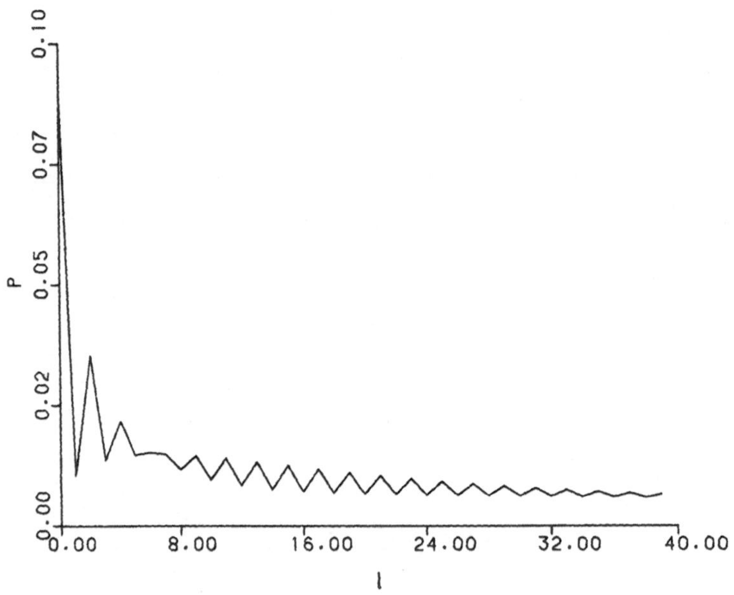

Figure 2. Photon number distribution $P_{snb}(\ell)$ of a squeezed negative binomial state when squeeze parameter r=2 and initial average photon number $\bar{n} = 2$, with w=4,$\bar{n} = 2$.

The Wigner function $W(\alpha)$ is defined in terms of the characteristics functions $C_W(\eta)$ by

$$W(\alpha) = \frac{1}{\pi^2} \int d^2\eta exp(\alpha\eta^* - \alpha^*\eta) C_W(\eta) \quad (6)$$

With the characteristics function defined as the expectation value of the Glauber translation operator

$$C_W(\eta) = T_r(\hat{\rho} exp(\eta \hat{a}^+ - \eta^* a)) \quad (7)$$

For squeezed binomial state the Wigner function $W_{sb}(\alpha)$ is

$$W_{sb}(\alpha) = (\frac{2}{\pi}) exp[\frac{1}{2}(\alpha - \alpha^*)^2 e^{-2r} - \frac{1}{2}(\alpha + \alpha^*)^2 e^{2r}]$$

$$\text{x} \sum_{n=0}^{N} \frac{N!}{n!(N-n)!} p^n (1-p)^{N-n} (-1)^n \mathcal{L}_n(\chi) \quad (8)$$

where,

$$\mathcal{L}_n(\chi) = \sum_{m=0} (-1)^m \binom{n}{m} \frac{\chi^m}{m!} \quad \text{(Laguerre polynomial).}$$

For the squeezed negative binomial state we find the Wigner function $W_{sb}(\alpha)$ as

$$W_{snb}(\alpha) = \frac{2}{\pi}(1-q)^W (1+q)^{-W} exp(-\frac{1}{2}\chi) {}_1F_1[{}^{W;}_{1;} \frac{\chi q}{1+q}] \quad (9)$$

where,

$$\chi = [(\alpha + \alpha^*)^2 e^{2r} - (\alpha - \alpha^*)^2 e^{-2r}] \quad (10)$$

and hypergeometric function

$${}_1F_1(a, b, z) = \sum_{n=0}^{\infty} \frac{(a)_n z^n}{(b)_n n!} \quad (11)$$

The Wigner function for both squeezed binomial and squeezed negative binomial states are plotted in Fig. 3 and 4 respectively. We find that both these functions are seen to be stretched by the action of squeezing. The Wigner function of a squeezed binomial state has an inverted peak which slowly reduces its depth as one moves away from a number state to a coherent state limit, finally becoming a Gaussian near coherent limit. Interestingly, we find a trough in the Wigner function of a squeezed negative binomial state for some value of the parameters (Fig. 4).

With the knowledge of the Wigner function for these two-states, the pairwise oscillatory behaviour of a photon number distribution could be easily explained in terms of the Wigner function phase space method.

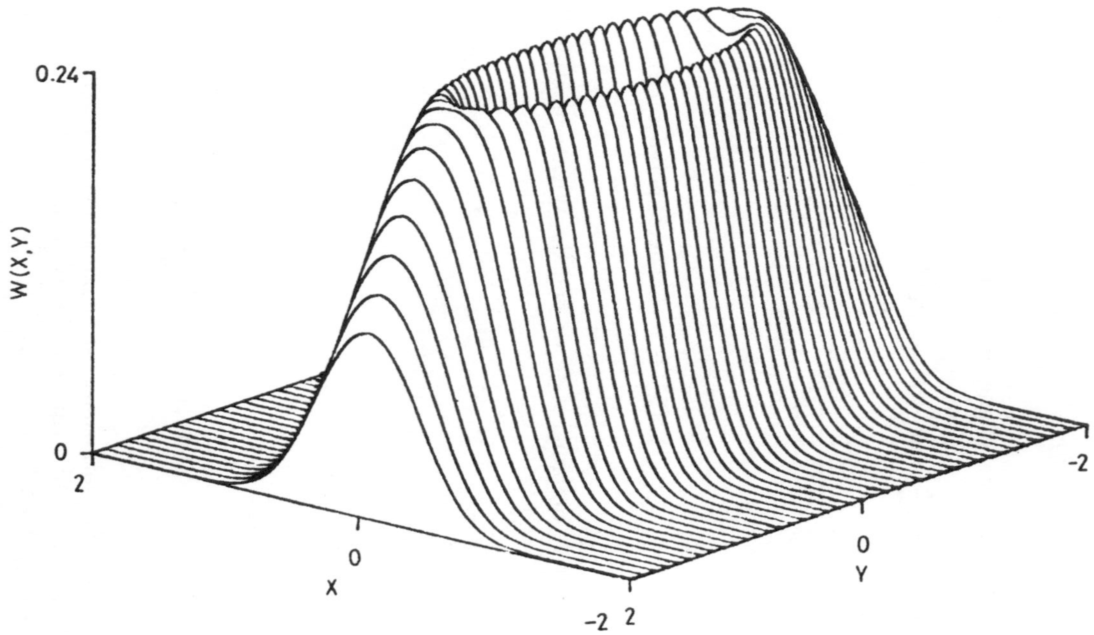

Figure 3. Wigner function $W(x,y)$ for a squeezed binomial state with squeeze parameter r=0.5 and $\bar{n} = 1$. Here, x=Re(α) and y=Im(α) and N=2,p=0.5.

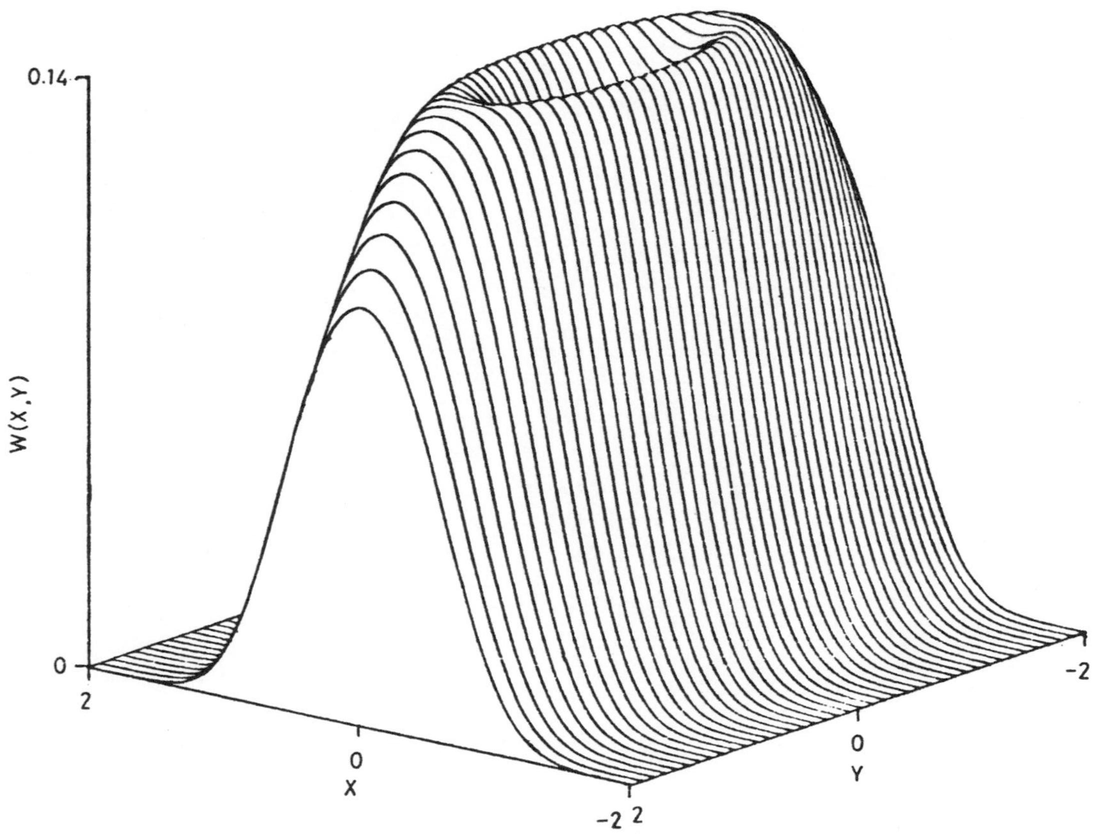

Figure 4. Wigner function $W(x, y)$ for a squeezed negative binomial state with squeeze parameter r=0.5 and $\bar{n} = 1$. Here, x=Re(α) and y=Im(α) and w=5,$\bar{n} = 1$.

REFERENCES

1. F. A. M. DeOliviera and P. L. Knight, *Phys. Rev. Lett.* **61**, 830 (1988).
2. C. K. Hong and L. Mandel, *Phys. Rev. Lett.* **56**, 58 (1986).
3. G. Dattoli, J. Gallardo, A. Torre, *J. Opt. Soc. Am.* **B4**, 185 (1987).
4. D. Stoler, B. E. A. Saleh, M. C. Teich, *Optica Acta* **32**, 345 (1985).
5. B. L. Schumaker and C. M. Caves, *Phys. Rev.* **A31** 3093 (1985).

OPTICAL MULTISTABILITY AT THE SURFACE OF A MEDIUM OF TWO-LEVEL ATOMS

M.B. Pande and S. Dutta Gupta

School of Physics
University of Hyderabad
Hyderabad 500134, India

We study a symmetric layered structure consisting of two identical metal films of thickness d_1 and dielectric constant ε_i separated by a distance of d_2. The space between the films is uniformly filled by two-level atoms exhibiting a nonlinearity given by the dielectric function[2]

$$\varepsilon_2 = \varepsilon_{20} + \frac{\alpha|\vec{E}|^2}{1 + (\alpha/\beta)|\vec{E}|^2} \tag{1}$$

where ε_{20} is the linear dielectric constant, α is the nonlinearity constant and β is the saturation parameter. Note that, for very large β, Eq. 1 represents a Kerr-type nonlinearity. The whole structure is embedded between two high index prisms of dielectric constant ε_i. Let a p-polarized wave be incident on the structure at an angle θ. We present exact numerical results for the reflectivity of the plane wave from the structure under conditions when coupled surface plasmons are excited in it.

We show optical bistability and multistability in reflectivity from such structures. Moreover, we investigate the effect of saturation on bistability to show that saturation effects inhibit multivalued output.

Assuming the x dependence of the fields to be $\sim e^{ik_x x}, k_x = (2\pi/\lambda)\sqrt{\varepsilon_i} sin\theta$ and with no z variation, Maxwell's equations for p-polarized field components for the nonlinear medium become

$$\dot{\tilde{E}}_x' = -F\tilde{E}_y'' \tag{2}$$

$$\dot{\tilde{E}}_x'' = F\tilde{E}_y' \tag{3}$$

$$\dot{\tilde{E}}_y' = (A_1A_2 - B^2)^{-1}[A_2(C\tilde{E}_x'' + D\tilde{E}_y') - B(-C\tilde{E}_x' + D\tilde{E}_y'')] \tag{4}$$

$$\dot{\tilde{E}}_y'' = (A_1A_2 - B^2)^{-1}[-B(C\tilde{E}_x'' + D\tilde{E}_y') + A_1(-C\tilde{E}_x' + D\tilde{E}_y'')] \tag{5}$$

Recent Developments in Quantum Optics, Edited
by R. Inguva, Plenum Press, New York, 1993

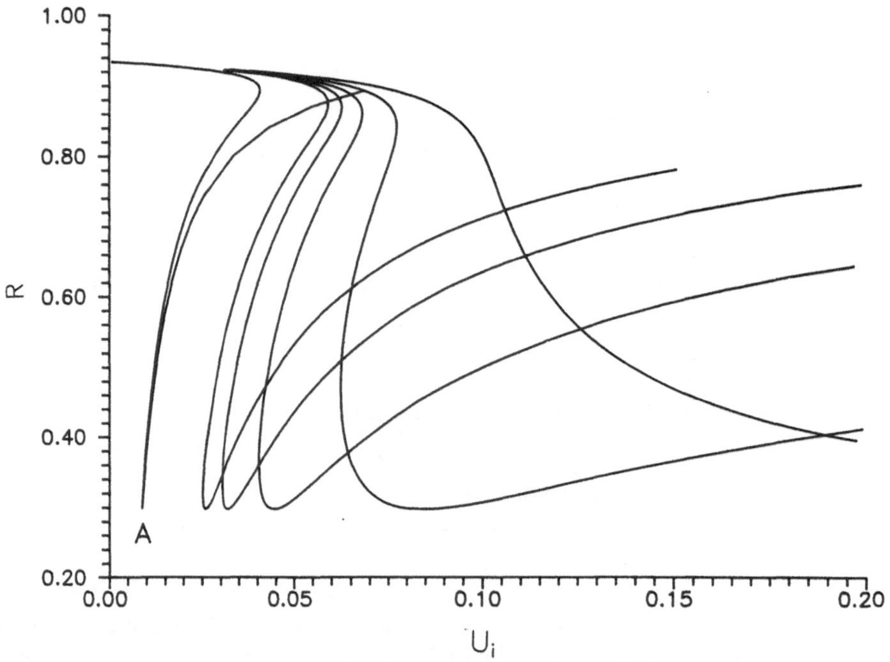

Figure 1. R as a function of U_i for $d_2 = 1\mu m$ and $\theta = 40.5°$. Curves from left to right are for $\beta = 10^5$, 0.1, 0.09, 0.08, 0.07 and 0.06.

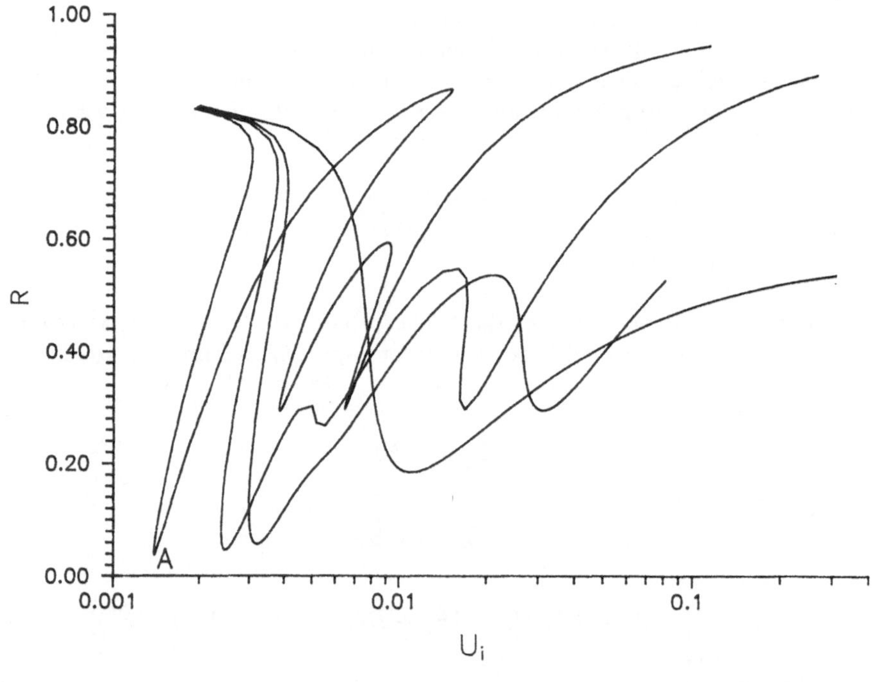

Figure 2: R as a function of U_i for $d_2 = 1.8$ μm and $\theta = 41.3°$. Curves from left to right are for $\beta = 10^5$, 0.08, 0.06, and 0.03.

where, the dots represent derivatives with respect to the dimensionless length $2\pi y/\lambda$. The quantities with tilda denote the dimensionless field amplitudes and the prime and two primes denote, respectively, the real and imaginary parts. The other quantities are

$$A_1 = (\varepsilon_2 M + 2\tilde{E}_y'^2),\ A_2 = (\varepsilon_2 M + 2\tilde{E}_y''^2),\ B = 2\tilde{E}_y'\tilde{E}_y'',$$

$$C = \varepsilon_2 \eta M,\ D = (\eta - \varepsilon_2/\eta),\ F = 2D(\tilde{E}_x'\tilde{E}_y'' - \tilde{E}_x''\tilde{E}_y')$$

$$M = [1 + (1/\beta)(\tilde{E}_x'^2 + \tilde{E}_x''^2 + \tilde{E}_y'^2 + \tilde{E}_y''^2)]^2,\ \eta = k_x\lambda/2\pi \quad (6)$$

The equations (2)-(5) for the nonlinear medium were integrated numerically, whereas propagation in the linear media was handled using the standard characteristic matrix formalism.

We define the dimensionless intensities $U_t = |\tilde{H}_t|^2$, $U_i = |\tilde{H}_i|^2$ and $U_r = |\tilde{E}_r^2|$, where the subscripts i, r and f refer to the incident, reflected and transmitted waves respectively. In what follows we present the results of our numerical calculations. The power dependence of the reflectivity is shown in Figs. 1 and 2. The parameters chosen were as follows: $\lambda = 1.06\mu m$, $d_1 = 0.045\mu m$, $\varepsilon_i = 6.145$, $\varepsilon_1 = -67.03 + 2.44i$, $\varepsilon_{20} = 2.54$. In Fig. 1 we have shown $R(= U_r/U_i)$ as a function of U_i for $d_2 = 1.0\mu m$ and $\theta = 40.5°$ where the long and the short range resonances are well separated and the value of the angle of incidence corresponds to the excitation of the short range mode of the linear system. It is clear from Fig. 1 that one can have bistability using the short range mode. Curves from left to right are for decreasing values of β. Curve A is for $\beta = 10^5$ and is seen to be close to that for a Kerr-type nonlinearity[3]. We see that as β decreases the bistable behaviour vanishes as a result of saturation. In Fig. 2 we study R as a function of U_i for $d_2 = 1.8\mu m$ and $\theta = 41.3°$, for various β. Curve A is for $\beta = 10^5$ and is seen to be very close to that for a Kerr-type nonlinearity. As β decreases strong saturation effects result in a broadening of the resonances leading to a loss of the finer structures in the resonances.

In conclusion we have shown the possibility of optical multistability with coupled surface plasmons and have studied the saturation effects in such structures.

REFERENCES

1. For a detailed study of the linear properties of such structures, see K. R. Welford and J. R. Sambles, *J. Mod. Opt.* **35**, 1467 (1988).
2. Th. Peschel, P. Dannberg, U. Langbein and F. Lederer, *J. Opt. Soc. Am.* **B5**, 29 (1988).
3. M. B. Pande and S. Dutta Gupta, *Opt. Lett.* **15**, 944 (1990).

SQUEEZED AND NONCLASSICAL STATES DUE TO FREQUENCY CHANGES IN HARMONIC OSCILLATORS

S. Arun Kumar*

School of Physics
University of Hyderabad
Hyderabad - 500 134, India

1. INTRODUCTION

In the recent years there has been a lot of interest in the generation of squeezed and nonclassical light.[(1-4)] We present in this paper our results on the squeezing of fluctuations in the quadrature operators of a harmonic oscillator (of unit mass), which is represented by the Hamiltonian

$$H = \frac{1}{2}p^2 + \frac{1}{2}\omega^2(1+\beta(t))x^2 \tag{1.1}$$

where

$$\begin{aligned} \beta(t) &= 0 && \text{for} \quad -\infty < t < 0 \\ &= \beta_o \tfrac{t}{T} && \text{for} \quad 0 \le t \le T \\ &= \beta_o && \text{for} \quad T < t < +\infty. \end{aligned} \tag{1.2}$$

The fluctuations in the quadratures are given by the variances σ_x^2 and σ_p^2. We discuss the two limiting cases, viz., sudden and adiabatic changes in β. For sudden changes[5] in β there is maximum squeezing, whereas for the adiabatic case[6] there is no squeezing.

2. EFFECTS OF $\beta(t)$ ON SQUEEZING

We consider the time dependent Hamiltonian (1.1). We define dimensionless operators X and P, called quadrature operators:

$$X = \sqrt{\frac{\omega}{2}}x, \quad P = \sqrt{\frac{1}{2\omega}}p, \quad [X, P] = \frac{i}{2}, \tag{2.1}$$

and dimensionless time $\tau = \omega t$. In terms of these quadrature operators, eq (1.1) takes the form,

*Work done in collaboration with Prof. G.S. Agarwal

Recent Developments in Quantum Optics, Edited
by R. Inguva, Plenum Press, New York, 1993

$$H = \omega[P^2 + (1+\beta(\tau))X^2), \quad h = 1. \tag{2.2}$$

We work in the Heisenberg picture. The Heisenberg equations of motion for the operators X and P are

$$\frac{dX}{d\tau} = P, \quad \frac{dP}{d\tau} = -(1+\beta(\tau))X. \tag{2.3}$$

The Heisenberg equations can be solved and the solution written in the form

$$\begin{pmatrix} X(\tau) \\ P(\tau) \end{pmatrix} = \begin{pmatrix} U & V \\ \dot{U} & \dot{V} \end{pmatrix} \begin{pmatrix} X(0) \\ P(0) \end{pmatrix}. \tag{2.4}$$

where the functions U and V satisfy the differential equation

$$\frac{d^2}{d\tau^2}\phi + (1+\beta(\tau))\phi = 0 \tag{2.5}$$

with the initial conditions

$$U(0) = 1, \quad \dot{U}(0) = 0, \quad V(0) = 0, \quad \dot{V}(0) = 1. \tag{2.6}$$

For $\beta(\tau)$ given by (1.2) the general solution of (2.5) in the domain $0 \le \tau \le \omega T$ can be expressed in terms of Bessel functions of order 1/3

$$\phi = C_1\sqrt{\tau'}J_{1/3}(z) + C_2\sqrt{\tau'}Y_{1/3}(z), \tag{2.7}$$

where

$$\tau' = \tau + \frac{\omega T}{\beta_o}, \quad z = \frac{2}{3}\left[\frac{\beta_o}{\omega T}\tau'^3\right]^{1/2}. \tag{2.8}$$

Thus the functions U and V can be fixed by using (2.6) in (2.7).

One can also directly determine the variances by numerical integration. We rewrite equation (2.3) in terms of the mean values of the operators:

$$\dot{\varphi}_1 = \varphi_2, \quad \dot{\varphi}_2 = -(1+\beta(\tau))\varphi_1 \tag{2.9}$$

where

$$\varphi_1 = \langle X \rangle \text{ and } \varphi_2 + \langle P \rangle. \tag{2.10}$$

Defining

$$\Psi_1 = \langle X^2 \rangle, \; \Psi_2 = \langle XP + PX \rangle /2 \text{ and } \Psi_3 = \langle P^2 \rangle \tag{2.11}$$

we have

$$\begin{aligned} \dot{\Psi}_1 &= 2\Psi_2 \\ \dot{\Psi}_2 &= -(1+\beta(\tau))\Psi_1 + \Psi_3 \\ \dot{\Psi}_3 &= -2(1+\beta(\tau))\Psi_2 \end{aligned} \tag{2.12}$$

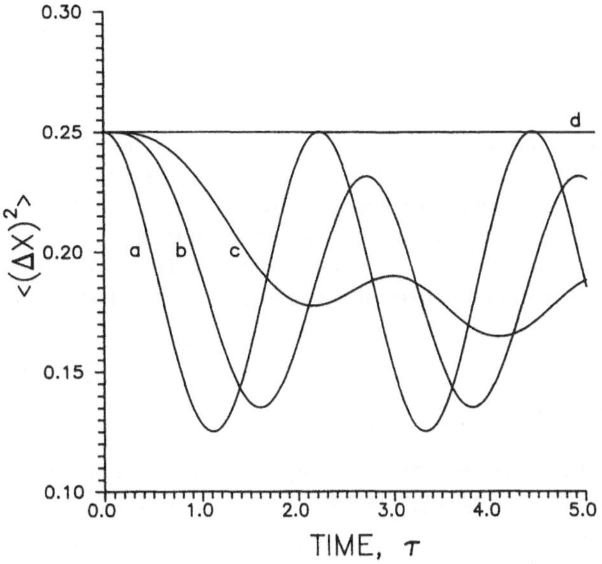

Figure 1. $\langle(\Delta X)^2\rangle$ as a function of τ with $|0\rangle$ as initial state. The parameters chosen are $\beta_o = 1$ and $\omega T = 10^{-3}$(a), 1(b), 3(c) and 10^3(d).

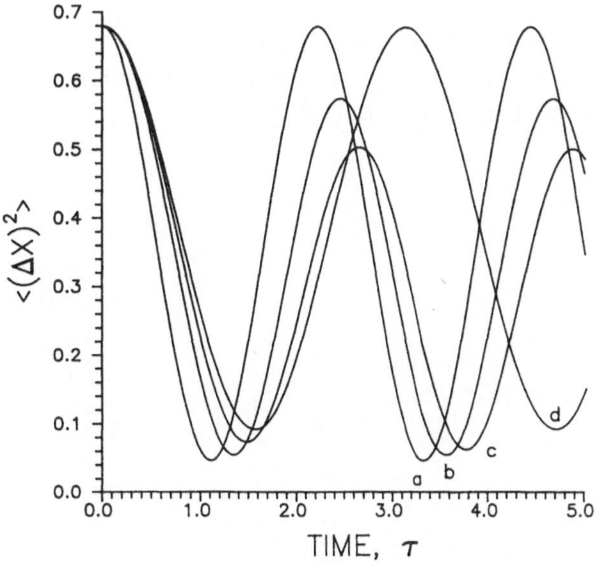

Figure 2. $\langle(\Delta X)^2\rangle$ as a function of τ with $|0,\zeta\rangle$ as initial state. The parameters chosen are $\beta_o = 1$, $\omega T = 10^{-3}$(a), 1(b), 3(c) and 10^3(d) and $\zeta = 0.5\exp(-i\pi)$.

Thus we have two sets of coupled differential equations. By numerically integrating them one could determine the variances as

$$\sigma_x^2 = \Psi_1 - \varphi_1^2 \text{ and } \sigma_p^2 = \Psi_3 - \varphi_2^2. \tag{2.13}$$

3. RESULTS

In figure 1, we have plotted the variance σ_x^2 as a function of τ for the vacuum as initial state. We note that a sudden change in the frequency of the oscillator amounts to large squeezing, whereas for slower changes in the frequency it leads to lesser amounts of squeezing. In figure 2, we have plotted σ_x^2 with squeezed vacuum as initial state having squeezing parameter $\zeta = 0.5\exp(i\pi)$. We see that the frequency changes result in the enhancement of squeezing which is already present in the initial squeezed state.

ACKNOWLEDGEMENTS

The author would like to thank the Department of Science and Technology, Government of India for supporting this work and Prof. V.I. Man'ko for many interesting discussions.

REFERENCES

1. Walls, D.F., *Nature*, Lond., **306**, 141 (1983).
2. Loudon, R. and Knight, P.L., *J. Mod. Opt.* **34**, 709-759 (1987).
3. Teich, M.C. and Saleh, B.E.A., *Quantum Opt.* **1**, (1989) 153-191.
4. See special issues on squeezed and nonclassical light: (i) *J. Mod. Opt.* **34**, Nos. 6 and 7 (1987), (ii) *J. Opt. Soc. Am.* B**4** (1987).
5. Janzsky, J. and Yushin, Y.Y., *Opt. Commun.* **59** (1986).
6. Graham, R., *J. Mod. Opt.* **34**, 873 (1987).

CAVITY QUANTUM ELECTRODYNAMICS: FUNDAMENTAL THEORY OF THE MICROMASER

R.K. Bullough, N. Nayak* and B.V. Thompson

Department of Mathematics
UMIST, P.O. Box 88
Manchester M60 1QD, U.K.

1. INTRODUCTION

The paper given to the meeting in Hyderabad, January 2–8, under the present title was followed by a closely related one to the NATO ARW at Cortina, January 20–25, 1991 now published.[1] Circumstances have delayed the writing of the present paper for these Proceedings of the Hyderabad meeting, and in the meantime we have obtained further results. This paper therefore amplifies the Cortina paper[1] but is intended to be self-contained.

As in Ref. 1, the paper reports on aspects of the action of the one ^{85}Rb atom-one mode micromaser currently in operation[2–5] at the Max Planck Institute, Garching by Munich, Germany.

The report concerns a new theoretical investigation, numerically based, of the action of the micromaser. In the recent experiments[2,3] there is at most one ^{85}Rb atom in the maser cavity during one atomic transit time $t_{int} \sim 35\mu$sec. But, even at *precise* atomic velocity selection, atoms still enter stochastically with an exponential distribution parameterized by a mean repetition rate T_p (say): $T_p \sim 7,000\mu$sec in the experiments – so the cavity is empty of atoms with probability ≈ 1 for some 99% of the time. The main conclusion of this paper then supports the conclusion of Ref. 1 namely that if the empty-of-atoms time $t_e \equiv T_p - t_{int}$ is reduced towards the values of t_{int} itself, the system acts more like a maser with infinite cavity Q even at the now realized Q and temperature T[2,3] $Q = 3\text{x}10^{10}$, $T = 0.5°$K. Thus, in distinction to previous discussions, eg. that by Meystre *et al.*,[6] we suggest that it may be possible to create Fock states in the micromaser at this Q and T. However, as T_p is reduced, the theoretical problem becomes complicated by the possibility of more than one atom being in the cavity at the same time – the effects of which on Fock state evolution we are still investigating.[1]

The conclusion that Fock states will emerge within single atom theories is based on the numerical results already reported in Ref. 1. These results are amplified in the present paper by further studies *in the current regime of micromaser operation* which

* Now at S.N. Bose National Centre for Basic Sciences, DB-17, Sector-1, Salt Lake, Calcutta 700064, India

Recent Developments in Quantum Optics, Edited by R. Inguva, Plenum Press, New York, 1993

is $T_p \approx t_e >> t_{int}$.[2,3] Even in this regime the conclusion is that, as T_p moves towards t_{int}, even though $T_p >> t_{int}$ always, the variance for the maser cavity field defined here by (Refs. 2,3 use Q_a for our symbol σ)

$$\sigma \equiv [< (a^\dagger a)^2 > -(< a^\dagger a >)^2]/ < a^\dagger a > \tag{1}$$

systematically falls. The natural unit of time is g^{-1} where g is the real valued coupling constant between the atom and the field: g =44kHz in the experiments[2,3] but we shall not need this number as such in this paper. Thus for $gt_{int} = 1.047$ and $gT_p = 720$ with the $Q = 3\text{x}10^{10}$ and the mean black-body photon number $\bar{n}_{th} = 0.15$ at the temperature $T = 0.5°\text{K}$ of the experiments[2,3] we find $\sigma = 0.281$ becoming $\sigma = 0.204$ for $gT_p = 216$. The gt_{int} is chosen to *minimize* σ (more or less - see the Fig. 5 in §4 where $Q = 2\text{x}10^9$). The values of σ are lower than those observed [2,3] (the smallest σ observed[2,3] is $\sigma \approx 0.3$ for $gT_p = 308$). The fields are sub-Poissonian ($\sigma < 1$) but still differ substantially from the $\sigma = 0$ required by a cavity field in a pure Fock state (n-state, $|n>$) of n photons. The fall of σ with T_p in this large T_p regime is nevertheless consistent with the numerical evidence of Ref. 1 that Fock states, $\sigma = 0$, will emerge for the much smaller T_p where $gT_p \gtrsim gt_{int}$. The practical message for Fock state creation seems to be to reduce T_p *to some extent* - so that the probability of more than one atom in the cavity at the same time still remains small.

The material is presented in the following way. In §2 we sketch our methods for computing the effects of black-body radiation on single atoms occupying (microwave) cavities with Q in the range $10^3 < Q \leq 3\text{x}10^{10}$. In §3 we make a fundamental analysis of the canonically quantized micromaser, introducing the 'trapping states'[6-10] and also drawing a comparison with the ordinary laser much as this was done in Ref. 9. The §4 reports sets of numerical results which support the conclusions of Ref. 1, all but two of these evaluated in the $T_p >> t_{int}$ regime. The §5 summarizes the paper.

2. SINGLE ATOM CANONICAL CAVITY q.e.d. AT FINITE TEMPERATURES

The model is the Jaynes-Cummings model[7] in which the single mode is coupled to a heat bath at $T > 0$. The well known master equation is [11-14]

$$\frac{d\rho}{dt} = -i[\hat{H}, \rho] - (\bar{n}_{th} + 1)\kappa[a^\dagger a\rho - 2a\rho a^\dagger + \rho a^\dagger a] - \bar{n}_{th}\kappa[aa^\dagger\rho - 2a^\dagger\rho a + \rho aa^\dagger], \tag{2}$$

in which $\kappa = \frac{1}{2}\omega Q^{-1}$ and $\bar{n}_{th} = (e^{\beta\omega} - 1)^{-1}$ ($\hbar = 1$) the Planck function: $\hat{H}$ is the J-C Hamiltonian with number operator $\hat{N}$

$$\hat{H} = \omega_o S^z + \omega a^\dagger a + g(a^\dagger S^- + S^+ a), \quad \hat{N} = S^z + a^\dagger a + \frac{1}{2} \tag{3}$$

and $[\hat{H}, \hat{N}] = 0 : [a, a^\dagger] = 1$ and the single cavity mode provides one quantum degree of freedom: the single atom (frequency ω_o) with the spin-$\frac{1}{2}$ operators $S^z, S^\pm$ satisfying the su(2) Lie algebra

$$[S^\pm, S^z] = \mp S^z, \ [S^+, S^-] = 2S^z \tag{4}$$

but with total spin $S = \frac{1}{2}$, provides a second quantum degree of freedom. The system is therefore quantum completely integrable.[16] At zero temperature it is well known[7] the system can be solved by diagonalizing a sequence of 2x2 matrices characterized by the eigenvalues n of $\hat{N}$: $\hat{N}|n,\pm> = n|n,\pm>, n = 0,1,2,...,$; $\hat{H}|n,\pm> = \{(n-\frac{1}{2})\omega \pm \frac{1}{2}[(\omega-\omega_o)^2 + 4g^2n]^{\frac{1}{2}}\}|n,\pm>$. However at finite temperatures $\hat{N}$ is no longer a good quantum number.

The calculational problem at finite temperatures rests in the infinite number of photon states $|n>$ spanning the Hilbert space: a complete basis is $|e>|n-1>$, $|g>|n>$ with $n = 1,2,...$, and $|g>|0>$. Our procedure has been[11-15,17,18] to calculate the four one-body quantities $P_n \equiv <a^{\dagger n}a^n>$, $Q_n \equiv <a^{\dagger n}a^nS^z>$ (and $n \geq 0$), $S_n \equiv <S^+a^{\dagger n-1}a^n + a^{\dagger n}a^{n-1}S^->$, $R_n \equiv i<S^+a^{\dagger n-1}a^n - a^{\dagger n}a^{n-1}S^->$ (and $n > 0$), which satisfy a closed coupled system of equations infinite in number: $Q_o \equiv <S^z>$ and $P_1 = <a^\dagger a>$, $P_2 = <a^\dagger a^\dagger aa>$. The problem reduces[11] to the solution of the matrix equations

$$\frac{d\psi_n}{dt} = \mathcal{A}_n\psi_n + B_n\psi_{n+1} + C_n\psi_{n-1},\ n \geq 0;\ C_o = 0; \tag{5}$$

in which $\psi_n = [P_n, Q_n, R_n, S_n]^T$ and $\mathcal{A}_n$, B_n and C_n are 4x4 matrices. The Laplace transform on t yields a 3-term matrix recurrence relation solved by matrix continued fraction techniques introduced by Risken.[20] We also calculate the 2-body correlation functions in a similar way.[12-14] These correlations at t and $t+\tau$ satisfy a closed system of the form (5) with larger matrices and τ replacing t.[12-14] Moreover[12-14], we have calculated the fluorescence and cavity field spectra for $10^3 \lesssim Q \leq 10^8$ and watched their *qualitative* changes as Q increases: vacuum field Rabi splitting is first discernible for $Q's \sim 10^6$. One can also expect to see the breakdown of the Born and Markov approximations with increasing Q.[12]

We turn to the micromaser.

An instructive result found by the methods just described is shown in the Fig. 4 of Ref. 1. This shows the evolution of the Fock state $|10>$ in the presence of the single atom for $Q = 2\text{x}10^{10}$ and $\bar{n}_{th}$=0.15: $\kappa g^{-1} = 0.0000767$, and $63P_{3/2} \rightarrow 61D_{5/2}$ transitions of ^{85}Rb ($\omega_o = 2\pi$ x 21.456 GHz) are assumed: detuning $\Delta = \omega_o - \omega = 0$. The scale of gt is that 9 units $\sim$ one atomic transit time for crossing the cavity (gt_{int}, §1). The figure shows that after 8 such transit times the amplitude of the inversion has fallen some 10%. In the empty-of-atoms cavity time $gt_e = gT_p - gt_{int}$ of the micromaser (§1 and Refs. 2,3) this amplitude would be $\sim (0.9)^{36}$=0.02. Thus if the Fock state $|10>$ were created it would decay substantially in each period T_p. This suggests that a smaller T_p is needed for the creation and maintenance of Fock states in the micromaser. We investigate this suggestion in the next sections §§3 and 4. We look first of all at the fundamental quantum theory of the micromaser.

3. THE CANONICALLY QUANTIZED MICROMASER AS A FUNDAMENTAL QUANTUM DEVICE

For long repetition rates $T_p >> t_{int}$ there is, to a probability ≈ 1, at most one atom in the maser cavity. Moreover, during most of the time T_p the maser cavity is empty of atoms. Theories of the micromaser have been developed for $Q = \infty$ throughout all of the motion[7,10,21] and for the micromaser coupled to a heat bath and $Q < \infty$ when, but only when, there is no atom in the cavity.[8,9] When the atom is in the cavity during the short time t_{int}, the atom and field are supposed to evolve under the *unitary* transformation $\hat{U} = exp[-it\hbar^{-1}\hat{H}]$ where $\hat{H}$ is the J-C Hamiltonian eqn. (2), i.e. the evolution during t_{int} is for $Q = \infty$.[8,9] If an atom enters the maser cavity at $t = 0$ in its upper state $|e>$ with the cavity field in the state $|n>$ at $t = 0$, the unitary operator $\hat{U}$ transforms this state to $|e>|n>$ with probability[21]

$$a(n) = \cos^2 \sqrt{n+1} g t_{int}, \tag{6}$$

and to $|g> \; |n+1>$ with probability

$$b(n) = \sin^2 \sqrt{n+1} g t_{int}; \tag{7}$$

and $a(n) + b(n) = 1$. In the *sequence* (at $Q = \infty$) in which atoms enter the maser cavity successively the probability p_n of finding n photons in the cavity will depend on the number of atoms k which have passed through the cavity i.e. $p_n = p_n(k)$. Thus[10]

$$p_n(k+1) = a(n)p_n(k) + b(n-1)p_{n-1}(k). \tag{8}$$

Consequently, if $b(n-1) = 0$, $p_n(k+1) = a(n)\, p_n(k)$ and if initially $p_n(0) = 0$ for all $n > n_0$, $p_n(k) = p_n(k+1) = 0$ for all $n > n_0$ and the p_n with $n > n_0$ never build up in the cavity field. In effect a barrier is created[10] at $n = n_0$. A crucial result demonstrated in Ref. 10 is then, that if $n = n_0$ is the zero of (7), and initially $p_n = 0$ for all $n > n_0$, then $p_{n0}(k) = p_{n0}(k+1) \to 1$, i.e. a pure *Fock* state $|n_0>$ is created in the cavity. In the present analysis we shall be primarily concerned to discover in detail under what conditions (if any) a Fock state could build up in a $Q = \infty$ cavity which initially contains black-body radiation and, more to the point for an experiment, under what conditions (if any) a Fock state could build up in a real maser cavity with $Q < \infty$ which is coupled all the time to a heat bath at a finite temperature T.

The zeros of $b(n)$, eqn (7), define 'trapping' states of the field.[6–10] It is convenient to consider these trapping states as the zeros of (7) labelled by n_o and r : $(n_o+1)^{\frac{1}{2}} g t_{int} = r\pi$ (Ref. 6 uses q for our r). Thus, as example, $g t_{int} = \frac{1}{2}\pi$ and $r = 1$, $n_o = 3$ defines a 'first' trapping state; then there are the second, third, etc trapping states with $n_o = 15$, 35, 63, ... There are trapping states for *any* $g t_{int} = r\pi(n_o+1)^{-\frac{1}{2}}$, i.e. for $g t_{int}$ the trapping states have the measure of the irrationals, the $g t_{int}$ for trapping states are therefore mathematically 'dense', and there is a very large number of such states. Without attempting to make this argument rigorous one can therefore expect that the cavity field of the micromaser 'almost always', i.e. with probability 1, evolves to a Fock state. The Fock state is characterized by the variance σ, defined by eqn. (1), equal to zero. The experiments[2,3] show however that such evolution does not necessarily occur: *sub*-Poissonian fields with $0 < \sigma < 1$ were detected, and the observations conform to the very different predictions of the theoretical work[8,9] for $Q < \infty$ and finite temperatures.

At $Q = \infty$ one finds in any case[10] that evolution to a pure Fock state $|n_o>$ will not occur if the cavity initially contains black-body radiation. If $g t_{int}$ defines a first trapping state (n_o, r), then if (n_1, r), (n_2, r) etc are the second, third, etc. trapping states, each of these will evolve as long as the initial state of the cavity field is $\sum_{n=0}^{N_{max}} \alpha_n |n>$ $(\alpha_{N_{max}} \neq 0)$ and the n_i of the trapping states have $n_i < N_{max}$. In the case of black-body radiation, $N_{max} = \infty$ so all of the sequence of trapping states (n_i, r) will evolve, *and no steady state will ever be attained!* However in Ref. 10 no time scale is given for this evolution. From a computer calculation with $Q = \infty$ and $g t_{int} = \frac{1}{2}\pi$, we find that after 91 inverted atoms have entered and excited successively, p_3 =0.9997105, p_{15} =0.2659983x10^{-3} and p_{35} =0.197007x10^{-14}. After 500 inverted atoms this becomes p_3 =0.9997105, p_{15} =0.2894501x10^{-3} and p_{35} =0.701391x10^{-14}.

After 2000 inverted atoms only p_{35} changes (to 0.7019316x10^{-14}). All the figures then remain unchanged as 100,000 atoms have passed!

The second reason why Fock states do not occur at the values $Q = 3\text{x}10^{10}$ and $T = 0.5°\text{K}$ of the experiments[2,3] is that the finite but small value $\bar{n}_{th}$=0.15 of the black-body radiation induced by coupling to a heat bath at that T *qualitatively* (not just quantitatively) changes the action of the micromaser compared with its action at $Q = \infty$. The observations[2,3] which use $T_p(\sim 7,000\mu\text{sec}) >> t_{int}(\sim 35\mu\text{sec})$, are consistent with the theoretical predictions.[8,9] We sketch the nature of these particular theories[8,9] which use unitary evolution by $\hat{U}$ in the short time t_{int} and couple in the heat bath only during the long time $t_e = T_p - t_{int}$.

The result (8) means that

$$p_n(k+1) - p_n(k) = a(n)p_n(k) + b(n-1)p_{(n-1)}(k) - p_n(k). \tag{9}$$

If a dynamically steady state can be reached repeating each T_p then, with the average *rate* $R \equiv T_p^{-1}$, the change $\Delta\rho^f_{n,n}$ in the field density matrix during a time Δt in $t_{int} << \Delta t << T_p$ is

$$\Delta\rho^f_{n,n} = R\Delta t[p_n(k+1) - p_n(k)] \tag{10}$$

since the square bracket is the change for a single atom after k atoms have passed. But providing the situation *is* steady, so that we need not distinguish the number of atoms k which have passed through.

$$\begin{aligned} \Delta\rho^f_{n,n}|_{gain} &= R\Delta t[(a(n)-1)\rho^f_{n,n} + b(n-1)\rho^f_{n-1,n-1}] \\ &\equiv \Delta t[-A(n+1)F(n+1)\rho^f_{n,n} + AnF(n)\rho^f_{n-1,n-1}] \end{aligned} \tag{11a}$$

where $F(n) \equiv n^{-1}\sin^2\sqrt{n}gt_{int}$ and $A \equiv R$, and the $\rho^f_{n,n}$ do not distinguish the passage of atoms. Then by coupling in the heat bath during the long time t_e, one finds from the master equation (2) that

$$\begin{aligned} \Delta\rho^f_{n,n}|_{loss} = \Delta t\{(\bar{n}_{th}+1)C(n+1)\rho^f_{n+1,n+1} + \bar{n}_{th}Cn\rho^f_{n-1,n-1} \\ -C[(n+1)\bar{n}_{th} + n(\bar{n}_{th}+1)]\rho^f_{n,n}\} \end{aligned} \tag{11b}$$

where $C = 2\kappa = T_{cav}^{-1}$. The total change $\Delta\rho^f_{n,n}$ is then given *additively* (i.e. 'by additive gain and loss')

$$\frac{\Delta\rho^f_{n,n}}{\Delta t} = \left.\frac{\Delta\rho^f_{n,n}}{\Delta t}\right|_{gain} + \left.\frac{\Delta\rho^f_{n,n}}{\Delta t}\right|_{loss}, \tag{12}$$

and the coarse grained expression $\Delta\rho^f_{n,n}/\Delta t$ is identified with the actual derivative $\partial\rho^f_{n,n}/\partial t$ for the field (*N.B.* $\Delta t >> t_{int}$). One then has the coarse grained master equation for the field density matrix $\rho^f_{n,n}$

$$\frac{\partial \rho^f_{n,n}}{\partial t} = -A(n+1)F(n+1)\rho^f_{n,n} + AnF(n)\rho^f_{n-1,n-1}$$

$$+(\bar{n}_{th}+1)C(n+1)\rho^f_{n+1,n+1} + \bar{n}_{th}Cn\rho^f_{n-1,n-1}$$

$$-C(n+1)\bar{n}_{th}\rho^f_{n,n} - Cn(\bar{n}_{th}+1)\rho^f_{n,n} \tag{13}$$

and (compare Ref. 21 for the full density matrix ρ) *this is diagonal throughout the whole motion.* In the steady state $\partial \rho^f_{n,n}/\partial t = 0$, detailed balance means

$$Cn(\bar{n}_{th}+1)\rho^f_{n,n} = [AnF(n) + Cn\bar{n}_{th}]\rho^f_{n-1,n-1} \tag{14}$$

so

$$\rho^f_{n,n} = f(\bar{n}_{th})\Big[\frac{A}{C}\frac{F(n)}{\bar{n}_{th}} + 1\Big]\rho^f_{n-1,n-1} \tag{15}$$

where $f(\bar{n}_{th}) \equiv \bar{n}_{th}/(1+\bar{n}_{th})$. Without detailed balance (13) depends on $\rho^f_{n+1,n+1}$, $\rho^f_{n,n}$, $\rho^f_{n-1,n-1}$ and in the steady state yields a 3-term recurrence relation (§2). Detailed balance yields (15) which is two term and easily solved to

$$\rho^f_{n,n} = C_o\Big(\frac{\bar{n}_{th}}{1+\bar{n}_{th}}\Big)^n \prod_{k=1}^{n}\Big[1 + \frac{A}{C}\frac{F(k)}{\bar{n}_{th}}\Big] \tag{16}$$

in which C_o is a normalization constant such that $\sum_{n=0}^{\infty} \rho^f_{n,n} = 1$: by definition $AC^{-1} = R\,T_{cav}$. The expression (16) for the micromaser was first obtained by Meystre[8] by a somewhat different route. The route *we* have used has been developed in such a way that A and C and $F(n)$ can be interpreted in terms of ordinary single mode laser theory: if $A \equiv 2R\Omega^2\gamma_{eg}^{-2}$, $C = T_{cav}^{-1}$ and $nF(n) = n[1 + 4n\Omega^2\gamma_{eg}^{-2}]^{-1}$ in which Ω is the Rabi frequency and γ_{eg}^{-1} is the spontaneous emission time, we regain the essentials of the Scully-Lamb laser theory[22] as extended to include the contribution of black-body radiation at finite temperatures[23]: γ_{eg}^{-2} is replaced by a combination of γ_g, γ_e and γ_{eg} in Ref. 22 and in effect $f(\bar{n}_{th})$ replaces $\bar{n}_{th}$ inside the square bracket in Ref. 23 (however at sub-Kelvin temperatures $f(\bar{n}_{th}) \approx \bar{n}_{th}$). The threshold for laser operation is [22] $A = C$, and for $A > C$ the photon distribution $p(n) = \rho^f_{n,n}$ has a peak at $n \neq 0$ (coherence). Moreover for large AC^{-1} (well above threshold) $\rho^f_{n,n} \approx \frac{|\alpha|^2}{n}\rho^f_{n-1,n-1}$ where (for large enough Ω^2) $F(n) = [4n\Omega^2\gamma_{eg}^{-2}]^{-1}$ and $|\alpha|^2 = AC^{-1}F(n)n = T_{cav}2R\Omega^2\gamma_{eg}^{-2}/4\Omega^2\gamma_{eg}^{-2} = \frac{1}{2}T_{cav}R$.

Lugiato et al[9] in effect gave the route from (10) to reach (16) for the micromaser identically, with $AC^{-1} = RT_{cav}$. They make the same point that (13) for the micromaser is also conventional laser theory. The new feature in the micromaser theory is that instead of the spontaneous emission time γ_{eg}^{-1} controlling the theory, so that $\gamma_{eg}^{-1} << \Delta t << T_{cav}$, it is t_{int}, satisfying $t_{int} << \Delta t << T_{cav}$, which does so. This has the important consequence that, unlike laser theory where there is essentially only one parameter, AC^{-1} with $AC^{-1}nF(n) \sim RT_{cav}$, the micromaser theory yielding (15) has the *two* parameters RT_{cav} and gt_{int}. It is this which enables the micromaser to perform as a strictly *quantum* device producing *sub*-Poissonian light in the cavity field (see below). Meystre[8] defined $RT_{cav} = N_{ex}$, the mean number of atoms traversing the cavity in a cavity decay time $T_{cav} = (2\kappa)^{-1}$, and gave a natural 'tipping angle' Θ for the atom in the cavity

$$\Theta \equiv (RT_{cav})^{1/2} g t_{int} = (N_{ex})^{1/2} g t_{int}. \tag{17}$$

The formula (15) with $AC^{-1} = N_{ex}$ will presumably be good for $T_p >> t_{int}$ as was assumed providing the temperature T is large enough. Evidently, as $T \to 0$, $\bar{n}_{th} \to 0$ and (15) yields

$$\rho^f_{n,n} = C_o N_{ex} \prod_{k=1}^{n} \frac{\sin^2 \sqrt{k} g t_{int}.}{k} \tag{18}$$

This is the exact solution of $C\rho^f_{n,n} = AF(n)\rho^f_{n-1,n-1}$ which derives from (14) and (13) in the steady state at $T = 0$: moreover $p_n = \rho^f_{n,n} = 0$ for $n \geq (n_o + 1)$ where $\sin(n_o + 1)^{1/2} g t_{int} = 0$. But now the theory is incomplete compared with that of (8), because no Fock state $p_{n_o} = 1$ is predicted; nor does the theory show that the higher trapping states will emerge in the presence of black-body radiation at $T > 0$ and that there is no steady state – (18) is still a steady state theory. At $Q = \infty$ (no damping) $C = 0$ in (13) and for a steady state

$$-A(n+1)F(n+1)\rho^f_{n,n} + AnF(n)\rho^f_{n-1,n-1} = 0. \tag{19}$$

This is

$$-(1 - \cos^2 \sqrt{n+1} g t_{int})\rho^f_{n,n} + (\sin^2 \sqrt{n} g t_{int})\rho^f_{n-1,n-1} = 0, \tag{20}$$

namely (8) with no distinction made in the $\rho^f_{n,n}$ for the number of atoms k which have passed through the cavity ($\rho^f_{n,n}$ is independent of k). For the creation of Fock states through the 'trapping state' (n_o, r) the state of the field does depend significantly on the number k of atoms which have passed (as the Fig. 1 above makes clear). *Apparently sequential evolution is fundamental to the creation of Fock states in the maser cavity.*

The formula (16) for the micromaser, with $AC^{-1} = N_{ex} = RT_{cav}$, is sensitive to the value of $\bar{n}_{th}$. It is intuitive that for $T_p >> t_{int}$ it is a good approximation to permit unitary evolution under $\hat{U}$ in the short time t_{int} and to couple in the heat bath during the long time t_e. Predictions based on (16)[8,9] agree with the observations.[2,3] The conclusion is that this description of the effects of black-body radiation changes the theory *qualitatively* compared with the $Q = \infty$ theory based on (8).

To achieve Fock states in the cavity it seems[6] from (16) that temperatures $T \lesssim 0.2°$K and the current $N'_{ex}s \sim 50$[2,3] will be needed. However it seems we must give up (16) for $T = 0$ and use (8). Alternatively we can look again at the more exact equations for the total density matrix ρ for which (13) is both a projection to ρ^f and an approximation. The Fig. 4 of Ref. 1 based on the exact equations for ρ^f suggests that if we wish to create Fock states in the micromaser at finite $Q < \infty$ and $T > 0$ we should reduce T_p to $T_p \gtrsim t_{int}$. Evidently it is then necessary to couple in the heat bath throughout the motion. This coupling introduces an intrinsically *nonlinear* gain and loss mechanism: linear additive gain and loss eqn. (12) does not apply.

A simple minded approach to see this, which however is still concerned with t_e, the empty-of-atoms cavity time is as follows: assume there *is* a steady state varying during

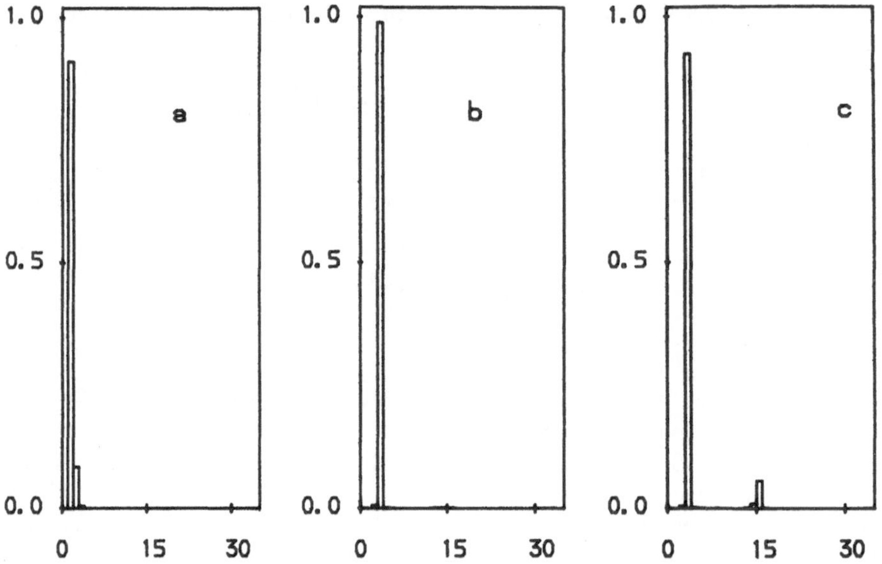

Figure 1. Calculated histogram of $p_n = \rho_f(n)$ after 1(a), 40(b) and 500(c) atoms have passed through the cavity: $\omega = \omega_o$ and $t_{int} = \pi/2g$. See end this §3 for details.

T_p but repeating each T_p. Suppose $<\hat{n}>\equiv<a^\dagger a>$, the mean number of photons in the cavity, is n_1 as the inverted atom enters, is n_2 when it leaves after time t_{int}, and is n_3 before the next atom enters at T_p. For the steady state $n_3 = n_1$, and

$$\bar{n}_{th} + (n_2 - \bar{n}_{th})e^{-2\kappa(T_p - t_{int})} = n_3 = n_1 \tag{21}$$

where the case $n = 1$ of the exact expression (23) below has been used. Since $n_2 \leq n_1+1$ (only one photon can be dumped) then, for small enough $\bar{n}_{th}$ compared with n_1, n_2, n_3 (a numerically valid assumption in the present context though this removes $\bar{n}_{th}$ from the theory)

$$n_2 \leq [1 - e^{-2\kappa(T_p - t_{int})}]^{-1}. \tag{22}$$

It follows that for $t_{int} << T_p$ (the assumption of Refs. 8,9) and $2\kappa T_p << 1$ (an assumption of Ref. 9), $n_2 \leq (2\kappa T_p)^{-1} = N_{ex}$. The equality is the equality of linearly additive gain and loss $R = T_p^{-1} = 2\kappa n_2$. More generally this simple minded analysis shows the gain-loss mechanism is nonlinear even in the empty-of-atoms time t_e. But it tells us nothing about the evolution during t_{int}, only that n_2 could become large for $T_p \approx t_{int}$.

We have used the methods sketched in the previous section to investigate this nonlinear gain and loss regime: we solve (2) for the total density operator ρ during t_{int} when the atom is in the cavity: the heat bath is thus coupled in. We then solve the same equation with the atom omitted during the time $t_e = T_p - t_{int}$ and the heat bath coupled in. This is done numerically or we can make use of the exact analytic formula derived from (2) when no atom is present: we define $P_n = (n!)^{-1} < a^{\dagger n} a^n >$ and find (by solving equation (26) below) that

$$P_n(\tau) = (\bar{n})^n + \sum_{r=1}^{n} e^{-r\tau} M_{nr}(\bar{n})^{n-r} \sum_{j=1}^{r} (-1)^{r-j} M_{rj}(\bar{n})^{r-j}\{P_j(0) - (\bar{n})^j\} \tag{23}$$

in which $M_{ij} = i!(j!(i-j)!)^{-1}$, $i \geq j$; $M_{ij} = 0$, $i < j$: κ enters through $\tau = 2\kappa t_e$ and (23) generalizes (21) with $\bar{n} \equiv \bar{n}_{th}$.

So far in these calculations both t_{int} and t_e repeat regularly with fixed T_p (a stochastic input of the atoms is not yet included). The empty of atoms cavity has initial condition $\rho_f(0) = Tr_a\rho(t_{int})$, where $\rho(t_{int})$ is ρ as the atom of current interest leaves the cavity and $\rho_f(0)$ evolves to $\rho_f(t_e)$ in the empty of atoms cavity time, either by numerical solution for $\rho_f(t)$ from (2) with no atom present or through (23): $\rho_f(t_e)$ now serves as initial condition for the cavity field as the next atom enters so $\rho = |e> \rho_f(T_p - t_{int}) < e|$ is the initial condition for ρ as the next atom enters (we here conveniently use ρ_f for the cavity field density operator not ρ^f).

The equations of motion during the interaction time t_{int} can be reduced to a 3-term recurrence relation solved once again by matrix continued fraction methods. Details will be published elsewhere.[15,24] A significant point is that although off-diagonal elements of ρ develop, $\rho_f(t_e)$ is diagonal if $\rho_f(-t_{int}) = Tr_a\rho(0)$ is initially diagonal. But this is the case since the first atom enters a cavity containing only black-body radiation. This feature also arises in the unitary transformation of ρ during the time t_{int},[21] but the equations (13) for ρ_f are always diagonal.

Obviously the smallest empty-of-atoms time is zero, i.e. $T_p - t_{int} = 0$ and $T_p = t_{int}$. The Fig. 1 shows the evolution of probabilities p_n for finding n photons in the cavity as a function of the number of inverted atoms which have passed through the cavity in successive intervals $gt_{int} = \frac{1}{2}\pi$. This is the condition for evolution to a first trapping state $n_o = 3$, $r = 1$ at $Q = \infty$ and the next trapping state has $n_o = 15$. However, in these calculations $Q = 2\text{x}10^{10} < \infty$ and $T = 0.5°\text{K}$ much as in the experiments.[2,3] It will be seen that after 40 atoms have passed p_3 evolves to 0.988, while p_{15} now evolves and after 500 atoms have passed p_3 and p_{15} have become 0.924 and 0.056 respectively. The Fig. 2 of Ref. 18 also shows comparable results for $gt_{int} = \frac{1}{4}\pi$ and the first trapping state is at $n_o = 15$: $p_{15} = 0.774$ after 400 atoms have passed and is still growing. Thus at this finite Q and T the system displays the action predicted for $Q = \infty$ in Ref. 10. Presumably no steady state is reached as the higher trapping states at n_o =35, 63,... begin significantly to grow. Notice the dependence on the numbers of atoms which have passed: the theory is very different from the steady state theories based on eqns. (13).

Thus the major *conclusion must be that as* $t_e = T_p - t_{int}$ *becomes smaller the finite* Q *finite* T *system behaves more and more like a* $Q = \infty$ *system as described in Refs 7,10.*

We support this conclusion by the further work described in the next section.

4. ACTION OF THE MICROMASER AS THE EMPTY OF ATOMS CAVITY TIME CHANGES

The numerical calculations described in §§2 and 3 are slow even on the Amdahl 5890 system available in Manchester. The matrix continued fraction methods used offer the very high accuracy and good stability necessary for the detailed investigation of black-body dynamics displayed in Fig. 1, but calculations are very long. We have developed faster, less accurate and less stable, methods for the PC. We described some of the results found so far through these in Ref. 1. We amplify those results in Ref. 1 by the further results in the $T_p >> t_e$ regime described here. These particular calculations determine the moments P_n, Q_n, R_n and S_n mentioned above eqn. (5). On resonance we need P_n, Q_n and R_n only, and the equations of motion derived from (2), with the atom present in the cavity, are

$$\begin{aligned}
\dot{Q}_o &= -R_1 \\
\dot{P}_n &= nR_n - 2\kappa g^{-1} n(P_n - \bar{n}P_{n-1}) \\
\dot{Q}_n &= -\tfrac{1}{2}nR_n - (n+1)R_{n+1} - 2n\kappa g^{-1}(Q_n - \bar{n}Q_{n-1}) \\
\dot{R}_n &= 4Q_n + P_{n-1} + 2Q_{n-1} - \kappa g^{-1}\{(2n-1)R_n - 2(n-1)R_{n-1}\bar{n}\}.
\end{aligned} \tag{24}$$

Conveniently we use $\bar{n}$ for $\bar{n}_{th}$ and have rescaled $P_n \rightarrow (n!)P_n$, etc. so with time t for gt

$$\begin{aligned}
P_n(t) &\equiv Tr\{\rho(t)(a^\dagger)^n a^n\}/n! \\
Q_n(t) &\equiv Tr\{\rho(t)(a^\dagger)^n a^n S^z\}/n! \\
R_n(t) &\equiv Tr\{\rho(t)S^+(a^\dagger)^{n-1}a^n - \rho(t)(a^\dagger)^n a^{n-1}S^-\}/n! \quad .
\end{aligned} \tag{25}$$

The empty of atoms cavity field evolves as

$$\dot{P}_n = -2n\kappa g^{-1}(P_n - \bar{n}P_{n-1}) \tag{26}$$

and the analytical solution of this is (23). When the atom is in, we close the system by $R_{m+1} \approx (m+1)^{-1}P_1R_m$. We need $m >> < a^\dagger a > \equiv P_1$ and there are $3m+1$ equations.

The Figs. 2 and 3 show results achieved this way for $Q \doteq 1.5\text{x}10^8$, $\bar{n}_{th} = 0.15$, $\omega = \omega_o$ (exact resonance) and different gt_{int}. In the Fig. 3 there are pronounced sub-Poissonian regions where $\sigma < 1$ and there is a general qualitative agreement in this respect with the results predicted in Refs. 8 and 9 and confirmed in the experiments.[2,3] Qualitative agreement over the plotted range $gt_{int} < 2.5$ can be reached through $\Theta = (N_{ex})^{\frac{1}{2}}gt_{int} = (2\kappa T_p)^{-\frac{1}{2}}gt_{int} = \sqrt{10}gt_{int} \doteq 3.162gt_{int}$. Thus there is also the super-Poissonian region in Figs. 2, 3 which occurs after the threshold and where $\Theta \approx 5.69$ ($\sim 2\pi$), and the $<$ is satisfied in (22) with $N_{ex} = 10$. The significance of these results compared with previous work [2,3,8,9] however is that $gt_{int} \lesssim 2.5$ and $gT_p = 5$, so the empty of atoms cavity time reaches up to as little as 50%. Agreement with Refs. 2, 3, 8, 9 is thus reached by the rather low value ($1.5\text{x}10^8$) chosen for Q. Figs. 4 and 5 depict the steady state when the quality factor Q is improved so that $\kappa g^{-1} = 7\text{x}10^{-4}$ and $< gt_e > = gR^{-1} = 72$ (so $N_{ex} \approx 10$ as before). The error made in regarding t_e as a Poisson process rather than $(t_e + t_{int})$ is very small. For the stochastic average over the t_e (23) it suffices to replace the exponentials $e^{-2\kappa r t_e}$ by $(1+2\kappa r R^{-1})^{-1}$ and all of Figs. 4 to 10 are calculated on this basis. The two sets of figures 2, 3 and 4, 5 are very similar for $0 < gt_{int} < 1.7$ although only Figs. 4, 5 approach the $T_p >> t_{int}$ regime of the experiments.[2,3] In neither set is there any evidence of the trapping state $n_o = 3$ at $gt_{int} = \frac{1}{2}\pi$ (for example) in agreement with the ideas of Ref. 6. Instead the super-Poissonian region at $\Theta \sim 5.69$ occurs. Thus there are *three* important parameters: the empty-of-atoms cavity time t_e, $\bar{n}_{th}$, and Q.

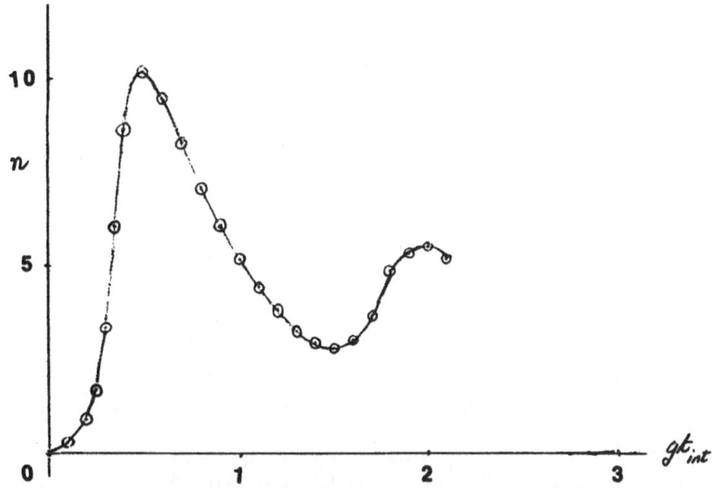

Figure 2. Plot from the PC of the 'long time' value $<\hat{n}>=<a^{\dagger}a>$ on atomic exit as a function of gt_{int} for Q =1.5x10^8 (κg^{-1} = 0.01) and $\bar{n}_{th}$ = 0.15: gT_p = 5 (N_{ex} =10) and Θ = (3.162)gt_{int}.

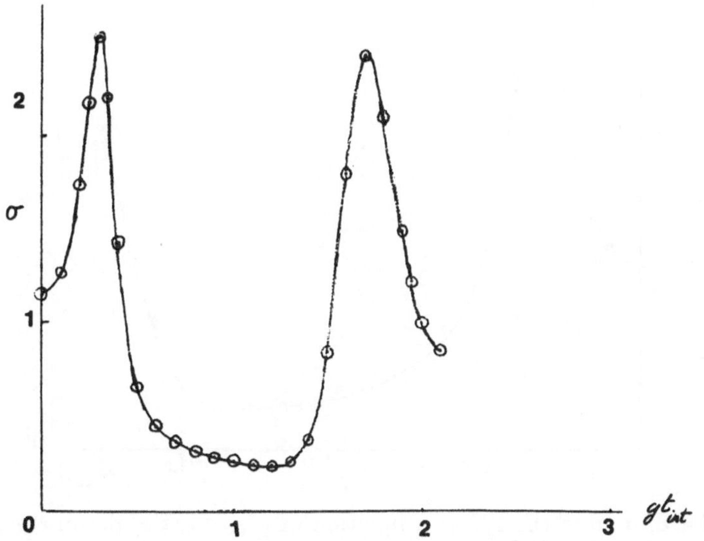

Figure 3. Plot of σ from the PC as a function of gt_{int} for the parameters of Fig. 2. Note the super-Poissonian region were $gt_{int} \approx 1.8$ ($\Theta \approx 5.69 \sim 2\pi$).

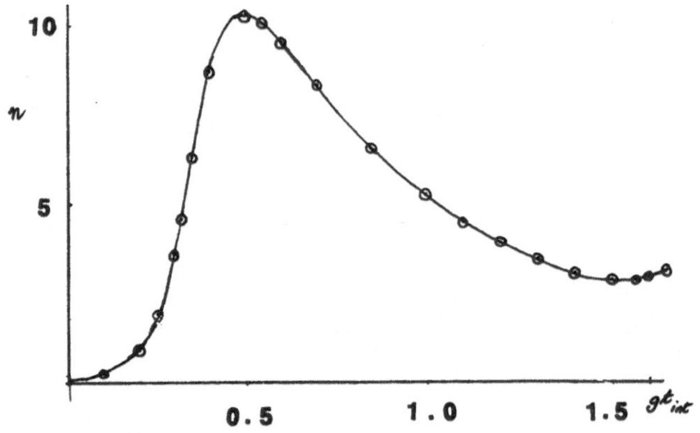

Figure 4. Evolution of the cavity photon number: $Q \approx 2\text{x}10^9$ ($\kappa g^{-1} = 7\text{x}10^{-4}$) and $< gt_e >_{Av} = gR^{-1} = 72$ (so $N_{ex} \approx 10$ as in Fig. 3): $\bar{n}_{th} = 0.15$.

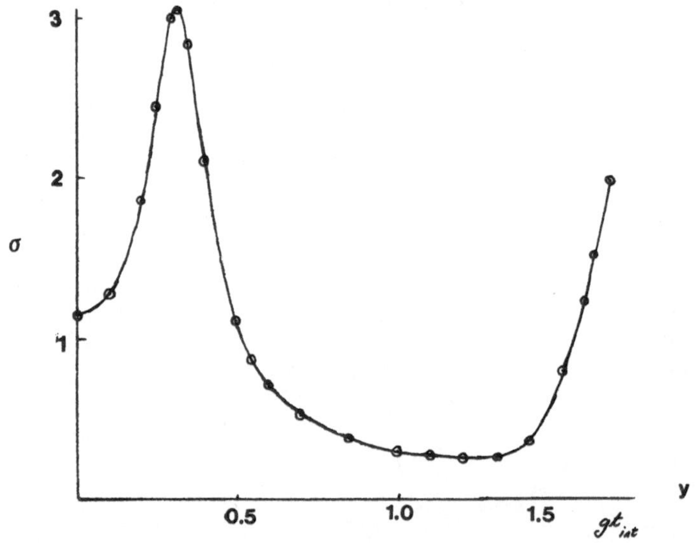

Figure 5. Plot of σ from the PC as a function of gt_{int} for the parameters of Fig. 4. Note the super-Poissonian region again occurring where $gt_{int} \approx 1.8$ ($\Theta \approx 5.69$).

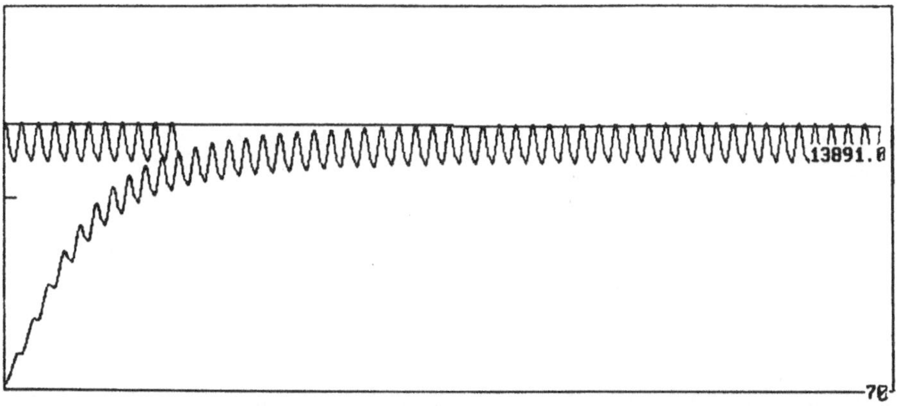

Figure 6. Evolution of cavity photon number for $gt_{int} = 1.047$ chosen to minimize σ on exit: $gR^{-1} = 216$, $\kappa g^{-1} \sim 7\text{x}10^{-5}$ ($Q = 2\text{x}10^{10}$); $\bar{n}_{th} = 0.15$. For convenience gt (mod 72) is plotted rather than gt. After 65 atoms $< \hat{n} > = 6.16$, $\sigma = 0.204$, $< S^z >= 0.332$.

Notice that by setting $N_{ex} = 10$ for both Figs. 2, 3 and Figs. 4, 5 they show the very similar behaviors. The Figs. 8-11 in Ref. 1 show that, on the contrary for large $Q \simeq 2\text{x}10^{10}$ and gT_p small ($gT_p = gt_{int}$ and then gT_p increasing) plots of $< \hat{n} >=< a^\dagger a >$ as a function of time show that for $gt_{int} = \frac{1}{2}\pi$ the trend is that as $gt_e \to 72$, σ on final exit has increased from 0.053 (almost a Fock state) for $gt_e = 1$ ($gT_p = 1 + gt_{int} = 1 + \frac{1}{2}\pi$) to $\sigma = 1.106$ which is the super-Poissonian ($\sigma > 1$) value much as in the observations[2,3] – though very many more atoms were used in the steady state for these observations.

Our conclusion from this is[1] that a relatively short empty of atoms cavity time t_e will help in the creation of Fock states at finite Q and T. Evidently the nonlinear gain and loss mechanism enhances the cavity Q towards $Q = \infty$ (as of course does the reduction $T_p << T_{cav}$).

The Figs. 6-10 support this conclusion. All of these figures are in the regime of large $gT_p >> gt_{int} : gt_{int} = 1.047$ and gT_p =216, 360, 504, 720 and 1008. The σ at exit follow the trend σ =0.204, 0.216, 0.242, 0.281, 0.332. These figures compare with $\sigma \geq 0.3$ as observed in the experiments[2,3] with $gT_p \sim 308$. Again very many more atoms were used in these experiments to achieve a steady state. At zero temperature, with $Q = \infty$, $gt_{int} = 1.047$ should give a steady oscillation between the two number states $|8>$ and $|9>$. Although, of the Figs. 6-10, Fig. 6 represents conditions where thermalization effects are least (T_p is smallest), the expectation value $< \hat{n} >= n$ on exit is much less than 8 and $\sigma \approx 0.2$ is well above zero. Nevertheless, we think Figs. 6-10 support our view that reducing T_p towards t_{int} will increase the trend towards pure Fock state evolution in the micromaser cavity field.

We recognize[1] decreasing T_p increases the likelihood of more than one atom in the cavity at the same time and this considerable complication is being explored.

5. SUMMARY AND CONCLUSIONS

The paper makes a fundamental investigation of the action of the Garching micromaser both analytically (for large recurrence times T_p) and numerically (for both small and large recurrence times T_p). The aim is to investigate the possibilities for Fock state creation in the cavity field of the micromaser. The §2 reviews our numerical methods

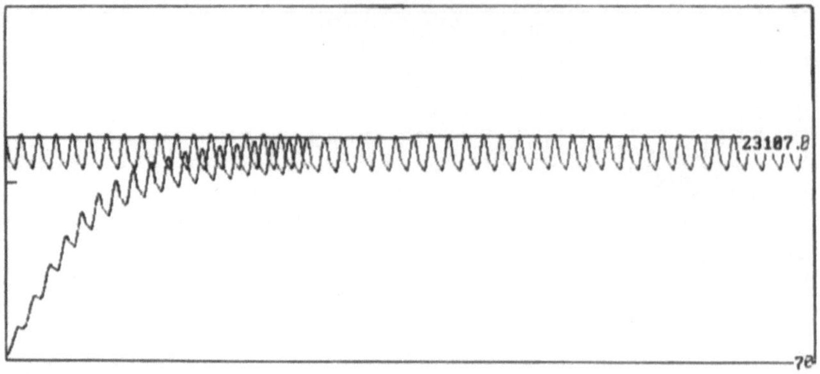

Figure 7. Evolution of cavity photon number. Parameters as for Fig. 6 but $gR^{-1} = 360$. AFter 65 atoms have passed $< \hat{n} >= 5.68$, $\sigma = 0.216$, $< S^z >= 0.221$.

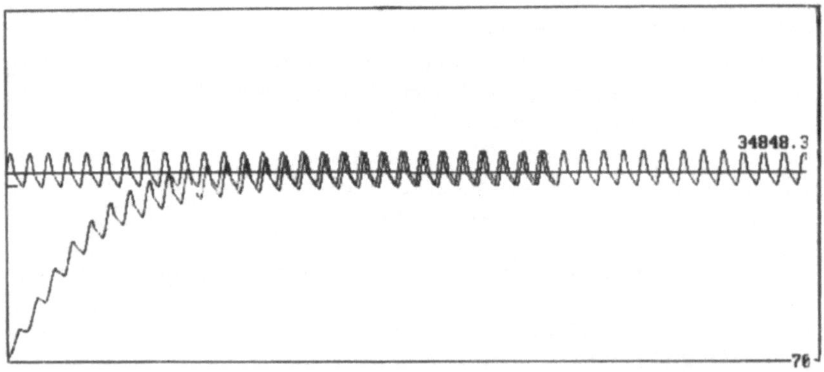

Figure 8. Evolution of cavity photon number. Parameters as for Fig. 6 but $gR^{-1} = 504$. After 70 atoms have passed $< \hat{n} >= 5.31$, $\sigma = 0.242$, $< S^z >= 0.1720$.

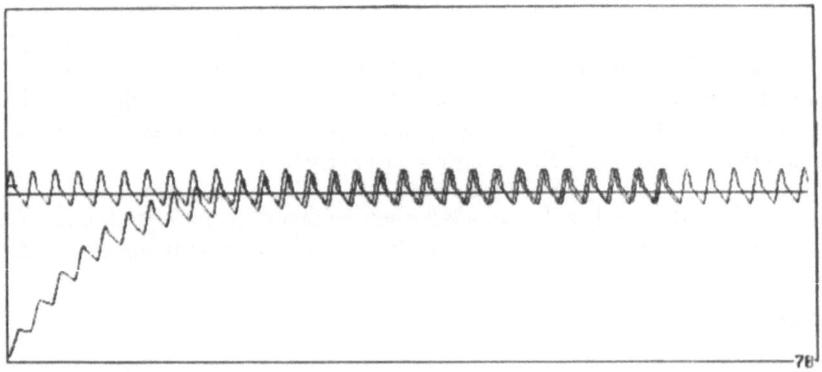

Figure 9. Evolution of cavity photon number. Parameters as for Fig. 6 but $gR^{-1} = 720$. After 65 atoms have passed $< \hat{n} >= 4.92$, $\sigma = 0.281$, $< S^z >= 0.0510$.

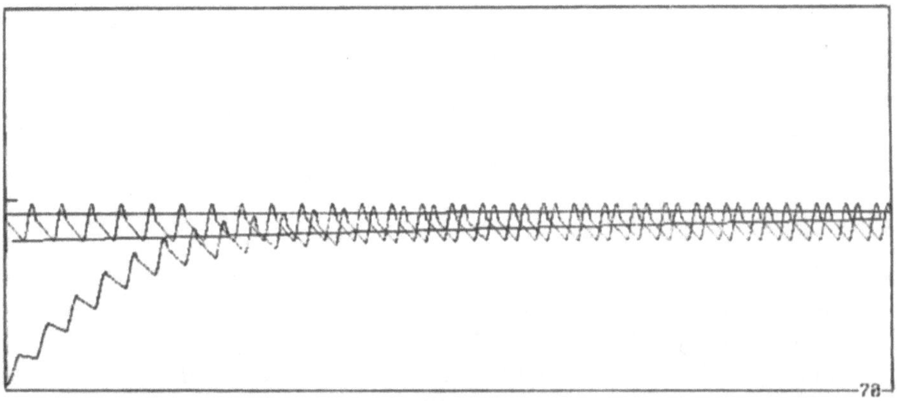

Figure 10. Evolution of cavity photon number. Parameters as for Fig. 6 but $gR^{-1} = 1008$. After 60 atoms have passed $< \hat{n} >= 4.47$, $\sigma = 0.332$, $< S^z >= -0.0379$.

for calculating effects of black-body radiation on single atoms in cavities of arbitrary Q: the §3 develops fundamental theory for the canonically quantized micromaser; it is related to conventional laser theory much as in Ref. 9 under conditions of long empty-of-atoms times $t_e = T_p - t_{int}$, and under which at finite temperatures T it seems to make sense to envisage a steady state. However, it is shown that at $T = 0$, and $Q = \infty$, there is no steady state: evolution towards a Fock state is both typical and possible, but a *sequential* description of the atoms passing through the cavity becomes a necessary feature of the description of the dynamics.

Thus at long empty-of-atoms times $t_e >> t_{int}$ there is a *qualitative* change between $Q = \infty$, $T = 0$ theories and $Q < \infty$, $T > 0$ theories even when as in the recent experiments[2,3] T's as low as 0.5°K are under consideration. For long enough cavity lifetimes T_c the two other times t_{int} and T_p are important: t_{int} is the interaction time of single atoms with the cavity field, and T_p $(= R^{-1})$ is the mean repetition rate of atoms entering the cavity. Then $t_e \equiv T_p - t_{int}$. The main result of §3 is to show following Ref. 1 that, as $T_p >> t_{int}$ becomes $T_p \approx t_{int}$ (and $T_c \gtrsim T_p$ becomes $T_c >> T_p$), the micromaser becomes more like a $Q = \infty$ system. Evidently $T_c >> T_p$ enhances $Q = \infty$ behavior. But the Fig. 1 shows there is no steady state even[2,3] at the $Q = 3\text{x}10^{10}$ and $T = 0.5$°K of the experiments when $T_p \approx t_{int}$. Moreover, the timescale for the evolution in Fig. 1 is very short compared with the timescale for such evolution at $Q = \infty$.

The *qualitative* change in behavior is induced by the advent of an intrinsically non-linear gain-loss mechanism as T_p approaches t_{int} and the sequential evolution of the system as the atoms pass through the cavity becomes important. The §4 reports new results in the regime of linear gain and loss, $T_p >> t_{int}$, which tend to confirm the experimental observations[2,3] in this regime. Nevertheless, even in this regime, lower values of σ, the cavity field variance, are achieved for the smaller T_p.

The conclusion is that reached previously[1]: the possibility of Fock state creation in the micromaser will be enhanced by reducing T_p towards t_{int} bearing in mind, however, that as $T_p \rightarrow t_{int}$ the increasing probability of more than one atom in the cavity at the same time could substantially change and complicate predictions.

The message for Fock state creation therefore seems to be: – Reduce T_p *to some extent* and reduce T. But it may then be necessary to follow a sequential evolution of the cavity field as the inverted atoms enter since no steady state is apparently reached in the presence of black-body radiation (of whatever small amount).

Further detailed studies of trapping state evolution at realistic T_p and T's $< 0.2°K$ are in hand.

REFERENCES

1. R. K. Bullough, N. M. Bogoliubov, N. Nayak and B. V. Thompson, in: Proc. NATO ARW "Quantum Measurements in Optics," P. Tombesi and D. F. Walls, eds., Plenum, New York, (1992), pp. 129-149.
2. G. Rempe, F. Schmidt-Kaler and H. Walther, *Phys. Rev. Lett.* **44**, 2783 (1990).
3. Gerhard Rempe and Herbert Walther, *Phys. Rev. A* **42**, 1650 (1990).
4. D. Meschede, H. Walther and G. Müller, *Phys. Rev. Lett.* **54**, 551 (1985).
5. G. Rempe, H. Walther and N. Klein, *Phys. Rev. Lett.* **58**, 353 (1987).
6. P. Meystre, G. Rempe and H. Walther, *Opt. Lett.* **13**, 1078 (1988).
7. E. T. Jaynes and F. W. Cummings, *Proc. IEEE* **51**, 89 (1963).
8. P. Filipowicz, J. Javainen and P. Meystre, *Phys. Rev. A.* **34**, 3077 (1986).
9. L A. Lugiato, M. O. Scully and H. Walther, *Phys. Rev. A* **36**, 740 (1987).
10. F. W. Cummings and A. K. Rajagopal, *Phys. Rev. A* **39**, 3414 (1989).
11. N. Nayak, R. K. Bullough, B. V. Thompson and G. S. Agarwal, *IEEE J. Quant. Electronics* **QE-24**, 1331 (1988).
12. G. S. Agarwal, R. K. Bullough and G. P. Hildred, *Optics Commun.,* **39**, 23 (1986).
13. G. S Agarwal, R. K. Bullough, G. P. Hildred and N. Nayak, "Cavity quantum electrodynamics at arbitrary Q and temperature T." To be published.
14. G. S. Agarwal, R. K. Bullough and N. Nayak, *Optics Commun.* **85**, 202 (1991).
15. N. Nayak, B. V. Thompson and R. K. Bullough, "Cavity quantum electrodynamics: finite temperature theory of the ^{85}Rb atom maser." To be published.
16. R. K. Bullough and J. Timonen, "Quantum groups and quantum complete integrability: theory and experiment" in: Proc. 19th Intl. Conf. on Diff. Geom. Methods in Theor. Phys., C. Bartocci, U. Bruzzo and R. Cianci, eds., Springer Lecture Notes in Physics **375**, Springer-Verlag, Berlin (1991), p. 69.
17. N. Nayak, R. K. Bullough and B. V. Thompson in: "Coherence and Quantum Optics," J. H. Eberly, L. Mandel, and E. Wolf, eds., Plenum, New York, (1990), pp. 809-813.
18. N. Nayak, B. V. Thompson and R. K. Bullough in: "ECOOSA'90 Quantum Optics," Inst. of Phys. Conf. Ser. 115: Section 1, M. Bertolotti and E. R. Pike, eds. (Inst. of Phys. Bristol), pp. 81-84.
19. G. P. Hildred, R. R. Puri, S. S. Hassan and R. K. Bullough, *J. Phys. B: At Mol. Phys.* **17**, L535 (1984).
20. H. Risken "Fokker-Planck Equations," Springer-Verlag, Berlin (1984).
21. J. Kruse, M. O. Scully and H. Walther, *Phys. Rev. A* **36**, 4547 (1987).
22. M. O. Scully and Willis E. Lamb, Jr., *Phys. Rev.* **159**, 203 (1967).
23. M. O. Scully, Dae M. Kim and Willis E. Lamb, Jr., *Phys. Rev. A* **2**, 2529 (1970).
24. N. Nayak, B. V. Thompson and R. K. Bullough. To be published.

THE JAYNES-CUMMINGS MODEL GENERALIZED: THREE-LEVEL ATOMS INTERACTING WITH TWO OSCILLATORS

Fritz Haake and Klaus-Dieter Harms

Fackbereich Physik
Universität-Gesamthochschule Essen
4300 Essen 1, Deutschland

A collection of N identical atoms with three equidistant levels is resonantly coupled to two harmonic oscillators according to the following scheme[1]:

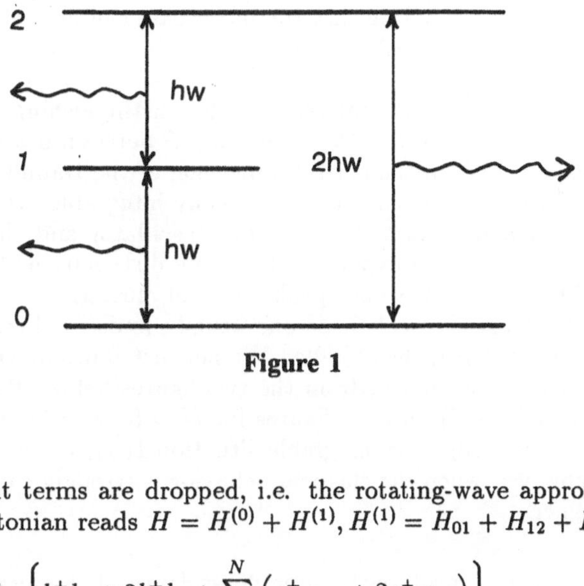

Figure 1

If all antiresonant terms are dropped, i.e. the rotating-wave approximation (rwa) is made, the Hamiltonian reads $H = H^{(0)} + H^{(1)}$, $H^{(1)} = H_{01} + H_{12} + H_{02}$ with

$$H^{(0)} = \hbar\omega\left\{ b_1^+ b_1 + 2b_2^+ b_2 + \sum_{\mu=1}^{N} \left(a_{\mu 1}^+ a_{\mu 1} + 2a_{\mu 2}^+ a_{\mu 2} \right) \right\}$$

$$H_{01} + H_{12} = (\hbar g_1/\sqrt{N}) b_1 \sum_{\mu=1}^{N} (a_{\mu 1}^+ a_{\mu 0} + a_{\mu 2}^+ a_{\mu 1} + H.c.$$

$$H_{02} = (\hbar g_2/\sqrt{N}) b_2 \sum_{\mu} a_{\mu 2}^+ a_{\mu 0} + H.c. \tag{1}$$

Recent Developments in Quantum Optics, Edited
by R. Inguva, Plenum Press, New York, 1993

Here b_i and b_i^+ annihilate and create photons in the ith oscillator ($i = 1, 2$) while $a_{\mu j}$ and $a_{\mu j}^+$ annihilate and create the μth atom in its jth level ($\mu = 1, 2, \ldots N$; $j = 0, 1, 2$). Due to the rwa the four additive parts of the Hamiltonian, $H^{(0)}$, H_{01}, H_{02}, H_{12}, are separately conserved. Furthermore, the Hamiltonian is symmetric under exchange of atom labels. We may therefore use basis vectors symmetric in all atoms, $|N_0, N_1, N_2; n_1, n_2>$; the quantum numbers labeling these states are the photon numbers $n_{1,2}$ for the two oscillators and the global population numbers $N_{0,1,2}$ of the three atomic levels. Two combinations of these quantum numbers are conserved, the total number of atoms $N = N_0+N_1+N_2$ and the total number of quanta of energy $\hbar\omega$, $Z = N_1+2N_2+n_1+2n_2$. With N and Z fixed, the number dim (N, Z) of independent states $|N_0, N_1, N_2; n_1 n_2>$ is given by a polynomial of third order in N and Z. We have diagonalized H in that space of dimension dim (N, Z).

The model in question acquires classical behavior in the limit $N \to \infty$, $Z \to \infty$. A classical description naturally employs as observables two complex mode amplitudes, three complex polarizations, and three real occupation numbers. Even after accounting for the conservation laws mentioned, one is left with sufficiently many independent observables for classical chaos to be possible. Indeed, Shepelyanski[2] has verified chaos to arise.

The possibility of chaos is an interesting difference to the simpler Jaynes Cummings model which consists of two-level atoms coupled to a single oscillator, provided antiresonant terms are excluded from the Hamiltonian. That model is contained in our as the special case $g_1 = 0$. Incidentally, the case $g_2 = 0$ corresponds to a cascade of two Jaynes-Cummings models and is also integrable in the classical limit. When both coupling constants are of comparable magnitude classical chaos is most strongly pronounced. It may be worth noting that the latter case can be realized in atoms only with the help of strong parity-breaking fields.

A well known quantum mechanical criterion for distinguishing chaos and regular motion is based on the distribution $P(S)$ of spacings S between nearest-neighbor levels in the energy spectrum.[3] When the matrix dimension of the Hamiltonian is large that distribution becomes reasonably smooth. Classically integrable systems with at least two degrees of freedom, if generic (the harmonic oscillator and the Kepler problem are exceptions), tend to have an exponential spacing distribution, $P(S) = e^{-S}$. Since zero is the most likely value of S one speaks of level clustering. On the other hand, under conditions of global classical chaos one usually finds level repulsion, according to $P(S) = (\pi/2)exp(-\pi S^2/4)$, the so-called Wigner distribution. Our model displays the expected behavior, as is obvious from the two figures below. We plot the spacing staircase $I(S) = \int_0^S dS'\ P(S')$, in both figures for $N = 8$, $Z = 20$, dim $(N, Z) = 305$. For $g_2/g_1 = 0.05$, a classically near integrable situation $I(S)$ is indeed close to $1 - e^{-S}$. For $g_2/g_1 = 0.5$, however, when the classical behavior is strongly chaotic, we find level repulsion à la Wigner.

The final pair of figures displays the time dependence of the mean photon number in the oscillator with eigenfrequency ω, $< n_1(t) >$. In both cases the initial state is the same, $|N_0, N_1, N_2; n_1, n_2 >= |8, 0, 0; 20, 0>$. As for the preceding figures, we thus have $N = 8$, $Z = 20$, dim $(N, Z) = 305$. Both graphs below display quasiperiodic behavior in time, as is indeed characteristic of finite series of harmonic functions. However, in the near integrable case $g_2/g_1 = 0,05$ quasiperiodicity takes the form of a sequence of collapses and revivals,

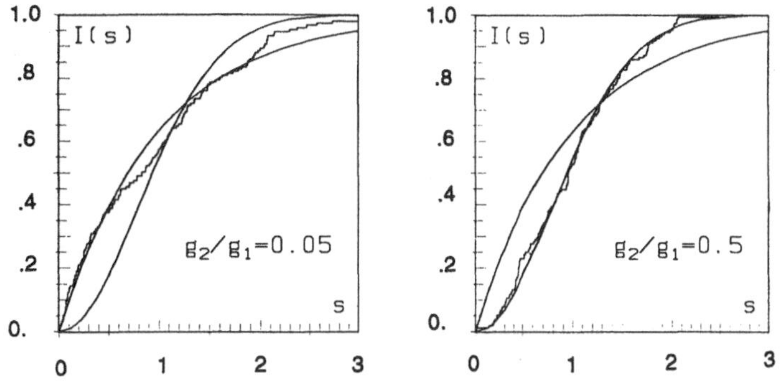

Figure 2

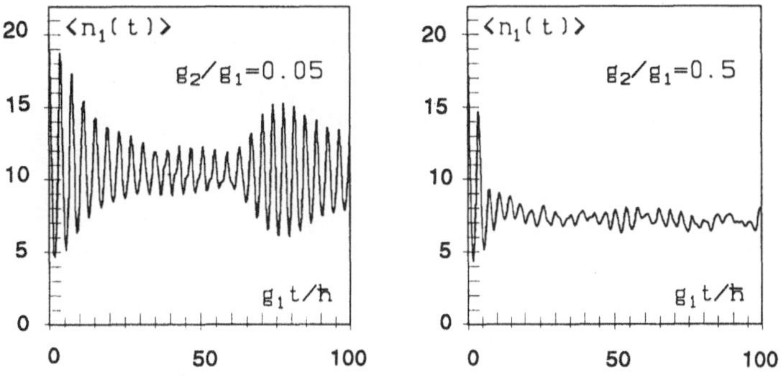

Figure 3

while for $g_2/g_1 = 0.5$ the time dependence is more reminiscent of dissipative behavior; sizeable recurrence events, bound to happen sooner or later, appear to be rarer events under conditions of classical chaos than in near integrable situations. What is important is the difference of the two behaviors. Clearly, we are facing a quantum counterpart of the classical distinction between regular and chaotic motion. For an explanation and another example of this phenomenon the reader is referred to the literature.[3]

ACKNOWLEDGEMENT

We gratefully acknowledge support by the Deutsche Forschungsgemeinschaft, especially the Sonderforschungsbereich "Unordnung und groBe Fluktuationen."

REFERENCES

1. K. D. Harms and F. Haake, *Z. Physik* B **79**:159 (1990).
2. D. L. Shepelyanski, *Phys. Rev. Lett.* **57**:1857 (1986).
3. F. Haake, "Quantum Signatures of Chaos," Springer, Berlin (1991).

QUANTUM NOISE REDUCTION IN LASERS THROUGH NONLINEAR INTRACAVITY DYNAMICS

H. Ritsch* and P. Zoller†

*Institute for Theoretical Physics
University of Innsbruck, A-6020 Innsbruck (Austria)

†Joint Institute for Laboratory Astrophysics & Dept. of Physics
University of Colorado, Boulder, CO 80309 (USA)

INTRODUCTION

As can be found in almost any textbook on laser physics, standard theories for a single mode laser predict, that far above threshold the laser generates nearly a coherent state (with a phase randomly varying in time). Hence the counting statistics for the output photons is Poissonian and the intensity fluctuation spectrum is shot noise limited (this defines the so-called standard quantum limit (SQL)). A closer investigation reveals three main sources of the laser quantum intensity fluctuations, namely pump noise, noise through cavity losses and spontaneous emission noise. The last contribution, which turns out to be less important far above threshold, can be minimized by proper choice of the laser transition. Recently there has been increased interest in developing a so called quiet laser, i.e. a laser with sub-shotnoise intensity fluctuations. In this context several authors have proposed a laser with an external sub-Poissonian *pump*[1–6] (i.e. pumping with amplitude squeezed light or a sequence of regularly spaced short pump pulses, as well as injection of a regular beam of excited atoms, electrons), which leads to an intensity noise reduction below the standard quantum limit. In the best case the variance of the laser photon number distribution is reduced by a factor of 2 compared to the Poissonian case and the slow output intensity fluctuations around zero frequency are completely suppressed. In contrast to an external sub-Poissonian pump, in this work we will present two alternative mechanisms to quench the quantum noise in a laser, which do not rely on some externally injected regularity, but are based on nonlinear dynamic self regularization.[7–11]

The first proposed scheme relies on a modification of the laser loss noise by an **intracavity nonlinear absorber.**[8,10] We will show that by coupling a nonlinear absorber (e.g. an (unsaturated) N-photon transition) to the laser mode, the photon number variance gets below the value corresponding to a coherent state. It turns out that the photon number distribution gets the narrower the higher the order N of the nonlinear process and reaches the same minimum width of 1/2 of the coherent state width as for a perfectly regular pump. However, the bandwidth of the intensity noise reduction can be significantly enhanced. If we modify our absorbers to allow for a nonlinear feedback, this behaviour can even be improved further.

Recent Developments in Quantum Optics, Edited
by R. Inguva, Plenum Press, New York, 1993

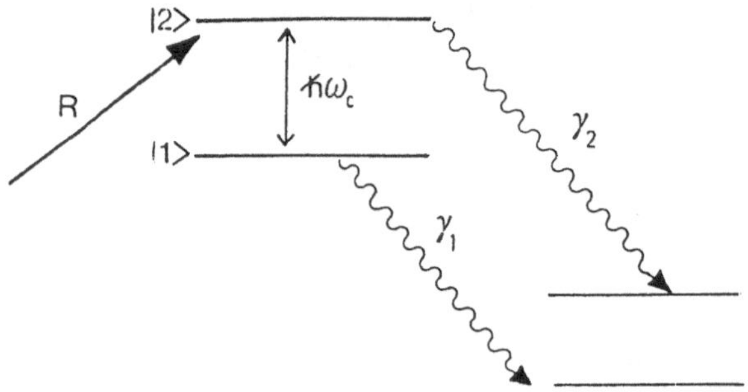

Figure 1. Schematic level diagram of laser active atoms.

In the second part of this work we identify a new mechanism of pump noise suppression in multilevel laser systems, based on the **nonlinear dynamics of the active laser atoms.**[9] Surprisingly to our knowledge this mechanism, which is closely related to the origin of antibunching in atomic resonance fluorescence, has not been discovered in previous treatments of the single mode laser. In the best case we find (at least in principle) a complete suppression of the low frequency laser intensity noise. Note that in order to achieve this result, we do not have to make any fundamental changes in the laser setup but rather choose suitable operating conditions and laser atoms.

LASER WITH INTRACAVITY NONLINEAR ABSORBER

Following the laser treatment developed by Scully and Lamb[12] (SL), we consider a single quantized lasing mode inside a high quality optical cavity. The cavity losses are induced by a continuum of output modes outside the laser cavity. The laser pumping prepares model atoms in an excited state $|2>$ at given times $t(i)$ with a mean rate R. These atoms then subsequently undergo transitions to a lower level $|1>$, where the transition frequency ω_{12} is assumed to be resonant with the cavity eigenfrequency ω. Both atomic levels are assumed to decay with rates γ_1, γ_2 into some other atomic levels, where they do not interact with the laser mode any more. The level scheme of such a model is depicted in Fig. 1. Note that this assumption implies that each atom contributes either one or no photon to the lasing mode, independent of all other atoms. Assuming fast atomic decay rates $\gamma_1, \gamma_2 >> \kappa$ with κ the cavity damping rate, one can adiabatically eliminate the atoms to get a master equation for the laser mode dynamics. Assuming a Poissonian distribution of atomic injection times $t(i)$ one finds the following set of coupled differential equations for the photon number distribution $p_n := < n|\rho_f|n >^{2}$ [12],:

$$(\dot{p}_n)_{SL} = -\kappa(n\ p_n - (n+1)p_{n+1}) + R(\beta_n p_{n-1} - \beta_{n+1}p_n), \tag{1}$$

with ρ_f being the internal laser mode density matrix and $|n>$ an n-photon eigenstate of the internal photon number operator. Here $\beta_n = \beta_o \frac{n/n_o}{1+n/n_o}$, with $\beta_o = \gamma_1/(\gamma_1+\gamma_2)$ and n_o is proportional to the saturation intensity of the lasing transition. As Marte

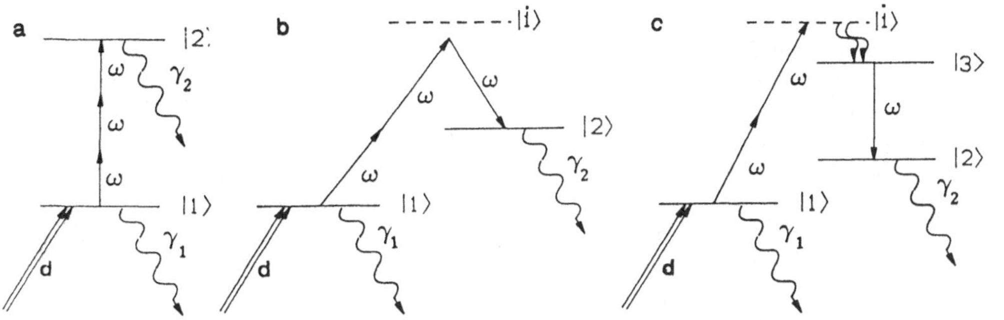

Figure 2. Schematic energy level diagram for a standard nonlinear absorber atom (a), an absorber with a Raman-type transition (b) and an absorber with nonlinear feedback (c).

and Zoller[1] have shown, a sub-Poissonian distribution of pump events $t(i)$ leads to the additional terms (labelled Q):

$$(\dot{p}_n)_Q = RQ_p(\beta_{n+1}^2 p_n - \beta_n(\beta_n + \beta_{n+1})p_{n+1} + \beta_n\beta_{n-1}p_{n-2}) \tag{2}$$

where Q_p is a parameter associated with the time distribution of pump events[2] (for a perfectly regular pump $Q_p = -\frac{1}{2}$; for a Poissonian distribution of pump events one has $Q_p = 0$).

In order to include the effect of a nonlinear absorbing medium inside the cavity, we assume an additional independent kind of atoms to be prepared at times $t'(j)$ in some atomic state $|1'>$. This eigenstate is coupled by an N-photon transition to some other level $|2'>$, changing the number of photons in the mode by M. Note that M is not necessarily equal to N. A similar averaging over the injection times $t'(j)$ and again assuming that the atomic decay time scales $1/\gamma_1'$, $1/\gamma_2'$ are much faster than $1/\kappa$, we get the following additional terms in the master equation for the p_n :[8]

$$(\dot{p}_n)_{NL} = d(c_n^N p_{n+M} - c_{n-M}^N p_n). \tag{3}$$

Here d is the mean injection rate and the 'transition properties' $c(n)$ are given by

$$c_n := c_o \frac{(n/n_{01})^N}{1 + (n/n_{01})^N}. \tag{4}$$

n_{01} is the saturation photon number associated with the nonlinear transition and c_o a constant.[12] The types of nonlinear transitions we will consider here are schematically depicted in Fig. 2a-c. Fig. 2a shows a standard N-photon absorber with $N = M$. Fig. 2b shows the three photon 'Raman'-type process ($N = 3, M = 1$), where we assume that a direct one-photon transition $|1'> \rightarrow |2'>$ is prohibited by e.g. angular momentum selection rules. Another possibility is a nonlinear feedback process (see Fig. 2c).

The atom is raised by an N-photon transition from the initially prepared state $|1'>$ to some rapidly decaying intermediate state $|i'>$, from which it decays into a lasing level $|3'>$. Subsequently it can thus re-emit one photon into the laser mode by a transition to $|2'>$. We call this a nonlinear feedback mechanism, as via a nonlinear transition the laser populates an upper lasing level and thus contributes to the pump events itself. Processes of this sort have been observed e.g. in semiconductors[13] where electrons in the conduction band are generated by a two-photon transition, which through intraband transitions relax very fast to the conduction band edge, where the laser transition takes place.

Knowing the distribution p_n one can of course easily calculate the mean and the variance of the laser photon number

$$\bar{n} :=< a^\dagger a >= \sum_{n=0}^{\infty} n p_n \tag{5}$$

$$(\Delta n)^2 =< (a^\dagger a - \bar{n})^2 >= \sum_{n=0}^{\infty} (n - \bar{n})^2 p_n \tag{6}$$

and thus obtain the corresponding Mandel Q parameter

$$Q = ((\Delta n)^2 - \bar{n})/\bar{n}. \tag{7}$$

Another important quantity characterizing the laser field statistics is the output field fluctuation spectrum:[15]

$$S(\omega) = 1 + 2\eta\kappa \frac{Re\left\{ \int_o^\infty d\tau e^{i\omega\tau} (G_2(\tau) - \bar{n}^2) \right.}{\bar{n}},$$

with

$$G_2(\tau) :=< a^\dagger(0) a^\dagger(\tau) a(\tau) a(0) > \tag{8}$$

being the normally and time ordered second order intensity correlation function. We have defined $S(\omega)$ in a way that the flat shot noise contribution is normalized to one. η is the quantum efficiency of the detector, which for simplicity we will set to $\eta = 1$ in the following. Note that for a coherent field the integrand in Eq. (8) vanishes and only the shot noise contribution remains. For nonclassical fields the integral in Eq. (8) can become negative and we get less fluctuations than the shot noise level.

Let us now assume a large mean photon number $\bar{n} >> 1$, so that $\bar{p}_n$ can be treated as a continuous function $p(n)$. Using the identity

$$e^{-m\frac{\partial}{\partial n}} p(n) = p(n - m) \tag{9}$$

we can convert the difference equations (1-3) into a Fokker Planck equation

$$\frac{\partial}{\partial t} p(n,t) = (-\frac{\partial}{\partial n} A(n) + \frac{1}{2}\frac{\partial^2}{\partial n^2} B(n)) p(n,t), \tag{10}$$

with drift

$$A(n) := -(\kappa n - R\beta_n + dMc_n^N) \tag{11}$$

and diffusion

$$B(n) := \kappa n + R\beta_n + dM^2 c_n^N + 2RQ_p\beta_n^2 + 2dQ_a M^2 (c_n^N)^2. \tag{12}$$

The stationary solution of Eq. (10) is given by:

$$p_s(n) = \frac{\mathcal{N}}{B(n)} exp\{-2\int_{n_o}^{n} dn' \frac{A(n')}{B(n')}\} \tag{13}$$

with $\mathcal{N}$ a normalization constant. Expanding the exponent in Eq. (13) up to second order in $\delta n = (n- << n >>)$, we find the following expression for the stationary photon number distribution:

$$p^s(n) := \frac{\mathcal{N}}{B(\bar{n})} exp\{-\frac{(n-\bar{n})^2}{B(\bar{n})/a(\bar{n})}\}, \tag{14}$$

with

$$a(n) := \frac{d}{dn} A(n).$$

We then can easily read off the Mandel Q parameter

$$Q = -1 + \frac{\Delta n^2}{\bar{n}} = -1 + \frac{B(\bar{n})}{2\bar{n}\ a(\bar{n})} \equiv$$

$$= \frac{d\ Mc_{\bar{n}}^N(1 - 2N + M) + R\beta_n 2Q_p\beta_{\bar{n}}^2 + d\ M^2 c_{\bar{n}}^N 2Q_a(c_{\bar{n}}^N)^2}{2\bar{n}(\kappa - R\frac{d}{dn}\beta_{\bar{n}} + dM\frac{d}{dn}c_{\bar{n}}^N)} - 1. \tag{15}$$

for the laser mode. Equation (15) is our central result, as we now have derived a rather simple analytic expression for the Q parameter of a laser with a very general type of a nonlinear intracavity absorber. The above formula also includes the effects of a regular pumping. We will now discuss this general expression for various special cases:

a) In the usual laser with a saturated laser transition, we have $d = 0$, $Q(p) = Q(a) = 0$ and $\beta(n)$=const., so that $V = 1$ and the Mandel Q-parameter is equal to zero as expected.

b) Adding a nonlinear non-saturated absorber to the usual laser we find

$$Q = -1 + \frac{\kappa\bar{n} + R\beta_{\bar{n}} + dM^2(\bar{n}/n_{sa})^N}{2(\kappa\bar{n} + dMN(\bar{n}/n_{sa})^N)} \overrightarrow{\kappa n << R} \frac{1}{2}\frac{M+1}{N} - 1. \tag{16}$$

For a high quality cavity, where we have $\kappa\bar{n} << R\beta$ Q gets negative for $N > 1$. Thus, quite generally, a nonlinear absorber can lead to a noise reduction below the coherent state level. For a simple N-photon absorption process we have $N = M$ and hence $Q = -\frac{1}{2} + 1/N$. In the case of a two photon absorber we have $N = 2$ and thus find $Q = -\frac{1}{4}$, which agrees with the result obtained by Golubev and Sokolov[17] using a different method. We see that the noise reduction is the better the higher the order of the absorption process. In the limit of high N the best achievable noise reduction amounts to $Q = -\frac{1}{2}$, which is just 50% reduction compared to the lowest possible value of $Q = -1$. For $M < N$, however, we can get beyond this limit, and in principle achieve almost perfect noise reduction, i.e. $Q \to -1$, for $N >> M$. In particular for the 3-photon Raman process as depicted in Fig. 2b we find $Q = -2/3$. The nonlinear feedback type absorber (Fig. 2c) leads to $Q = -1/2$. Physically the improved behaviour of the latter types of absorbers compared to the usual N-photon absorber can be attributed to the fact, that they show the same intensity dependence of the transition probability, but remove less photons from the field leading to a higher mean value of the photon number. Hence we have two advantages at the same time, a higher laser intensity and less fluctuations.

c) As Marte et Zoller[2] have shown, reduction of pump fluctuations in a laser also leads to sub-Poissonian photon number distribution in the laser mode. For a perfectly regular pump (corresponding to $Q(p) = -\frac{1}{2}$) one obtains $Q = -\frac{1}{2}$ for the laser mode, as can be seen from Eq. (15), if we set $Q(p) = -\frac{1}{2}$, $d = 0$ and $\beta(n) = 1$. Combining a regular pump and a nonlinear absorber, we now set the nonlinear absorber injection rate $d \neq 0$ and $Q(p) = -\frac{1}{2}$. Again assuming a good cavity $\kappa\bar{n} >> R$, we find:

$$Q = \frac{1}{2}\frac{M}{N} - 1. \tag{17}$$

Note that in comparison to Eq. (16) the one in the numerator, connected to the pump fluctuations, is now missing. Nevertheless, for the standard nonlinear absorber $M = N$, we find no further improvement beyond the limit for a regular pump, $Q = -\frac{1}{2}$. For $M < N$, however, the noise can be reduced further. For instance the nonlinear feedback type absorber leads to $Q = -3/4$ (compared to $Q = -\frac{1}{2}$ without regular pump) and the three-photon Raman process results in $Q = -5/6$ (compared to $Q = -2/3$). Again, using the higher order processes, Q can in principle acquire values close to the optimum possible value $Q \simeq -1$. Thus also in this case the types of absorbers with $M < N$ are more favorable than a simple nonlinear absorption process.

d) **Output power spectrum:** For the output power spectrum $S(\omega)$, we will again use the linearized form of the Fokker Planck equation (10), which leads to :[15,16]

$$S(\omega) := 1 + 2\kappa Q \frac{a(\bar{n})}{\omega^2 + a(\bar{n})^2} \geq 0, \tag{18}$$

Obviously, for a negative Q we get a Lorentzian dip of width

$$a(\bar{n}) = \kappa + dM(\frac{d}{dn}c_n^N)|_{\bar{n}}$$

below the shot noise level. For a weak nonlinearity the frequency range of the noise reduction is mainly given by the cavity width κ, whereas in the opposite case of a strong nonlinear absorber a much larger frequency band carries reduced fluctuations. Of course this is at the expense of a smaller amount of noise reduction at $\omega = 0$:

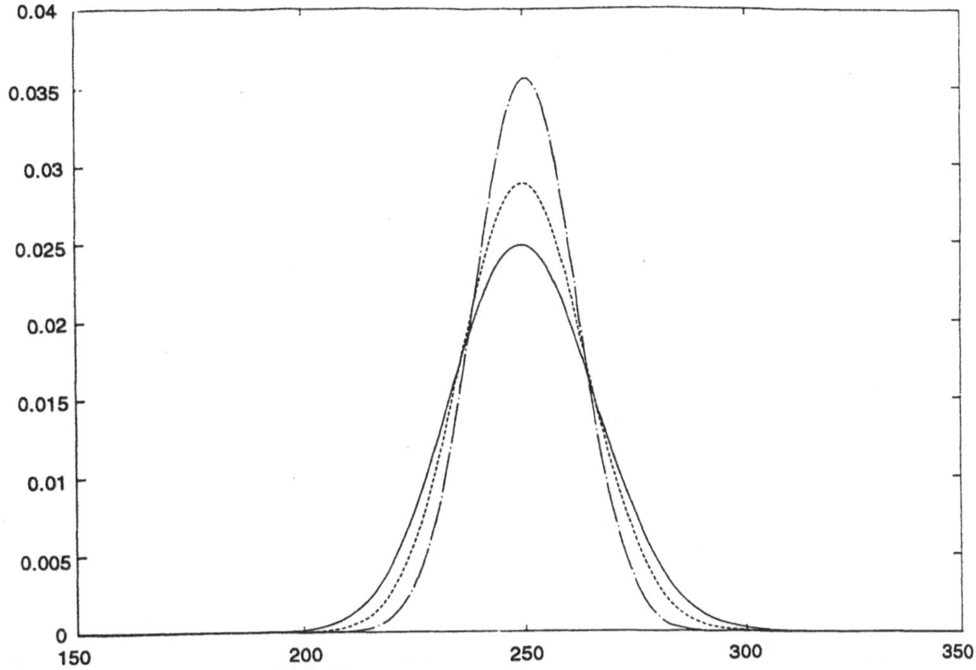

Figure 3. Photon number distribution for a laser with a two photon absorber for fixed mean photon number =250. The different curves correspond to three different values of pump noise strength $Q(p) = 1/2$ (solid line), $Q(p) = 0$ (dashed line) and $Q(p) = -\frac{1}{2}$ (dot-dashed line).

$$S(0) = 1 + \frac{2Q}{(1 + dMNc_o\bar{n}^{N-1}/\kappa)}, \tag{19}$$

where we have assumed a nonlinear absorber in the unsaturated limit. In the sub-Poissonian pump limit $d = 0$, and $Q_p = -\frac{1}{2}$, we have $S(0) = 0$ in agreement with Ref. 2, whereas for the nonlinear feedback case $M = 1$, $N = 2$ we get the same Q, but the dip in the spectrum is much broader and shallower, i.e.

$$S(0) = 1 - \frac{1}{1 + 2d\bar{n}} > 0. \tag{20}$$

e) **Examples**: In the following we will illustrate some of the above results by a series of figures. In Fig. 3 we plot the photon number probability distribution $p(n)$ calculated by a numerical solution of Eq. (1-3) for a mean photon number $n = 250$ and three different values of the pump noise strength $Q(p) = +\frac{1}{2}$ (noisy pump, solid line), $Q(p) = 0$ (standard laser, dashed line) and $Q(p) = -\frac{1}{2}$ (regular pump, dot- dashed line). The dependence of the laser Q parameters on the nonlinear absorption rate is shown in Fig. 4, where we have plotted Q as function of the nonlinear absorption rate for a fixed value of the intensity $n = 250$. The three curves correspond to different values of the pump noise

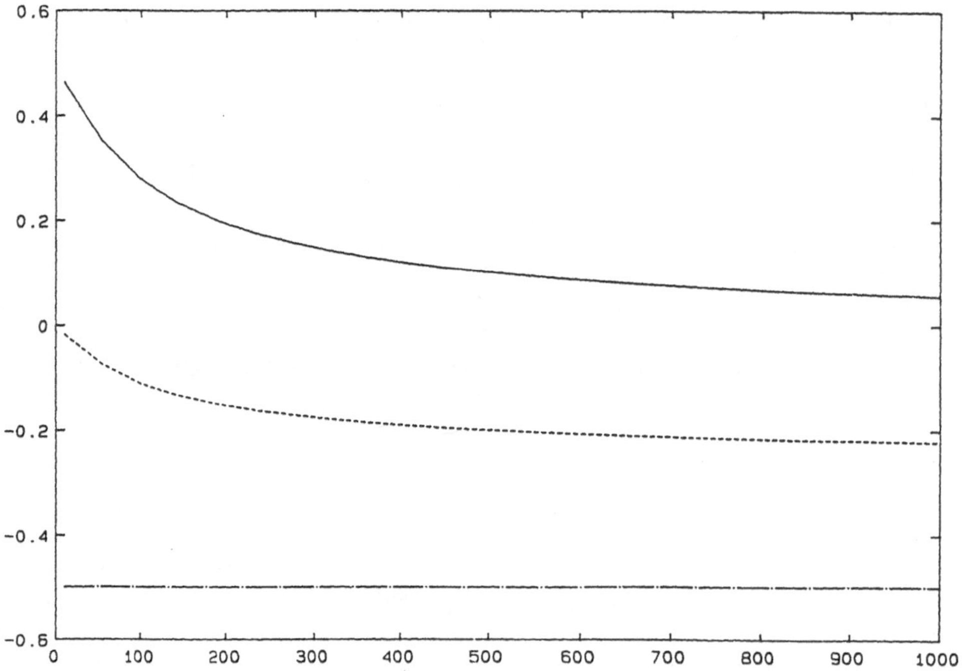

Figure 4. Mandel Q parameter of the laser output as a function of the nonlinear absorption rate with fixed mean output intensity $n = 250$. The three curves correspond to different values of the pump noise strength as in Figure 3.

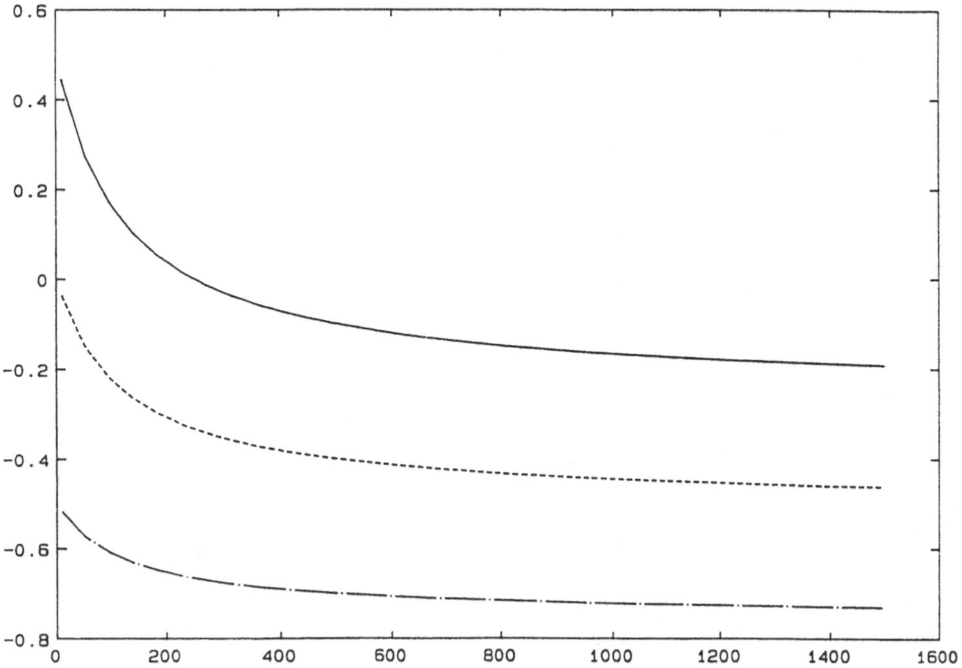

Figure 5. Same as Fig. 4 for a laser with nonlinear feedback.

parameter $Q(p) = +\frac{1}{2}$ (solid line), $Q(p) = 0$ (dashed line) and $Q(p) = -\frac{1}{2}$ (dot-dashed line). Clearly we see that for a regularly pumped laser we find no improvement of the minimum achievable $Q = -\frac{1}{2}$, whereas for the standard laser Q is reduced from $Q = 0$ $(d = 0)$ to $Q = -\frac{1}{4}$ (large d).

For a nonlinear feedback mechanism as depicted in Fig. 2c, the intensity dependence of the absorption rate is the same as for the two-photon absorber treated above, but we have $M = 1$. In Fig. 5 we plot the Mandel Q parameter for $n = 250$ and three different values of the pump noise strength $Q(p) = \frac{1}{2}$ (solid line), $Q(p) = 0$ (dashed line) and $Q(p) = -\frac{1}{2}$ (dot-dashed line). In this case even for a perfectly regular pump we find a further reduction of the laser Q from $Q = -\frac{1}{2}$ to $Q = -3/4$ as a function of absorber injection rate d. This is in agreement with the analytical result given by Eqs. (15,16).

In Fig. 6, we plot the output fluctuation power spectrum $S(\omega)$ for the case of a perfectly regular pump $Q(p)$, with (dashed line) and without (solid line) the nonlinear absorber. In the first case we have perfect noise reduction $S(0) = 0$ for zero frequency $\omega = 0$, whereas in the second case the nonlinear absorber leads to a much broader dip in the fluctuation spectrum, but we have less noise reduction for $\omega = 0$. Hence we see that the negative Mandel Q in the two cases is related to two different kinds of noise reduction. A nonlinear absorber mainly diminishes the fast fluctuations of the laser, i.e. we have antibunching of

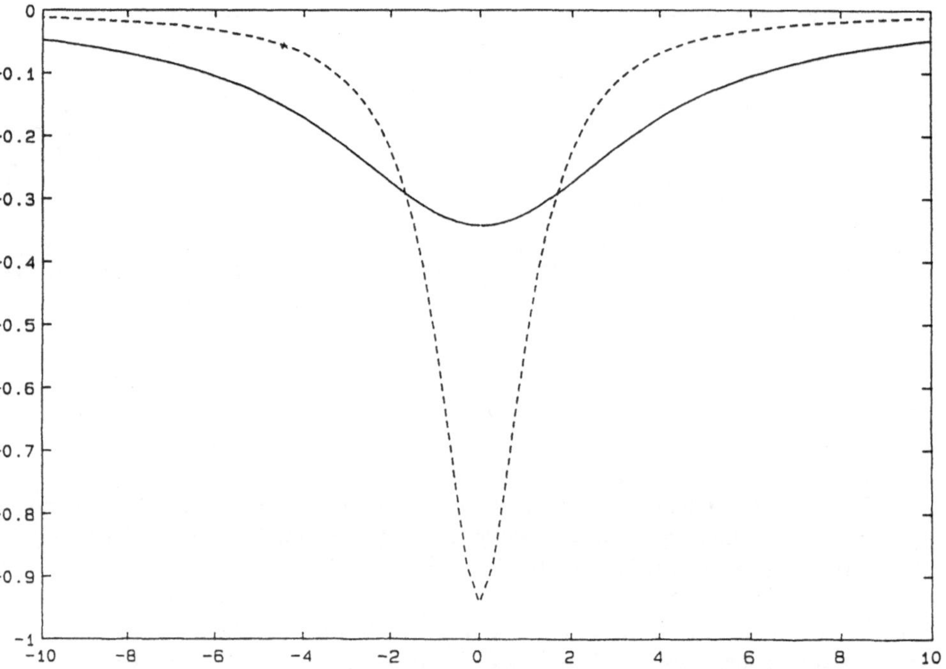

Figure 6. Output intensity fluctuation spectrum for a laser with a regular pump (- - - line) and a laser with a nonlinear feedback type absorber (— line).

the photons at short times, whereas it introduces additional fluctuations in the long term stability of the laser intensity. By a regular pump we can efficiently stabilize the amplitude of the laser on a long time scale, whereas fast fluctuations are hardly quenched.

LASER WITH DYNAMIC PUMP NOISE REDUCTION

Similar to the previous section, we assume that the laser active medium consists of a fixed large number N of m-level atoms, with one pair of levels ($|1> - |2>$) resonantly coupled to a single lasing mode. However, in contrast to before, we now explicitly include the repumping of the atom from the lower lasing level of the electron into the upper lasing level via the (m-2) remaining levels. The corresponding series of transitions from $|j> \rightarrow |i>$ are described by transition rates w_{ij}. As an important difference to the previous section, we now cannot assume a short atom field interaction time anymore, as the atom can undergo my cycles of emitting a laser photon and being re-excited to the upper lasing level. For simplicity we still treat the whole problem in the good cavity limit. This allows us again to eliminate the atoms adiabatically. The corresponding Langevin equation for the electric field amplitude is[8,16]

$$\dot{\alpha} = -\frac{1}{2}\gamma\alpha[1 - \frac{Cd_o}{1 + \alpha\alpha^+/n_s}] + \zeta_\alpha(t) \tag{21}$$

A similar equation holds for α^+ obtained by the replacement $\alpha \leftrightarrow \alpha^+$. In (21) the first term is the cavity damping while the second term describes gain. γ is the damping constant, $C = 2\mu^2 N/\Gamma_{12}\gamma$ the cooperativity parameters, with μ the dipole moment matrix element of the lasing transition and Γ_{12} the transverse damping rate. Furthermore d_o denotes the zero cavity-field population inversion and n_s the saturation photon number. $\zeta_\alpha(t)$ is a Langevin force with diffusion coefficients

$$D_{\alpha\alpha^+} = N\mu^2 \int_o^\infty d\tau(< \sigma^+, \sigma^-(\tau) > + < \sigma^+(\tau), \sigma^- >), \tag{22}$$

$$D_{\alpha\alpha} = N\mu^2 \int_o^\infty d\tau < \sigma^-(\tau), \sigma^- >, D_{\alpha^+\alpha^+} = N\mu^2 \int_o^\infty d\tau < \sigma^+, \sigma^+(\tau) > . \tag{23}$$

Here $\sigma^\pm$ refers to the atomic raising (lowering) operator on the laser transition. The atomic correlation functions in Eq. (22) have to be evaluated with the help of the Quantum Regression Theorem from the Optical Bloch Equations for the multilevel system, where the interaction of the atom with the cavity mode is represented by c-number amplitudes α, α^+.

A simple, but rather realistic, example for a laser atom is a four level system ($m = 4$) as depicted in Fig. 7. We have solved Eqn. (21) for this case and derived the following analytical expression for the Mandel Q-Parameter defined by $(< \hat{n}\hat{n} > - < \hat{n} >^2)/ < \hat{n} > = 1 + Q$ (with $\hat{n}$ the photon number operator of the cavity mode):

$$Q = \frac{-w_{23}w_{34}w_{41}(2w_{23} + 2w_{34} + w_{41} + 2w_{43})}{(2w_{23}w_{34} + w_{23}w_{41} + w_{34}w_{41} + w_{41}w_{43})^2}. \tag{24}$$

Fig. 8 shows Q as a function of the pump rate w_{34} for the laser operated well above threshold $Cd_o << 1$ and various combinations of the atomic decay rates. Note that

Q is negative, exhibiting nonclassical sub-Poissonian statistics and shows a minimum for some intermediate value of the pump rate w_{34}. Neglecting spontaneous emission, a short calculation yields an optimum Q value of $Q = -2/7$ in the case of $w_{42} = (4/3)w_{34}$ and $w_{13} = w_{34}$. Including the spontaneous emission rate w_{21} on the lasing transition to lowest order in w_{21}/w_{42} we find:

$$Q = -\frac{2}{7} + \frac{5}{7}\frac{w_{21}}{w_{42}}, \tag{25}$$

which for a typical laser, where we have a fast decay out of the lower lasing level, is very small.

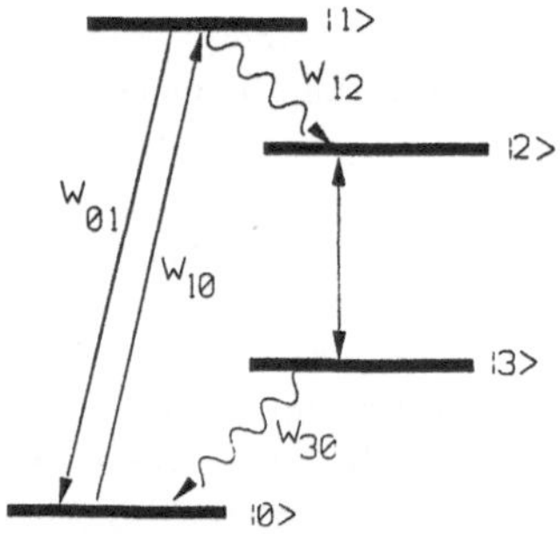

Figure 7. Four level laser scheme. $|2> -|1>$ and $|4> -|3>$ are the laser and pump transition, respectively.

The physical origin of this sub-Poissonian behaviour is closely related to the well known antibunching in the atomic resonance fluorescence. Once one of the laser active atoms has emitted a photon, its wavefunction is projected into the lower of the two lasing levels. Hence it cannot reemit a second photon until it is re-excited to the upper lasing level. This delay between two successive emissions introduces a regularity in the pump process, which leads to the negative Q as calculated above. By increasing the number of intermediate steps in this recycling process this noise reduction can be enhanced even further. To understand the dependence of the Mandel Q parameter on the number of atomic levels involved in this recycling process of the active electron, we consider the m-level scheme shown in Fig. 8.

Again $|2> \rightarrow |1>$ is the laser transition and electrons are recycled back from $|1>$ to the upper state $|2>$ via the levels $|m>$, $|m-1>$, ... $|3>$. Each of these steps is modeled by a unidirectional incoherent rate process. To identify the mechanism

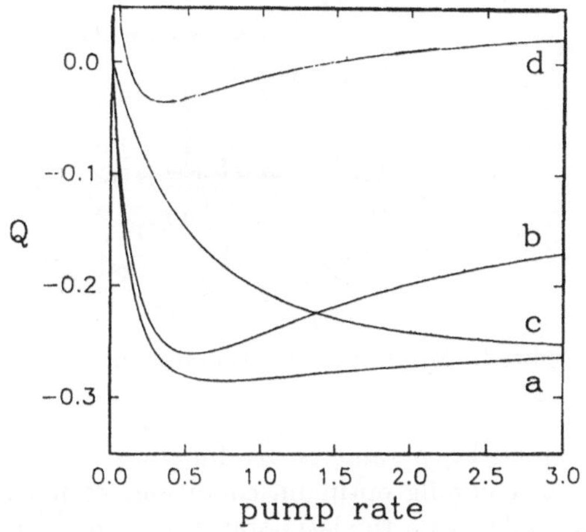

Figure 8. Mandel Q parameter as function of the pump rate $w_{34} \equiv B$. The parameters for the various curves are $w_{23} = 3/2$, $A = w_{12} = 0$ (solid line), $w_{23} = 10$, $A = w_{12} = 0$ (dashed line), $w_{23} = 3/2$, $A = 5$ (dot-dashed line), $w_{23} = 3/2$, $A = 0$, $w_{12} = 1/4$ (dotted line). All rates are in units of w_{14} (compare Fig. 7). A and B are the Einstein coefficients on the pump transition.

of noise suppression we note that for a single step $|j> \rightarrow |j-1>$ with rate r the conditional probability for the electron jumping to $|i>$ in the time interval $[t, t+dt]$, provided it was prepared in $|j>$ at time $t=0$, is given by an exponential decay law $\tilde{c}(t) = re^{-rt}$. On the other hand $m-1$ consecutive rate steps lead to

$$\tilde{c}(t) = r(rt)^{m-1}e^{-rt}/m! \tag{26}$$

Thus the recycling of the electron through these steps leads to anticorrelation in the sense that $\tilde{c}(t=0) = 0$ for $m > 2$. For a fixed mean jump time $\tilde{c}(t)$ approaches a delta function when m tends to infinity. In this limit there are no more fluctuations and the stochastic process becomes deterministic.

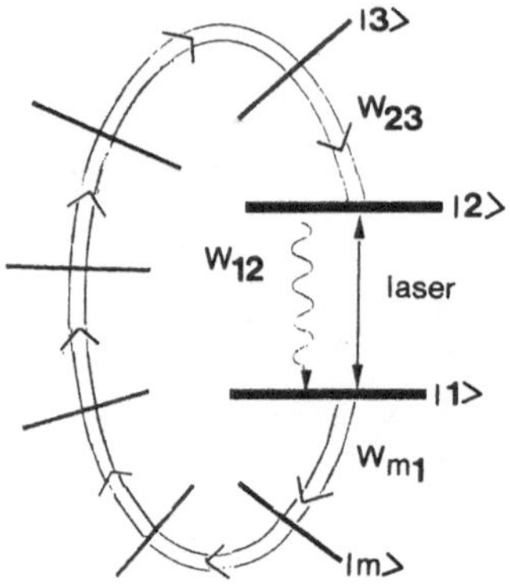

Figure 9. Symbolic laser scheme with m levels.

Fig. 9 shows the Mandel Q parameter as a function of the cooperativity parameter C for $m = 3, 4, ..., 10$ assuming optimum conditions of matched rates $w_{23} = ... = w_{m-1m} = \frac{1}{2}w_{m1}$ and no decay on the laser transition, $w_{12} = 0$. We see that for large C and increasing number of atomic levels Q decreases. We have been able to show that for $C >> 1$ the Mandel Q parameter is given by

$$Q = \frac{1}{2}\frac{m-2}{m-1} \quad (c >> 1) \tag{27}$$

which predicts $Q = 0$ for the two-level system, $Q = -1/4$ and $Q = -1/3$ for the three and four-level system, respectively, and $Q \rightarrow -1/2$ for $m >> 1$ which corresponds to the ideal limit of complete noise suppression. In this limit the results are identical to the case of a laser pumped by a regular injection of excited atoms. However, in our case no external regularity is assumed from the outside, but it is internally generated by the nonlinear dynamics of the laser atoms themselves.

ACKNOWLEDGEMENTS

The authors thank M. Marte for many helpful discussions. This work was supported by the Fonds zur Foerderung der wissenschaftlichen Forschung under grant Nr. P7295.

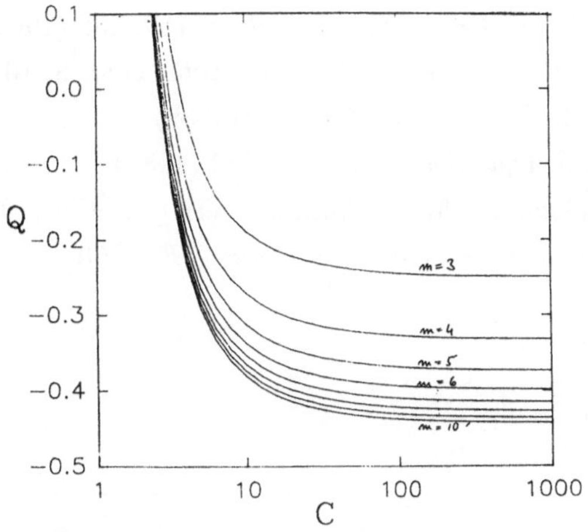

Figure 10. Mandel Q parameters as a function of the cooperativity parameter C for $m = 3, 4, ..., 10$ levels.

REFERENCES

1. Y. Yamamoto, S. Machida and O. Nilson, *Phys. Rev.* A34, 4025 (1986).
2. M. A. M. Marte and P. Zoller, *Phys. Rev.* A40, 5774 (1989).
3. M. A. M. Marte and P. Zoller, *Quantum Opt.* 2, 229 (1990).
4. C. Benkert, M. O. Scully, J. Bergou, L. Davidovich, M. Hillary and M. Orszag, *Phys. Rev.* A41, 2756 (1990).
5. F. Haake, S. M. Tan and D. F. Walls, *Phys. Rev.* A40, 7121 (1989); ibid. 41, 2808 (1990).
6. H. Ritsch, P. Zoller and C. W. Gardiner, *Phys. Rev. A*, submitted.
7. T. A. B. Kennedy and D. F. Walls, *Phys. Rev.* A40, 6366 (1989).
8. H. Ritsch, *Quantum Opt. 1*, 189 (1990).
9. H. Ritsch, P. Zoller, C. W. Gardiner and D. F. Walls, *Phys. Rev. A*, submitted.
10. M. J. Collett, A. S. Lane and D. F. Walls, *Phys. Rev. A*, to be published.
11. Gorbachev, *Opt. Comm.*, to be published (1990).
12. M. O. Scully and W. E. Lamb Jr., *Phys. Rev.* 159, 208 (1967).
13. K. J. McNeil and D. F. Walls, *J. Phys. A: Math, Gen.* 8, 104 (1975).
14. K. W. Delong et al. *J. Opt. Soc. Am.* B6, (1989).
15. H. J. Carmichael, *J. Opt. Soc. Am.* B4, 1588 (1987).
16. C. W. Gardiner, *Quantum Noise*, Springer Verlag, Berlin (1991).
17. Y. M. Golubev and I. V. Sokolov, *Sov. Phys. JETP* 60, 234 (1984).

DYNAMICS OF A MULTIMODE LASER WITH NONLINEAR, BIREFRINGENT INTRACAVITY ELEMENTS

Rajarshi Roy, C. Bracikowski and G.E. James

School of Physics
Georgia Institute of Technology
Atlanta, GA 30332

I. INTRODUCTION

Many laser systems have been developed that contain intracavity nonlinear crystals for harmonic generation. Often, these crystals are birefringent in nature. If the laser cavity does not contain any polarization or mode selecting elements such as Brewster windows and Fabry-Perot etalons, such a laser may operate in many modes with orthogonal polarizations. A diode laser pumped Nd:YAG laser with a birefringent intracavity KTP crystal was studied by Baer [1], who found the output of the laser exhibited large scale amplitude fluctuations when the intracavity doubling crystal was present. He developed a rate equation model for the laser, including the effect of the crystal nonlinearity, but neglecting its birefringence, and showed that the model predicted unstable behavior for multimode operation.

As first pointed out by Oka and Kubota [2], a complete analysis of this laser system must include the polarizations of the axial cavity modes. These polarizations are given by the eigenvectors of the round trip Jones matrix M for this laser cavity. This matrix describes how a given polarization of light is affected after one round trip in the laser cavity which contains birefringent elements. The round trip matrix M is the product of the individual Jones matrices that describe how each element in the cavity affects the polarization of light [3]. Here, we will present an analysis of a resonator with birefringent elements (section II).

In a series of papers [4-8], we have shown that the instability observed in this laser system is chaotic in nature, demonstrated a technique for the elimination of this chaos to obtain stable steady state operation, and also observed the existence of "antiphase" states in this laser system. In this paper, we provide an analysis of the dynamics of a multimode laser with nonlinear, birefringent intracavity elements (sections III and IV). The laser cavity is taken to be high Q at the fundamental wavelength (1.06 μ) and low Q at the frequency doubled wavelength (0.532 μ). Two birefringent elements in the cavity are considered; the highly birefringent KTP crystal, and the slightly birefringent YAG crystal. Normally, YAG is an isotropic material, but it can display stress and thermally induced birefringence (due to mechanical forces and heating from pump laser absorption), and we have found this to be an important element of the model that was

Recent Developments in Quantum Optics, Edited
by R. Inguva, Plenum Press, New York, 1993

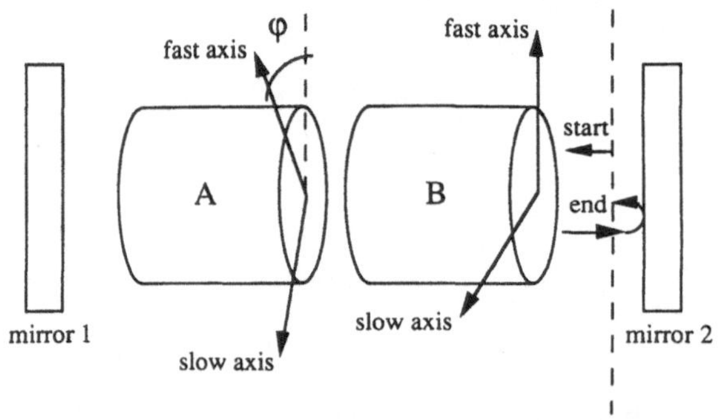

Figure 1. Schematic of a laser cavity with two birefringent intracavity elements. The dashed line marks the starting and end point of the round trip Jones matrix calculation.

not included in earlier work [1,2]. We finally focus on two physical aspects of the laser system behavior; relaxation oscillations (section V) and polarization dynamics (section VI). The laser dynamics is modelled accurately by the coupled differential equations of section IV.

II. LASER CAVITY WITH BIREFRINGENT ELEMENTS

Consider the laser cavity with two birefringent elements shown in Fig. 1. A detailed description of the cavity elements has been given in ref [5,8]. Let ξ and δ be the phase delays (phase delay $= 2\pi L(n_e - n_O)/\lambda$) of the birefringent elements A (the YAG crystal) and B (the KTP crystal) respectively, and φ be the angle between the fast axes of the two elements. The cavity high reflector, mirror 1, is coated directly onto the YAG crystal and mirror 2 is the output coupler. Consider a round trip in this cavity starting just prior to entering element B (position labelled "start") and ending just after a reflection from mirror 2 (position labelled "end"). The round trip Jones matrix M is then given by

$$M = I\ C(\delta)R(\varphi)C(\xi)I\ C(\xi)R(-\varphi)C(\delta) \tag{1}$$

$$M = \begin{bmatrix} e^{i\delta/2} & 0 \\ 0 & e^{-i\delta/2} \end{bmatrix} \begin{bmatrix} \cos\varphi & -\sin\varphi \\ \sin\varphi & \cos\varphi \end{bmatrix} \begin{bmatrix} e^{i\xi/2} & 0 \\ 0 & e^{-i\xi/2} \end{bmatrix} \begin{bmatrix} e^{i\xi/2} & 0 \\ 0 & e^{-i\xi/2} \end{bmatrix} \begin{bmatrix} \cos\varphi & \sin\varphi \\ -\sin\varphi & \cos\varphi \end{bmatrix} \begin{bmatrix} e^{i\delta/2} & 0 \\ 0 & e^{-i\delta/2} \end{bmatrix} \tag{2}$$

and after all the matrices are multiplied together,

$$M = \begin{bmatrix} a & iy \\ iy & a^* \end{bmatrix} \tag{3}$$

where

$$a = e^{i(\delta+\xi)}\cos^2(\varphi) + e^{i(\delta-\xi)}\sin^2(\varphi) \tag{4}$$

$$y = \sin(2\varphi)\sin(\xi) \tag{5}$$

and

$$C(\delta) = \begin{bmatrix} e^{i\delta/2} & 0 \\ 0 & e^{-i\delta/2} \end{bmatrix} \tag{6}$$

$$R(\varphi) = \begin{bmatrix} \cos(\varphi) & -\sin(\varphi) \\ \sin(\varphi) & \cos(\varphi) \end{bmatrix} \tag{7}$$

$$I = \begin{bmatrix} 1 & 0 \\ 0 & 1 \end{bmatrix} \tag{8}$$

The matrix $C(\delta)$ (equation (6)) describes passage through a birefringent element which introduces a relative phase delay δ between the two orthogonal field components. The vector basis for any Jones matrix is the fast and slow axes of the particular birefringent element the matrix describes. Since the fast and slow axes of elements A and B are not parallel, the rotation matrix $R(\varphi)$ (equation (7)) is needed to account for the angle φ between the fast axes of the two birefringent elements. The matrix $R(\varphi)$ rotates the basis set. In this calculation it is assumed that the mirrors are lossless such that reflection from them can be described by the identity matrix I (equation (8)).

The eigenvalues and corresponding unnormalized eigenvectors of M for $y \neq 0$ are given by

$$\lambda_{\pm} = \left\{Re(a) \pm i\sqrt{1-[Re(a)]^2}\right\} \tag{9}$$

$$w_{\pm} = \begin{bmatrix} Im(a) \pm \sqrt{1-[Re(a)]^2} \\ y \end{bmatrix} \tag{10}$$

Note that this form of the eigenvalues was chosen because $\{1-[Re(a)]^2\} \geq 0$ since

$$Re(a) = \cos(\delta+\xi)\cos^2(\varphi) + \cos(\delta-\xi)\sin^2(\varphi) \tag{11}$$

and, incidentally,

$$Im(a) = \sin(\delta+\xi)\cos^2(\varphi) + \sin(\delta-\xi)\sin^2(\varphi) \tag{12}$$

The eigenvalues and corresponding normalized eigenvectors of M for $y = 0$ are given by

$$\lambda_{+} = a \tag{13}$$

$$w_{+} = \begin{bmatrix} 1 \\ 0 \end{bmatrix} \tag{14}$$

$$\lambda_- = a^* \tag{15}$$

$$w_- = \begin{bmatrix} 0 \\ 1 \end{bmatrix} \tag{16}$$

The round trip Jones matrix M (equation (3)) is unitary, i.e. $\det(M) = 1$, since all the matrices that describe the individual cavity elements are unitary. This implies $aa^* + y^2 = 1$, the eigenvalues have unit magnitude ($|\lambda|^2 = 1$) and the eigenvectors are orthogonal. These properties are easily confirmed. The eigenvalues have unit magnitude as a result of the assumption that the cavity mirrors were lossless, and since no other loss or gain elements were considered in this calculation. The two orthogonal eigenvectors of the round trip matrix M define the two polarizations of light that will lase in the cavity because they define the only two polarization directions that are unchanged after one round trip in the cavity. It can be seen that the components of the eigenvectors of the matrix M given above are real. That is, the laser light in the cavity will at most consist of two linearly polarized components along the two orthogonal eigenvector directions.

In the calculation of M, it was assumed that both cavity mirrors were lossless. If mirror 2 is assumed to be the output coupling mirror which reflects equal fractions r of the two orthogonal components of the incident field, then the new round trip Jones matrix M' is

$$M' = \begin{bmatrix} r & 0 \\ 0 & r \end{bmatrix} M = rM \tag{17}$$

the new eigenvalues λ'_+ and λ'_- are

$$\lambda'_+ = r\lambda_+ \tag{18}$$

$$\lambda'_- = r\lambda_- \tag{19}$$

where

$$|\lambda'_+|^2 = r^2 \tag{20}$$

$$|\lambda'_-|^2 = r^2 \tag{21}$$

and the eigenvectors are identical to those of the original matrix M. The magnitude of the eigenvalues reflects the loss of the output coupling mirror. If it is assumed that the loss affects the two field vector components equally, then the location of the loss element in the cavity and hence in the cascaded matrix calculation is irrelevant since only a unit matrix is involved. If the loss is polarization dependent, then the position of the loss element(s) would be significant and the matrix calculation would have to be repeated from scratch.

Since the eigenvectors of M and M' are real and orthogonal, the laser light that leaks out of the cavity is linearly polarized along one of two orthogonal directions. Experimental observations confirm that the laser modes are indeed linearly polarized along two orthogonal directions. It is not necessary, however, that the output light have components in both directions. The analysis described above does not allow us to predict the number of modes that will lase in the two polarization directions. Experimentally, different numbers of linearly polarized modes ranging from one to ten have been observed with various combinations of modes in the two orthogonal directions. Note that the gain of the laser medium and the nonlinearity of the KTP crystal have not been incorporated into this geometrical analysis; the final lasing mode configuration is crucially dependent on these considerations

III. EFFECT OF NONLINEARITY

Green light is produced in the KTP crystal both by second harmonic generation from a single cavity mode and by sum frequency generation between pairs of modes. In second harmonic generation two photons from the same cavity mode at the fundamental frequency ω combine to create one photon of green at frequency 2ω. In sum frequency generation one photon from a cavity mode at frequency ω_1 and one photon from a different mode at frequency ω_2 combine to create one photon of green at frequency $\omega_3 = (\omega_1 + \omega_2)$. Now that the lasing polarizations are known, as found above, the amount of green light generated in the KTP crystal can be calculated. It will turn out that the sum frequency generated green light produced by two parallel polarized cavity modes and that produced by two orthogonally polarized modes are not the same.

In the KTP crystal, the doubled intensity is produced with Type II phase matching. For Type II phase matching, one fundamental field polarized along the extraordinary axis of the KTP and another fundamental field along the ordinary axis combine to produce green light polarized along the extraordinary axis [9].

Consider two linearly polarized fields in this laser cavity, i.e. two cavity modes. One field $\underline{E_1}$ is at the fundamental frequency ω_1 and is polarized along the direction $\underline{u}$. The other field $\underline{E_2}$ is at the fundamental frequency ω_2 and is polarized along the direction $\underline{v}$, where the properly normalized vectors $\underline{u}$ and $\underline{v}$ are each along one of the two orthogonal eigenvector directions of M. That is,

$$\underline{E_1}(\omega_1, t) = |E_1(t)| e^{i(\omega_1 t + \vartheta_1)} \begin{bmatrix} u_1 \\ u_2 \end{bmatrix} \tag{22}$$

$$\underline{E_2}(\omega_2, t) = |E_2(t)| e^{i(\omega_2 t + \vartheta_2)} \begin{bmatrix} v_1 \\ v_2 \end{bmatrix} \tag{23}$$

where ϑ_1 and ϑ_2 are arbitrary initial phases. The polarization vectors u and v are relative to the extraordinary and ordinary axes of the KTP crystal since it is the birefringent element closest to the starting and ending position for the Jones matrix calculation. The nonlinear polarization P_d induced in the Type II phase matched KTP crystal by these two fundamental fields can be written as [9]

$$
\begin{aligned}
P_d(\omega_1,\omega_2,t) &= d_{eff}E_o(\omega_1,\omega_2,t)E_e(\omega_1,\omega_2,t) \\
&= d_{eff}[|E_1(t)|u_1e^{i(\omega_1t+\vartheta_1)} + |E_2(t)|v_1e^{i(\omega_2t+\vartheta_2)}] \\
&\text{x} \quad [|E_1(t)|u_2e^{i(\omega_1t+\vartheta_1)} + |E_2(t)|v_2e^{i(\omega_2t+\vartheta_2)}] \\
&= d_{eff}[|E_1(t)|^2u_1u_2e^{i2(\omega_1t+\vartheta_1)}] \\
&+ d_{eff}[|E_2(t)|^2v_1v_2e^{i2(\omega_2t+\vartheta)}] \\
&+ d_{eff}[|E_1(t)||E_2(t)|(u_1v_2+u_2v_1)e^{i((\omega_1+\omega_2)t+(\vartheta_1+\vartheta_2))}]
\end{aligned} \tag{24}
$$

where E_o and E_e are the components of the total field in the KTP crystal along its ordinary and extraordinary axes respectively and d_{eff} is the effective nonlinear coefficient for the KTP. As can be seen from equation (24), the induced polarization P_d has terms that oscillate at frequency $2\omega_1$, $2\omega_2$ and $(\omega_1+\omega_2)$. These oscillations produce the green light by dipole radiation. The parameter d_{eff} describes the efficiency of the conversion of fundamental into green light; d_{eff} describes the degree of polarizability of the media. For KTP, due to the large differences between n_z and n_x or n_y and the small difference between n_x and n_y, an approximate expression for d_{eff} is given by [9-12]

$$
d_{eff} \approx (d_{24}-d_{15})\sin(2\theta)\sin(2\phi) - (d_{15}\sin^2(\phi) + d_{24}\cos^2(\phi))\sin(\theta) \tag{25}
$$

where θ is the angle relative to the z-axis and ϕ is the angle relative to the x-axis in the x-y plane that define the propagation direction, $\theta = 90°$ and $\phi \sim 26°$ for Type II phase matching in KTP, $d_{15} = 6.1$ pm/V, $d_{24} = 7.6$ pm/V and the d_{ij} are elements of the second order polarizability tensor d_{ijk}. The subscripts ijk represent the rectangular coordinates x, y and z respectively. However, because KTP has mm2-orthorhombic crystal symmetry, d_{ijk} can be written as [9,13,14]

$$
d_{ijk} = \begin{bmatrix} 0 & 0 & 0 & 0 & d_{15} & 0 \\ 0 & 0 & 0 & d_{24} & 0 & 0 \\ d_{31} & d_{32} & d_{33} & 0 & 0 & 0 \end{bmatrix} \tag{26}
$$

The doubled intensity I_d is then given by

$$
\begin{aligned}
I_d(t) &= P_d(\omega_1,\omega_1,t)P_d^*(\omega_1,\omega_1,t) \\
&= d_{eff}^2[|E_1|^4(u_1u_2)^2 + |E_2|^4(v_1v_2)^2 + |E_1|^2|E_1|^2(u_1v_2+u_1v_2)^2] \\
&+ 2d_{eff}^2|E_1|^2|E_2|^2u_1u_2v_1v_2\cos\{2[(\omega_1-\omega_2)t + (\vartheta_1-\vartheta_2)]\} \\
&+ d_{eff}^2|E_1|^3|E_2|u_1u_2(u_1v_2+u_2v_1)e^{i[(\omega_1-\omega_2)t+(\vartheta_1-\vartheta_2)]} \\
&+ d_{eff}^2|E_1||E_2|^3v_1v_2(u_1v_2+u_2v_1)e^{-i[(\omega_1-\omega_2)t+(\vartheta_1-\vartheta_2)]}
\end{aligned} \tag{27}
$$

The doubled intensity I_d contains terms which oscillate at the intermode beat frequency $|\omega_1 - \omega_2|$ and at twice this frequency. If it is assumed that changes in the envelope of the electric field, i.e. $|E_1|$ and $|E_2|$, occur on a much slower time scale than that defined by the beat frequency, the explicit time dependence in equation (27) can be eliminated, and an approximate expression for the doubled intensity time averaged over many cycles of the beat frequency can be written as

$$\begin{aligned} I_d(t) &\approx < P_d(\omega_1,\omega_1,t)P_d^*(\omega_1,\omega_1,t) > \\ &= d_{eff}^2[I_1^2(t)u_1^2u_2^2 + I_2^2(t)v_1^2v_2^2 + I_1(t)I_2(t)(u_1v_2 + u_2v_1)^2] \end{aligned} \tag{28}$$

where

$$< P_d(\omega_1,\omega_1,t)P_d^*(\omega_1,\omega_1,t) >= \frac{1}{\tau}\int_0^\tau P_d(\omega_1,\omega_1,t)P_d^*(\omega_1,\omega_1,t)dt \tag{29}$$

The approximation is validated by experimental observations which reveal that the intermode beat frequency is of the order of a few GHz, whereas the intensity fluctuations occur at a few tens of kHz.

From equation (28) it can be seen that only the first term in equation (27) remains after the time average is taken. It should be noted that the initial phases ϑ_1 and ϑ_2 canceled out naturally as the result of a multiplication of oppositely signed exponentials in equation (27) even before the time average was taken. The terms in I_1^2 and I_2^2 represent the creation of green light due to second harmonic generation at $2\omega_1$ and $2\omega_2$ respectively. The I_1I_2 term represents sum frequency generation at $(\omega_1 + \omega_2)$. Both types of green generation processes have arisen naturally from Type II phase matching as is evidenced by the presence of terms in u_1u_2, v_1v_2, u_1v_2 and u_2v_1 which represent all possible combinations of products of the orthogonal vector components. Recall that in Type II phase matching fundamental field components polarized along the extraordinary and ordinary axes combine to create green light polarized along the extraordinary axis. The expression for the doubled intensity (equation (28)) tells nothing of its polarization. An interesting point to note here is the form of the factor $(u_1v_2 + u_2v_1)^2$ such that a term $2u_1v_2u_2v_1$ will occur. This represents the coherent addition of sum frequency generated fields at frequency $(\omega_1 + \omega_2)$. Such coherent addition yields an extra term over what incoherent addition $(u_1v_2)^2 + (u_2v_1)^2$ would produce.

Let

$$g_1 \equiv 4u_1^2u_1^2 \tag{30}$$

$$g_2 \equiv 4v_1^2v_1^2 \tag{31}$$

$$\sigma \equiv 4(u_1v_2 + u_2v_1)^2 \tag{32}$$

Using these definition, equation (30) can be rewritten as

$$I_d(t) = \left(\frac{d_{eff}^2}{4}\right)[g_1I_1^2(t) + g_2I_2^2(t) + \sigma I_1(t)I_2(t)] \tag{33}$$

To put I_d in units of Watts, define the parameter ε as [1,4]

$$\varepsilon \equiv \frac{d_{eff}^2}{4\varepsilon_o c D^2} \tag{34}$$

where D is an appropriate atomic length scale for the doubling crystal. Using equations (34), equation (33) can be rewritten as

$$I_d(t) = \varepsilon[g_1 I_1^2(t) + g_2 I_2^2(t) + \sigma I_1(t) I_2(t)] \tag{35}$$

Since $\boldsymbol{u}$ and $\boldsymbol{v}$ are unit vectors, an angle η can be defined such that $u_1 = \cos\eta$ and $u_2 = \sin\eta$ and

i) if $\boldsymbol{v} \parallel \boldsymbol{u}$ then $v_1 = u_1 = \cos\eta$, and $v_2 = u_2 = \sin\eta$
ii) if $\boldsymbol{v} \perp \boldsymbol{u}$ then $v_1 = \sin\eta$ and $v_2 = -\cos\eta$

This implies that

i) if $\boldsymbol{v} \parallel \boldsymbol{u}$ then $g_1 = g_2 = g$ and $\sigma = 4g$
ii) if $\boldsymbol{v} \perp \boldsymbol{u}$ then $g_1 = g_2 = g$ and $\sigma = 4(1-g)$

and equation (37) can be rewritten as

$$I_d(t) = g\varepsilon I_1^2(t) + g\varepsilon I_2^2(t) + \sigma\varepsilon I_1(t) I_2(t) \tag{36}$$

At first glance it might seem odd that the coefficients of the I_1^2 and I_2^2 terms (the second harmonic terms) are identical even if the two modes are orthogonally polarized. However, this is just a consequence of the orthogonality itself and the fact that Type II phase matching is used. If Type I phase matching was considered instead, the two coefficients of the second harmonic terms would be the same if the modes were parallel polarized and different if the modes were orthogonally polarized.

The parameter g defined above is a purely geometrical factor whose value depends on the phase delays of the YAG and KTP crystals, ξ and δ respectively, and the angle φ between the YAG and KTP fast axes. The value of the parameter g varies between 0 and 1.

This analysis was carried out for only two cavity modes. It may be easily extended to the case of N modes such that

$$I_d(t) = g\varepsilon \sum_{j=1}^{N} I_j^2(t) + 4\varepsilon \sum_{\substack{j,k=1 \\ j>k}}^{N} \mu_{jk} I_j(t) I_k(t) \tag{37}$$

where $\mu_{jk} = g$ if the j-th and k-th modes are parallel polarized, and $\mu_{jk} = (1-g)$ if they are orthogonally polarized.

Since the green light is produced at the expense of the fundamental, the second harmonic generation and the two types of sum frequency generation (whether between parallel or orthogonal pairs of modes) must be included into the laser rate equations for the fundamental intensity as losses. The losses due to sum frequency generation are divided equally between the two participating modes.

IV. DYNAMICAL EQUATIONS FOR THE LASER

The dynamical equations necessary to describe a multimode YAG laser must contain the dynamics of both the population inversion and the mode amplitudes or intensities. We will outline a simple derivation of the equations that have been used by Baer and ourselves [1,4-8], and then incorporate the effect of the birefringent and nonlinear elements to obtain a dynamical description of the laser system. We begin with the well-known coupled equations for the population inversion and mode intensities of a standing wave, multimode laser, given in the early work of Tang, Statz and deMars [15]

$$\frac{d}{dt}n(z,t) = -\frac{(n-\bar{n})}{\tau_f} - \sum_i Dg_i I_i(t)[1-\cos(2m_i\pi z/L)]n(z,t) \tag{38}$$

$$\frac{dI_i}{dt} = -\frac{\alpha_i I_i}{\tau_c} + \frac{Dg_i I_i}{\tau_c}\int_0^L n(z,t)[1-\cos(2m_i\pi z/L)]dz \tag{39}$$

In these equations $n(z,t)$ is the spatially varying population inversion between the upper and lower lasing levels, $\bar{n}$ is the steady state value of $n(z)$ in the absence of stimulated emission and τ_f is the upper level fluorescence decay time. Dg_i is a stimulated emission coefficient appropriate for the i-th mode. I_i is the intensity of the i-th mode, α_i is the associated loss coefficient and τ_c is the cavity round trip time. Our notation is slightly different from that of Tang et al for convenience. Many different schemes have been devised to solve these coupled equations, involving different approximation techniques to eliminate the integral in equation (39). Tang et al [15] approximated $n(z,t)$ by a sinusoidal expansion and obtained a closed set of coupled equations for the expansion coefficients and the mode intensities. Fleck [16] and Svelto [17] have described other approximation methods. Here, we show that equations of the form used by Baer are readily obtained from equations (38) and (39), which involve coupled equations for the gain associated with each mode and its intensity.

Define the gain of the i-th mode

$$G_i = Dg_i\int_0^L n(z,t)[1-\cos(2m_i\pi z/L)]dz \tag{40}$$

We may then obtain the equation for the gain G_i, using equation (38)

$$\frac{d}{dt}G_i = -\frac{G_i}{\tau_f} + \frac{L\bar{n}Dg_i}{\tau_f} - \sum_j Dg_j I_j G_i$$
$$+\sum_j D^2 g_j I_j g_i\int_0^L n(z,t)[1-\cos(2m_i\pi z/L)]\cos(2m_j\pi z/L)dz \tag{41}$$

We now define a mode-coupling coefficient

$$K_{ij} = \frac{\int_0^L n(z,t)[1-\cos(2m_i\pi z/L)]\cos(2m_j\pi z/L)dz}{\int_0^L n(z,t)[1-\cos(2m_i\pi z/L)]dz}. \tag{42}$$

This coefficient is truly a constant if $n(z,t)$ factors into separate time and space dependent components; this is probably a good assumption for a standing wave cavity. Equation (41) may then be rewritten

$$\tau_f \frac{d}{dt} G_i = L\bar{n}Dg_i - (1 + Dg_i I_i \tau_f (1 - K_{ii}) + \sum_{j \neq i} Dg_j I_j \tau_f (1 - K_{ij})) G_i \tag{43}$$

Introducing the notation

$$\gamma_i = L\bar{n}Dg_i \tag{44a}$$

$$\beta_i = Dg_i \tau_f (1 - K_{ii}) \tag{44b}$$

$$\beta_{ij} = Dg_j \tau_f (1 - K_{ij}) \tag{44c}$$

we can rewrite equation (43) as

$$\tau_f \frac{d}{dt} G_i = \gamma_i - (1 + \beta_i I_i + \sum_{j \neq i} I_j \beta_{ij}) G_i \tag{45}$$

while equation (41) becomes

$$\tau_c \frac{dI_i}{dt} = (G_i - \alpha_i) I_i \tag{46}$$

The losses due to the doubling crystal may now be included in the above intensity equations. As has been shown above, each cavity mode can exist in one of two orthogonal eigenpolarization directions, which can be labelled x and y. Let m and n be the number of modes polarized in the x- and y- directions respectively where $N = m + n$ is the total number of lasing modes. The rate equations for the fundamental intensities I_k and gains G_k are then [5-8]

$$\tau_c \frac{dI_k}{dt} = \Big(G_k - \alpha_k - g\varepsilon I_k - 2\varepsilon \sum_{j \neq k} \mu_{jk} I_j\Big) I_k \tag{47}$$

$$\tau_f \frac{dG_k}{dt} = \gamma_k - \Big(1 + \beta_k I_k + \sum_{j \neq k} \beta_{kj} I_j\Big) G_k \tag{48}$$

where τ_c and τ_f are the cavity round trip time and the fluorescence lifetime of the Nd^{3+} ion respectively; I_k and G_k are respectively the intensity and gain associated with the k-th longitudinal mode; α_k is the cavity loss parameter for the k-th mode, γ_k is the small signal gain which is related to the pump rate, β_k is the self saturation parameter, β_{jk} is the cross saturation parameter and g is the geometrical factor whose value depends on the angle φ between the YAG and KTP fast axes, as well as on the phase delays ξ and δ due to their birefringence. For modes having the same polarization as the k-th mode, $\mu_{jk} = g$, while $\mu_{jk} = (1 - g)$ for modes having the orthogonal polarization. The parameter μ_{jk} represents the different amounts of sum frequency generated green light produced by pairs of parallel polarized modes or by pairs of orthogonally polarized modes. Here, ε is a nonlinear coefficient whose value depends on the crystal properties of the KTP and describes the conversion efficiency of the fundamental intensity into doubled intensity. The terms $g\varepsilon$ and $\mu_{jk}\varepsilon$ in equation (47) result in different effective nonlinear conversion efficiencies for the various harmonic generating processes. It is interesting to note that since $0 \leq g \leq 1$ and $\mu_{jk} \leq 1$ these effective conversion efficiencies are always less than or equal to ε; harmonic conversion cannot be enhanced by changing the value of g.

In equations (47) and (48) the I_k are actually intracavity powers in units of Watts which are proportional to the intracavity photon numbers and to the intracavity intensity which is in units of Watts/cm^2. The output power from the laser is I_k times the transmission of the output coupler. Experimentally the laser output is detected by a photodiode which produces a voltage signal proportional to the incident intensity.

In utilizing these rate equations we have often made the simplifying approximation [4-8] that the gains γ_k and cross saturation parameter β_{kj} are the same for all modes, and scaled the self saturation coefficients to unity. The individual mode losses are assumed to differ only slightly, with $\alpha_k \sim 0.01$. The parameter values given above represent typical experimental operating conditions.

The cross saturation parameter β_{jk} is a measure of the competition among the various longitudinal modes for a given population inversion. It reflects the amount of spatial overlap of the two given standing wave cavity modes inside the gain medium and the extent of homogeneous broadening of the laser transition [18,19]. In the cavity configuration used here, since the gain medium is located at one of the cavity mirrors, the amount of spatial overlap among the various lasing modes is large. This occurs because all the mode fields must go to zero at the mirror in order to satisfy the cavity boundary conditions. If the gain medium was located farther away from the mirror, in the middle of the cavity for example, the amount of spatial overlap of the modes and the mode competition would both be reduced. In general the cross saturation parameter β_{jk} will be different for each pair of cavity modes.

V. RELAXATION OSCILLATIONS

A characteristic frequency for exchange of energy between the light and the active medium is specified by the relaxation oscillations which may be observed in the laser output with or without the doubling crystal. These oscillations occur when the laser is perturbed away from stable steady state due to noise, and may be observed with an rf spectrum analyzer. The signal produced by a photodiode served as the input to the spectrum analyzer. The magnitude of the oscillations is less than a fraction of a percent of the average steady state intensity. This magnitude was found to decrease as the cavity length was increased. Detailed calculations of the relaxation oscillation frequency for our laser system show that the doubling crystal has a very small effect (less than a percent) on the observed relaxation oscillation frequency for typical parameter values. We have experimentally observed relaxation oscillation frequencies in the range of 20-150 kHz for pumping up to five times above threshold. The frequency increases as the square root of the pump power, as is well known. In the rest of this thesis the large periodic and chaotic intensity fluctuations that occur with similar characteristic frequencies will be described.

The damping rate κ_{rel} and the angular frequency ω_{rel} of the relaxation oscillation for a single mode laser are given by [19]

$$\kappa_{rel} = \frac{r\rho_f}{2} \tag{49}$$

$$\omega_{rel} = \sqrt{\rho_f \frac{\alpha}{\tau_c}(r-1)} \tag{50}$$

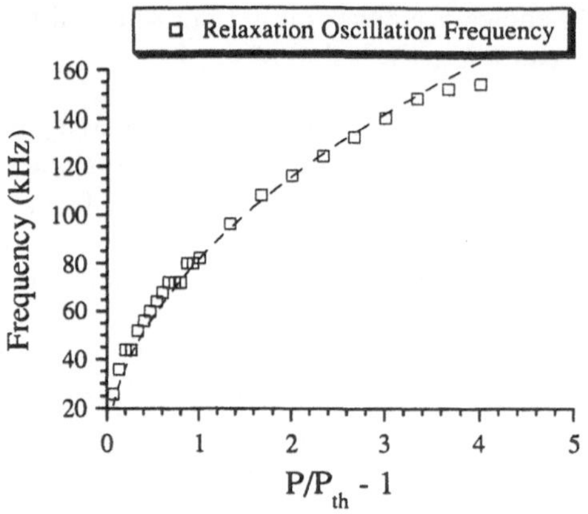

Figure 2. Experimentally measured relaxation oscillation frequency with no KTP crystal in the cavity. The dashed line is the fit to the data using equation (50). The fit was best for a total cavity loss of $\sim$ 1.2%.

Here $\rho_f = 1/\tau_f$, and $r = \gamma/\gamma_{th}$, where γ_{th} is the pump rate to reach laser threshold. For fixed pumping, as the cavity decay rate increases the relaxation oscillation frequency increases. An increase in the cavity decay rate could be produced by shortening the laser cavity or increasing the total loss. Changing the damping κ_{rel} rate for fixed pumping r is more difficult since the fluorescence lifetime of the active medium would have to be altered. Notice that the relaxation oscillation frequency varies as the square root of the pump power.

In Fig. 2 the experimentally measured relaxation oscillation frequency in the Nd:YAG laser (no intracavity KTP crystal) is plotted vs pump power. The output from the laser was filtered to remove any residual pump light and was incident upon a photodiode. The relaxation oscillations were observed directly on an rf spectrum analyzer. The frequency was also measured manually or using the intrinsic FFT function on a digital oscilloscope. The laser cavity length was 2.9 cm and a threshold pump power of $P_{th} = 8$ mW was experimentally measured.

The plot of the relaxation frequency in Fig. 2 shows a square root dependence on pump power as expected. The dashed curve in Fig. 2 is a fit to the experimental data using the equation (50) for the relaxation oscillation frequency with the only adjustable parameter being the total cavity loss. The best fit occurs for a total cavity loss of around 1.2%. It is seen from these results that the relaxation oscillations of the total intensity have the same frequency dependence on pump excitation as a single mode laser.

To discover how the relaxation oscillation frequency is affected by the nonlinearity of an intracavity doubling crystal, a linear stability analysis was carried out on the laser rate equations (47) and (48) derived above, assuming only single mode operation. The appropriate rate equations are

$$\tau_c \frac{dI(t)}{dt} = (G - \alpha - g\varepsilon I)I \tag{51}$$

$$\tau_f \frac{dG(t)}{dt} = \gamma - G(1 + \beta I) \tag{52}$$

The results are as follows:

$$\kappa'_{rel} = \kappa_{rel}\Big[1 + \Big(\frac{g\varepsilon\tau_f}{\beta\tau_c}\Big)\Big(\frac{r-1}{r}\Big)\Big] \tag{53}$$

$$\omega'_{rel} = \omega_{rel}\sqrt{1 + \Big(\frac{g\varepsilon}{\alpha\beta}\Big)\Big(\frac{3}{2}r - 1\Big)} \tag{54}$$

For the typical laser parameter values of $\varepsilon = 5\text{x}10^{-6}\ \text{W}^{-1}$, $g = 0.1$, $\beta = 1\ \text{W}^{-1}$, $r = 5$ (i.e. pumping five times threshold), $\alpha = 0.01$, $\tau_c = 0.29$ nsec, $\tau_f = 230\ \mu$sec

$$\kappa'_{rel} = (1.3)\kappa_{rel} \tag{55}$$

and

$$\omega'_{rel} \approx (1.0003)\omega_{rel} \approx \omega_{rel} \tag{56}$$

For these parameter values the nonlinearity of the doubling crystal has a negligible effect on the relaxation oscillation frequency, but it increases the decay rate appreciably. In general, the presence of the doubling crystal in a single mode laser will only significantly affect the decay rate of the relaxation oscillations and not their frequency.

VI. CHAOTIC POLARIZATION DYNAMICS

After this description of relaxation oscillations, the reader may get the impression that insertion of the doubling crystal into the multimode laser cavity will have no more effect than to damp out the relaxation oscillations faster than in the basic laser system. This conclusion would be quite true if we had a set of uncoupled modes independently obeying equations similar to equations (50) and (51). Equations (47) and (48) for the multimode laser, however, contain the nonlinear mode-coupling terms due to cross-saturation and sum-frequency generation. Interestingly, if one omits the nonlinear terms due to the intracavity doubling crystal and includes the cross-saturation terms, the laser model still leads to stable, steady state operation. It is only when the sum-frequency generation terms are included that the model predicts unstable behavior. Instability can occur irrespective of the mode polarizations, though the conditions for the onset of instability depend on the details of the polarization configuration as has been discussed in ref [5]. It was further shown there that it is possible to eliminate the chaotic fluctuations in the output intensity by proper geometric orientation of the KTP and YAG crystals (adjustment of the value of g). In references [7,8] a detailed studies of the chaotic intensity fluctuations was reported, and it was shown that these fluctuations are approximately Gaussian distributed.

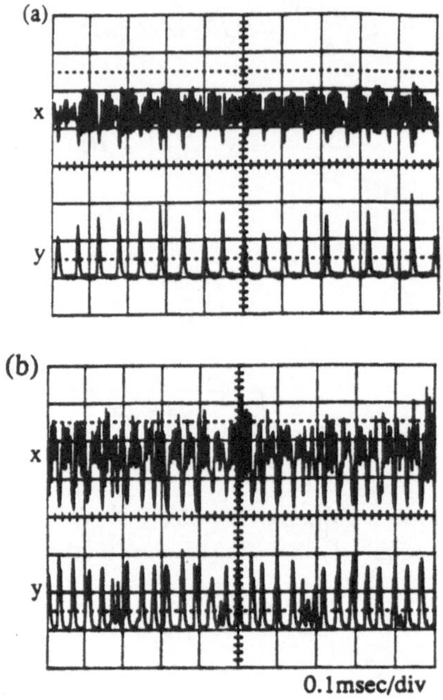

Figure 3. Experimental time traces showing x- and y- polarized intensities measured simultaneously at pump powers of a) 40 mW and b) 90 mW.

It is very interesting to examine the dynamics of the laser output in the two orthogonal polarization directions. The laser was adjusted to produce a chaotic total intensity with only one mode polarized in the y-direction; the orthogonal x-polarization direction had many modes and no effort was made to control their number. The chaotic time trace was obtained by tilting the output coupling mirror and adjusting the relative orientation angle between the YAG and KTP crystals.

The harmonic wavelength was filtered from the laser output. A polarizing prism was placed in the fundamental beam and was adjusted to pass only the one y-polarized mode onto a photodiode. The photodiode signal was observed on a digital oscilloscope. The digitized data values were transferred to the microcomputer where the ON and OFF dwell time probability distributions were calculated. The one y-polarized mode was considered to be ON when its intensity increased past a specified reference level. Similarly, the mode was considered OFF when its intensity decreased below the reference level. The reference level was a set voltage level above ground on the digital oscilloscope. The pump power was varied from 40 mW to 130 mW ($P_{th} \sim 39$ mW), and at each pump power an ON dwell time distribution and an OFF dwell time distribution was accumulated from 500 time traces of length 1 msec each. Each 1 ms interval typically contained more than 15 switching events. This number increased as the pump power increased, i.e., the switching rate increased with pump power. Figures 3 show characteristic time traces of the x- and y- polarized intensities at two sample pump powers. Notice the anticorrelation of the orthogonally polarized intensities. Figures 4 show histograms for the ON and OFF dwell times for the same pump powers as the time traces included in Figs. 3. The mean ON and OFF dwell times are plotted against pump power in Figs. 5. The error bars are smaller than the symbols used in

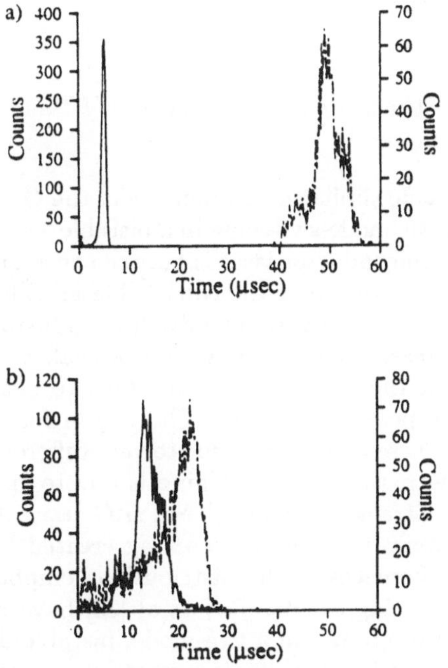

Figure 4. ON (solid lines) and OFF (dashed lines) dwell time histograms for pump powers of a) 40 mW and b) 90 mW.

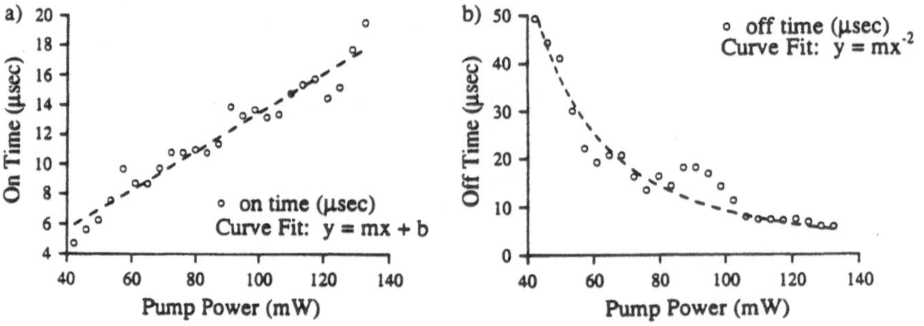

Figure 5. (a) Mean ON dwell time as a function of pump power. (b) Mean OFF dwell time as a function of pump.

Figs. 5. The mean OFF times are seen to vary as $1/P^2$ and the mean ON times vary linearly with the pump power P.

Gray and Roy have made similar measurements of the ON and OFF dwell time distributions for the stochastic mode-switching in a bistable semiconductor laser [20]. The mode-switching in the semiconductor laser is between two longitudinal modes and not between two polarizations as occurs in the Nd:YAG laser. The dwell time distributions for the two lasers look quite different. The dwell time histograms acquired from the semiconductor laser decrease monotonically from a peak at very near $t = 0$ whereas the dwell times found for the Nd:YAG laser are seen to peak at some nonzero value and often show a double peaked structure. The reason for this difference is that the switching that occurs in these two lasers is due to very different phenomena. In the case of the semiconductor laser, the mode-switching is due to spontaneous emission noise [20-22], whereas in the intracavity doubled Nd:YAG laser the polarization-switching is due to the global, deterministic mode coupling created by sum frequency generation. The qualitative differences in the distributions emphasize the different nature of the sources of the switching events. Roy et al [23] have measured the mean dwell times as a function of pump power in a two-mode, bistable dye laser. Here, as for the semiconductor laser, the mode-switching is caused by quantum spontaneous emission noise. The histograms measured for the ON and OFF dwell times in the dye laser are similar to those obtained from the semiconductor laser and can be fit very well by an exponential distribution. The mean dwell times for the dye laser were found to increase exponentially with pump excitation.

The differences in the dwell time measurements for these other laser systems and for the polarization switching described here reflect the difference between stochastic (due to spontaneous emission noise) and deterministic (due to global mode coupling created through sum frequency generation) fluctuations.

The average frequency of the polarization switching in the intracavity doubled Nd:YAG laser is related to the relaxation oscillation frequency in that they both have the same relative dependence on the system parameters, although the two frequencies are not equal. This is to be expected, since the x- and y- polarized intensities are sums of the longitudinal mode intensities for the correspondingly polarized modes.

Table 1. Dependence of the polarization switching frequency on the cavity loss α, pumping r, nonlinear coefficient ε and cross-saturation β.

Parameter Value Increased	Effect on Polarization Switching Frequency
α	increased
r	increased
ε	decreased
β	decreased

From the results of integration of the numerical model it has been observed that as the pumping or the cavity loss increases, or as the nonlinear coefficient ε or the cross saturation parameter β decreases, this polarization switching frequency increases. This is the same parameter dependence exhibited by the relaxation oscillation frequency (equation (50)).

The numerical model was integrated with four modes in the x- polarization direction ($m = 4$) and one mode in the y-polarization direction ($n = 1$) with various values for the loss α, pump excitation r, cross saturation β and nonlinear coefficient ε. The goal of these integrations was to find the right combination of parameters in order to produce a polarization switching frequency of about 40 kHz as was measured experimentally for a similar mode configuration at an excitation level of $r = 5$. In addition, time traces for the total, x- polarized and y-polarized intensities that resembled those measured experimentally were desired.

The frequency of the polarization switching in the intracavity doubled Nd:YAG laser is related to the relaxation oscillation frequency in that they both have the same relative dependence on the system parameters, although the two frequencies are not equal. After a few initial integrations were performed with $\tau_c = 0.29$ nsec and $\tau_f = 230$ μsec, distinct patterns were observed in the dependence of the polarization switching frequency on the parameters α, β, ε and r that are summarized in Table 1.

The results of this integration are shown in Figs. 6. The parameters used in the integration of the numerical model were $m = 4$ (four modes in the x-polarization direction), $n = 1$ (one mode in the y-polarization direction), $\beta = 0.7$ W^{-1}, $r = 5$, $\varepsilon =$ 5x10^{-6} W^{-1}, equal mode losses $\alpha_j = \alpha = 0.016$ and $g = 0.1$. The differential pumping used here was 2% yielding individual mode pump values of 1.06*r, 1.04*r 1.02*r, r(four modes x- polarized) and 1.02*r (one mode y- polarized). In this case the frequency of the spikes in the intensity of the one y-polarized mode, and hence the polarization switching frequency, was about 40 kHz.

VII. CONCLUSION

If one were to just observe the irregular switching between the x- and y- polarized intensities, one may be tempted to conclude that the switching was of a noise induced origin. The shapes of the dwell time histograms and the dependence of the mean dwell times on the pump excitation are, however, quite different from those observed in noise induced mode-hopping phenomena in other lasers. The dwell times do not

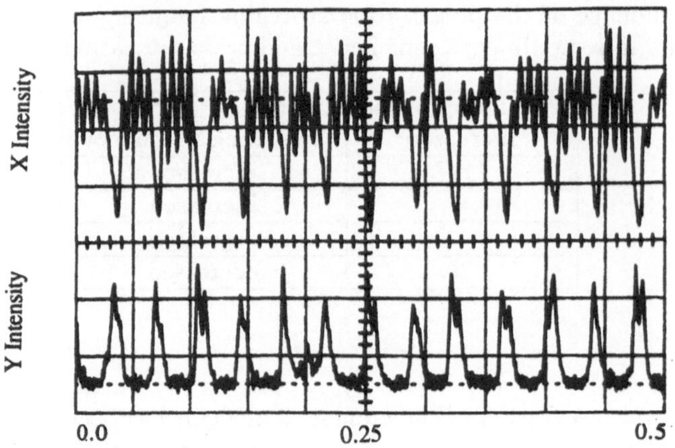

Figure 6(a). Time traces of the x- and y- polarized intensities measured experimentally.

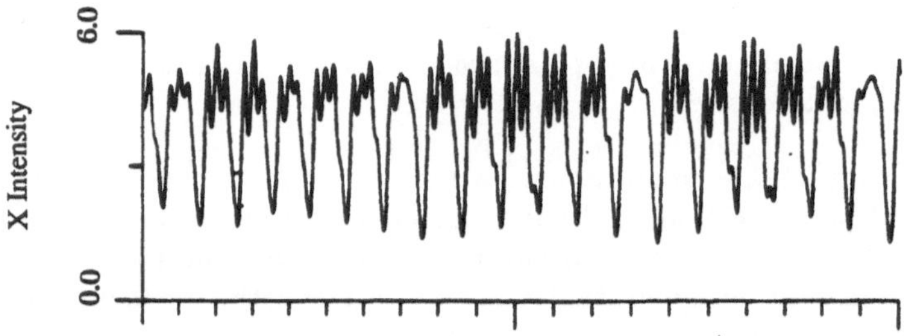

Figure 6(b). Time trace of the x-polarized intensity produced by the numerical model for (m,n)=(4,1).

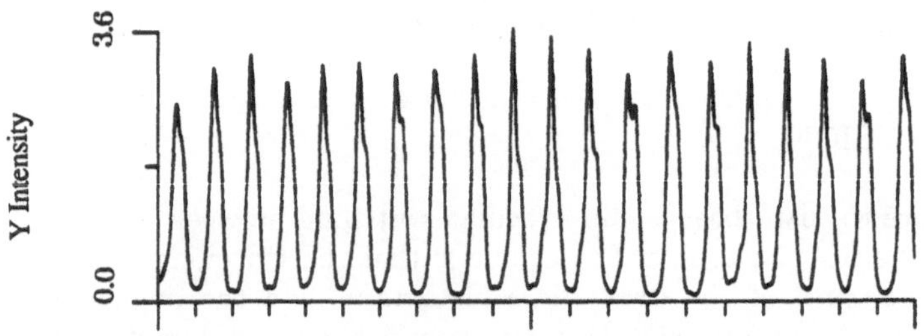

Figure 6(c). Time trace of the y-polarized intensity produced by the numerical model. The total time is 0.5 msec.

increase exponentially with pumping; there is a linear increase of the ON dwell time and quadratic decrease of the OFF dwell time for the y- polarized intensity. The deterministic numerical model presented here is able to account quite well for many aspects of the laser dynamics, but there is much work still to be done to obtain a deep intuitive understanding of the nature of chaotic instabilities and the statistical characteristics of the fluctuations exhibited by chaotic systems.

ACKNOWLEDGEMENTS

This research was supported by NSF grant ECS-8722216 (R.R. and C.B) and by the U.S. Air Force Academy Department of Mathematical Sciences (G.J.). We acknowledge collaborations and many helpful discussions with E. Harrell and K. Wiesenfeld.

REFERENCES

1. T. Baer, *J. Opt. Soc. Am* B**3**, 1175 (1986).
2. M Oka and S. Kubota, *Opt. Lett* **13**, 805 (1988).
3. E. Hecht and A. Zajac. Optics (Addison-Wesley Publishing Company, Reading, Massachusetts, 1979).
4. (a) G. E. James, E. M. Harrell II and R. Roy, *Phys. Rev.* A**41**, 2778 (1990); (b) G. E. james, Ph.D. thesis, Georgia Institute of Technology (1990).
5. G. E. James, E. M. Harrell II, C. Bracikowski, K. Wiesenfeld, and R. Roy, *Opt. Lett.* **15** 1141 (1990).
6. K. Wiesenfeld, C. Bracikowski, G. E. James, and R. Roy, *Phys. Rev. Lett.* **65**, 1749 (1990).
7. C. Bracikowski and R. Roy, *Phys. Rev.* **A43**, 6455 (1991).
8. (a) C. Bracikowski and R. Roy, Chaos **1**, 49 (1991); (b) C. Bracikwoski, Ph.D. thesis, Georgia Institute of Technology (1991).
9. H. Ito, H. Naito and H. Inaba, *J. Appl. Phys.* **46**, 3992 (1975).
10. J. D. Bierlein and H. Vanherzeele, *J. Opt. Soc. Am.* B**6**, 622 (1989).
11. T. Y. Fan, C. E. Huang, B. Q. Hu, R. C. Eckardt, Y. X. Fan, R. L. Byer and R. S. Feigelson, *Appl. Opt.* **26**, 2390 (1987).
12. R. C. Eckardt, H. Masuda, Y. X. Fan and R. L. Byer, *IEEE J. Quantum Electron.* **26**, 922 (1990).
13. J. Q. Yao and T. S. Fahlen, *J. Appl. Phys.* **55**, 65 (1984).
14. D. A. Kleinman, *Phys. Rev.* **126**, 1977 (1962).
15. C. L. Tang, H. Statz and G. deMars, *J. App. Phys.* **34**, 2289 (1963).
16. J. A. Fleck and R. E. Kidder, *J. App. Phys.* **36**, 2327 (1965).
17. O. Svelto, Principles of Lasers, p. 465-470 (Third Edition, Plenum Press, New York, 1989).
18. M. Sargent, M. O. Scully and W. E. Lamb, Jr. Laser Physics (Addison-Wesley, Reading, Massachusetts, 1974).
19. A. Siegman, Lasers (University Science Books, Mill Valley, California, 1986).
20. G. R. Gray and R. Roy, *J. Opt. Soc. Am.* B**8**, 632 (1991).
21. G. Gray and R. Roy, p. 441, Coherence and Quantum Optics VI, J. H. Eberly, L. Mandel and E. Wolf, eds. (Plenum, New York, 1990).

22. M. Ohtsu, Y. Otsuka and Y. Teramachi, *Appl. Phys. Lett* **46**, 108 (1985).
23. R. Roy, R. Short, J. Durnin and L. Mandel, *Phys. Rev. Lett.* **45**, 1486 (1980).

QUANTUM NOISE IN LASERS AND PARAMETRIC OSCILLATORS: INSIDE VERSUS OUTSIDE

Sudhakar Prasad

Department of Physics and Astronomy and
Center for Advanced Studies
University of New Mexico
Albuquerque, NM 87131

1. INTRODUCTION

Resonant cavities have been employed in a large variety of extremely useful optical applications, ranging from interference filters and etalons[1] to ordinary lasers and masers[2] to exotic devices like the single-atom maser[3] and two-photon lasers[4] to devices based on nonlinear interactions like parametric generation[5] that produce squeezed states[6] of light. In many of these applications, the principal role of the cavity is to provide frequency and mode discrimination. Yet such discrimination cannot be perfect since light must necessarily be allowed to exit as the useful output through partially transmitting end mirrors, which also permit the incoherent external vacuum field to leak in and contaminate the cavity field. The functioning of optical devices which generate quantum mechanical states of light with properties like squeezing, photon antibunching, etc. is especially sensitive to such contamination.

Early treatments of mirror transmission involved a reservoir of fictitious oscillators or atoms to which the cavity radiation mode is coupled. Such coupling to an infinite reservoir gives rise to a damping of the cavity mode as well as fluctuations, mimicking mirror transmission. However, Lang, Scully, and Lamb[7] - and later Ujihara[8] - adopted a unified approach in which by modeling the mirror by a dielectric bump and introducing an auxiliary perfect mirror at infinity, they could quantize the field everywhere at once in terms of well-defined modes. In this approach, the same mode annihilation and creation operators describe intracavity, output, and input fields, depending on which traveling piece of the quantum mechanical field operator and where (inside or outside the leaky cavity) it is observed.

Later, Collett and Gardiner[9] employed a simpler approach based on the mirror boundary conditions in which the intracavity and extracavity fields are taken to be independent quantum observables. These authors demonstrated the striking difference in the quantum character of the intracavity and extracavity fields from an optical device like the parametric oscillator, in affirmation of the results of Yurke who found perfect extracavity squeezing without perfect intracavity squeezing. Subsequent work of Carmichael[10] and others[11] strengthened the foundation of this inside-outside connection. However, from a fundamental point of view, this approach is unsatisfactory,

Recent Developments in Quantum Optics, Edited
by R. Inguva, Plenum Press, New York, 1993

for in a leaky cavity the field inside cannot be quantized independently of the field outside.

The principal objectives of this paper are two-fold. The first is to review our work[12,13] on squeezing and quantum noise motivated by the work of Scully, Lang, and Lamb. For a nearly perfect cavity, although this approach yields the same dynamical equations as those of Collett and Gardiner, it does more by providing a simple physical picture of the connection of quantum fields and their noise inside and outside the cavity. We shall consider as illustrative examples quantum noise in ordinary lasers and parametric oscillators.

The second important objective concerns an alternative formulation predicated on the boundary conditions that linearly couple the intracavity field to the externally located output and input fields at the output mirror(s). We shall see that this non-modal approach is the more suitable of the two approaches discussed here when the out-coupling of the cavity is arbitrarily large. Although this approach may appear merely to be an extension of Collett and Gardiner's somewhat adhoc approach, that is not really so, since no attempt is made to quantize the inside and outside fields separately here. It is perfectly consistent with our first approach and therefore entirely rigorous.

This non-modal approach will be illustrated by considering a sub-threshold degenerate parametric oscillator with arbitrary out-coupling, or cavity-Q. We shall see that for high output transmission the quadrature noise reduction, or squeezing, characteristic of parametric sub-harmonic generation becomes very small. On the other hand, in the limit of a perfect cavity, the maximum available squeezing assumes the usual values of 50% and 100% for the intracavity and output fields, respectively. We shall derive expressions for noise reduction that are uniformly valid for all values of out-coupling and therefore also govern the intermediate regime of imperfect cavity below threshold.

2. REVIEW OF THE UNIVERSE MODES[7,12]

We construct a one-dimensional universe by adding an unphysical perfectly reflecting mirror to the far left - at $z = -L$- of the leaky physical cavity at hand, which is assumed to have a single output end with a perfectly reflecting mirror at $z = \ell$ and a partially transmitting output mirror at $z = 0$ (Fig. 1). The universe is then the larger cavity between the perfectly reflecting mirrors at positions $z = -L$ and $z = \ell$. The unphysical mirror at $z = -L$ must be taken away by setting $L = \infty$ before any physically meaningful statements can be made. For fields normally incident from the right, the output mirror is described by real amplitude reflection and transmission coefficients $-\tilde{r}$ and $\tilde{t} = (1 - \tilde{r}^2)^{1/2}$.

Let $U_k(z)$ represent the electric field amplitude at frequency $\Omega_k = ck$. The vanishing of the field at the perfect mirrors dictates the following form for $U_k(z)$:

$$U_k(z) = \begin{cases} \xi_k \sin k(z + L) & \text{for } z < 0 \\ M_k \sin k(z - \ell) & \text{for } z > 0. \end{cases} \tag{2.1}$$

The boundary conditions at the leaky mirror at $z = 0$ imply a discrete set of allowed values of k, separated one from the next by $\Delta k = \pi/L$ in the limit $L >> \ell$, and determine the ratio of M_k and ξ_k via the relations

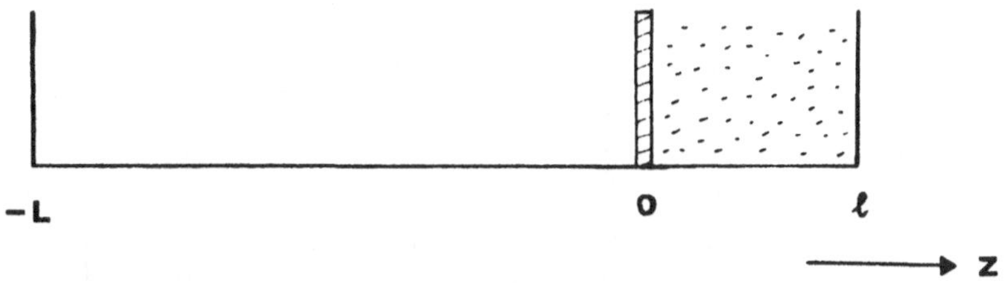

Figure 1. The leaky cavity is bounded by the perfectly reflecting mirror at $z = \ell$ and a partially transmitting mirror at $z = 0$. The space between the mirror at $z = \ell$ and the unphysical perfectly reflecting mirror at $z = -L$ is the "universe" for our purposes.

$$\xi_k \tilde{r} e^{ikL} - M_k \tilde{t} e^{ik\ell} = -\xi_k e^{-ikL}, \quad \xi_k \tilde{t} e^{ikL} + M_k \tilde{r} e^{ik\ell} = M_k e^{-ik\ell}, \tag{2.2a}$$

By letting ξ_k alternate between 1 and -1 as k increases, we may show that

$$M_k = \left(\frac{p}{p^2 \cos^2 k\ell + \sin^2 k\ell}\right)^{1/2}, \quad p = \frac{1-\tilde{r}}{1+\tilde{r}}, \tag{2.2b}$$

which is a periodically peaked function of k. In the good cavity limit, $\tilde{r} \to 1$, the behavior of M_k^2 around each peak may be approximated by a Lorentzian form with half-width at half maximum being $\Gamma = c(1-\tilde{r})/2\ell$, the amplitude decay rate.

Different modes $\{U_k\}$ are orthogonal and therefore rigorously independent. We may therefore quantize the electric field of radiation in terms of these modes of the universe:

$$\begin{aligned} E(z,t) &= E^{(+)}(z,t) + E^{(-)}(z,t) \\ &= \textstyle\sum_k \sqrt{\frac{\hbar\Omega_k}{\epsilon_0 \mathcal{A} L}} a_k(t) U_k(z) + H.C. \end{aligned} \tag{2.3}$$

where a_k are the universe mode annihilation operators defining the positive frequency part, $E^{(+)}(z,t)$, of the field in the Heisenberg picture, $\mathcal{A}$ is the cross sectional area of the cavity, and $H.C.$ stands for the phrase 'Hermitian Conjugate.' We can identify *intracavity* and *extracavity* fields by considering the regions $z > 0$ and $z < 0$, respectively. Inside the leaky cavity of interest, $z > 0$, it is the periodically peaked behavior of M_k's which defines the cavity quasimode structure (Fig. 2), the width of each peak being determined by the reflectivity $\tilde{r}$. Thus, for instance, the positive frequency component of the left-traveling field at $z = 0^+$, which we call the intracavity field, is given by

$$E_{cav}^{(+)} = -\frac{1}{2i} \sum_k \sqrt{\frac{\hbar\Omega_k}{\epsilon_0 \mathcal{A} L}} M_k a_k e^{ik\ell}. \tag{2.4a}$$

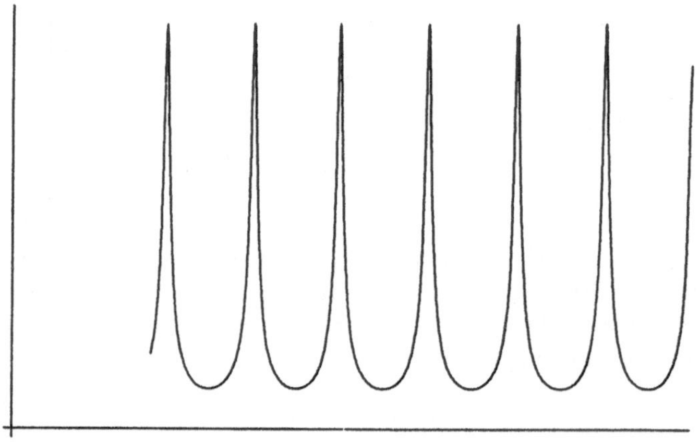

Figure 2. The cavity quasimode structure represented by plotting M_k^2 as a function of k.

Similarly, the positive-frequency parts of the input and output fields at $z = 0^-$ have the following expressions:

$$E_{in}^{(+)} = \frac{1}{2i}\sum_k \sqrt{\frac{\hbar\Omega_k}{\epsilon_0 \mathcal{A}L}}\xi_k a_k e^{ikL}, \tag{2.4b}$$

$$E_{out}^{(+)} = -\frac{1}{2i}\sum_k \sqrt{\frac{\hbar\Omega_k}{\epsilon_0 \mathcal{A}L}}\xi_k a_k e^{-ikL}. \tag{2.4c}$$

The relations between M_k and ξ_k that follow from the boundary conditions at $z = 0$ enable one to write[12] $\tilde{t}E_{cav}^{(t)}$ in terms of $E_{in}^{(+)}$ and $E_{out}^{(+)}$ as

$$E_{out}^{(+)} = \tilde{r}E_{in}^{(+)} + \tilde{t}E_{cav}^{(+)}, \tag{2.5}$$

which has an obvious physical interpretation in terms of the reflection and transmission of the mirror at $z = 0$. The preceding decomposition has far-reaching consequences for the case an active medium is in the cavity $[0, \ell]$. In this case, although the output field is again a sum of the reflected input field and transmitted cavity field, there is a qualitative distinction between the two: it is only the cavity field that undergoes interaction with the active medium.

In the good-cavity limit, the width of each peak in the quasimode structure becomes sufficiently small compared to the separation of successive peaks. In this limit and if one has a narrow enough gain profile for the active medium driving the cavity, one may assume that the coherent field inside the cavity excites a single (quasi)mode. One can make contact with the usual single-mode theories by truncating the sum in Eq. (2.4a) to run over a single quasimode centered at wavevector k_0, namely over the range $[k_0 - \pi/2\ell, k_0 + \pi/2\ell]$ and equating the truncating sum, denoted by a prime superscript, to

$$E_{cav}^{(+)\prime}(t) = \chi a(t)e^{ik_o\ell - i\Omega t}, \tag{2.6}$$

in which a represents an effective single-mode intracavity annihilation operator in the sense $[a, a^\dagger] = 1$. Consistently, the normalization constant χ must obey the constraint

$$|\chi|^2 = [E_{cav}^{(+)\prime}(z,t), E_{cav}^{(-)\prime}(z,t)] = \sum_k{}' \frac{\hbar\Omega_k}{4\epsilon_0\mathcal{A}L}|M_k|^2.$$

In the limit $L \to \infty$, the preceding sum over k may be converted to an integral over k leading to the result $|\chi| = \sqrt{\hbar\Omega/4\epsilon_0\mathcal{A}\ell}$. These considerations enable one to write, up to a phase factor, the expression for $a(t)$ directly in terms of the annihilation operators a_k of the modes of the whole space,

$$a(t) \simeq i\sqrt{\frac{\ell}{L}}\sum_k{}' M_k a_k(t)e^{ik\ell + i\Omega t}. \tag{2.7}$$

As an application of the formalism developed in this section, we may derive the operator equation for the decay of the intracavity field inside a good cavity when the cavity is empty. In that case, the a_k's in Eq. (2.7) have the trivial time dependence $a_k(t) = a_k(0)\exp(-i\Omega_k t)$ and so we can write

$$a(t + 2\ell/c) = i\sqrt{\frac{\ell}{L}}\sum_k M_k a_k(0)e^{-ik\ell - i\delta\Omega_k t},$$

where we have defined $\delta\Omega \equiv c(k - k_0)$. Then by Eq. (2.2a), we obtain

$$\begin{aligned} a(t + 2\ell/c) &= \tilde{r}i\sqrt{\tfrac{\ell}{L}}\textstyle\sum_k M_k a_k(0)e^{ik\ell - i\delta\Omega_k t} \\ &\quad + \tilde{t}i\sqrt{\tfrac{\ell}{L}}\textstyle\sum_k \xi_k a_k(0)e^{ikL - i\delta\Omega_k t} \\ &= \tilde{r}a(t) - \tilde{t}2\sqrt{\tfrac{\epsilon_0\mathcal{A}L}{\hbar\Omega_k}}\mathcal{E}_{in}^{(+)}(t), \end{aligned} \tag{2.8}$$

in which $\mathcal{E}_{in}^{(+)}(t)$ is the slowly varying amplitude of the input field:

$$E_{in}^{(+)}(t) = \mathcal{E}_{in}^{(+)}(t)e^{-i\Omega t}. \tag{2.9}$$

Now the approximation

$$a(t + 2\ell/c) \approx a(t) + \frac{2\ell}{c}\frac{da}{dt} \tag{2.10}$$

enables us to get

$$\frac{da}{dt} = -\Gamma a - \tilde{t}\frac{c}{\ell}\sqrt{\frac{\epsilon_0\mathcal{A}\ell}{\hbar\Omega}}\mathcal{E}_{in}^{(+)}(t). \tag{2.11}$$

The equation for da/dt is of the familiar form

$$\frac{da}{dt} = -\Gamma a + F(t), \tag{2.12}$$

where the Langevin noise operator $F(t)$ represents fluctuations arising from the incoming external field and guarantees the commutation relations. Such an equation was derived in a more formal way by Collett and Gardiner.[9] It shows that as the field inside the cavity is damped it is replaced by the incoming field. Hence in steady state the cavity field and the incoming field show strong time correlations, a fact that is critical to the relationship of squeezing inside vs. outside.

When the cavity contains a gain medium with a small single-pass gain, we may treat the gain as being separate from the losses. That is, we may simply add some terms to Eq. (2.12) to account for the way in which the field is changed by the gain medium alone. In an ordinary laser whose gain medium consists of inverted two-level atoms, these extra terms on the right-hand side of Eq. (2.12) are of form $\alpha a + G(t)$, where α is the gain coefficient and $G(t)$ is a noise term arising from the spontaneous emission from the inverted medium. We shall not discuss the ordinary laser further, but analyze the optical parametric oscillator in the remainder of this paper.

Degenerate Optical Parametric Oscillator Below Threshold

In a degenerate parametric oscillator driven by a classical pump field at frequency 2Ω, the additional term in Eq. (2.12) is $\epsilon a^\dagger$, where ϵ is proportional to the nonlinear susceptibility $\chi^{(2)}$ and the pump amplitude:

$$\frac{da}{dt} = -\Gamma a - \tau \mathcal{E}_{in}^{+}(t) + \epsilon a^\dagger, \tag{2.13}$$

where $\tau \equiv (4\Gamma / \overset{t}{\sim})(\epsilon_0 A\ell/\hbar\Omega)^{1/2}$. Although the same equation was the starting point of Collett and Gardiner's treatment, we follow a different solution procedure leading, of course, to the same end results as they obtained. Note that we have neglected noise due to spontaneous emission from the nonlinear medium.

By superposing linearly Eq. (2.13) and its adjoint, one may easily establish that if θ is the phase of the generally complex pump parameter ϵ, then the quadrature combinations

$$\begin{aligned} X^\theta &\equiv \left(\tfrac{1}{\chi} X^\theta_{cav}\right) = \tfrac{1}{2}(a e^{-i\theta/2} + a^\dagger e^{i\theta/2}) \\ Y^\theta &\equiv X^{\theta+\pi/2} = \tfrac{1}{2i}(a e^{-i\theta/2} - a^\dagger e^{i\theta/2}) \end{aligned} \tag{2.14}$$

obey the decoupled equations

$$\begin{aligned} \tfrac{d}{dt} X^\theta &= -(\Gamma - |\epsilon|) X^\theta - \tau \mathcal{E}^\theta_{in}, \\ \tfrac{d}{dt} Y^\theta &= -(\Gamma + |\epsilon|) Y^\theta - \tau \mathcal{E}^{\theta+\pi/2}_{in}. \end{aligned} \tag{2.15}$$

The X and Y quadratures, $\mathcal{E}^\theta_{in}$ and $\mathcal{E}^{\theta+\pi/2}_{in}$, of the input field are defined according to the same rule (2.14) that was used for the quadratures of the cavity field.

Before proceeding further, we remark that since Eq. (2.13) (or (2.15)) is linear in the field, it can only describe the parametric oscillator below threshold, i.e., for $\Gamma > |\epsilon|$. Keeping this in mind, one may solve the two equations (2.15) in the long-time (steady-state) limit as

$$X^{\theta}(t) = -\tau \int_{-\infty}^{t} \mathcal{E}_{in}^{\theta}(t') e^{-(\Gamma-|\epsilon|)(t-t')} dt',$$

and

$$Y^{\theta}(t) = -\tau \int_{-\infty}^{t} \mathcal{E}_{in}^{\theta+\pi/2}(t') e^{-(\Gamma+|\epsilon|)(t-t')} dt'. \tag{2.16}$$

One may now calculate the second-order noise moments of X_θ and Y_θ in terms of the two-time input-field noise correlations $< \Delta\mathcal{E}_{in}^{\theta}(t')\Delta\mathcal{E}_{in}^{\theta}(t'') >$. To compute the latter, we employ the rules (2.14) for forming the quadratures of the input field and Eqs. (2.4b) and (2.9). This procedure yields[12]

$$< \Delta\mathcal{E}_{in}^{\theta}(t')\Delta\mathcal{E}_{in}^{\theta}(t'') >= \frac{\hbar\Omega}{8\epsilon_0 Ac}\delta(t'-t''), \tag{2.17}$$

whose application to Eqs. (2.16) establishes the following noise moments for the cavity field X and Y quadratures:

$$< (\Delta X^{\theta})^2 >= \tau^2 \frac{\hbar\Omega}{8\mathcal{E}_0 Ac} \int_0^{\infty} e^{-2(\Gamma-|\epsilon|)t} dt = \frac{\Gamma}{4(\Gamma-|\epsilon|)}$$

and

$$< (\Delta Y^{\theta})^2 >= \frac{\Gamma}{4(\Gamma+|\epsilon|)}. \tag{2.18}$$

As the threshold is approached from below, i.e. as $|\epsilon| \to \Gamma$, the fluctuations in the X-quadrature of the cavity field become infinitely large while those in the Y-quadrature are suppressed by a factor of 2 relative to the vacuum state, in accordance with the results of Collett and Gardiner.[9]

We now turn to the output field which, unlike the cavity field, is amenable to a spectral analysis. In fact, the conventional technique of balanced homodyning to detect the squeezing of a signal provides a measurement of a generalized spectral quadrature, defined by the Fourier transform of an output-field quadrature like

$$X_{out} \equiv \frac{1}{2}\left(\mathcal{E}_{out}^{(+)}(t)e^{-i\theta/2} + \mathcal{E}_{out}^{(-)}(t)e^{i\theta/2}\right). \tag{2.19}$$

We note that the quadratures X_{out}, X_{in}^{θ}, and X_{cav}^{θ} are related formally via Eq. (2.5). If $\delta\omega$ denotes the spectral variable (detuning from the central frequency Ω) and if an overhead tilde sign represents Fourier transforms, then it follows from a use of the Fourier transform of relation (2.15) that

$$\begin{aligned}\tilde{X}_{out}^{\theta}(\delta\omega) &= \tilde{r}\tilde{X}_{in}^{\theta}(\delta\omega) + \tilde{t}\tilde{X}_{cav}^{\theta}(\delta\omega) \\ &= \tilde{X}_{in}^{\theta}(\delta\omega)\left\{\tilde{r} - \tfrac{\chi\tau\tilde{t}}{\Gamma-|\epsilon|-i\delta\omega}\right\},\end{aligned}$$

in which use was also made of the fact $X^\theta = (X^\theta_{cav}/\chi)$ (see Eq. (2.14)). Now noting that $\chi\tau\tilde{t} = 2\Gamma$, we have

$$\tilde{X}^\theta_{out}(\delta\omega) = \tilde{X}^\theta_{in}(\delta\omega)\left[\tilde{r} - \frac{2\Gamma}{\Gamma - |\epsilon| - i\delta\omega}\right]. \tag{2.20a}$$

Similarly,

$$\tilde{Y}^\theta_{out}(\delta\omega) = \tilde{Y}^\theta_{in}(\delta\omega)\left[\tilde{r} - \frac{2\Gamma}{\Gamma + |\epsilon| - i\delta\omega}\right]. \tag{2.20b}$$

In the good-cavity limit, $\tilde{r} \to 1$ and for the central frequency ($\delta\omega = 0$), it follows that as the threshold is approached from below, $|\epsilon| \to \Gamma$, the spectral quadrature noise has components

$$< (\Delta\tilde{X}^\theta_{out}(\delta\omega = 0))^2 >= \infty, \qquad < (\Delta\tilde{Y}^\theta_{out}(\delta\omega = 0))^2 >= 0 \tag{2.21}$$

Thus, the central-frequency component of the X-quadrature of the output field is infinitely noisy while the noise in its Y-quadrature is perfectly squeezed. Of course, for nonzero detuning, the results are intermediate between ∞ and 0.

Before proceeding to a discussion of the second, nonmodal approach, it is worth pointing out that interesting results like perfect intracavity squeezing with no output squeezing may be obtained in correlated-emission lasers. We shall, however, not review them here. We have also not dwelt upon the beautiful physical picture, in which to understand the inside-outside connection, that emerges in our formalism. We refer the reader for these discussions to Refs. 12 and 13.

3. THE NON-MODAL APPROACH TO THE PARAMETRIC OSCILLATOR PROBLEM

We saw in the previous section how the introduction of an unphysical mirror at $-L(L \to \infty)$ aided in rendering the traditional treatments of cavity damping rigorous. However, from a practical standpoint, that approach is at best cumbersome when cavity damping is large. One may argue that large cavity damping is of no interest, since some of the most fascinating results in quantum optics, for example, the states of a single-atom maser, arise in the highest-Q cavities. But that is not so for a class of devices, namely the semiconductor laser, in which the high intrinsic gain of the active medium obviates the need for highly polished end faces or mirrors. Thus, it has seemed to us that a general method, which is easy to apply to problems involving arbitrarily large or small mirror out-coupling, must be devised.

The non-modal approach that we discuss in this section is based on the simple connections of the input, cavity, and output fields at the output mirror in terms of the (arbitrarily large or small) reflection and transmission coefficients. In this approach there is no longer any need for the unphysical mirror at $z = -L$. In a general sense, one first propagates the optical field through a complete round-trip of interaction with the active medium, starting from and ending at the immediate right of the output mirror, which involves one reflection at the output mirror. The cavity damping due to mirror out-coupling is manifested in the output field, which via relation (2.5) drains the cavity energy. Just as in Sec. II, we shall for simplicity consider only single-ended cavities, although the extension to double-ended cavities is straightforward. We demonstrate this approach here by computing for the degenerate parametric oscillator its quantum noise and squeezing characteristics. Very detailed treatments of this problem and another problem concerning the intrinsic linewidth of a laser with arbitrary out-coupling appear elsewhere.[14,15]

We embark here on an effective-Hamiltonian treatment of a degenerate parametric oscillator. Although this will suffice for our purposes, a full microscopic analysis will be needed to include the subtleties of spontaneous emission. In the Heisenberg picture, the Hamiltonian for such a process may be written as[16,17]

$$H = \frac{1}{2}\int_{universe} [\epsilon(\vec{r})E^2(\vec{r},t) + \mu_0 H^2(\vec{r},t)]d\vec{r}, \tag{3.1}$$

where $\epsilon(\vec{r})$ is given by

$$\epsilon(\vec{r}) = \begin{cases} \epsilon_0 + \chi^{(2)}E(\vec{r},t), & z > 0 \\ \epsilon_0, & z < 0. \end{cases}$$

In view of this nonlinearity, the full Hamiltonian H can be split up into the usual unperturbed Hamiltonian H_0 and an interaction term H' arising purely from the nonlinearity:

$$H' = \frac{\mathcal{A}}{2}\chi^{(2)}\int_{cavity} E^3_{cav}(z,t)dz \tag{3.2}$$

where it has been assumed for simplicity that the electric field has no transverse $z-y$ dependence. The electric field $E_{cav}(z,t)$ is the sum of the quantized field (populated by photons in modes centered in frequency at Ω_0) and the intense classical pump field,

$$\begin{aligned} E_{pump}(z,t) &= \mathcal{E}_{pump}(z)e^{-2i\Omega_0 t} + \text{complex conjugate} \\ &= e_{pump}\sin(2k_0 z + \phi)e^{-i2\Omega_0 t} + \text{complex conjugate.} \end{aligned} \tag{3.3}$$

Here, the pump phase ϕ will be taken to be $\pi/2$. The general treatment for arbitrary ϕ is carried out in Ref. 14.

Substitution of the full cavity field into Eq. (3.1) generates several terms, but the particular nonlinear process of interest, namely the conversion of 1 pump photon into 2 signal photons and vice-versa, is described by only one of these terms. We ignore all other terms as being of no value here as well as all fast-varying terms, which amounts to the following form for the parametric interaction part of (3.2):

$$H'_{opo} = \frac{3\mathcal{A}}{2}\chi^{(2)}e^{2i\Omega_0 t}\int_0^{\ell} \mathcal{E}^{(+)*}_{pump}(z)E^{(+)^2}(z,t)dz + H.C. \tag{3.4}$$

We now separate the cavity field into its right and left traveling pieces. This separation is necessary, since not only are the boundary conditions specifically dependent on these particular fields, but also the output of the cavity is related to the transmitted part of the left traveling field alone. Let $e_+(z,t)$ and $e_-(z,t)$ denote the envelopes of the right and left traveling fields, i.e.,

$$E^{(+)}(z,t) \equiv [e_+(z,t)e^{ik_0 z} + e_-(z,t)e^{-ik_0 z}]e^{-i\Omega_0 t}. \tag{3.5}$$

In terms of the traveling pieces of the cavity field and e_{pump} (from Eqs. (3.1) and (3.4)), the Hamiltonian simplifies further into the form:

$$H'_{opo} = \frac{3\mathcal{A}}{4}\chi^{(2)}e^*_{pump}\int_0^\ell [e_+^2(z,t) + e_-^2(z,t)]dz + H.C., \tag{3.6}$$

in which spatial integrals of fast exponentials $\exp(\pm 2ik_0 z)$ were ignored.

The propagation of the traveling fields $e_\pm$ is described by the Maxwell's wave equation, which in the slowly-varying envelope approximation reduces to the following form:

$$\left(\frac{\partial}{\partial z} \pm \frac{1}{c}\frac{\partial}{\partial t}\right)e_\pm(z,t) = \pm i\frac{k_0}{2\epsilon_0}p_\pm^{NL}(z,t), \tag{3.7}$$

where $p_\pm^{NL}(z,t)$ are the slowly varying envelopes of the rightward and leftward traveling pieces of the nonlinear polarization field that drives the parametric amplification process. For Hamiltonians like H'_{opo} which are at most quadratic in the quantum field, the calculation of $p_\pm^{NL}(z,t)$ may be done by the classical procedure of functional differentiation with respect to the corresponding electric field:

$$p_\pm^{NL}(z,t) = -\frac{1}{\mathcal{A}}\frac{\delta}{\delta e_\pm^\dagger(z,t)}H'_{opo}.$$

The net result of this procedure is the following equations for $e_\pm(z,t)$:

$$\left(\frac{\partial}{\partial z} \pm \frac{1}{c}\frac{\partial}{\partial t}\right)e_\pm(z,t) = \pm q e_\pm^\dagger(z,t), \tag{3.8}$$

in which q is merely a constant

$$q \equiv -\frac{3i}{4}\chi^{(2)*}\frac{k_0}{\epsilon_0}e_{pump}.$$

The procedure from this point on is similar to that used in Sec. II. One may transform each of the two equations in (3.8) into two decoupled equations, if the following Hermitian quadrature combinations $X_\pm(z,t)$ and $Y_\pm(z,t)$ are used:

$$X_\pm(z,t) = \frac{e_\pm(z,t)e^{-\theta/2} + e_\pm^\dagger(z,t)e^{i\theta/2}}{2}; \tag{3.9a}$$

$$Y_\pm(z,t) = \frac{e_\pm(z,t)e^{-\theta/2} - e_\pm^\dagger(z,t)e^{i\theta/2}}{2}; \tag{3.9b}$$

in which θ is the phase of the complex number q. The following four equations are obtained for the quadratures:

$$\left(\frac{\partial}{\partial z} \pm \frac{1}{c}\frac{\partial}{\partial t}\right)X_\pm(z,t) = |q|X_\pm(z,t); \tag{3.10a}$$

$$\left(\frac{\partial}{\partial z} \pm \frac{1}{c}\frac{\partial}{\partial t}\right) Y_{\pm}(z,t) = -|q|_{\pm}(z,t). \tag{3.10b}$$

By reformulating these equations in terms of retarded time variables, $\tau_+ = t - z/c$ (for the e_+ field) and $\tau_- = t - (\ell - z)/c$ (for the e_- field), we may solve them in terms of simple exponentials:

$$X_+(z,t) = e^{|q|z} X_+(0, t - z/c), \quad X_-(z,t) = e^{|q|(z-\ell)} X_-(\ell, t - (\ell - z)/c);$$

$$Y_+(z,t) = e^{-|q|z} Y_+(0, t - z/c), \quad Y_-(z,t) = e^{-|q|(z-\ell)} Y_-(\ell, t - (\ell - z)/c). \tag{3.11}$$

Equations (3.11) reveal how the various quadratures evolve during a single rightward or leftward pass through the medium. The boundary conditions at the mirrors couple the left and right traveling fields and thereby enable us to determine the effect of propagation on either traveling field through an entire round trip. The π phase shift at the perfect mirror and the finite reflection and transmission coefficients of the output mirror give rise to boundary conditions for the fields at $z = \ell$ and $z = 0$:

$$e_+(\ell,t) = -e_-(\ell,t) \quad \text{and} \quad e_+(0,t) = -\tilde{r}e_-(0,t) + \tilde{t}e_+^{vac}(0,t). \tag{3.12}$$

Since $\tilde{r}$ and $\tilde{t}$ are real, the same connections hold for the quadratures (3.11) as well:

$$\begin{pmatrix} X_+(\ell,t) \\ Y_+(\ell,t) \end{pmatrix} = -\begin{pmatrix} X_-(\ell,t) \\ Y_-(\ell,t) \end{pmatrix}; \quad \begin{pmatrix} X_+(0,t) \\ Y_+(0,t) \end{pmatrix} = -\tilde{r}\begin{pmatrix} X_-(0,t) \\ Y_-(0,t) \end{pmatrix} + \tilde{t}\begin{pmatrix} X_+^{vac}(0,t) \\ Y_+^{vac}(0,t) \end{pmatrix}, \tag{3.13}$$

where X_+^{vac} and Y_+^{vac} are the quadratures of the rightward traveling vacuum field defined formally via (3.9). Given the single-pass evolutions (3.11) and boundary conditions (3.13), the full round-trip evolution of a quadrature, say X_+ is computed in the following four steps: First propagate X_+ from $z = 0$ to $z = \ell$ via Eq. (3.11). Then apply the boundary condition (3.13) at $z = \ell$ which transforms X_+ to X_-. Then propagate X_- backward from $z = \ell$ to $z = 0$. Finally apply the boundary condition (3.13) at $z = 0$ to get back X_+, albeit advanced from the starting X_+ by the round-trip time, $2\ell/c$. The result of this procedure is the following round-trip evolution equation:

$$\begin{pmatrix} X_+(0,t+2\ell/c) \\ Y_+(0,t+2\ell/c) \end{pmatrix} = \tilde{r}\begin{pmatrix} e^{2|q|\ell} X_+(0,t) \\ e^{-2|q|\ell} Y_+(0,t) \end{pmatrix} + \tilde{t}\begin{pmatrix} X_+^{vac}(0,t+2\ell/c) \\ Y_+^{vac}(0,t+2\ell/c) \end{pmatrix}. \tag{3.14}$$

The nominal threshold condition is obtained by first setting the vacuum field to zero in (3.14), and then finding the condition for which the gain X_+ experiences as it passes through the medium in time $2\ell/c$ compensates exactly for the loss at the output mirror. The nominal threshold condition is given by

$$\tilde{r}e^{2|q|\ell} = 1. \tag{3.15}$$

We now turn to the steady-state expectation values of cavity-field noise operators of interest. To determine the squeezing properties of the cavity field, we evaluate the moments $< \Delta X_+^2(z=0) >$, $< \Delta Y_+^2(z=0) >$, and $< \Delta X_+(z=0)\Delta Y_+(z=0) >$. In steady state these moments remain unchanged from one round trop to the next. Since the vacuum field is perfectly isotropic, they may all be expressed in terms of the variance of any quadrature of the vacuum field, say $< \Delta X_+^{vac}(z=0)^2 >$ which we denote by N for brevity. Thus, for example,

$$< \Delta X_+^2(z=0) >= \tilde{r}^2 e^{4|q|\ell} < \Delta X_+^2(z=0) > +\tilde{t}^2 < \Delta X_+^{vac}(z=0)^2 >,$$

so that

$$< \Delta X_+^2(z=0) >= \frac{(1-\tilde{r}^2)}{(1-\tilde{r}^2 e^{4|q|\ell})} N. \tag{3.16a}$$

Similarly,

$$< \Delta Y_+^2(z=0) >= \frac{(1-\tilde{r}^2)}{(1-\tilde{r}^2 e^{4|q|\ell})} N. \tag{3.16b}$$

Finally, since orthogonal quadratures of the vacuum field are uncorrelated, the $X-Y$ covariance of the cavity field vanishes:

$$< \Delta X_+(z=0)\Delta Y_+(z=0) >= 0 \tag{3.16c}$$

We note from Eqs. (3.16a) and (3.16b) that the X-quadrature of the cavity field has a larger fluctuation and its Y-quadrature has a smaller fluctuation than the fluctuation of any quadrature of the vacuum field. In other words, the Y-quadrature of the cavity field is squeezed. In the special limit that the threshold is approached from below, $\tilde{r}\exp(2|q|\ell) \to 1$, it is easy to see that

$$< \Delta Y_+^2(z=0) >= \frac{1}{1+\tilde{r}^2} N. \tag{3.17}$$

In the limit of a perfect cavity $\tilde{r} \to 1$, this reproduces the well-known answer of 50% squeezing for the cavity field.

We now turn to the spectral noise characteristics of the output field, which is related to the cavity field via (2.5). Note that the cavity field in that equation is really the left-going part of it, proportional to $e_-(0,t)$. So we first need the spectral characteristics of the X_- and Y_-quadratures, which are related to the X_+ and Y_+ quadratures by relation (3.13). By use of relations (2.5), (3.9), and (3.13) in the Fourier domain, it may be seen that the central Fourier component ($\delta\omega = 0$) of the X and Y quadratures of the output field are given as

$$\begin{pmatrix} \tilde{X}_{out}(\delta\omega=0) \\ \tilde{Y}_{out}(\delta\omega=0) \end{pmatrix} = -\frac{\tilde{t}}{\tilde{r}} \begin{pmatrix} \tilde{X}_+(\delta\omega=0) \\ \tilde{Y}_+(\delta\omega=0) \end{pmatrix} + \frac{1}{\tilde{r}} \begin{pmatrix} \tilde{X}_+^{in}(\delta\omega=0) \\ \tilde{Y}_+^{in}(\delta\omega=0) \end{pmatrix}, \tag{3.18}$$

This relation is written solely in terms of the incoming vacuum field by the use of the Fourier transform of Eq. (3.14):

$$\tilde{X}_{out}(\delta\omega+0) = \left(\frac{\tilde{r}-e^{2|q|\ell}}{1-\tilde{r}e^{2|q|\ell}}\right) \tilde{X}_{vac}(\delta\omega=0). \tag{3.19a}$$

A similar relation is obtained for the Y quadrature:

$$\tilde{Y}_{out}(\delta\omega = 0) = \left(\frac{\tilde{r} - e^{-2|q|\ell}}{1 - \tilde{r}e^{-2|q|\ell}}\right)\tilde{Y}_{vac}(\delta\omega = 0). \qquad (3.19b)$$

From these relations and the fact that any quadrature of the vacuum field has the same variance as any other, it follows that

$$< \Delta\tilde{X}^2_{out}(\delta\omega = 0) = \left(\frac{\tilde{r} - e^{2|q|\ell}}{1 - \tilde{r}e^{2|q|\ell}}\right)^2 < \Delta\tilde{X}^2_{vac}(\delta\omega = 0) > . \qquad (3.20a)$$

and

$$< \Delta\tilde{Y}^2_{out}(\delta\omega = 0) >= \left(\frac{\tilde{r} - e^{-2|q|\ell}}{1 - \tilde{r}e^{-2|q|\ell}}\right)^2 < \Delta\tilde{X}^2_{vac}(\delta\omega = 0) > . \qquad (3.20b)$$

It is instructive to specialize to the threshold condition (3.15). At threshold, it may be seen that the central Fourier component of the X-quadrature of the output field has infinitely large fluctuations, while Y-quadrature is fully squeezed. This is in conformity with the good-cavity results of Collett and Gardiner,[9] but the surprising new result here is that this in fact hold for an arbitrary-Q cavity.

4. CONCLUSIONS

In this paper, we have presented two self-consistent and rigorous formalisms in which to analyze any matter-field interaction problem inside a cavity which can have arbitrary mirror out-coupling. We have illustrated both approaches by calculating the quantum noise in a degenerate parametric oscillator below threshold. The first approach based on the introduction of rigorously defined modes, although valid for arbitrary output coupling, is well-suited to calculations concerning a good cavity, while the second, nonmodal approach based on mirror boundary conditions is suitable for arbitrary out-coupling.

A new result of our calculations based on the second approach is the detailed dependence of the degree of squeezing of the cavity and output fields on the output mirror transmission. At threshold, as the cavity out-coupling increases from 0 to 100%, the squeezing of the quantum noise in the cavity field decreases from 50% to 0, while the output field has its central frequency component perfectly squeezed regardless of the out-coupling. In our simple discussions, spontaneous emission from the active nonlinear medium has been ignored. Its inclusion will render our results experimentally more meaningful.

ACKNOWLEDGMENTS

The work reported here was partially done in collaboration with B. Abbott, J. Gea-Banacloche, N. Lu, L. Pedrotti, M Scully, and K. Wódkiewicz. The author gratefully acknowledges the kind invitation of Professor G.S. Agarwal and the receipt of travel funds from the U.S. Government that made this work possible.

REFERENCES

1. See for example, M. Born and E. Wolf, "Principles of Optics," Pergamon Press, Oxford (1975), Chapter 7.
2. See, for example, M. Sargent, M. Scully, and W. Lamb, Jr., "Laser Physics," Addison-Wesley, Reading (1974).
3. D. Meschede, H. Walther, and G. Müller, *Phys. Rev. Lett.* **54**:551 (1985).
4. M. Brune, J. Raimond, P. Goy, L. Davidovich, and S. Haroche, *Phys. Rev. Lett.* **59**:1899 (1987).
5. See, for example, Y. Shen, "The Principles of Nonlinear Optics," Wiley, New York, (1984).
6. See, for example, D. Walls, *Nature*, **306**:141 (1983); *ibid.*, 324:210 (1986); R. Loudon and P. Knight, it J. Mod. Opt., **34**:709 (1987); G. Milburn and D. Walls, *Opt. Commun.*, **39**:401 (1981).
7. R. Lang, M. Scully, and W. Lamb, Jr., *Phys. Rev. A*, **1**:1788 (1973); R. lang and M. Scully, *Opt. Commun.*, **9**:331 (1973).
8. K. Ujihara, *Phys. Rev. A*, **12**:148 (1975); *Jpn. J. Appl. Phys.*, **15**:1529 (1976); *Phys. Rev. A*, **16**:562 (1977).
9. M. Collett and C. Gardiner, *Phys. Rev. A*, **30**:1386 (1984); C. Gardiner and M. Collett, *ibid.*, **31**:3761 (1985).
10. H. Carmichael, *J. Opt. Soc. Am. B*, **4**:1588 (1987).
11. M. Collett and D. Walls, *Phys. Rev. A*, **32**:2887 (1985); D. Holm and M. Sargent, *Phys. Rev. A*, **35**:2150 (1987).
12. J. Gea-Banacloche, N. Lu, L. Pedrotti, S. Prasad, M. Scully, and K. Wódkiewicz, *Phys. Rev. A*, **41**:369 (1990).
13. J. Gea-Banacloche, N. Lu, L. Pedrotti, S. Prasad, M. Scully, and K. Wódkiewicz, *Phys. Rev. A*, **41**:381 (1990).
14. B. Abbott and S. Prasad, *Phys. Rev. A*, **45**:5039 (1992).
15. S. Prasad, *Opt. Commun.*, **85**:227 (1991); *Phys. Rev. A*, to be published.
16. W. Louisell, A. Yariv, and A. Siegman, *Phys. Rev.*, **124**:1646 (1961).
17. B. Mollow and R. Glauber, *Phys. Rev.*, **160**:1076 (1967).

CHAOS IN A SINGLE MODE LASER WITH INTRACAVITY PARAMETRIC AMPLIFICATION*

S. Dutta Gupta

School of Physics
University of Hyderabad
Hyderabad 500134, India

We investigate the modifications in the chaotic behaviour of a bad cavity single mode laser[1] when a parametric amplifying medium is inserted in the cavity.[2] We assume that the amplifying medium possesses a $\chi^{(2)}$ type nonlinearity and is pumped at twice the laser frequency. The nonlinear material will convert the pumping radiation at 2ω into radiation at ω. This down conversion process is known to lead to the generation of squeezed radiation. The equations for the dynamic variables in the presence of the parametric amplifier can be written as follows:

$$\dot{x} = \sigma(y - x) - |\alpha| e^{ie\psi} x^*$$

$$\dot{y} = -y - xz + \xi x$$

$$\dot{z} = -bz + (1/2)(x^* y + x y^*) \tag{1}$$

where σ is the ratio of the cavity decay and transverse relaxation rate and b is the ratio of the longitudinal and transverse relaxation rates. The parameter ξ depends on laser pumping and $|\alpha| e^{ie\psi}$ is determined by the nonlinearity and down conversion pump amplitude of the $\chi^{(2)}$ material. Note that the set of Eqs. (1) is reduced to Lorenz equations if we set the nonlinearity parameter $\alpha = 0$. If $\alpha \neq 0$, then Eqs. (1) represent a set of five coupled nonlinear equations with respect to the real and imaginary parts of x and y and the real variable z. It is easy to verify that the divergence of the flow is negative and like in Lorenz equations the z-axis remains invariant. The origin is a stationary point for all the parameter values and there are two bifurcations at $\xi = \xi_m = 1 - |\alpha|/\sigma$, and at $\xi = \xi_p = 1 + |\alpha|/\sigma$. These bifurcations lead to a pair of stationary points $Q_{1\pm}$ and $Q_{2\pm}$, respectively. The stability of these stationary points depends on the following parameters

$$\xi_a = \frac{(\sigma - |\alpha|)(\sigma - |\alpha| + 5)}{(\sigma - 3/\xi_m)}, \xi_b = \frac{(\sigma + |\alpha|)(\sigma + |\alpha| + 5)}{(\sigma - 3/\xi_p)} \tag{2}$$

*Work done in collaboration with G.S. Agarwal.

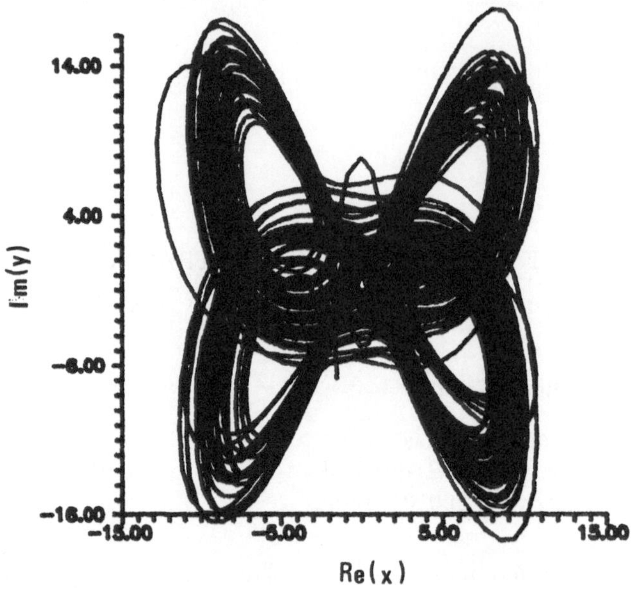

Fig. 1. Projection of the trajectory on Re(x)-Im(y) plane for parameter values $\sigma = 5, \xi = 30, |\alpha| = 0.1, \psi = 89°, b = 2$.

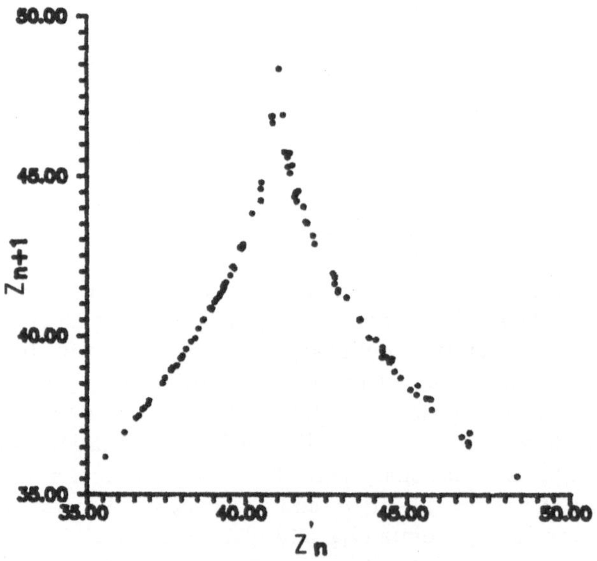

Fig. 2. Plot of successive local maxima Z_{n+1} against the previous one Z_n.

For $\xi > max(\xi_a, \xi_b)$ the system shows a rich chaotic dynamics. The presence of the parametric amplifier leads to the birth of an additional attractor. Out of the two Lorenz type attractors one is unstable. For suitable initial conditions the flow apparently settles down for a finite time around the unstable attractor $Q_{2\pm}$. Since the attractor given by $Q_{2\pm}$ is unstable (it has one additional eigenvalue compared to $Q_{1\pm}$) the flow departs from $Q_{2\pm}$ and after some transients settles down around the stable attractor $Q_{1\pm}$. The time span of the flow around the unstable attractor is determined by $|\alpha|$ and the initial conditions. We have shown this behaviour in Fig. 1 where Im(y) is plotted against Re(x). The first(second) and third(fourth) quadrant shows the unstable(stable) attractor. The plot of the successive maxima of z one against the previous one (Fig. 2), resembles the logistic map.

In conclusion we have shown how the presence of a parametric amplifier can change the chaotic behaviour of a strongly pumped bad cavity single mode laser, leading to the existence of two attractors in contrast to the single Lorenz attractor in its absence.[3] A similar effect was observed in a single mode laser with a phase conjugate mirror.[4] Birth of additional attractors was also noticed in loss modulated laser rate equations in presence of the parametric amplifier.[5] Thus the general effect of the presence of phase sensitive elements (parametric amplifier, phase conjugate mirror etc.) in the laser cavity leads to an increase in the dimension of phase space and results in the lifting of the phase degeneracy.

REFERENCES

1. H. Haken, *Phys. Let.* A **53**, 77 (1975).
2. The dynamics of the laser field in the good cavity limit in presence of the parametric amplifier was studied by Yu. M. Golubev, *Sov. Phys. JETP* **66**(2), 265 (1987).
3. S. Dutta Gupta and G. S. Agarwal "Instability and chaos in a single mode laser with intracavity parametric amplification." *J. Opt. Soc. Am.* **B** 8, 1712 (1991).
4. B. H. W. Hendriks, M. A. M. de Jong and G. Nienhuis *Opt. Commun.* **77**, 435 (1990).
5. S. Dutta Gupta and M. B. Pande, to be published.

CAVITY QED WITH A SINGLE MORSE OSCILLATOR

Gautam Gangopadhyay and Deb Shankar Ray

Indian Association for the Cultivation of Science
Jadavpur, Calcutta-700032
India

We consider the finite temperature cavity QED with a single Morse oscillator and its spectral aspects. It has been shown that the single peak structure of the emission spectrum in the presence of weak coupling splits up into a doublet in the limit of strong coupling. In the finite temperature case, for a small average number of photons $\bar{n}$ in the cavity, the lineshape first shows doublet structure as the coupling increases. However, these merge into a single peak structure with increasing $\bar{n}$ at values as small as 3-4 photons.

Experimental studies of Rydberg atoms in microwave cavities have demonstrated that the radiation emitted by the excited atoms in a cavity gets significantly modified by the strong coupling of the atoms with the cavity field mode(s).[1] This modification leads to a number of interesting fundamental cavity electrodynamic phenomena,[1-6] of which a very important one is vacuum field Rabi splitting.[2-5] The phenomenon is amenable to theoretical understanding in terms of the atom-cavity field mode coupled oscillator model. While these studies are based on systems with two-level atoms, it may be worthwhile to consider in this connection the multilevel systems with non-equidistant states and to examine to what extent the coupled oscillator model is generic at the fundamental cavity QED level. We have shown recently[7] that the Morse oscillator, the simplest molecular system with nonequidistant multilevel states, offers good opportunity for studying the aspect of quantum optical coherent interaction process. In the present paper we investigate the spectral characteristics of radiation scattered by a single weakly nonlinear Morse oscillator coherently interacting with a quantized cavity mode pumped by an external classical field.

We consider a Morse oscillator interacting with a cavity mode with frequency ω. The cavity is driven by an external field E(t) and damped at a rate γ_f. The energy decay rate of the oscillator is γ_a. The master equation for the reduced density operator ρ in the rotating wave, Born-Markov approximations is given by

$$\frac{d\rho}{dt} = \frac{-i}{\hbar}[H_o + H_d, \rho] + L_f[\rho] + L_a[\rho]. \tag{1}$$

The Hamiltonian H_o for the Morse oscillator ($\epsilon = 1$)[7,8], the cavity field with frequency ω, and the interaction with each other is

Recent Developments in Quantum Optics, Edited
by R. Inguva, Plenum Press, New York, 1993

$$H_o = \hbar f(S_o + \varepsilon S_- S_+) + \hbar\omega a^+ a + \hbar g(aS_+ + a^+ S_-). \tag{2}$$

g is the oscillator-cavity mode coupling constant. For $\varepsilon \neq 1$ one finds dipole-dipole interaction between two-level atoms.[9] S_+, S_-, and S_o are the generators of $SU(2)$ Lie algebra.[7,8]

The driving of the cavity by a classical field $E(t)$ of frequency ω_o is described by

$$H_d = \hbar[a^+ E(t) + aE^*(t)]. \tag{3}$$

The Liouvillians representing the interactions of the field (L_f) and the oscillator (L_a) with the heat baths are given by

$$L_f(\rho) = (1+\bar{n})\gamma_f[2a\rho a^+ - \rho a^+ a - a^+ a\rho] + \bar{n}\gamma_f[2a^+\rho a - aa^+\rho - \rho aa^+],$$

and

$$L_a(\rho) = \gamma_a[2S_-\rho S_+ - \rho S_+ S_- - S_+ S_- \rho],$$

respectively, where $\bar{n}$ is the thermal average number of photons in the cavity.

Using the standard operator disentanglement technique we derive the following Langevin stochastic differential equations for c-numbers α, α^* (corresponding to a and a^+ respectively) associated with the field and α_+, α_-, and α_o (corresponding to S_+, S_-, and S_o respectively) associated with the oscillator;

$$\dot{\alpha} = [-i\omega\alpha - \bar{\gamma}_f\alpha - iE(t)] + G_\alpha(t),$$

$$\dot{\alpha}^* = [i\omega\alpha^* - \bar{\gamma}_f\alpha^* + iE^*(t)] + G^*_\alpha(t). \tag{4}$$

$$\dot{\alpha}_+ = [-if(1-2\varepsilon)\alpha_+ - 2if\varepsilon\alpha_+\alpha_o - 2ig\alpha^*\alpha_o + 2\gamma_a\alpha_o\alpha_+] + G_{\alpha_+}(t),$$

$$\dot{\alpha}_- = [+if(1-2\varepsilon)\alpha_- + 2if\varepsilon\alpha_-\alpha_o + 2ig\alpha\alpha_o + 2\gamma_a\alpha_o\alpha_-] + G_{\alpha_-}(t),$$

$$\dot{\alpha}_o = [-ig\alpha\alpha_+ + ig\alpha^*\alpha_- - 2\gamma_a\alpha_+\alpha_-] + G_{\alpha_o}(t).$$

Here the G_{α_i}'s are the independent Langevin forces with zero reservoir averages as follows; $< G_{\alpha_i} >_R = 0$. The nonzero second moments of the Langevin forces are given by

$$< G_{\alpha_+}(t)G_{\alpha_+}(s) >_R = (-if\varepsilon\alpha_+^2 - ig\alpha^*\alpha_+ + \gamma_a\alpha_+^2)\delta(t-s),$$

$$< G_{\alpha_-}(t)G_{\alpha_-}(s) >_R = (if\varepsilon\alpha_-^2 + ig\alpha\alpha_- + \gamma_a\alpha_-^2)\delta(t-s),$$

$$< G_{\alpha_o}(t)G_{\alpha_o}(s) >_R = [-\frac{ig}{2}(\alpha\alpha_+ - \alpha^*\alpha_-) - \gamma_a\alpha_+\alpha_-]\delta(t-s)$$

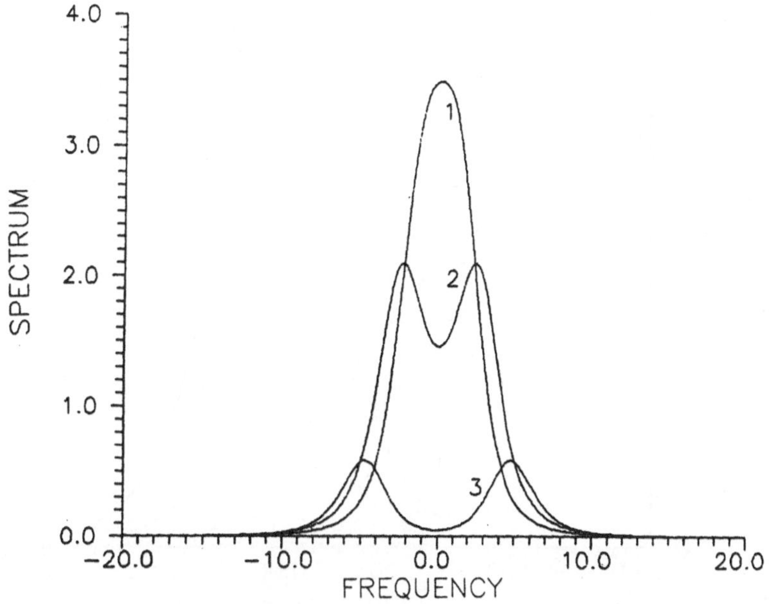

Figure 1. The power spectrum $S(\Omega - \omega)\gamma_a$ as a function of frequency $(\Omega - \omega)/\gamma_a$ for $\bar{n} = 0.1$ and $\varepsilon = 1$ (Morse Oscillator) but varying $g\gamma_f^{-1}$ = (1) 0.4, (2) 0.6, (3) 1.1 (both scale arbitrarily).

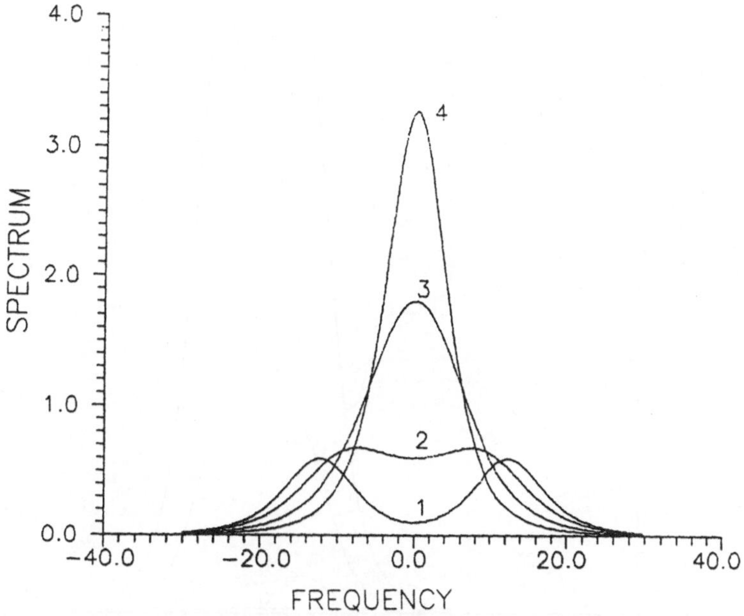

Figure 2. As in Figure 1 but $g\gamma_f^{-1} = 1.67$ and $\varepsilon = 1$ and $\bar{n}$ varies as (1) 1.6, (2) 2.4, (3) 4.0, (4) 6.4 (both scale arbitrarily).

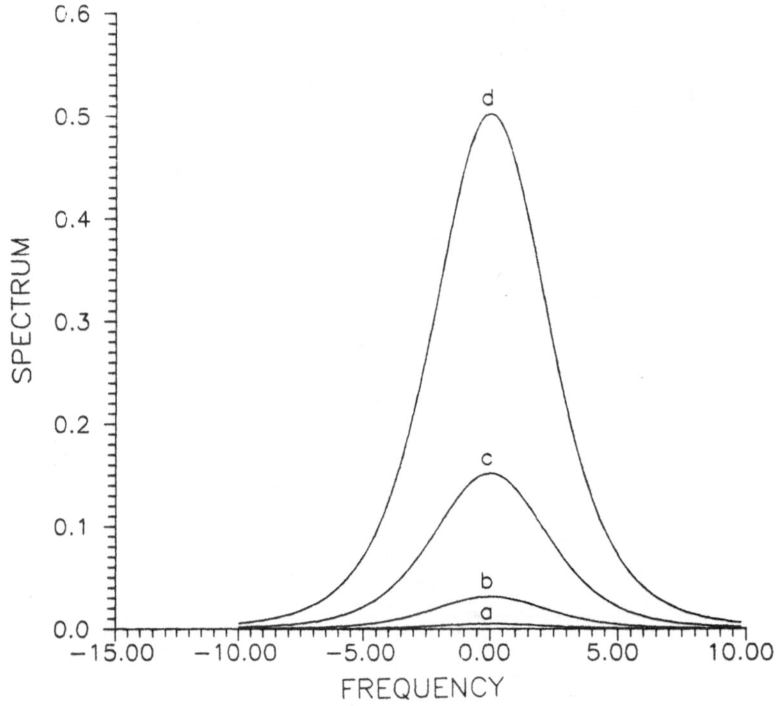

Figure 3. As in Figure 1 but $g\gamma_f^{-1} = 0.83$ and $\bar{n} = 2.0$ and ε varies as (a) 0.1, (b) 0.25, (c) 0.55, (d) 1.0 (both scale arbitrarily).

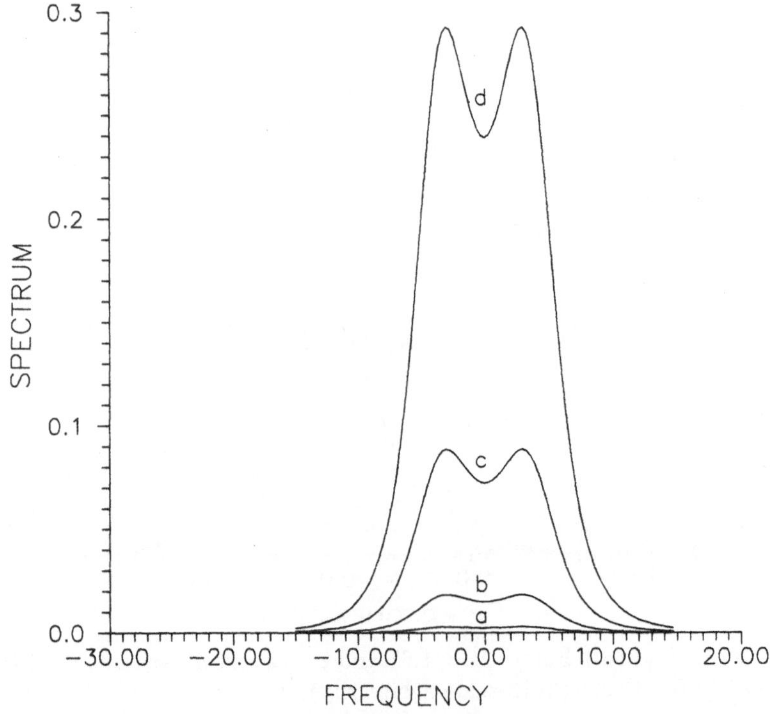

Figure 4. As in Figure 1 but $g\gamma_f^{-1} = 1.0$ and $\bar{n} = 1.0$ and ε varies as (a) 0.1, (b) 0.25, (c) 0.55, (d) 1.0 (both scale arbitrarily).

and

$$< G_\alpha(t)G_{\alpha^*}(s) >_R = 2\gamma_f\bar{n}\delta(t-s).$$

$\bar{\gamma}_f$ is defined as $\bar{\gamma}_f = (1+\bar{n})\gamma_f$.

Removing undesirable fast time dependence from $E(t)$ and $\alpha - s$ in eqn(4), one can calculate the Fourier transform of the relevant correlation function to obtain the fluctuation spectra in the usual way. First we consider the case of Morse oscillator ε=1. In the strong coupling limit, the spectrum is a superposition of two Lorentzian functions, where the separation of the two maxima is a function of the cavity-oscillator coupling g. For a weak coupling we find, however, a single peak structure at $\Omega = \omega$. As an illustration, we have displayed these spectra of a Morse oscillator for different values of g/γ_f in Fig. 1. The doublet structure is characteristic of the vacuum field Rabi oscillations of a single oscillator in a cavity where a single quantum of energy is transferred back and forth between the oscillator and the cavity.

Next we present the finite temperature effect in Figure 2. For a small enough average number of photons $\bar{n}$ in the cavity ($\sim$ 0.1), the line shape first shows a doublet structure as the coupling increases. However, these merge to a single peak with increasing $\bar{n}$ as small as 3-4 photons. This is a clear demonstration of the strong competition between the vacuum field Rabi oscillations and the thermally induced processes.

For $\varepsilon \neq 1$ we depict the situations displayed in Figs. 3 and 4. With the increase in ε, i.e. the strength of dipole-dipole interaction, the intensity increases. We find no further splitting in the spectra.

It is pertinent to note that notwithstanding the nonequidistant multilevel nature of the Morse oscillator and its nonlinearity, the results are in close correspondence with that for a single two-level atom QED in a cavity. This leads us to suspect that the system-field mode coupled oscillator is generic at the fundamental QED level for a very few number of photons for weak nonlinearity.

ACKNOWLEDGEMENT

Partial financial support from the Council of Scientific and Industrial Research under Grant No. 5/137/88 EMR II is thankfully acknowledged.

REFERENCES

1. See, for example, S. Haroche and J.M. Raimond, in Advances in Atomic and Molecular Physics, eds, D. Bates and B. Bederson (Academie, NY. 1985) Vol 20, p.347.
2. J.J. Sanchez-Mondragon, N.B. Narozhky and J.H. Eberly, Phys. Rev. Letts. 51 550 (1985).
3. G.S. Agarwal, Phys. Rev. Letts. 51 1732 (1984).
4. G.S. Agarwal, R.K. Bullough and G.P. Hildred, Opt. Commun. 59 23 (1986).
5. D. Meschede, H. Walther and G. Miller, Phys. Rev. Letts. 54 551 (1985); G. Rempe, H. Walther and N. Klein, Phys. Rev. Letts. 58 353 (1987).

6. Y. Zhu, A. Lezana, T. Mossberg and M. Lewenstein, Phys. Rev. Letts. 61 1946 (1988).

7. D.S. Ray, J. Chem. Phys. 92 1145 (1990), G Gangopadhyay and D.S. Ray, Phys. Rev. A41 6429 (1990). G Gangopadhyay and D.S. Ray, Phys. Rev. A (to be published).

8. R.D. Levine and C.E. Wulfman, Chem. Phys. Letts. 60 372 (1979); R.D. Levine, in Intramolecular Dynamics, eds. J Jortner and B. Pullman (Reidel, Dordrecht, 1982).

9. D.S. Ray, Phys. Letts. A 111 25 (1985) and references therein.

EFFECTS OF ATOMIC COHERENCE ON THE COLLAPSE AND REVIVAL PHENOMENON IN A TWO-PHOTON JAYNES-CUMMINGS MODEL

Amitabh Joshi and R.R. Puri

Bhabha Atomic Research Center
Trombay, Bombay 400 085
India

We discuss the effects of atomic coherence on dynamical and field statistical properties of an atom undergoing two-photon process in a coherent field inside a lossless cavity. The behaviour here is in contrast with an atom undergoing one-photon process in which case the Rabi oscillations are qualitatively the same for all values of coherences between the two atomic levels.

The problem of a two-level atom interacting with a single mode of the field has been studied in considerable detail [1-2]. Experimentally such a system can be realized in a single atom Rydberg maser where interaction of a single two-level atom with a single mode of the field may be regarded as dominant process [3].

Recently, the attention on the atom undergoing non-resonant two-photon process (modeled as effective two-level system) has been given in which the study of collapse and revival phenomenon of Rabi oscillations is reported [4-5]. The revivals in the case of two-photon process are both compact and regular in contrast with the case of single photon transition [2] in which the revivals are partial in the coherent field. The nature of this phenomenon depends on the statistics but not on the phase of the field if atom is in one of its two-states. However, the excitation probability do exhibits dependence on the phase of atomic coherence as well as that of excitation field if the atom is in the superposition of its two-states[6]. The phase dependence of excitation probability in coherent field provides a useful means of testing the quantum theory of radiation as against semiclassical and neoclassical theories [7].

Here we study the effects of atomic coherence on the dynamics and field statistics of an atom undergoing a two-photon process in an infinitely high-Q cavity.

In our effective two-level model (see Ref. [5]), the lower state $|g>$ is coupled to an excited state $|e>$. The atom interacts with the field of frequency $\omega = (E_e - E_g)/2\hbar$, where $E_g(E_e)$ is lower (excited) state. If the frequencies $(E_i - E_g)/\hbar$ and $(E_e - E_i)/\hbar$ (where E_i is the energy of an intermediate state $|i>$) are sufficiently different from the field frequency ω, then it has been shown that the transitions to intermediate levels can be considered as virtual so that the atom acts as an effective two-level atom absorbing and emitting two-photon of frequency ω each at a time. The Hamiltonian for such a system is then given by

Recent Developments in Quantum Optics, Edited
by R. Inguva, Plenum Press, New York, 1993

$$H = 2\hbar\omega S_z + \hbar\omega a^+a + \hbar g(a^{+2}S_- + S_+a^2) \qquad (1)$$

where $a(a^+)$ is the cavity field annihilation (creation) operator and $S_+ = |e><g|, S_- = |g><e|$ and $S_z = (|e><e| - |g><g|)/2$.

Since we are interested in investigating the effects of coherence between the atomic levels on the dynamics of the system, we consider the initial atomic state $|\mu>$ to be a superposition of ground state and excited state

$$|\mu>= (1 + |\mu|^2)^{-\frac{1}{2}}[|e> + \mu|g>] \qquad (2)$$

where μ is a complex number.

Let the initial state of the field be

$$\rho_f(o) = \sum_{m,n} P_{mn}|m><n| \qquad (3)$$

where,

$$P_{mn} = e^{-|z|^2}|z|^{m+n}/[m!n!]^{\frac{1}{2}} \qquad (4)$$

for the coherent state of the field.

Using the general solution of Eq. (1) obtained in Ref. [5] under dressed state approximation, it can be shown that for the initial state given by the direct product of the states in Eqs. (2) and (3), the probability $P_e(t)$ of finding an atom in the excited state is given by

$$P_e(t) = \frac{1}{2} + \frac{1}{2(1+|\mu|^2)}\sum([P_{nn} + |\mu|^2 P_{n+2,n+2}]$$

$$+[P_{nn} - |\mu|^2 P_{n+2,\ n+2}]cos(2gt\sqrt{(n+1)(n+2)})$$

$$+2|\mu|sin(\phi)P_{n+2n}sin(2gt\sqrt{(n+1)(n+2)}))$$

with

$$P_{nn} = e^{-|z|^2}|z|^{2n}/n! \qquad (5)$$

and $\mu = |\mu|exp(i\phi)$. The second and the third term of the series of Eq. (5) cannot be summed exactly. Nevertheless, the maximum contribution to the sum in Eq. (5) comes from n near $|z|^2$ so that for $n \sim |z|^2 >> 1$ we have

$$[(n+1)(n+2)]^{\frac{1}{2}} \cong n + \frac{3}{2} \qquad (6)$$

and the expression for $P_e(t)$ is obtained as follows

$$P_e(t) = \frac{1}{2} + \frac{1}{2(1+|\mu|^2)} exp[-|z|^2(1-cos(2gt))]$$

$$\text{x}\{cos[|z|^2 sin(2gt) + 3gt] - |\mu|^2 cos[|z|^2 sin(2gt) - gt]$$

$$+ 2|\mu| sin(\phi) sin[|z|^2 sin(2gt) + gt]\} \quad (7)$$

An interesting case in studying dynamics of the system is when $|\mu| = 1$, i.e. when the two-levels are equally populated. Here, it is found that the oscillations for both $\phi = 0$ and $\phi = \pi/2$ case look not only qualitatively different but for $\phi = 0$, their amplitude is considerably reduced as compared to $\phi = \pi/2$. The revivals of $P_e(t)$ for $\phi = 0$ exhibit a "doublet" structure (see Fig. 1). In this case the envelope function of revivals vanishes at $t = t_r$ and has extremas at $t_r \pm t_r/2\pi|z|$. Hence each revival becomes a "doublet" oscillation. This analytic result is in good agreement with exact numerical summation of the series. This result is in contrast with the one-photon coherent excitation case where no such "doublet" structure is obtained for $|\mu| = 1$, $\phi = 0$ case [7]. However, for $|\mu| = 1$, $\phi = \pi/2$, the envelope of oscillations have maxima at $t = t_r$ for all the terms. This explains the appearance of singlets of revivals centered at $t = t_r$ for $|\mu| = 1$, $\phi = \pi/2$.

The intensity-intensity correlation function $g^{(2)}(t)$ defined as

$$g^{(2)}(t) = < a^{+2}a^2 > / < a^+ a >^2 - 1 \quad (8)$$

does show significantly different behaviour for $\phi = 0$ and $\phi = \pi/2$.

For $|\mu| = 1$, $\phi = 0$; we arrive at the following expression of $g^{(2)}(t)$:

$$g^{(2)}(t) = \frac{(|z|^4 + 2)}{(|z|^2)^2} + exp[-|z|^2(1-cos(2gt))] sin(2gt)$$

$$\text{x}\left[\left(\frac{|z|^4+2}{|z|^6} sin(|z|^2 sin(2gt) + gt) - \frac{1}{|z|^2} sin(|z|^2\right.\right.$$

$$\left. sin(2gt) + 3gt)\right) - \frac{1}{2|z|^4}[cos(|z|^2 sin(2gt) + 3gt)$$

$$\left. + 3cos(|z|^2 sin(2gt) + gt)] + ...\right] - 1 \quad (9)$$

Clearly the envelope of the dominating term (first square bracket) has minima at $t = t_r$ and has its maxima at $t_r + t_r/2\pi\sqrt{|z|}$. The magnitude of the envelope function is in between $1/|z|^4$ and $1/|z|^2$. So, we have to consider the next term in Eq. (9) (of approximately the same magnitude) which gives nonvanishing contribution at $t = t_r$. We thus find a triplet structure in revivals at $t = t_r$, $t = t_r \pm t_r/2\pi|z|$. The expression given in Eq. (9) is in extremely good agreement with numerical summation of the series for $g^{(2)}(t)$ at $|\mu| = 1$, $\phi = 0$ (Fig. 2).

Next, the contribution of ϕ-dependent terms to revivals ($|\mu| = 1$) may be shown to be equivalent to

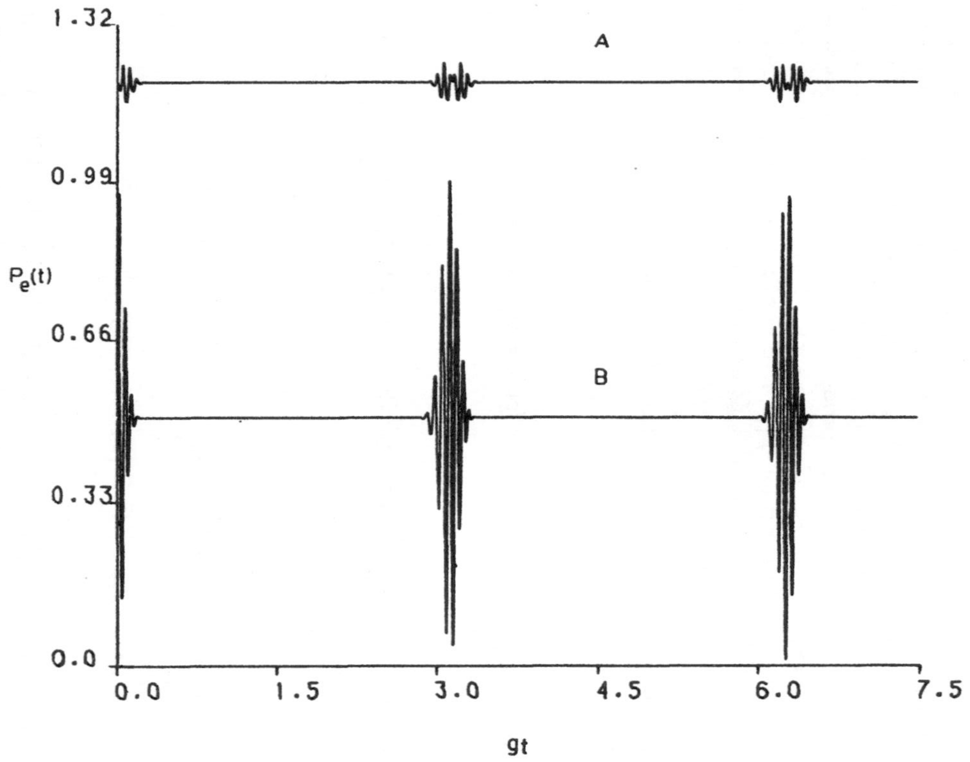

Figure 1. Probability $P_e(t)$ of finding an atom in the excited state as a function of time for initial coherent state with $|z|^2 = 50$. Curves $A(P_e(t)+0.7)$ and $B(P_e(t))$ are for $|\mu| = 1$, $\phi = 0$ and $|\mu| = 1$, $\phi = \frac{\pi}{2}$ respectively.

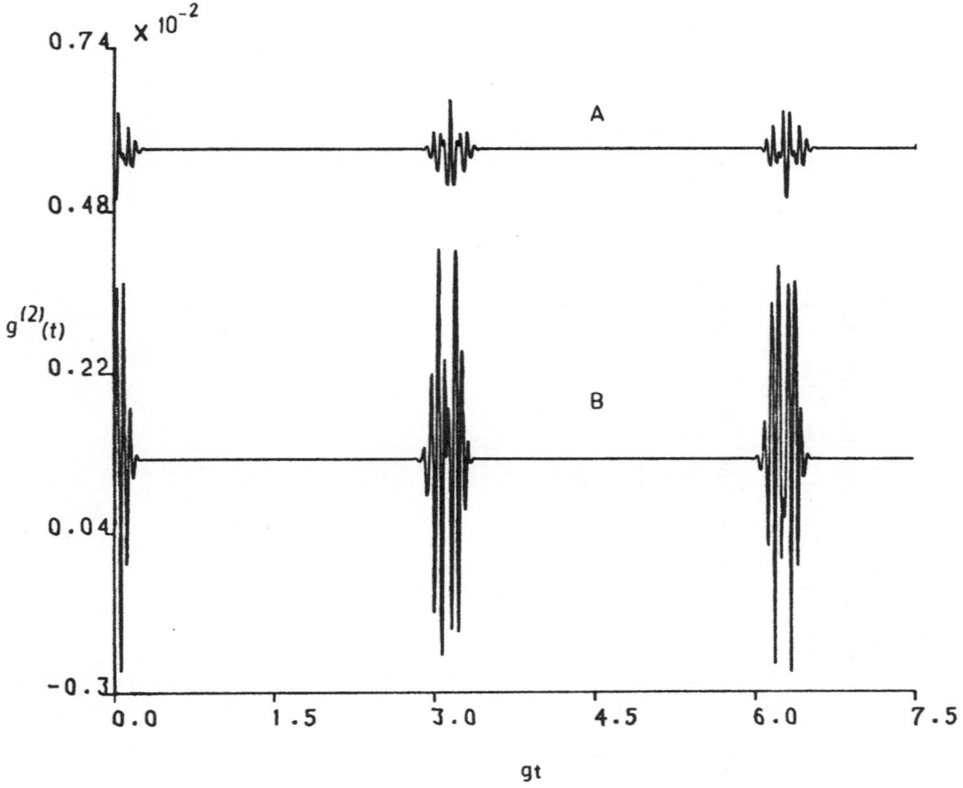

Figure 2. The intensity-intensity correlation function $g^{(2)}(t)$ as a function of time for initial coherent state with $|z|^2 = 50$. Curves $A(g^{(2)}(t)+0.005)$ and $B(g^{(2)}(t))$ are for $|\mu| = 1,\ \phi = 0$ and $|\mu| = 1,\ \phi = \frac{\pi}{2}$ respectively.

$$f_\phi(t) = (\frac{2}{|z|^2}) sin(\phi) exp[-|z|^2(1 - cos(2gt))] sin(gt) sin[|z|^2 sin(2gt) + gt] \quad (10)$$

Thus, the envelope function vanishes at $t = t_r$ and has maxima at $t = t_r \pm t_r/2\pi|z|$, resulting in the "doublets" of the revivals displaced by $\pm t_r/2\pi|z|$ from their centers when $|\mu| = 1$, $\phi = \pi/2$.

REFERENCES

1. E. T. Jaynes and F. W. Cummings. Proc. IEEE, 51, 89 (1963).
2. H. I. Yoo and J. H. Eberly, *Phys. Rep.* 118, 239 (1985).
3. D. Meschede, H. Walther, and G. Muller, *Phys. Rev. Lett.* **54**, 551 (1985); G. Rempe, H. Walther, and N. Kleing, ibid, **58**, 353 (1987).
4. P. Alsing and M. S. Zubairy, *J. Opt. Soc. Am.*, **B4** 117 (1987).
5. R. R. Puri and G. S. Agarwal, *Phy. Rev.* **37**, 3879 (1988).
6. A. Joshi and R. R. Puri, *J. Mod. Opt.*, **36**, 557 (1989).
7. G. S. Agarwal and R. R. Puri, *J. Opt. Soc. Am.*, **B5**, 1669 (1988).

LAMB SHIFT IN SQUEEZED THERMAL VACUUM[1]

P. Shanta

School of Physics
University of Hyderabad
Hyderabad-500 134, India

INTRODUCTION

In quantum optics several non-classical states of light have been discovered. Besides the coherent state, the Yuen squeezed states [1] and the Caves-Schumaker states have been produced in the laboratory. It is therefore very important to examine effects such as Lamb shift in quantum electrodynamics where vacuum plays an important role. The atomic level shifts for dipole interaction for a squeezed vacuum have already been calculated [2]. Here, we shall calculate the Lamb shift in hydrogen atom in a squeezed thermal vacuum. We shall use thermofield dynamics and Welton's method to calculate the Lamb shift.

THERMAL YUEN STATES

In thermofield dynamics the ensemble average of an observable A is

$$< A >= \frac{Tr\ e^{-\beta H} A}{Tr\ e^{-\beta H}} =< 0(\beta)|A|0(\beta) > \tag{1}$$

where $\beta = \frac{1}{kT}$ and $|0(\beta) >$ is the thermal vacuum. This thermal vacuum is defined in a direct product space $H \otimes \tilde{H}$ where H is spanned by a basis set of factors $\frac{a^{\dagger n}}{\sqrt{n!}}|0 >= |n >$. Every state $|\tilde{n} >$ and every observable A has its tilde counterpart. A commutes with its tildian $\tilde{A}$ and obeys the following tilde conjugation rules

$$(AB)^{\sim} = \tilde{A}\tilde{B}, (c_1 A)^{\sim} = C_1^* \tilde{A} \text{ and } (A^{\dagger})^{\sim} = \tilde{A}^{\dagger} \tag{2}$$

In particular the bosonic creation and annihilation operators $a, a^{\dagger}$ obey the following rules

$$[a, a^{\dagger}] = [\tilde{a}, \tilde{a}^{\dagger}] = 1 \text{ and } [a, \tilde{a}] = 0 \tag{3}$$

[1] In collaboration with S. Chaturvedi and V. Srinivasan, School of Physics, University of Hyderabad, Hyderabad-500134, India

Recent Developments in Quantum Optics, Edited
by R. Inguva, Plenum Press, New York, 1993

It is found that

$$|0(\beta)>= e^{-iG}|0,0> \tag{4}$$

where

$$G = i\theta(\beta)[a^\dagger\tilde{a}^\dagger - a\tilde{a}] \text{ and } tanh\theta(\beta) = e^{-\beta\omega\hbar}. \tag{5}$$

$|0(\beta)>$ is the vacuum at finite temperature. It can be easily verified that it satisfies Eq. (1). We refer to some of the earlier papers for a good review of TFD [3], [4], [5]. The thermal counterparts of various non-classical ground states of quantum optics have been recently constructed [6], [7]. In particular the thermal counterpart of Yuen squeezed (YS) vacuum is

$$|0(\beta)>_{YS}= D(\alpha)\tilde{D}(\alpha)S(z)\tilde{S}(z)e^{-iG}|0,0> \tag{6}$$

where

$$D(\alpha) = exp(\alpha a^\dagger - \alpha^* a), \quad \tilde{D}(\alpha) = exp(\alpha^*\tilde{a}^\dagger - \alpha\tilde{a}),$$

$$S(z) = exp\frac{z}{2}(a^{\dagger 2} - a^2), \quad \tilde{S}(z) = exp\frac{z}{2}(\tilde{a}^{\dagger 2} - \tilde{a}^2) \tag{7}$$

The following results are useful in later calculations:

$$e^{-iG}ae^{iG} = cosh\theta(\beta)a - sinh\theta(\beta)\tilde{a}^\dagger \tag{8}$$

$$S(z,\tilde{z})aS^\dagger(z,\tilde{z}) = coshz\ a - sinhz\ a^\dagger \tag{9}$$

$$D(\alpha,\tilde{\alpha}), aD^\dagger(\alpha,\tilde{\alpha}) = a - \alpha \tag{10}$$

where $S(z,\tilde{z}) = S(z)\tilde{S}(z)$, $D(\alpha,\tilde{\alpha}) = D(\alpha)\tilde{D}(\alpha)$, $\mu = coshz$ and $\nu = sinhz$.

LAMB SHIFT

We shall evaluate the Lamb shift for the vacuum $|0(\beta)>_{YS}$ using a method due to Welton [8], [9]. The electron in the hydrogen atom is subjected to a fluctuating force because of the vacuum fluctuations of the em field. Under the action of this fluctuating force the electron undergoes a displacement $\delta\vec{x}$ about its position $\vec{x}$ in orbit. The nuclear potential at $\vec{x}+\delta\vec{x}$ is expanded in a Taylor series about $\vec{x}$. The average potential seen by the electron when the em field is in the vacuum state $|0>$ is $<0|\phi(\vec{x}+\delta\vec{x})|0>$. The effect of vacuum fluctuations can be treated as a perturbation of the nuclear potential. The energy shifts are calculated using perturbation theory

The radiation field quantized in the transverse gauge is

$$\vec{E} = \sum_{\vec{k},\lambda} i[\frac{ch}{V}]^{\frac{1}{2}} k^{\frac{1}{2}} [a_\lambda(\vec{k})e^{i\vec{k}\cdot\vec{x}-i\omega t} - h.c.]\hat{e}_\lambda(\vec{k}) \tag{11}$$

where the $\vec{k}$ are the discreet eigenmodes of the field in a bound volume V. The $\hat{e}_\lambda(\vec{k})$ are the two unit polarization vectors orthogonal to $\vec{k}$. Also,

$$[a_{\lambda_1}(\vec{k}), a^\dagger_{\lambda_2}(\vec{k}')] = \delta_{\lambda_1\lambda_2}\delta_{\vec{k},\vec{k}'}, \; and \; a_\lambda(\vec{k})|0>= 0 \tag{12}$$

Using the Lorentz equation for force on the electron we find $\delta\vec{x}$ in terms of $\vec{E}(\vec{x},t)$:

$$\delta\vec{x} = \frac{-e}{m}[\frac{ch}{V}]^{\frac{1}{2}} \sum_{\vec{k},\lambda} i\frac{k^{\frac{1}{2}}}{k^2c^2}\hat{e}_\lambda(\vec{k})[a_\lambda(\vec{k})e^{i\vec{k}\cdot\vec{x}-i\omega t} - h.c.] \tag{13}$$

Using (12) we find that the perturbing potential is

$$\frac{1}{6} < 0|\delta\vec{x}\cdot\delta\vec{x}|0 > \cdot\nabla^2\phi(x) = \frac{4\pi e^2}{6} < 0|\delta\vec{x}\cdot\delta\vec{x}|0 > \delta(\vec{x}) \tag{14}$$

for a Coulomb potential. The energy shift of the atomic states ψ_{nlm} are

$$\Delta E_{nlm} = \frac{4\pi e^2}{6} < 0|\delta\vec{x}^2|0 > |\psi_{nlm}(0)|^2 \tag{15}$$

LAMB SHIFT IN THERMAL SQUEEZED VACUUM

To calculate the Lamb shift for Yuen thermal vacuum we replace $|0>$ by $|0(\beta)\rangle_{YS}$ of Eq. (6). The thermal transformation is given by

$$G = -i\sum_{\vec{k}} \theta_k(\beta)[a^\dagger(\vec{k})\tilde{a}^\dagger(\vec{k}) - a(\vec{k})\tilde{a}(\vec{k})] \tag{16}$$

while the squeezing and displacement transformations of Eq. (7) are applied to a single mode $a(\vec{K})$.

$$_{YS} < 0(\beta)|\delta\vec{x}^2|0(\beta) >_{YS} = \frac{e^2h}{m^2c^3\pi^2}\int\frac{1}{k}(1 + 2sinh^2\theta_k(\beta))dk$$
$$+ \frac{4e^2h}{m^2c^3V}\cdot\frac{1}{K^3}[|\alpha|^2 + sinh^2z(1 + 2sinh^2\theta_K(\beta))]. \tag{17}$$

In arriving at (17), we have made use of Eqs. (8), (9), (10), and (12) and dropped oscillatory time dependent terms [10]. The first term in the integral gives rise to the usual Lamb shift and the second gives the temperature dependent correction. We find that the α, z dependent correction factor is equal to $< n_{YS} >$ and hence the correction is proportional to the intensity of the Yuen mode. Also, as $sinh^2\theta_k(\beta) = \frac{1}{e^{\beta\omega\hbar}-1}$, we see that the temperature dependent correction does not have an ultraviolet divergence and that it vanishes at $T = 0$. The divergent integral in (17) is calculated using the upper cutoff $k = \frac{mc^2}{\hbar c}$ and the lower cutoff $k = \frac{E_<}{\hbar c}$. The thermal correction is approximately $\frac{4e^2h}{m^2c^3\pi^2}\cdot\frac{kT}{E_<}$. For $E_< = 17E_{100}$ and $T = 300K$ this is 3x10^{-5} times the uncorrected shift, which is found to be 1047 MHz. For $K = 10^5 cm^{-1}$ and $T = 0$, the α and z dependent correction is found to be $10^{-15}I_{YS}$ times the shift (I_{YS} in cgs units). This correction is greatly enhanced by continuous squeezing. By assumed uniform squeezing between the modes K_{max} and K_{min} we find from Eq. (17) that the correction factor is $[ln\frac{K_{max}}{K_{min}}](\alpha^2 + sinh^2 z)x10^{-1}$. This additional shift could be measured by placing the hydrogen atoms in a microwave cavity filled with squeezed coherent radiation.

ACKNOWLEDGEMENT

I thank the Council of Scientific and Industrial Research (CSIR) for financial support.

REFERENCES

1. H. P. Yuen, *Phys. Rev.* **A13**, 2226 (1976).
2. G. J. Milburn, *Phys. Rev.* **A34**, 4882 (1986). G. W. Ford and R. F. O'Connel, *J.O.S.A.* **B4**, 1710 (1987).
3. S. Chaturvedi and V. Srinivasan (unpublished).
4. Y. Takahasi and H. Umezawa, Collective Phenomena, 2, 55 (1975).
5. H. Umezawa, H. Matsumoto and M. Tachiki, Thermofield Dynamics and Condensed States (North Holland, Amsterdam, 1982).
6. S. Chaturvedi, R. Sandhya, V. Srinivasan and R. Simon, *Phys. Rev.* **A41**, 3969 (1990).
7. H. Fearn and M. J. Collett, *Journal of Mod. Optics,* **35**, 553 (1988).
8. T. Welton, *Phys. Rev.* **74**, 1157 (1948).
9. H. Umezawa, Quantum Field Theory (North Holland, Amsterdam, 1956).
10. S. Chaturvedi, P. Shanta, and V. Srinivasan, *Phy. Rev.* **A43**, 521, (1991).

REFLECTION OF ELECTROMAGNETIC SIGNAL FROM A MIRROR WITH LARGE QUANTUM MECHANICAL POSITIONAL UNCERTAINTY

V. P. Bykov

The General Physics Institute
Vavilova, 38, Moscow
USSR

Now it is interesting to transfer the major ideas of the squeezed light to other, nonoptical fields of physics, e.g. to acoustics or even to mechanics. Such transfer is possible to any field where oscillators are quantized as bosons. In this paper a mechanical oscillator which is in the squeezed vaquum state is investigated. The moments of time when its coordinate uncertainty attains maximum, i.e. the oscillator is squeezed in momentum canonically conjugated to the coordinate are of greatest interest.

The paper is devoted to the investigation of the reflection of an electromagnetic pulse from a mirror which is a part of the mechanical oscillator being in the squeezed state mentioned above (Fig. 1). In general form this problem is mathematically very difficult. So we'll simplify it without changing its physical essence. Firstly let the mirror have only two degrees of freedom.

One is the transversal oscillator, it describes the motion of the negative charges (bound electrons) along the mirror (displacement - Q, canonically conjugated momentum - P, surface mass density - ρ, surface charge density - σ). Just the motion of the negative charges provide the reflection of the electromagnetic pulse. The second degree of freedom is the longitudinal oscillator, it describes the motion of the mirror along the radiation propagation direction (coordinate - q, canonically conjugated momentum - p, surface mass density - μ). Secondly we take into account only waves incident normally to the mirror and polarized along the displacement Q.

We also let the mirror be infinitely thin and oriented perpendicularly to z-axis which is the direction of the propagation of the electromagnetic waves incident to the mirror. The polarization of waves and the displacement Q are directed along the X-axis. The transversal form of the field is the same as of the mirror, the area of the mirror is equal to s. Then the quantum system mirror+field can be described by the Hamiltonian

$$H = \frac{1}{2\rho s}\left(P - \frac{\sigma s}{c}A(q)\right)^2 + \frac{1}{2}sKQ^2 + \frac{1}{2}s \cdot \int dz \left[\frac{1}{4\pi}\left(\frac{\partial A}{\partial z}\right)^2 + 4\pi c^2 \Pi^2\right] + \frac{1}{2\mu s}p^2 + \frac{1}{2}s\kappa q^2, \tag{1}$$

Recent Developments in Quantum Optics, Edited by R. Inguva, Plenum Press, New York, 1993

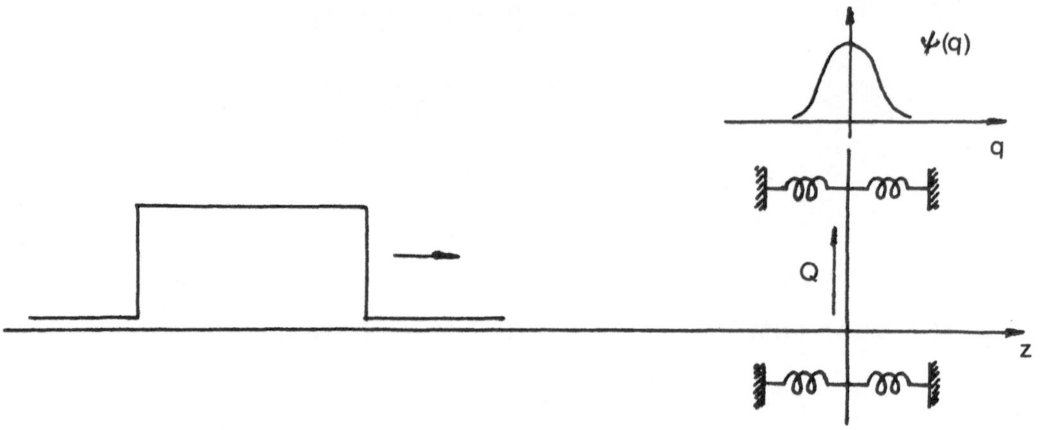

Figure 1. Signal reflection from the quantum mirror.

where K and κ are the elasticity coefficients of the transversal and longitudinal oscillators, $A(z)$ is the electromagnetic field vector-potential and $\Pi(z)$ is its conjugated momentum.

Let the field oscillators introduced as usual be initially in coherent states with such phases and amplitudes that the field forms a rectangular pulse filled in with high-frequency oscillations

$$< E_{in}(z,t) >= E_o sin\omega_o\tilde{t} \quad \Pi(\tilde{t}),$$

where

$$\tilde{t} = t - \bar{t} - z/c, \qquad \Pi(\tilde{t}) = \begin{cases} 1 & \text{when} \quad -\tau_o < \tilde{t} < \tau_o, \\ 0 & \text{when} \quad \tilde{t} < -\tau_o \text{ or } \tilde{t} > \tau_o, \end{cases}$$

$\bar{t}$ is the pulse arrival time at the coordinate origin where the mirror is located, $2\tau_o = 4\pi n_o/\omega_o$ is the pulse duration ($2n_o$ - the field periods number in the pulse). Let the transveral oscillator be in the ground or vacuum state. Let the longitudinal oscillator be in a squeezed state described by the wavefunction

$$\Psi(q) = (2\pi\bar{q}^2)^{-1/4} exp(-q^2/4\bar{q}^2), \tag{2}$$

where

$$\bar{q}^2 = \hbar/(2s\mu\nu\sqrt{\kappa})$$

and κ is the squeezing coefficient. The additional parameter in the wavefunction (2) - the coefficient of squeezing - provides the possibility to change the longitudinal oscillator parameters (μ,ν) in future without changing its distribution; in particular the transition to the motion in the free space $(\nu \rightarrow 0)$ will become possible.

The Heisenberg picture is convenient in our case. Then the Hamiltonian (1) gives the following Heisenberg equations

$$\frac{d^2Q}{dt^2} + \Omega^2 Q = \frac{-\sigma}{c\rho}\frac{dA(q,t)}{dt}, (\Omega^2 = K/\rho),$$

$$\frac{\partial^2 A}{\partial z^2} - \frac{1}{c^2}\frac{\partial^2 A}{\partial t^2} = -\frac{4\pi\sigma}{c}\frac{dQ}{dt}\delta(z-q),$$

$$\frac{d^2q}{dt^2} + \nu^2 q = \frac{\sigma}{c\mu}\frac{dQ}{dt}\frac{\partial A}{\partial t}|_{z=q}, (\nu^2 = \kappa/\mu). \tag{3}$$

The parameters μ and ν are only in the last Heisenberg equation (3). So in the limit $\nu \to 0$ and $\mu \to \infty$ the natural approximation is $q = const$. In this approximation we neglect a recoil essentially. Then the first two Heisenberg equations become linear and can be solved easily.

Our results are as follows. The field average value in the reflected signal is the stationary conditions and in the resonance ($\Omega = \omega_o$) is

$$< E_R(z,t) >= \frac{1}{2}E_o e^{-2\bar{q}^2/\lambda^2} sin\Omega(\tau - \bar{t}),$$

where $\lambda = \lambda/2\pi = c/\Omega$ and $\tau = t + z/c$. The average value of the squared field in the reflected signal is

$$< E_R^2(z,t) >= \frac{1}{8}E_o^2\left[1 - e^{-8\bar{q}^2/\lambda^2} cos2\Omega(\tau - t)\right];$$

this value is proportional to the electrical energy density. On the whole the reflected signal is shown in fig. 2. As can be seen when the position uncertainty of the mirror is small ($\bar{q} << \lambda$) the reflected signal keeps the properties of the coherent state; in particular $< E_R >^2 \simeq < E_R^2 >$. When the position uncertainty is large ($\bar{q} >> \lambda$) the field average value is close to zero but the average value of the squared field is not close to zero and keeps the finite value; only double frequency oscillations of energy density are close to zero. One can say that the amplitude reflection coefficient goes to zero when the position uncertainty grows, but the intensity reflection coefficient keeps finite value in the same case. It means that the reflected signal being a macroscopical one is in essentially a quantum state, as only in such states the relaxation ($< E >^2 < (E^2 >$) is possible. Consequently the distributed state of the mirror is an effective mean to transform the coherent light state into the quantum one, that is into the phaseless light state with average field strength close to zero.

An interesting consequence of the above results is the conclusions that it is possible to find out experimentally that the macroscopical body (mirror) is in a distributed quantummechanical state as a result of one pulse reflection from this body without essential change of its state. Indeed the reflected pulse, as we could see, bears the information about the distributed state of the mirror and being the macroscopical

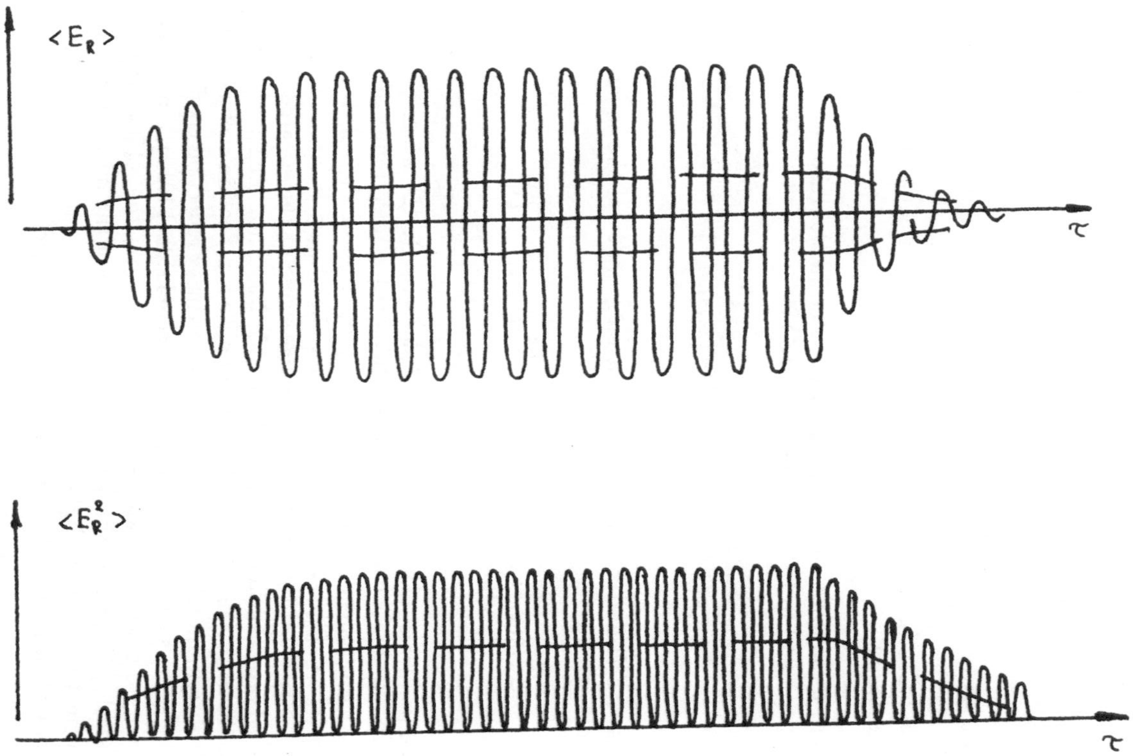

Figure 2. Reflected signal; continuous line is the signal reflected from the gathered mirror ($\bar{q} << \lambda$), dotted line is the amplitude of the signal reflected from the distributed mirror ($\bar{q} >> \lambda$).

one can be analyzed with the existing experimental means such as used for example to analyse the squeezed light [1,2]. At the same time the reflection process does not change the mirror state essentially. Such probability contradicts the usual interpretation of the distributed state as a state describing an ensemble of objects [3].

REFERENCES

1. R.E. Slusher, L. W. Hollberg, B. Yurke, J. C. Mertz, J. F. Valley *Phys. Rev. Lett.* **55**, 2409 (1985)
2. Ling-An Wu, H. J. Kimble, J. L. Hall, Huifa Wu *Phys. Rev. Lett.* **57**, 2520 (1986)
3. A. Messiah, Quantum Mechanics, Amsterdam, 2 vols, 1961-1962

TOWARDS SPECTROSCOPY OF PARTIALLY COHERENT SOURCES

Emil Wolf*

Department of Physics and Astronomy
University of Rochester
Rochester, NY 14627, USA

ABSTRACT

Traditional spectroscopy is concerned with spatially incoherent sources. In relatively recent times, after the invention of the lasers in 1960, spectroscopy of coherent sources has also been gradually developed. In the last few years it was found that spectroscopy of partially coherent sources i.e., those which are neither completely uncorrelated nor fully correlated, encounters certain problems. They arise from the fact predicted theoretically in 1986 and confirmed experimentally soon afterwards that, in general, the spectrum of light generated by a partially coherent source changes on propagation, even in free space. Such changes may take many different forms. For example a spectral line may be shifted or distorted or it may be split into several lines. In this talk the basic physical principles underlying this phenomenon will be described and some recent developments in this field will be reviewed.

Spectroscopy is based on the implicit assumption that the (relative) energy distribution in the spectrum of the electromagnetic radiation emitted by a source is entirely determined by the chemical composition of the source and that it is independent of the location, where the spectrum of the emitted radiation is measured. In a paper published in 1986 it was shown theoretically[1] that this assumption is incorrect and that, in general, the spectrum of the field depends not only on the spectrum of the source but also on some of its statistical properties, characterized by the so-called second-order degree of spectral coherence. It was also shown, that only for radiation generated by the most commonly occurring sources, for example blackbody sources and lasers, this effect may be absent under usual experimental conditions. This prediction has been verified experimentally soon afterwards[2] and has been demonstrated since then in many different laboratories.[3–8]

---

* Also at the Institute of Optics, University of Rochester.

Implications of this discovery for several areas of optical physics, optical engineering and astronomy are currently being explored. These investigations have already led to some interesting developments and are beginning to lay the foundations to a new discipline, namely spectroscopy of partially coherent sources. In this talk I will review very briefly the basic principles underlying these developments and I will describe some recent contributions to this subject.

Let me begin by stressing that any realistic optical source and, consequently, also the field which the source generates will in the course of time undergo random fluctuations. The fluctuations have to be characterized statistically, either in terms of appropriate probability distributions or in terms of various correlation functions. Suppose that $Q(\mathbf{r},t)$ represents a fluctuating source variable, say a Cartesian component of the current density, at a point whose position is specified by a vector $\mathbf{r}$, at time t, and that $V(\mathbf{r},t)$ represents the field which the source generates.† Assuming that the fluctuations are statistically stationary, i.e. that the underlying probabilities are invariant with respect to time translation, the simplest of the source correlation functions characterizes the "statistical similarity" of the source fluctuations at two space-time points, say $(\mathbf{r}_1,t)$ and $(\mathbf{r}_2,t+\tau)$. It is defined by the formula

$$\Gamma_Q(\mathbf{r}_1,\mathbf{r}_2,\tau) = < Q^*(\mathbf{r}_1,t)Q(\mathbf{r}_2,t+\tau) >, \tag{1}$$

where the angular brackets denote a statistical average. In a similar way we may define the correlation function of the field at two space-time points by the formula

$$\Gamma_V(\mathbf{r}_1,\mathbf{r}_2,\tau) = < V^*(\mathbf{r}_1,t)V(\mathbf{r}_2,t+\tau) > . \tag{2}$$

The functions Γ_Q and Γ_V characterize correlations in the *space-time domain.* For many purposes, particularly for the analysis of spectral properties of the field, it is more appropriate to employ functions which characterize correlations in the space-frequency domain. These correlation functions are just the Fourier transform of Γ_Q and Γ_V, viz.,

$$W_Q(\mathbf{r}_1,\mathbf{r}_2,\omega) = \frac{1}{2\pi}\int_{-\infty}^{\infty} \Gamma_Q(\mathbf{r}_1,\mathbf{r}_2,\tau)e^{i\omega\tau}d\tau, \tag{3}$$

$$W_V(\mathbf{r}_1,\mathbf{r}_2,\omega) = \frac{1}{2\pi}\int_{-\infty}^{\infty} \Gamma_V(\mathbf{r}_1,\mathbf{r}_2,\tau)e^{i\omega\tau}d\tau, \tag{4}$$

known as the *cross-spectral densities* of the source variable and of the field variable respectively. Actually the formulas (3) and (4) do not show that the cross-spectral densities themselves are correlation functions. This was demonstrated only relatively recently,[10] when it was shown that one may always introduce ensembles of random, frequency-dependent variables $\{U_Q(\mathbf{r},\omega)\}$ and $\{U_V(\mathbf{r},\omega)\}$ such that

$$W_Q(\mathbf{r}_1,\mathbf{r}_2,\omega) = < U_Q^*(\mathbf{r}_1,\omega)U_Q(\mathbf{r}_2,\omega) >_\omega \tag{5}$$

and

$$W_V(\mathbf{r}_1,\mathbf{r}_2,\omega) = < U_V^*(\mathbf{r}_1,\omega)U_V(\mathbf{r}_2,\omega) >_\omega . \tag{6}$$

†Throughout this paper we use the complex analytic signal representation ([9], Sec. 10.2) of the source variable and of the field variable. Also in order to bring out the essential features of correlation-induced spectral changes we ignore vector features of the source and of the field, i.e. we take $Q(\mathbf{r},t)$ and $V(\mathbf{r},t)$ to be scalar quantities.

In these two formulas the angular brackets, with the subscript ω, indicate averaging over these ensembles of frequency-dependent functions. The functions $U_Q(\mathbf{r},\omega)$ and $U_V(\mathbf{r},\omega)$ are *not* the Fourier transforms of the variables $Q(\mathbf{r},t)$ and $V(\mathbf{r},t)$ because, as is well-known, the sample functions of stationary random processes do not possess Fourier representation within the framework of ordinary function theory.[‡] The functions U_Q and U_V have to be introduced in terms of the eigenfunctions and the eigenvalues of a Fredholm integral equation, whose kernel is the cross-spectral density W_Q.

According to an obvious generalization of the Wiener-Khintchine theorem[11] the spectral density $S_Q(\mathbf{r},\omega)$ of the source distribution and the spectral density $S_V(\mathbf{r},\omega)$ of the field distribution are just the "*diagonal elements*" of the cross-spectral densities $W_Q(\mathbf{r}_1,\mathbf{r}_2,\omega)$ and $W_V(\mathbf{r}_1,\mathbf{r}_2,\omega)$, i.e.

$$S_Q(\mathbf{r},\omega) = W_Q(\mathbf{r},\mathbf{r},\omega), \tag{7}$$

and

$$S_V(\mathbf{r},\omega) = W_V(\mathbf{r},\mathbf{r},\omega). \tag{8}$$

In view of the formulas (5) and (6) the spectral densities may be expressed in intuitively more appealing forms as

$$S_Q(\mathbf{r},\omega) = < U_Q^*(\mathbf{r},\omega)U_Q(\mathbf{r},\omega) >_\omega \tag{9}$$

and

$$S_V(\mathbf{r},\omega) = < U_V^*(\mathbf{r},\omega)U_V(\mathbf{r},\omega) >_\omega . \tag{10}$$

Let us now consider light propagation in free space, from a localized source. Because the field variable and the source variable are then related by the inhomogeneous wave equations, one can show that the cross-spectral densities are rigorously related by the equation[12]

$$(\nabla_1^2 + k^2)(\nabla_2^2 + k^2)W_V(\mathbf{r}_1,\mathbf{r}_2,\omega) = (4\pi)^2 W_Q(\mathbf{r}_1,\mathbf{r}_2,\omega). \tag{11}$$

Here ∇_j^2, $(j = 1,2)$, is the Laplacian operator, acting with respect to the coordinates of the point $\mathbf{r}_j$ and

$$k = \omega/c, \tag{12}$$

c being the speed of light in vacuo. The solution of Eq. (11) for the cross-spectral density function of the field radiated by the source may readily be shown to be given by the formula

$$W_V(\mathbf{r}_1,\mathbf{r}_2,\omega) = \int_D\int_D W_Q(\mathbf{r}_1',\mathbf{r}_2',\omega)\frac{e^{ik(R_{22}-R_{11})}}{R_{11}R_{22}}d^3r_1'd^3r_2', \tag{13}$$

where

$$R_{11} = |\mathbf{r}_1 - \mathbf{r}_1'|, \qquad R_{22} = |\mathbf{r}_2 - \mathbf{r}_2'| \tag{14}$$

‡For discussion of this point see, for example, ref. [10a], Sec. 2.

and the integrations extend twice independently over the source domain D, assumed to be finite.

Suppose now that the field points are situated in the far zone of the source, at the same distance r from an origin O in the source region, in directions specified by unit vectors $\mathbf{u}_1$ and $\mathbf{u}_2$ respectively (Fig. 1). Then

$$R_{11} \sim r - \mathbf{u}_1 \cdot \mathbf{r}'_1, \qquad R_{22} \sim r - \mathbf{u}_2 \cdot \mathbf{r}'_2 \tag{15}$$

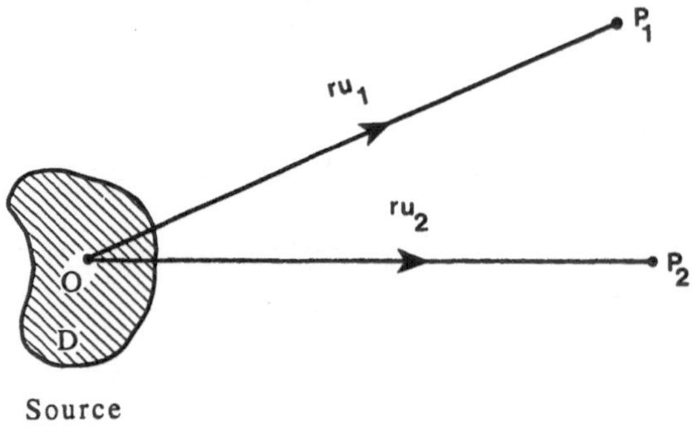

Figure 1. Illustrating the notation relating to the calculation of the cross-spectral density and of the spectrum of the field at point P_1 and P_2 in the far zone of a three-dimensional source.

($r = |\mathbf{r}|$) and the formula (13) becomes (with superscript "∞" indicating the far-zone value)

$$W_V^{(\infty)}(r\mathbf{u}_1, r\mathbf{u}_2, \omega) = \frac{1}{r^2}\tilde{W}_Q(-k\mathbf{u}_1, k\mathbf{u}_2, \omega), \tag{16}$$

where

$$\tilde{W}_Q(\mathbf{K}_1, \mathbf{K}_2, \omega) = \int\int W_Q(\mathbf{r}'_1, \mathbf{r}'_2, \omega) e^{-i(\mathbf{K}_1 \cdot \mathbf{r}'_1 + \mathbf{K}_2 \cdot \mathbf{r}'_2)} d^3r'_1 d^3r'_2. \tag{17}$$

Because $W_Q(\mathbf{r}'_1, \mathbf{r}'_2, \omega) = 0$ when either $\mathbf{r}'_1$ or $\mathbf{r}'_2$ are position vectors of points located outside the source region D, the integrals on the right-hand side of Eq. (17) may be taken formally to extend over the whole space. We see that $\tilde{W}_Q(\mathbf{K}_1, \mathbf{K}_2, \omega)$ is just the six-dimensional spatial Fourier transform of the cross-spectral density function of the source distribution. From Eqs. (8) and (16) it follows at once that the spectral density of the field in the far zone, at a point specified by the position vector $\mathbf{r} \equiv r\mathbf{u}(\mathbf{u}^2 = 1)$ is given by the formula

$$S_V^{(\infty)}(r\mathbf{u},\omega) = \frac{1}{r^2}\tilde{W}_Q(-k\mathbf{u},k\mathbf{u},\omega). \tag{18}$$

We will now show that this simple-looking formula has important physical implications regarding the spectra of fields produced by sources of different states of coherence. For this purpose we first introduce the degree of spectral coherence at frequency ω of the source distribution by the formula[13]

$$\mu_Q(\mathbf{r}_1',\mathbf{r}_2',\omega) = \frac{W_Q(\mathbf{r}_1',\mathbf{r}_2',\omega)}{\sqrt{S_Q(\mathbf{r}_1',\omega)}\sqrt{S_Q(\mathbf{r}_2',\omega)}}. \tag{19}$$

It can be shown that for all values of its variables,

$$|\mu_Q(\mathbf{r}_1',\mathbf{r}_2',\omega)| \leq 1. \tag{20}$$

The extreme value $|\mu_Q| = 1$ characterizes complete correlation (complete spatial coherence) and the other extreme, $\mu_Q = 0$, characterizes complete absence of correlation (complete spatial incoherence) of the source fluctuations at the points $\mathbf{r}_1'$ and $\mathbf{r}_2'$, at frequency ω.

Suppose now that the spectrum of the source is the same at every source point. We may then write $S_Q(\omega)$ in place of $S_Q(\mathbf{r},\omega)$ and Eq. (19) shows that in this case

$$W_Q(\mathbf{r}_1',\mathbf{r}_2',\omega) = S_Q(\omega)\mu_Q(\mathbf{r}_1',\mathbf{r}_2',\omega). \tag{21}$$

It will be useful to introduce the normalized far-zone spectrum

$$s_V^{(\infty)}(r\mathbf{u},\omega) = \frac{S_V^{(\infty)}(r\mathbf{u},\omega)}{\int_0^\infty S_V^{(\infty)}(r\mathbf{u},\omega)d\omega}, \tag{22}$$

whose integral, over all frequencies has evidently the value unity. On substituting from Eqs. (18) into Eq. (22) and making use of the formula (21) we finally obtain the following expression for the normalized far-zone spectrum:

$$s_V^{(\infty)}(r\mathbf{u},\omega) = \frac{1}{N(\mathbf{u})}S_Q(\omega)\tilde{\mu}_Q(-k\mathbf{u},k\mathbf{u},\omega), \tag{23}$$

where

$$\tilde{\mu}_Q(\mathbf{K}_1,\mathbf{K}_2,\omega) = \int\int \mu_Q(\mathbf{r}_1',\mathbf{r}_2',\omega)e^{-i(\mathbf{K}_1\cdot\mathbf{r}_1'+\mathbf{K}_2\cdot\mathbf{r}_2')}d^3r_1'd^3r_2' \tag{24}$$

is the six-dimensional spatial Fourier transform of the degree of spectral coherence of the source and

$$N(\mathbf{u}) = \int_0^\infty S_Q(\omega)\tilde{\mu}_Q(-k\mathbf{u},k\mathbf{u},\omega)d\omega \tag{25}$$

is a normalization factor.

The formula (23) is one of the main results relating to spectra of radiation fields produced by sources of any state of coherence. It shows that *the normalized far-zone spectrum depends not only on the source spectrum but also on the degree of spectral coherence of the source.* Moreover, as the formula also shows, the far-zone spectrum depends, in general, also on the direction $\mathbf{u}$ of observation.

It is clear from Eq. (23) that the coherence properties of the source may drastically affect the form of the spectrum of the radiated field. In particular, it is evident that if the source spectrum $S_Q(\omega)$ is a single spectral line centered on frequency ω_0 and if for some direction $\mathbf{u}$, $\tilde{\mu}_Q(-k\mathbf{u}, k\mathbf{u}, \omega)$ is sharply peaked around some frequency $\omega_1 \neq \omega_0$, the normalized far-zone spectrum $s_V^{(\infty)}(r\mathbf{u}, \omega)$ will no longer be centered on ω_0 but rather on a frequency that is closer to ω_1. In particular if $\omega_1 < \omega_0$ the spectrum $s_V^{(\infty)}$ will be centered at a frequency lower than ω_0, i.e. it will exhibit a *redshift*, whereas when $\omega_1 > \omega_0$ it will be centered on a frequency higher than ω_0 i.e. it will exhibit a *blueshift*. Such spectral changes[§] were discussed in refs. [14] and [15] and are illustrated by an example in Fig. 2.

The preceding analysis was concerned with influence of correlations involving three-dimensional primary sources. However, the phenomenon we are discussing is much more general and is manifested also in situations which involve secondary sources of various kinds, such as illuminated apertures[1,5,6,7,16] or illuminated scattering media whose macroscopic parameters vary randomly in space[17] and possibly also in time.[18–23] In some cases the spectral changes produced by scattering from suitably correlated fluctuating random media may imitate the Doppler effect, even though the source, the scattering medium and the observer are all at rest with respect to each other. An example is given in Fig. 3.

Perhaps the simplest situation which illustrates correlation-induced spectral changes involves two small apertures illuminated by partially coherent light.[¶] An example of this kind, due to Gori, Palma and Padovani,[25] is illustrated in Fig. 4. It makes use of two, mutually independent beams B_α and B_β, with nearly the same spectrum, illuminating pinholes P_1 and P_2. A phase difference is introduced between the beams by means of a beam splitter BS. By superposing the two beams the light at the two pinholes becomes partially coherent and its degree of coherence can be varied by changing the orientation of the beam splitter. With suitable orientations redshifts as well as blueshifts of spectral lines can be produced.

Another conceptually simple demonstration of the effects of source correlations on the spectra of radiation is by means of the Young interference experiment with partially coherent light, provided the bandwidth of the light incident on the pinholes is sufficiently broad or its degree of spectral coherence varies rapidly enough with frequency over the bandwidth.[26,27] For example when a uniform, circular, quasi-homogeneous,[28] secondary source illuminates two symmetrically placed pinholes P_1 and P_2 as shown in Fig. 5a, the light at the two pinholes will be partially coherent and its degree of coherence will depend on the pinhole separation.[‖] In consequence the spectrum of the

[§]First experimental demonstration of shifts of spectral lines due to source correlations was made with acoustical sources by M. F. Bocko, D. H. Douglass and R. S. Knox, *Phys. Rev. Let.* **58**, 2649-2651 (1987).

[¶]The possibility that when two beams with identical normalized spectra are superposed, the spectrum of light in the region of superposition may differ from that of the beams was first pointed out by L. Mandel [24] in his research on cross-spectral purity.

[‖]This fact follows from one of the reciprocity relations for fields generated by quasi-homogenous sources [28] which is analogous to the well-known van Cittert-Zernike theorem ([9], Sec. 10.4.2).

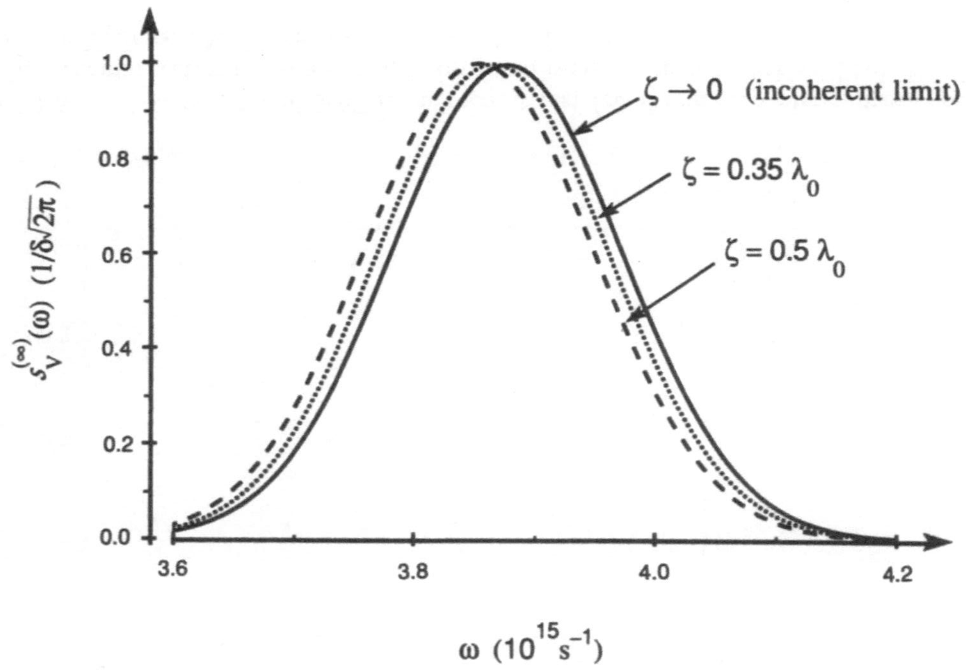

Figure 2. Normalized spectra $s_V^{(\infty)}(\omega)$, in units of $(1/\delta\sqrt{2\pi})$, of the far field generated by sources with the same normalized source spectrum $s_Q(\omega) = (\delta\sqrt{2\pi})^{-1} exp[-(\omega - \omega_0)^2/2\delta^2]$ and degree of spectral coherence $\mu_Q(\mathbf{r}', \omega) = exp[-\mathbf{r}'^2/2\zeta^2]$, with $\omega_0 = 3.887\text{x}10^{15}\text{s}^{-1}(\lambda_0 = 4861\text{Å})$ and $\delta = 9.57\text{x}10^{13}\text{s}^{-1}$, for several values of the effective source correlation length ζ. The solid curve $(\zeta \to 0)$ also represents the source spectrum (adapted from E. Wolf [14]).

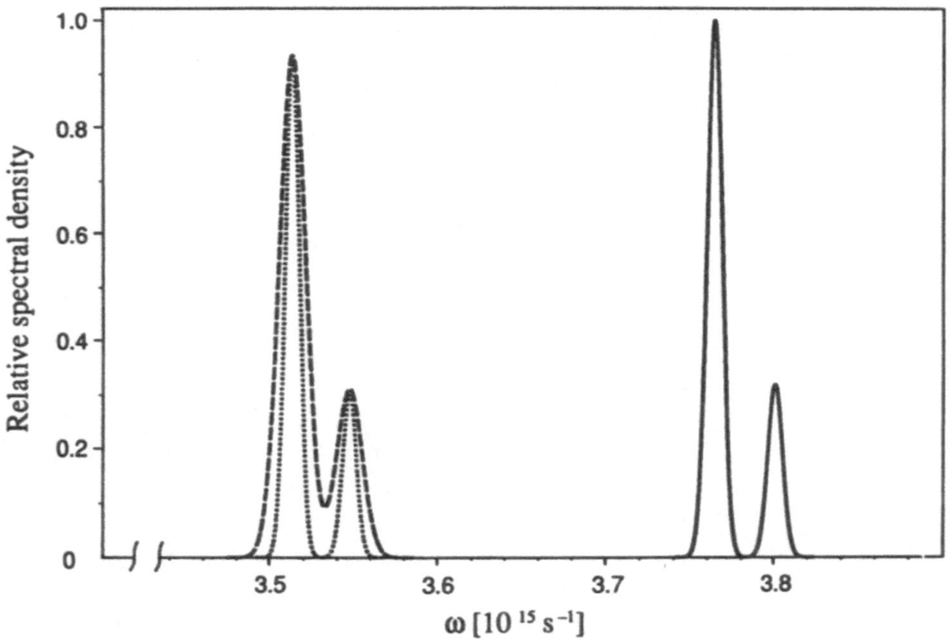

Figure 3. Two O III lines ($\lambda = 4959$Å and 5007 Å) as seen at rest (*solid line*), Doppler shifted (dotted line) and shifted by dynamic scattering on a suitably correlated random medium (dashed line). For details see the paper by D.F.V. James, M. Savedoff and E. Wolf [20], from which this figure is reproduced.

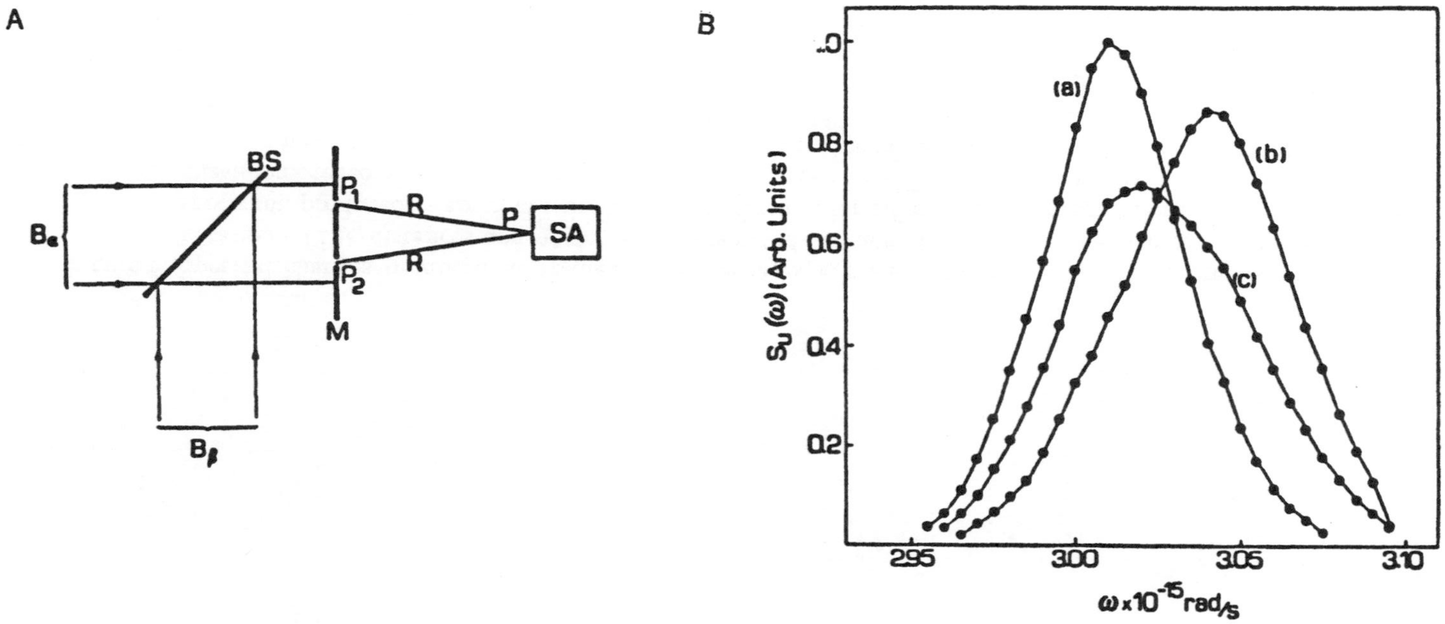

Figure 4. Experimental demonstration of shifts of spectral lines due to correlation between two secondary sources. (A) Layout of the experiment: B_α, B_β are mutually uncorrelated beams of essentially the same spectrum, BS is a beam splitter, P_1, P_2 are pinholes (secondary sources), SA is a spectral analyzer. (B) Results of observations shown by heavy dots. Redshifted spectrum (a) and blueshifted spectrum (b). The curve (c) shows the spectrum, multiplied by a factor 2, produced at the point of observation by light passing through only one of the pinholes. For the sake of clarity, the experimental points have been connected by lines. (After F. Gori, G. Guattari, C. Palma and C. Padovani [25]).

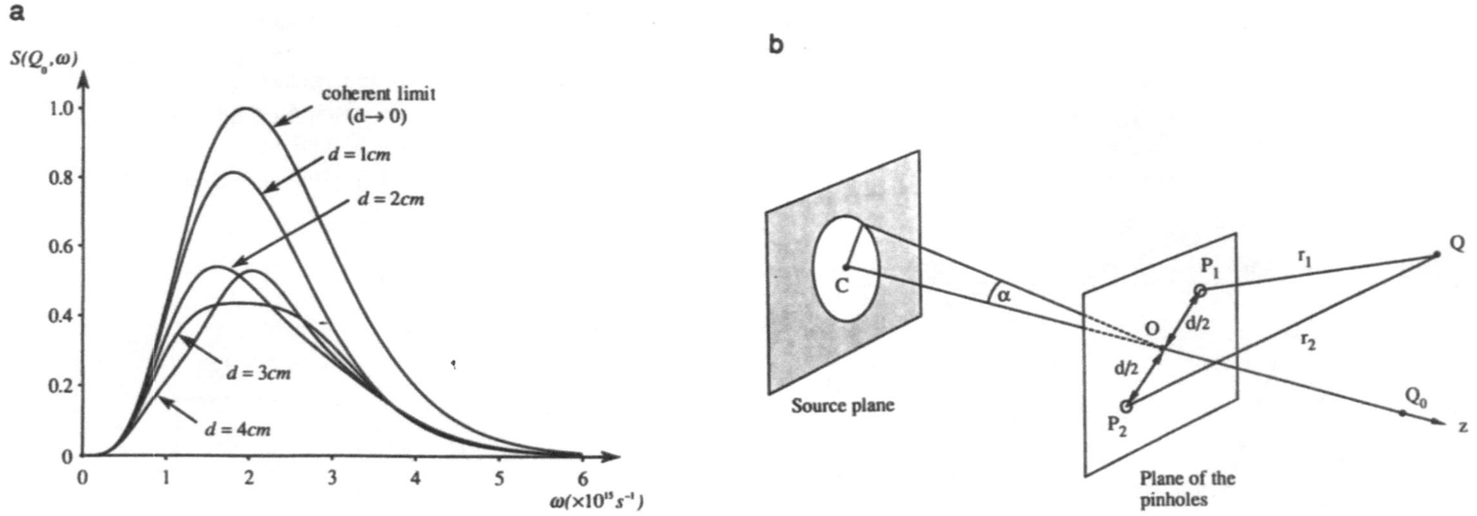

Figure 5. Spectral changes produced in Young's interference experiment. (a) Illustrating the geometry and the notation. (The distances and angles are exaggerated to show detail.) (b) The changes in the Planck spectrum, produced at an axial point Q_0 by interference, for different pinhole separations d. The circular, quasi-homogeneous source was assumed to be at temperature $T = 3000°$K and to subtend an angular semi-diameter $\alpha = 2.96\text{x}10^{-5}$ radians at 0. The units on the vertical axis are arbitrary, but are the same for all values of d. (After D. F. V. James and E. Wolf [26]).

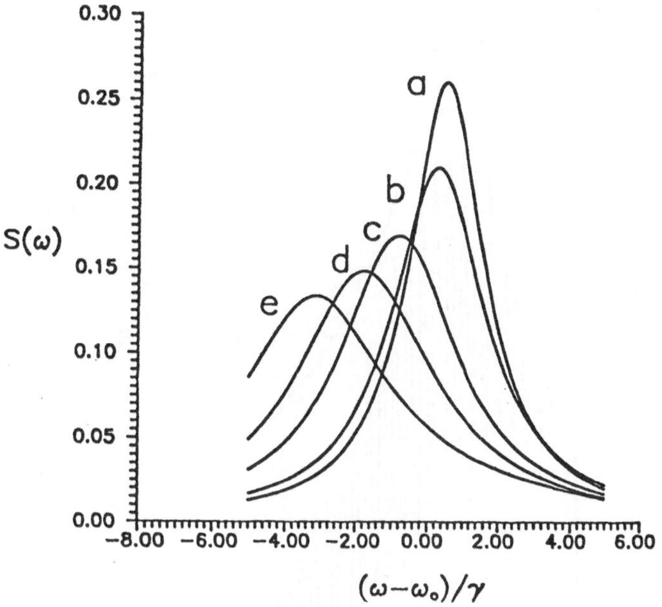

Figure 6. The spectra $S(\omega)$ of radiation emitted by a pair of identical two-level atoms, which are interacting with a thermal field at a fixed temperature and which are coupled by the dipole-dipole interaction. The direction of observation was chosen to be perpendicular to the line joining the two atoms and the mean number of thermal photons was taken as $\bar{n} = 0.1$. The different curves correspond to different values of the (dimensionless) interatomic distance $k_0 r_{12}$. The curves (a) and (b) correspond to $k_0 r_{12} = 3$ and 2 respectively (blueshifted lines), whereas (c), (d) and (e) correspond to $k_0 r_{12} = 1$, 0.7 and 0.5 respectively (redshifted lines). The normalization factor γ which appears in the variable that indicates the relative frequency shift (horizontal axis) is one-half of the Einstein A-coefficient. (After Varada and Agarwal [29]).

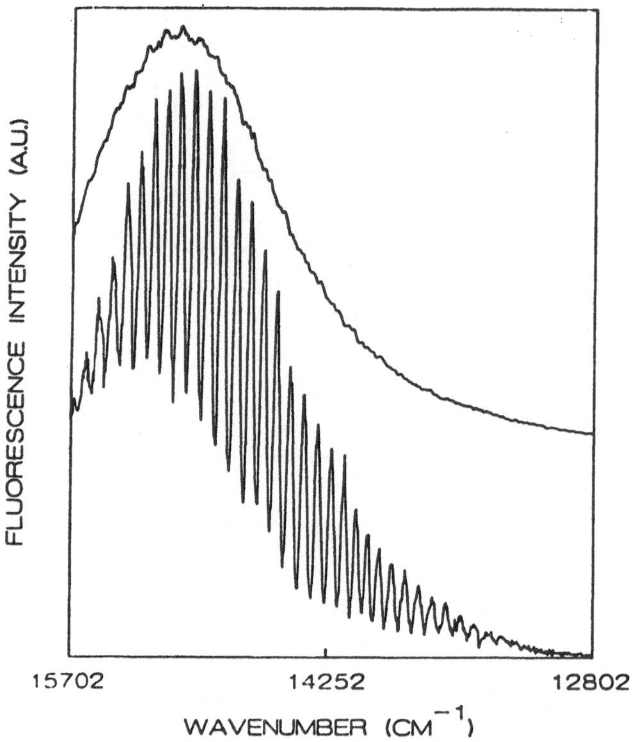

Figure 7: Fluorescence spectra observed at different points of radiating molecules placed in the neighborhood of a mirror. Both spectra were obtained from the same sample. (After H.K.E. Drabe, C. Cnossen, D.A. Wiersma, H.A. Ferwerda and B.J. Hoenders [30]).

light in the region of superposition will differ from the spectrum of the light incident on the two pinholes. In Fig. 5b spectra are shown which would be observed at an axial point Q_0, if the pinholes were illuminated with thermal radiation whose spectrum is given by Planck's law.**

So far we described correlation-induced spectral changes entirely on the basis of classical, statistical theory, but this effect is not restricted to the domain of classical physics. First quantum analysis of a phenomenon of this kind was developed by Varada and Agarwal[29] who studied frequency shifts due to cooperative effects at the microscopic level. They examined the spectrum of radiation emitted by an atomic system consisting of two identical two-level atoms which interact with a thermal field and which are coupled by the dipole-dipole interaction. They found that for directions of observation perpendicular to the line joining the two atoms, the steady-state spectrum consists of a single line which is, however, shifted with respect to the line that would be generated by each atom separately. The direction of the shift was found to depend on the sign of the dipole-dipole interaction term, which in turn is determined by the interatomic distance. With a mean number of thermal photons $\bar{n} = 0.1$ they found that when the distance between the atoms is greater than about half a wavelength, the spectral line is blueshifted, whereas for somewhat smaller separations it is redshifted (Fig. 6). For much smaller interatomic separation the spectrum breaks up into two distinct resonances.

Some experiments which illustrate correlation-induced spectral changes and which are most naturally analyzed quantum mechanically were performed by Drabe *et al.*[30] They found that the fluorescence spectra of radiating molecules change drastically when a mirror is placed in their vicinity and that the spectra are different at different points in space (Fig. 7). These spectral changes may be regarded as manifestation of correlations which exist between the molecules and their mirror image.

In this short review we were only able to present an account of the basic principles underlying the rapidly developing field of spectroscopy of partially coherent sources and to note some of the recent contributions to this field. Although the first papers on this subject appeared only six years ago, there are by now about sixty publications dealing with this topic and some applications of the underlying phenomenon of correlation-induced spectral changes have been considered. Some of these developments are briefly described in two recent review articles.[31,32]

ACKNOWLEDGEMENT

This paper was prepared under the sponsorship of the Department of Energy, under grant DE-FG02-90ER 14119. The views expressed in this article do not constitute an endorsement by the Department of Energy.

REFERENCES

1. E. Wolf, *Phys. Rev. Lett.* **56**, 1370 (1986).
2. G. M. Morris and D. Faklis, *Opt. Commun.* **62**, 5 (1987).
3. F. Gori, G. Guattari, C. Palma and C. Padovani, *Opt. Commun.* **67**, 1 (1988).
4. D. Faklis and G. M. Morris, *Opt. Lett.* **13**, 4 (1988).

**Since this talk was given, spectral changes produced in Young's interference experiment were observed in beautiful experiments performed by M. Santarsiero and F. Gori, (*Phys Lett A*, in press).

5. G. Indebetouw, *J. Mod. Opt.* **36**, 251 (1989).

6. H. C. Kandpal, J. S. Vaishya and K. C. Joshi, *Opt. Commun.* **73**, 169 (1989).

7. H. C. Kandpal. J. S. Vaishya and K. C. Joshi, *Phys. Rev.* A**41**, 4541 (1990).

8. H. K. E. Drabe, G. Cnossen, D. A. Wiersma, H. A. Ferwerda and B. J. Hoenders, *Phys. Rev. Lett.* **65**, 1427 (1990).

9. M. Born and E. Wolf, **Principles of Optics** (Oxford and New York, Pergamon Press, 6th Ed., 1980).

10. E. Wolf, (a) *J. Opt. Soc. Amer.* **72**, 343 (1982); (b) *ibid* A**3**, 76 (1986).

11. C. Kittel, **Elementary Statistical Physics** (New York, Wiley, 1958), Sec. 28.

12. W. H. Carter and E. Wolf, *Optica Acta* **28**, 227 (1981).

13. L. Mandel and E. Wolf, *J. Opt. Soc. Amer.* **66**, 529 (1976).

14. E. Wolf, *Nature* **326**, 363 (1987).

15. E. Wolf. *Opt. Commun.* **62**, 12 (1987).

16. J. T. Foley, (a) *Opt. Commun.* **75**, 347 (1990); (b) *J. Opt. Soc. Amer.* **A8**, 1099 (1991).

17. E. Wolf, J. T. Foley and F. Gori, *J. Opt. Soc. Amer.* A**6**, 1142 (1989); *ibid* **7**, 173 (1990).

18. J. T. Foley and E. Wolf, *Phys. Rev.* A**40**, 588 (1989).

19. E. Wolf, *Phys. Rev. Lett.* **63**, 2220 (1989).

20. D. F. V. James, M. Savedoff and E. Wolf, *Astrophys. J.* **359**, 67 (1990).

21. D. F. V. James and E. Wolf, *Phys. Letts.* A**146**, 167 (1990).

22. B. Cairns and E. Wolf, *J. Opt. Soc. Amer.* A**8**, 1992 (1991).

23. J. S. Vaishya, M. Chandra, H. C. Kandpal and K. C. Joshi, submitted to *Opt. Commun.*

24. L. Mandel, *J. Opt. Soc. Amer.* **51**, 1342 (1961).

25. F. Gori, G. Guattari, C. Palma and C. Padovani, *Opt. Commun.* **67**, 1(1988).

26. D. F. V. James and E. Wolf, *Opt. Commun.* **81**, 150 (1991).

27. D. F. V. James and E. Wolf, *Phys. Letts.* A**157**, 6 (1991).

28. W. H. Carter and E. Wolf, *J. Opt. Soc. Amer.* **67**, 785 (1977).

29. G. V. Varada and G. S. Agarwal, *Phys. Rev.* A**44**, 7626 (1991).

30. H K. E. Drabe, G. Cnossen, D. A. Wiersma, H. A. Ferwerda and B. J. Hoenders, *Phys. Rev. Lett.* **65**, 1427 (1990).

31. P. W. Milonni and S. Singh, in *Advances in Atomic, Molecular and Optical Physics*, D. Bates and B. Bederson, ed., (Academic Press, New York, 1991), vol. 28, chapt. VIII, p. 127.

32. E. Wolf, in **International Trends in Optics**, ed J. W. Goodman (Academic Press, San Diego, 1991), p. 221.

A MICROSCOPIC APPROACH TO WOLF EFFECT†

G. V. Varada

School of Physics
University of Hyderabad
Hyderabad-500 134, India

It has been shown by Wolf[1,2] that correlations between different parts of the source play an important role in determining the spectrum of the light emitted by the source. In particular, the frequency of the emitted light shows either a red shift or a blue shift as compared to its frequency when the source is assumed to be completely uncorrelated. Here, we present a microscopic model for the origin of this phenomenon.[3] For simplicity, we assume that the source consists of a pair of two- level atoms interacting with each other via the dipole-dipole interaction. We also include the effects of the finite temperature of the source. For this microscopic model, we calculate the spectrum observed in the far zone. We show how the correlations between the atoms induced by the dipole-dipole interaction affects the red and blue frequency shifts in the spectrum.

The Hamiltonian for the atomic system consisting of two identical two-level atoms of frequency ω_o interacting via the dipole-dipole interaction, in the interaction picture is

$$H = \Omega_{12}(S_1^+ S_2^- + h.c.), \tag{1}$$

where $S_i^{\pm}, i = 1,2$ are the spin 1/2 operators for each of the two level atoms. The dipole-dipole interaction coupling constant in Eq. (1) is given by

$$\Omega_{12} = -2\gamma cos(k_o r_{12})/k_o r_{12}, k_o = \omega_o/c \tag{2}$$

where we have averaged over the random orientation of the dipole moment $\vec{d}$ of the atoms with respect to $\vec{r}_{12}$ the line joining the two atoms. γ is the spontaneous decay rate of each atom.

The central variable required in the calculation of the steady state spectrum of the radiation emitted from the pair of interacting atoms in the far zone is the two time correlation function defined as

†Work done in collaboration with G.S. Agarwal.

$$G(t,t+\tau)=\sum_{ij=1}^{2} e^{i\vec{k}.(\vec{r}_i-\vec{r}_j)} < S_i^+(t+\tau)S_j^-(t) >, \vec{k}=k_o\vec{R}, \tag{3}$$

where $\vec{R}$ is the direction of observation. If the direction of observation is perpendicular the interatomic distance, then Eq. (3) gives

$$G(t,t+\tau)=\sum_{ij=1}^{2} < S_i^+(t+\tau)S_j^-(t) > . \tag{4}$$

In the original frame, the steady state spectrum is given by

$$S(\omega)=\lim_{t\to\infty}\ 2Real\int_o^\infty d\tau e^{-i(\omega-\omega_o)\tau}\Big\{ < S_1^+(t+\tau)S_1^-(t) >$$

$$+ < S_2^+(t+\tau)S_1^-(t) > + < S_1^+(t+\tau)S_2^-(t) > + < S_2^+(t+\tau)S_2^-(t) > \Big\} \tag{5}$$

Eq. (5) shows that the far zone spectrum has contributions arising from each atom independently as well as from the correlations between the atoms. In the absence of the dipole-dipole interaction, the atoms are uncorrelated. The far zone spectrum thus has contributions from just the first two terms in Eq. (5). As the atoms come closer to each other, the dipole- dipole interaction makes the atoms correlated. The contribution to the far zone spectrum arising from the correlations between the atoms is given by the last two terms in Eq. (5). These terms are responsible for the shift in the frequency of the spectrum.

We use the master equation approach to calculate the correlation functions in Eq. (5). The master equation for the evolution of the atomic density operator is given by

$$\frac{\partial\rho}{\partial t}+(1+\bar{n})\sum_{ij=1}^{2}\gamma_{ij}(S_i^+S_j^-\rho-2S_j^-\rho S_i^+ +\rho S_i^+S_j^-)+\bar{n}\sum_{ij=1}^{2}\gamma_{ij}(S_i^-S_j^+\rho-2S_j^+\rho S_i^- +\rho S_i^-S_j^+)$$

$$+i\sum_{\substack{ij=1\\ i\neq j}}^{2}\Omega_{ij}\left[S_i^+S_j^-,\rho\right]=0. \tag{6}$$

Here $\bar{n}$ is the mean number of thermal photons per mode at the optical frequency ω_o i.e.,

$$\bar{n}=\frac{1}{e^{\beta\omega_o}-1},\qquad \omega_o=k_oc, \tag{7}$$

The collective decay parameter γ_{12} obtained after averaging over the random orientation of the atomic dipole moment with respect to the interatomic distance is given by

$$\gamma_{12}=\frac{\gamma sin(k_or_{12})}{(k_or_{12})}. \tag{8}$$

Fig. 1 shows the far zone spectrum S(ω) for various values of the interatomic distance as a function of the frequency $(\omega-\omega_o)/\gamma$. For independent atoms (Ω_{12} =

0), the spectrum consists of a single peak at origin. At finite interatomic distances the atoms get coupled due to the dipole-dipole interaction and the spectrum shows a frequency shift. The direction of the shift is determined by the sign of the dipole-dipole interaction term (Eq. (2)) which in turn is determined by the distance between the atoms. Our results show that for $k_o r_{12} > \pi/2$ the spectral line is blue shifted from the origin. Conversely for $k_o r_{12} < \pi/2$ the spectrum is red shifted. This behaviour can be intuitively understood from the energy level diagram of the atomic system. The atomic system consisting of two identical two-level atoms is equivalent to a four level atomic system. The four levels can be described in terms of the eigen states associated

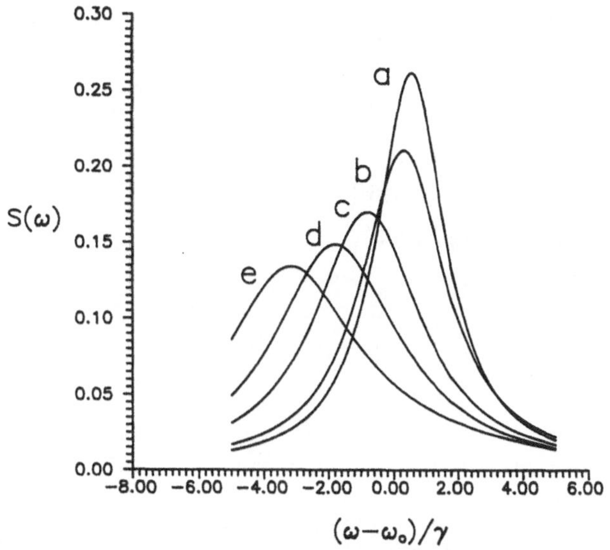

Figure 1. The red and blue shift of the spectrum $S(\omega)$ of the radiation emitted by a pair of identical coupled atoms, as a function of the frequency $(\omega - \omega_o)/\gamma$. The mean number of photons is $\bar{n} = 0.1$. Different curves correspond to various values of the interatomic distance $k_o r_{12}$. The curves (a) and (b) correspond respectively to $k_o r_{12} = 3$ and 2 (blue shifted lines). Curves (c), (d), (e) respectively correspond to $k_o r_{12} = 1, 0.7, 0.5$ (red shifted lines).

with the collective spin operator $\vec{S} = \vec{S}_1 + \vec{S}_2$. One then has three symmetric states corresponding to S=1 and one antisymmetric state corresponding to $S = 0$. Figure 2 shows the energy levels and their eigen values. Neglecting the weak transitions to the antisymmetric state, the population distribution in the symmetric states follows the Boltzmann distribution at thermal equilibrium. Thus the transition at frequency $\omega_o + \Omega_{12}$ from $|1,0>$ to $|1,-1>$ state dominates over the transition at frequency $\omega_o - \Omega_{12}$ from $|1,1>$ to $|1,0>$ state. Then, the spectrum should be blue shifted if Ω_{12} is positive and red shifted if Ω_{12} is negative. The sign of Ω_{12} is determined by the distance between the atoms and thus the far zone spectrum shows the red and blue frequency shift as the atoms come closer to each other in the presence of the dipole-dipole interaction.

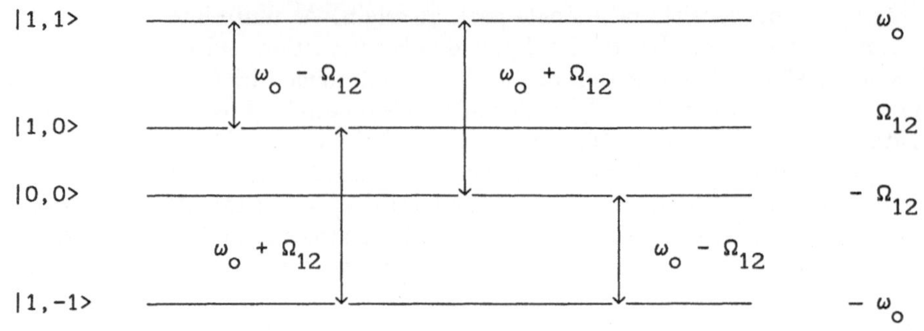

Figure 2. Schematic diagram of the energy levels of the two-atom system in the collective spin operator basis. The corresponding energy eigenvalues are shown.

REFERENCES

1. E. Wolf, *Nature* (London) **326**, 363 (1987)
2. E. Wolf, *Phys. Rev. Lett.* **58**, 2646 (1987)
3. G. V. Varada, and G. S. Agarwal.

STILES-CRAWFORD EFFECT OF SECOND KIND IN THE LIGHT OF SPECTRAL SHIFT DUE TO SOURCE CORRELATION

Mahesh Chander, H.C. Kandpal, J.S. Vaishya and K.C. Joshi

Standard Division
National Physical Laboratory
New Dehli - 110 012

ABSTRACT

Stiles-Crawford effect of second kind (SCE II) states that a monochromatic stimuli when excites the same retinal area from different angles, its perceived hue depends on its obliquity with respect of retina. SCE II is believed to be of retinal origin. The theories proposed to explain the phenomenon are based on the assumption that cone photoreceptors show directional sensitivity. On the contrary Alpern et al,[6–8] have demonstrated that underlying mechanism of SC I and SC II are not common. In other words it means that SCE I may be explained in terms a directional sensitivity concept of cone respectors but not SCE II. However, it still remains a big question - what causes SCE II.

In this communication we have endeavored to explain the cause of SCE II effect on the basis a physical phenomenon known as source correlation or spatial coherence existing between source points.[(10)] It has been theoretically predicted by Wolf, that source correlation causes spectral shift when light propagates through free space. More recently Foley[(12)] has shown that a certain correlation introduced by focusing optics and dispersing element at an aperture causes spectral shift. His theory supports the experimental observations already made.[(16)] On the basis of source correlation theory leading to spectral shift SCE II effect could be given an explanation that the source correlation developed during the passage of stimuli through occular media changes the spectral distribution. Detailed experimental analysis is also given.

INTRODUCTION

It is observed that the spectral distribution of a stimulus entering the pupil at its center (at normal incidence) appears different than when entering at a point away from the center (at oblique incidence). To explain such spectral changes called Stiles-Crawford effect of second kind[(1)] (SCE II), various physiological models[(2–5)] have been proposed in the past. But none of these models is able to explain successfully the observed spectral changes. Alpern et al,[(6–8)] through their detailed studies have inferred that the fundamental assumption of Enoch and Hiles[(3)] that Sc E-II in due to differences in directional response of three cone systems is not true.

Recent Developments in Quantum Optics, Edited
by R. Inguva, Plenum Press, New York, 1993

In recent years, theoretical studies[9-12] have shown that as a consequence of spatial correlations within the source, the spectrum of light is not invariant on propagation even in free space. Experimental studies[13-18] have confirmed that suitable source correlations can cause redshifts or blueshifts of stimulus. On account of an intimate relationship between radiation emanating from localized sources and scattering from the media of a finite extent, it has been shown theoretically[19-21] and confirmed experimentally[22] that spectral shifts may also be produced by scattering on static as well as temporarily varying media. These recent findings may influence the explanation given for the observed spectral changes in colour vision experiments.

In the present study it has been shown that the optical set-up used in colour vision experiments produces a correlation and the spectral shift during propagation of the stimulus through the occular media. It has been shown experimentally that the spectrum of the secondary source formed at the focal plane of the eye lens, i.e. retinal region, changes during its course of propagation. The spectral shift observed is different for normal and oblique incidence. Further change in the spectral shift may also arise as a result of scattering from occular media. In short, our study suggests that a part of the observed spectral change is due to source correlation induced by the combined effect of the optical set-up used in the experiment, the eye lens and cornea and scattering from other components of occular media e.g. aqueous humor and vitreous body.

SOURCE CORRELATION THEORY

Emil Wolf[2] has shown that the normalized spectrum of light produced by a planer, secondary, quasi-homogeneous source which has the same normalized spectrum at every source point, is the same throughout the far zone and across the source itself if the degree of spatial coherence of the light in the source plane has the functional form

$$\mu^{o}(P_2 - P_1, \omega) = f(k!P_2 - P_1!) \tag{1}$$

where $\omega = kc = 2\pi c/\lambda$

The eqn. (1) is called the scaling law and shows that the degree of spatial coherence μ^o of light at any two points P_2 and P_1 in the source plane is a function of a single variable $k|P_2 - P_1|$. If the source plane does not obey the scaling law, its normalized spectrum changes on propagation[9-11] i.e. the spectral distribution at the detector plane is different from that at the source plane.

Recently Foley[12] has shown that the spatial correlations of the radiated light from an incoherent source at a circular aperture kept in the focal plane of a lens causes spectral shift when detected at an on-axis points in the far zone. These theoretical findings supported earlier experimental results.[12-14] It has been shown that the magnitude of spectral shift due to source correlation depends considerably on the direction of detection in the far field and its magnitude is different for on-axis and off-axis points.[13-15,24]

EXPERIMENTAL PROCEDURE AND RESULTS

To demonstrate the effect of source correlation in SCE II, light from a 450 W tungsten halogen lamp was made incident on a monochromator without using any

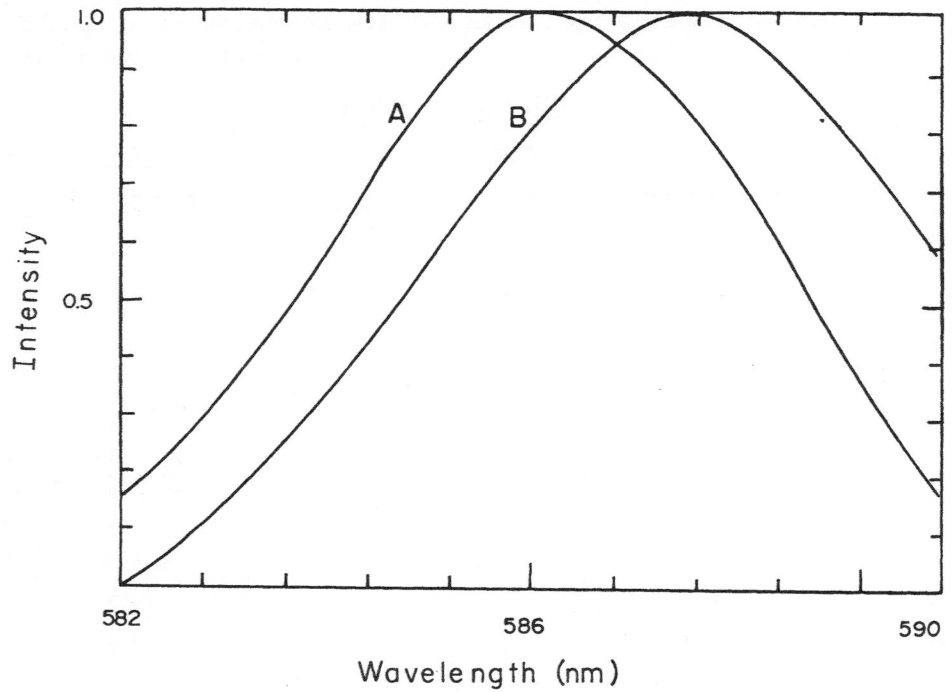

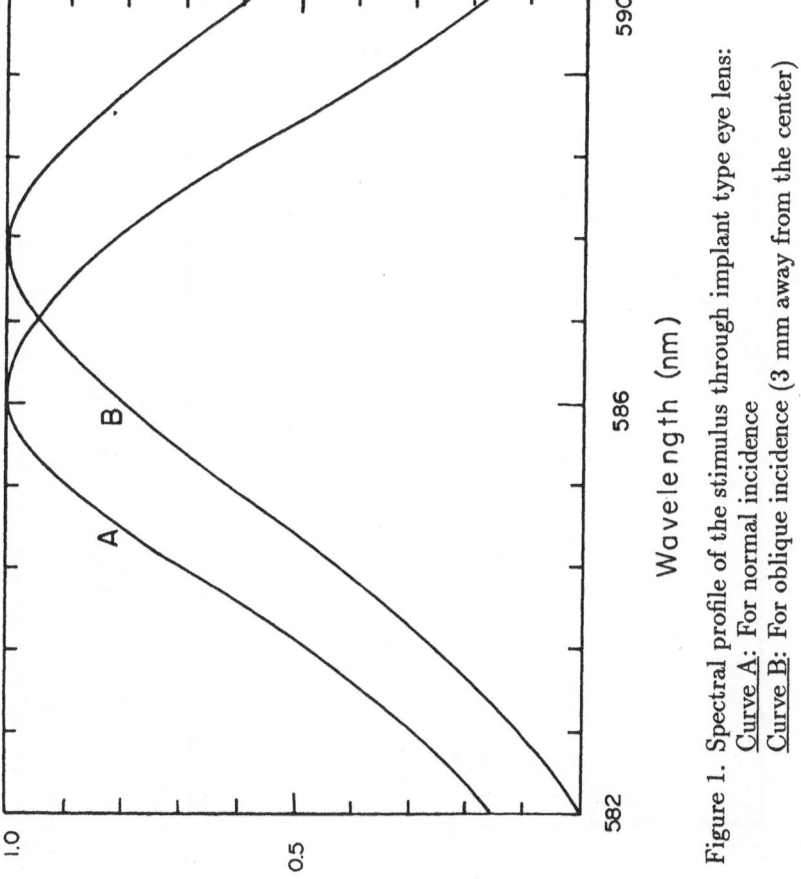

Figure 1. Spectral profile of the stimulus through implant type eye lens:
<u>Curve A</u>: For normal incidence
<u>Curve B</u>: For oblique incidence (3 mm away from the center)

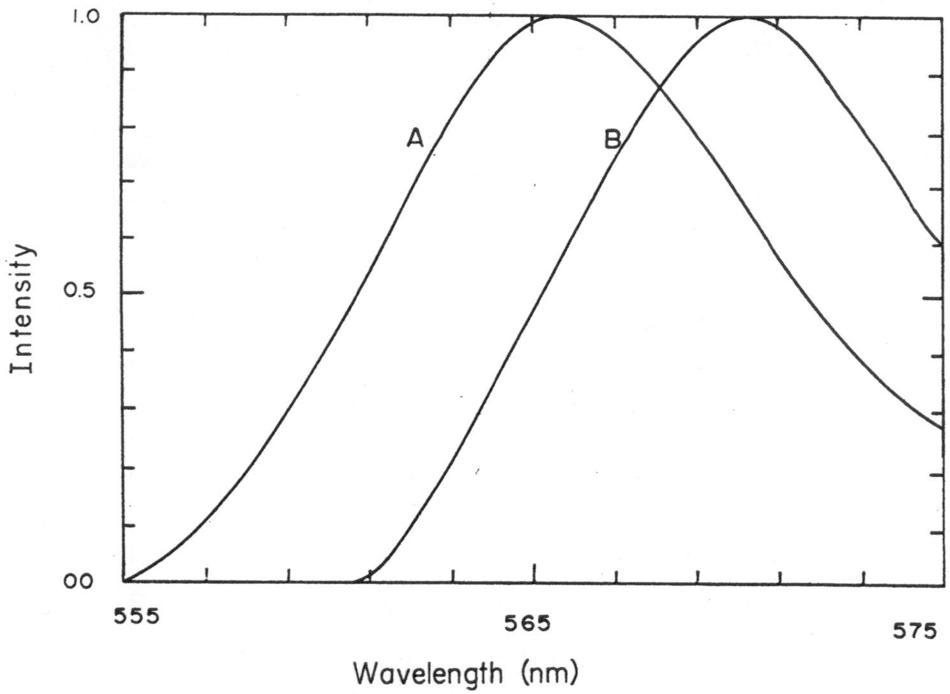

Figure 2. Spectral profile of the stimuls thorugh the lens:
<u>Curve A</u>: When the radiation was made incident at the center of the lens
<u>Curve B</u>: When the radiation was made incident at 10 mm away from the center of the lens

focusing optics. The monochromator was already calibrated *in situ*. The entrance and exits slits were opened to such an extent that a spectral band of approximately half band width of 5 nm at 585 nm was obtained from the exit slit. This radiation was focused at the center of a synthetic implant type eye lens of 8 mm dia and of power 14 diopter. The light from the lens was made incident at the entrance slit of a Spex 1404 double monochromator. The spectral distribution of the stimulus (through the eye lens) recorded is shown in Fig. 1 by curve A. When the radiation was made incident on the eye lense at a distance of 3 mm away from its center, the spectral distribution in the image plane was found to be modified. This is shown by curve B in Fig. 1. From Fig. 1, it can be seen that the peak of the curve B is shifted by about 3 nm towards the red end with respect to the peak of the curve A. It follows that any ray incident at the eye lens at a spot still away from the center will increase the shift at the peaks further towards the red end of the spectrum. The maximum shift observed in the experiment was about 5 nm, when the experiment was repeated with sources of different bandwidth. The experiment was repeated with the fine bandwidths which were obtained from discharge lamps like sodium and cadmium and also broad bandwidth from a tungsten helogen lamp with interference filters. The spectral shift in these cases was found to depend on the half bandwidth of the spectral light. For a half band width of 0.15 nm a spectral shift of approximately the same order was observed. For broader line (half bandwidth 10nm) a shift of 5 nm is observed. The results are shown in Fig. 2.

It may be realized that when the light enters the eye, it undergoes refraction at various surfaces. Consequently, the light stimuli bends as it passes from air to cornea, from aqueous humor to lens and from lens to vitreous body since at these interfaces significant changes in refractive indices occur. We believe that a definite amount of spectral shift may also occur during this process. This is because the light scattered from the space-time fluctuation in the occular media also contributes to spectral shift. The quantity of the shift on such account would again depend on the direction of incidence on the retina. However, the shift due to scattering may be small since in scattering experiments the spectral shift of the order of a fraction of nm are found.

Our above observations indicate that a significant part of the spectral shift observed in SCE II may arise due to physical reasons—the source correlation. However, it may be mentioned that the above phenomenon (source correlation) may not be the only cause for SCE II. The physical facts that we have described above supplement other physiological basis of SCE II. It is therefore emphasized that a commultative effect of all the phenomena—physical as well as physiological—are responsible for SCE II.

ACKNOWLEDGEMENT

The authors express their gratitude to Prof. S.K. Joshi, Director, National Physical Laboratory, New Delhi for his kind permission to publish this paper.

REFERENCES

1. W. S. Stiles, "The Luminous efficiency of monochromatic rays entering the eye pupil at different points and a new colour effect." *Proc. R. Soc.* **8** 90-118 (1937).
2. W. S. Stiles and B. H. Crowford. "The Luminous efficiency of rays entering the eye pupil at different points." *Proc., R. Soc.* **B-112** 428-450 (1933).

3. J. M. Enoch and W. S. Stiles, "The colour charges of monochromatic light with retinal angle of incidence optica." *Ac* **8** 329-350 (1961).

4. A. W. Synder and C. Pask, "The Stiles - Crowford effect - explanation and consequences," *Vis Res.* **13**, 1115-1157 (1973).

5. L. J. Bour and C. M. Verhoosel: "Directional Sensitivity of Photoreceptores for different degrees of coherence and directions of polarization of the incident light;" *Vis. Res.* **717**, 719 (1979).

6. M. Alpern and K. Kitahara: "The directional sensitivities of the Stiles, Colour mechanisms." *J. Physiol (Lond.)*, **338**, 627-649 (1983).

7. M. Alpern., K. Kitahara, and R. Takmi: "The dependence of the colour and brightness of a monochromatic light upon its angle of incidence on retina," *J. Physiol.* **338**, 651-668 (1983).

8. M. Alpern and R. Tamaki: "The saturation of monochromatic lights obliquely incident on the retina," *J. Physiol.*, **338**, 669-691 (1983).

9. Wolf "Invariance of spectrum of light on propagation" *Phys. Rev. Lett.* **56** 1370-1372 (1986).

10. E. Wolf, "Non-cosmological redshifts of spectral lines" *Nature* **326** (1987) 363-365.

11. E. Wolf, "Red shifts and blue shifts and spectral lines caused by source correlations" *Opt. Commun.*, **62** (1987) 2646-2648.

12. I. T. Foley "The effect of aperture on the spectrum of partially coherent light," *Optc. Commun.* **75** (1990) 347-352.

13. G. M. Morris and D. Faklis, "Effect of source correlation on the spectrum of light," *Opt. Commun.*, **62**, 5-9 (1987).

14. D. Faklis and C. M. Morris "Spectral shifts produced by source correlations," *Opt. Lett.* **13**, 4 (1988).

15. H. C. Kandpal, J. S. Vaishya and K. C. Joshi, "Simple experimental arrangement for observing spectral shifts due to source correlation" *Phys. Rev. A.* **41** 4541-4542 (1990).

16. H. C. Kandpal, J. S. Vaishya and K. C. Joshi "Wolf shift and its applications in spectroradiometry," *Opt. Commun.* **73** 169-172 (1989).

17. H. C. Kandpal, J. S. Vaishya and K. C. Joshi, "Spectral shift due to source correlation for parexial rays." *Opt. Commun.* **75**, 1-3 (1990).

18. Indebetouw, "Synthesis of polychromatic light sources with arbitrary degree of coherence some experiments" **36**, 251-269 (1989).

19. E. Wolf, J. T. Foley and F. Gori "Frequency shifts of spectral lines produced by scattering from specially random media" *J. Opt. Soc. Amer.* **A-6** 1142-1149 (1989) ibid A-7 173-179 (1990).

20. J. T. Foley and E. Wolf, "Frequency shifts of spectral lines generated by scattering from space time fluctuations," *Phys. Rev.* **A-40**, 588-598 (1989).

21. E. Wolf, "Correlation induced Doppler-like frequence shift of spectral lines," *Phys. Rev. Lett.* **63**, 220-223 (1989).

22. H. C. Kandpal. J. S. Vaishya and K. C. Joshi, "Spectral shift due to dynamic scattering experimental verification," *Opt. Commun.* (submitted for publication).

23. G. S. Brindley: "The effect on colour vision of adaptation to very bright light" *J. Phy.* (Lond) 332-350 (1953).

24. G. P. Agrawal and A. Gamliel, "Spectrum of partially coherent light: transition from near to far zone." *Optics Commun.* **78**, 1-6 (1990).

FREQUENCY SHIFTS OF SPECTRAL LINES BY SCATTERING FROM SPACE TIME FLUCTUATIONS

H.C. Kandpal, J.S. Vaishya and K.C. Joshi

Standards Division
National Physical Laboratory,
Dr. K.S. Krishnan Road
New Delhi - 110012 (India)

ABSTRACT

The spectral distribution of energy in scattered radiation is influenced by the physical parameters, like dielectric susceptibility of the scattering medium and their space-time correlation. For dynamic scatters where dielectric susceptibility is a random function of position with appropriate correlation properties, spectral shift is observed when the radiation is scattered by the medium. Such a spectral change imitates Doppler effect even though the source of radiation, the scattering medium and the observer (detector) are at rest with respect to each other.

It has been established that correlations in the fluctuations of a source distribution influence the spectrum of radiation when it propagates through free space. In particular, it has been shown theoretically[1-3] and subsequently confirmed by experiments[4-9] that suitable spatial correlations within the source or developed between field points on propagation can affect the distribution of energy in the spectrum of radiated fields resulting in frequency shifts of spectral lines. Recently it has been predicted theoretically[10-13] that similar changes may also occur on scattering of radiation by a medium whose physical parameters (e.g. the dielectric susceptibility) are random function of position but have an appropriate correlation property. Such spectral changes will imitate the Doppler effect, even though the source of radiation, the scattering medium and the observer are at rest with respect to each other.

In this communication we have experimentally verified the theoretical prediction[12-13] that frequency shift can be produced by scattering of radiation from spatially random medium like a dynamical scatterer formed by methyl benzyliden butyl aniline (MBBA) liquid crystal film or solutions of some ionic salts.

Scattering of polychromatic partially coherent light by a medium whose dielectric susceptibility varies randomly both in space and time but suitably correlated has been studied with (i) MBBA type liquid crystal film embedded between two conducting glass plate and (ii) solution of various ionic salts as dynamic scatterers. The experimental setup consists of an integrating sphere (0.5 m diameter and coated with $BaSO_4$ paint), a tungsten halogen lamp as a primary source, apertures of varying

Recent Developments in Quantum Optics, Edited
by R. Inguva, Plenum Press, New York, 1993

sizes, interference filters and a Spex double grating monochromator equipped with data acquisition and processing system. The lamp is kept at the center of the integrating sphere. Light from an aperture of 4 mm diameter on the surface of the integrating sphere is passed through apertures of varying size (1 mm to 3 mm) followed by an interference filter so that a narrow band of polychromatic partially coherent light is obtained. This light is focused on the dynamic scatterer. Since the degree of coherence produced at the focal plane of the lens depends on the aperture size and the focal length of the lens, it can be varied. The transmitted/scattered radiation is focused at the entrance slit of the grating monochromator. A schematic diagram of the experimental setup is shown in Fig. 1.

Curve A in Fig. 2 shows the transmitted spectrum of the interference filter in the absence of liquid crystal sample in the optical path. Curve B shows the spectrum when the liquid crystal film is put at the focal plane of the lens. It is noted that curve B is redshifted compared to curve A by about 0.05 nm. This frequency shift in curve B is essentially the effect of scattering by a medium whose physical properties are characterized by random functions of position while temporal fluctuations are slow enough to be ignored. Under such circumstances the process of scattering is influenced by the response of dipole oscillators of the scatterer and their correlation. Since the incident light is polychromatic, the different frequency components of the incident light are scattered in any particular direction with different strengths. Also the polarization induced in the scattering medium by the incident wave may be correlated over finite distances of the medium. As a result, the spectrum of the scattered light differs from that of the incident light.[10–12] Curve C shows the spectrum when varying electric field (of frequency 10 Hz - KHz) is applied across the liquid crystal film. It is noted that curve C is red shifted by about 0.3 nm compared to curve A which is obtained when the varying field reaches a value of 1 KHz. Application of electric field makes the dipoles of the liquid crystal orient themselves in the direction of the applied electric field. But due to the oscillating electric field the dipoles also vibrate. Interaction of the molecular dipole moments of the liquid crystal and the electric field of the incident light cause the scattering of the light and a shift is observed. The results are consistent with the theoretical prediction of James and Wolf.[13] Due to limitation in the selection of the dynamic scatterer and also its limited ability to get influenced by the varying electric field due to large molecular dimension of the liquid crystal desired frequency shift could not be obtained as predicted by James and Wolf.[13] However, further work is in progress.

Some results obtained by replacing the liquid crystal sample with solutions of different ionic salts are also reported here. Curve A in Fig. 3 and curve A in Fig. 2 represent the same spectrum of the transmitted light by the interference filter without any scatterer. Curve B in Fig. 3 represents the spectrum when the transmitted light by the interference filter is scattered by a cell containing the solution of ionic salt and the scattered light is focused at the entrance slit of the monochromator. The direct transmitted light through the cell is stopped and observations are made almost at on-axis points. A look at Fig. 3B shows a blue shift of about 0.3 nm with respect to the unscattered light (Fig. 3A). By changing the degree of coherence of the radiation falling at the solution, shift of varying magnitude from 0.1 nm to 0.3 nm could be observed. These experiments are repeated with plane polarized light and the magnitude of spectral shift observed is similar. The results obtained so far are consistent with the theoretical predictions. Further work is in progress.

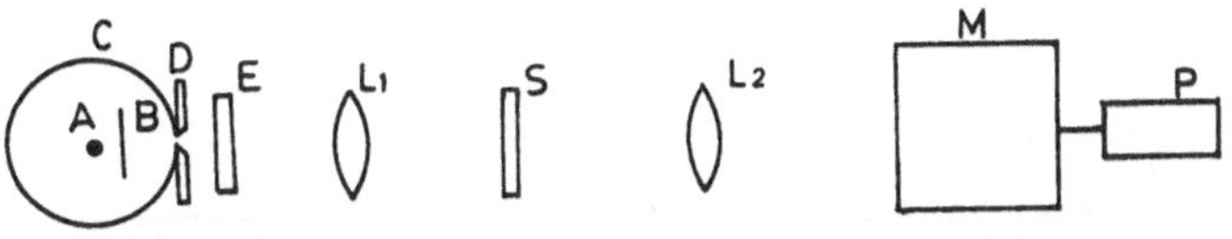

Figure 1. Schematics of experimental set up.

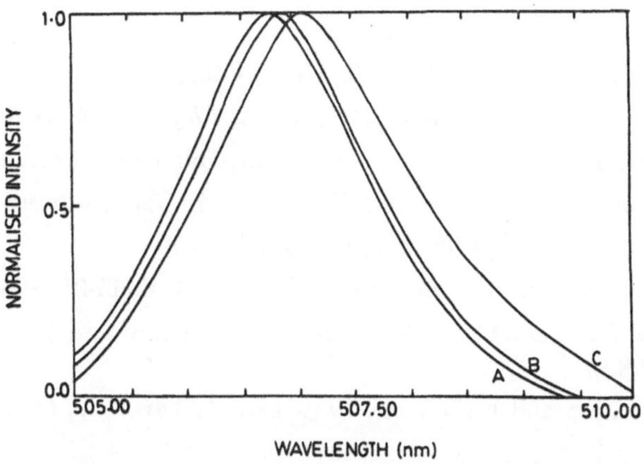

Figure 2. Spectrum of the light transmitted by the interference filter (A) in absence of sample S (B) when MBBa liquid crystal film embedded between glass plate is introduced at the focal plane of lens L_1 (C) when electric field of varying frequency is applied across MBBA liquid crystal film.

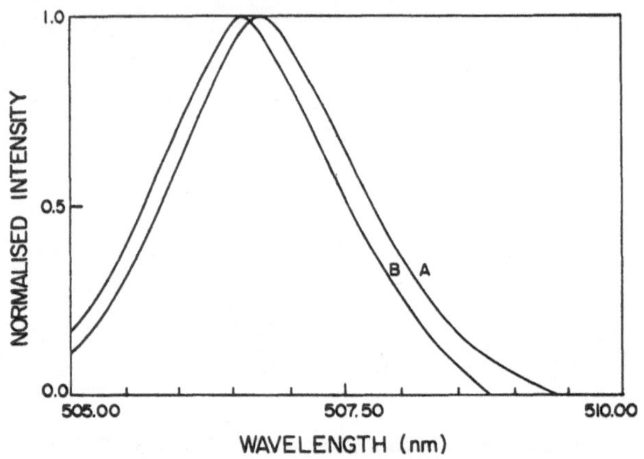

Figure 3. Spectrum of light transmitted by the interference filter (A). In absence of sample (B) when quartz cell containing ionic salt solution is kept at the focal plane of lens L_1.

REFERENCES

1. E. Wolf, *Phys Rev Lett* **56**, 1370, (1986).
2. E. Wolf, *Nature* **326**, 363, (1987).
3. E. Wolf, *Opt Commun* **62**, 12, (1987).
4. G. M. Morris and D. Faklis, *Opt Commun* **62**, 5, (1987).
5. D. Faklis and G. M. Morris, *Opt Lett*, **13**, 4, (1988).
6. H. C. Kandpal, J. S. Vaishya and K. C. Joshi, *Phys Rev A* **41** 4541, (1990).
7. H. C. Kandpal, J. S. Vaishya and K. C. Joshi, *Opt Commun*, **73**, 69 (1989).
8. F. Gori, G. Guattari, C. Palma and G. Padovani, *Opt Commun* **67**, 1 (1988).
9. G. Indebetouw, *J Mod Opt* **36**, 251, (1989).
10. E. Wolf, J. T. Foley and F. Gori, *J Opt Soc Am A*, **6**, 1142 (1989).
11. E. Wolf, *Phys Rev Lett*, **63**, 2220, (1989). J. T. Foley and E. Wolf, *Phys Rev A*, **40**, 588, (1989).
12. F. V. David, James and Emil Wolf, *Phys Lett A*, **146**, 167, (1990).

OVERVIEW OF RECENT ADVANCES IN SHORT-PULSE STRONG-FIELD IONIZATION OF ATOMS

J.H. Eberly

Department of Physics and Astronomy
University of Rochester
Rochester, NY 14627 USA

ABSTRACT

We sketch progress in high-power laser development and recent advances in theoretical understanding of high-intensity laser ionization of atoms. Phenomena of recent interest are atomic electron localization and an accompanying suppression of ionization. These are observed in new supercomputer simulations. They occur in short-pulse laser-atom interactions in a range of laser intensities near to and above 10^{16} watts/cm^2.

OVERVIEW OF STRONG-FIELD LASER DEVELOPMENT

The first anecdotal reports of atomic ionization in strong optical fields were based on observations of air breakdown at the focus of high-power lasers in the mid-1960's. Experimental results were difficult to interpret, but further ionization experiments were encouraged by the development of the first short-pulse lasers, typically Q-switched ruby or Nd:glass lasers. Systematic single-atom multiphoton ionization studies using low- pressure gas chambers were initiated by groups in Moscow [1] and in Paris [2] between 1965 and 1970. After the development of high-quality Nd:YAG crystals and mode locking, laser pulses at the YAG wavelength $\lambda \approx 1\mu$m could be obtained in the sub-nanosecond time domain with greatly increased output powers. Focused intensities in the neighborhood of 10^9- 10^{10} watts/cm^2 were achievable. However, a disadvantage of conventional very-high-power solid-state laser systems is their low repetition rate, mandated by the need to limit thermal distortions in amplifier rods.

Gas lasers are largely immune to thermal damage, and CO_2 lasers can be used for high-power ionization studies at $\lambda \approx 10\mu$m, but typically with longer pulse durations, larger focal spots and lower peak powers, than can be achieved with optical lasers. Excimer lasers are the best example of gas lasers that overcome some of these disadvantages and permit very high intensities, from 10^{16} to 10^{18} watts/cm^2, with pulse durations in the hundreds of femtoseconds range and tighter focal spots than infrared or optical lasers due to their shorter wavelength, $\lambda \approx 0.25\mu$m. One important disadvantage of high-power excimer systems is the cost and complexity of the amplifiers needed.

Recent Developments in Quantum Optics, Edited
by R. Inguva, Plenum Press, New York, 1993

The experimental situation has recently changed dramatically with the development of chirped pulse amplification (CPA) lasers [3]. These are solid state laser systems that operate by injecting a single very low-power mode-locked pulse through a dispersive optical element that imposes a wide chirp (frequency sweep) and stretches the pulse duration by several orders of magnitude. The stretched pulse has a much lower peak intensity. In this condition it can safely pass through a solid state amplifier without danger of overheating the rod, acquiring as much or more than 1 joule of optical energy. Upon removal from the amplifier or amplifiers the very energetic long pulse is de-chirped by a grating with a relatively high damage threshold. This shortens the pulselength, ideally to its bandwidth limit, giving an output power as high as 10^{12} watts. Focused CPA output intensities higher than 10^{17} watts/cm^2 have been reported [4]. With crystals such as alexandrite and Ti:sapphire, laser bandwidths exceeding those available in Nd:YAG and Nd:YLF are compatible with pulse durations in the same range as excimer lasers, well below 1 picosecond.

The rapid pace of laser development has been accompanied by rapid advances in multiphoton atomic physics. In the 1980's electron and photon spectral studies accompanying multiphoton ionization (MPI) became significant new subfields, in which new phenomena were first reported experimentally, with theoretical explanations coming later [5]. The main body of work in these fields has been carried out in the intensity range between 10^{12} and 10^{15} watts/cm^2, with laser pulse durations in the vicinity of 1-10 picoseconds.

SUPERCOMPUTER MODELING OF STRONG-FIELD PHENOMENA

Theoretical rather than experimental advances have been the motive force behind the most recent developments [6,7]. A variety of supercomputer programs able to solve Schrödinger's wave equation numerically now permit essentially exact analyses of the quantum mechanical response of an atomic electron exposed to both a strong time-dependent laser field and the static coulombic binding potential of the atom. There are no comparable (even approximate) analytic wave functions known. The new effects that these computer analyses have revealed are the main subject of the remainder of this paper.

First, the limitations of the existing computer programs should be sketched. Several approaches have been used to turn Schrödinger's equation into an equation suitable for numerical solution. In one way or another all approaches are based on discretization of space and time. For example, one replaces spatial and temporal derivatives by their well-known discrete versions, and replaces the electron wave function by a vector whose elements are the values of the wave function at the discrete grid points. Schrödinger's differential equation becomes a matrix equation of very high dimension. The matrix solution is presumed to provide an accurate replica of the actual wave function if the grid points are closely spaced. The achievement of convergence in the numerical sense can be tested by reducing grid spacing until results are insensitive to further reductions.

Non-numerical approximations are also typically incorporated into numerical solutions of Schrödinger's equation. For example, the electron is treated as non-relativistic and spinless, and the electric dipole approximation is used. These approximations can usually be justified rather easily. For example, the optical wavelength is much larger than any grid size in use, so the dipole approximation is natural. A further assumption commonly made is that the electron's motion is so nearly one-dimensional in a strong laser field [along the laser field's polarization direction] that the other two dimensions of Schrödinger's wave equation can be eliminated without harm. The same assumption is used in treating strong field ionization at microwave frequencies, and agreement between experiment and theory is good in that case. The main advantage of this extra assumption is to permit the use of the grid points that would have been needed for the y and z grids to be employed on the x grid. The main disadvantage is that intrinsically three-dimensional physical effects are automatically excluded. The most important of these are laser polarization effects and angular distributions of emitted photons and photoelectrons. Comparisons of 1-D with 3-D results and of 1-D results obtained by different methods with each other, and comparisons with experiments, have shown that the one-dimensional assumption is generally satisfactory for accurate semi-quantitative modeling of strong-field ionization dynamics.

The main constraint on numerical studies is not computer time, but computer memory. Very large arrays associated with wave function values at thousands of grid points must be stored and manipulated. The size of these arrays arises from the physical nature of the problem at hand. To give an example, the shape of the chosen binding potential places a requirement on the grid density, since the potential must be accurately represented by the grid. If a maximum photoelectron momentum k is to be resolved in the calculated photoelectron spectrum there must be a large number of spatial points per smallest electron wavelength, usually making a more severe demand on grid density. If, in addition, the ionization process takes time T to develop fully, the absolute size of the spatial grid must be large enough to keep the fastest photoelectron from leaving the grid before time T.

To take a specific instance we can consider 4-photon ionization of hydrogen, for which $\omega = 0.125$ a.u. is the threshold photon energy. An electron absorbing an additional 4 photons achieves a kinetic energy of 0.5 a.u. and a velocity $v = 1$ a.u. During 20 optical cycles ($T = 20\text{x}2\pi/\omega \approx 1000$ a.u.) the electron will travel a distance $\Delta\text{x} \approx \pm 1000$ a.u., and this establishes the need for a grid size $2L \approx 2000$ a.u. The electron's wavelength is $2\pi/v \approx 10$ a.u., so the use of 10-20 grid points per wavelength to permit good resolution of the electron wave function calls for $N \approx 2000 - 4000$ total points on the space grid. While there may be ways to side-step one or two specific constraints [an absorbing spatial grid boundary is sometimes employed], the inevitable result is the need for numerical wave function vectors with thousands of components.

ELECTRON LOCALIZATION AND ATOMIC STABILIZATION

Recently, one-dimensional simulations of ionization in strong fields have provided convincing evidence that ionization does not proceed in a conventional or familiar way if the intensity and frequency of the laser field are high enough [8]. For example, in Fig. 1 we show a sequence of graphs of ionization probability vs. time, for a series of increasing values of laser peak field strength. The first three graphs show a steadily increasing ionization probability, and in each case the ionization rate (slope of probability vs. time) is reasonably well defined. It has been determined [8] that these slopes are related to each other in the manner predicted by perturbation theory. A continuation of these graphs is shown in Fig. 2. Here the peak field strengths are much larger, up to a value corresponding to a laser intensity of $9\text{x}10^{17}$ watts/cm^2. It is clear that in no case is there a very well defined ionization rate (slope), and it is also clear that the final value of the ionized probability decreases with increasing field

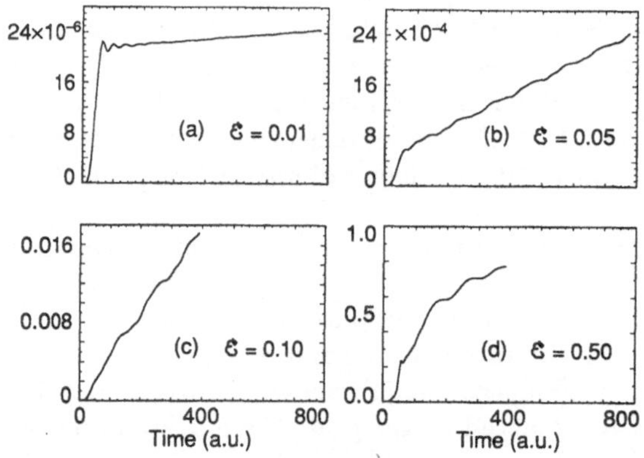

Figure 1. Probability of ionization vs. time (in dimensionless atomic units) for a sequence of increasingly stronger laser intensities. The peak laser field strength is given in atomic units, where $\varepsilon = 0.10$ corresponds to a laser intensity of $I \approx 3.6x10^{14}$ watts/cm^2. The laser frequency is $\omega = 0.52$ a.u., and the model atom's binding energy is 0.67. The laser field has been turned on smoothly over 5 cycles (about 60 a.u.) and then held constant. Note the greatly different vertical scales used.

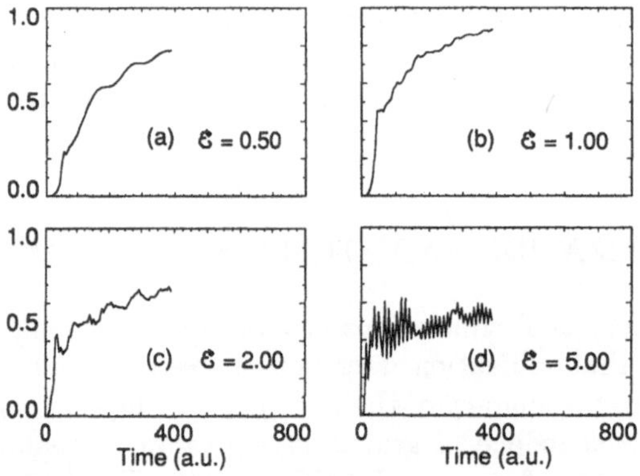

Figure 2. Probability of ionization vs. time, for greater laser intensities but otherwise the same parameters as in Fig. 1. Note that the vertical scales are all the same.

strength. This counter-intuitive decrease is the signature of what is called atomic stabilization or ionization suppression [9].

The results of Figs. 1 and 2 were obtained from numerical solutions of Schrödinger's equation in 1-D, using the soft-core coulombic potential [10]

$$V(x) = \frac{-1}{\sqrt{1+x^2}} \tag{1}$$

which has a binding energy of approximately 18 eV (0.67 a.u.), close to that of neon. The laser pulse used in these numerical simulations had peak field strengths ranging between $\mathcal{E} = 0.03$ and 5.0 a.u. (9×10^{17} watts/cm^2), and was smoothly turned on over 5.25 optical cycles. The photon energy was about 14 eV (0.52 a.u.), adequate for two-photon ionization of this model atom, but much higher than any available laser frequency. However, these results can be translated to lower frequencies by an appropriate scaling of the potential $V(x)$ [see Eberly, ref. 10].

Even more interesting, however, than the existence of stabilization, is the space-time behavior of the electron wave function. In Fig. 3 we show the electron probability wave packet $|\Psi(x,t)|^2$ as a function of time. The graphs show snapshots taken at $T = 0 - 15$ cycles in Fig. 3a, and during the 32nd cycle of laser radiation in Fig. 3b. In all cases the peak field strength is $\mathcal{E} = 5$ a.u., corresponding to the probability curve in Fig. 2d. The remarkable finding is that the electron sheds some probability during early times [see the irregular packet moving away from $x = 0$ during $T = 5-15$ cycles] and by $T \approx 30$ cycles the electron has become stably localized with a packet spread of $\Delta x \approx 20$ a.u. Since the initial width of the electron wave packet is just the ground state binding range, $\Delta x \approx 1$ a.u., the localization range is outside the nominal binding range, but very far inside the boundaries of the numerical grid that was used ($L \approx 1000$ a.u.).

Subsequent numerical studies [11-12] have amply confirmed the results shown here, and have extended the range of field strengths and frequencies at which stabilization and localization have been observed. It has become clear that the initial suggestions about stabilization [9] were too restrictive in requiring both frequencies and intensities to be much higher than one atomic unit, and it is clear that the effect is not an artifact of purely monochromatic fields. In fact both stabilization and localization survive under sufficiently rapid and smooth pulse turn-ons for frequencies near to or even below 1 a.u.

It is now believed that most aspects of stabilization and electron localization are best studied in the so-called Kramers-Henneberger average coordinate frame [13]. This frame oscillates with the amplitude and period of a free electron in the laser field. The electron wave function in this frame has most of the rapid oscillations of the lab frame removed. The residual ionization is a very slow smooth process, as shown in Fig. 4. Attention now needs to be given to this slow residual ionization rate and there are several analytic estimates for it [14]. Unfortunately, they fail by large factors (10-50) to agree with numerical experiments.

Finally, it is important to mention that no laboratory experiments have been able to study either stabilization or localization at optical frequencies to date. The requirements on laser frequency and turn-on are still apparently too severe. Stabilization with three-photon ionization has now been observed numerically [15], so the high-frequency requirement may be less important than currently understood. However,

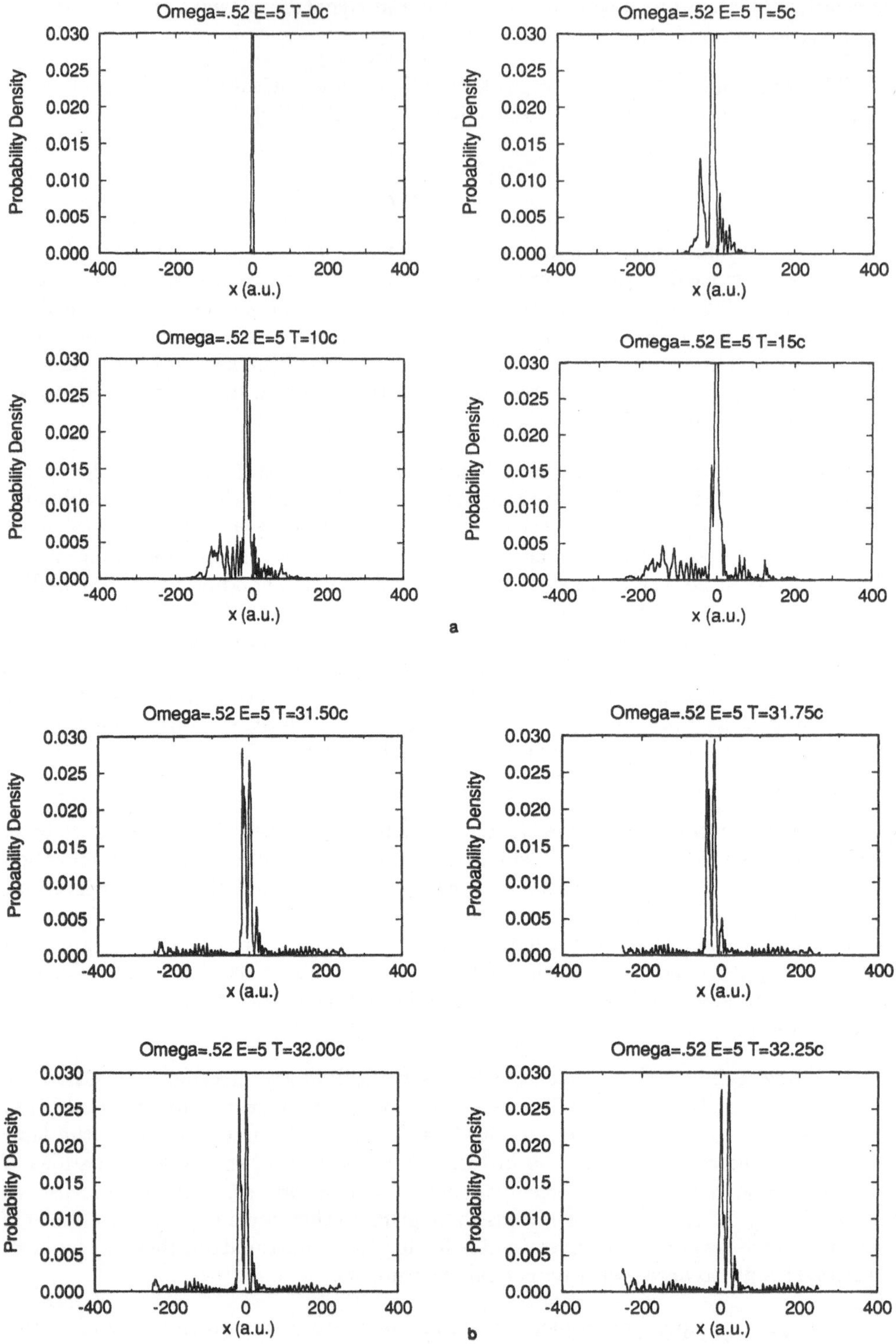

Figure 3. Space-time pictures of the ionization process, showing the electron probability density $|\Psi(x,t)|^2$ as a function of x for different times (a) t=0, 5, 10 and 15 cycles, and (b) t=31.50, 31.75, 32.00 and 32.25 cycles. In all cases $\mathcal{E} = 5$ a.u., and otherwise the same parameters as in Fig. 1.

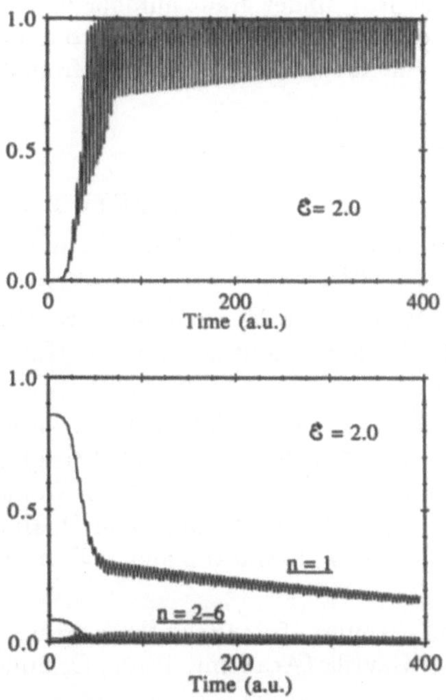

Figure 4. Ionization probability as a function of time, computed in the lab frame (top), and probability of occupying various low-lying bound states in the Kramers-Henneberger picture (bottom). Almost all probability is in the K-H ground state, and it leaks out slowly from the nearly fully stabilized atom. Compare the slope of the decay curve with the slope of the lower envelope of the large "jitter" oscillations of the lab frame. Here the laser has field strength $\mathcal{E} = 2.0$, frequency $\omega = 0.8$ and a turn-on of 10 cycles.

in order to confirm the specific numerical predictions made so far, there seems to be no way to avoid the need for a very strong pulse with a smoothly rising ultrashort (about 10 femtoseconds) leading edge followed by a constant-intensity plateau, and a laser with this capability is certainly not yet available.

ACKNOWLEDGEMENTS

Thanks are due to G.S. Agarwal for organizing the International Quantum Optics Conference in Hyderabad, January 1990, at which these results were presented, and to R. Inguva for organizing the US delegation with support from the National Science Foundation, Division of International Programs. The research reported here was partially supported by the NSF under grant number PHY-88 22730, and assisted by computer grants from the John von Neumann Supercomputing Center and the Pittsburgh Supercomputing Center, as well as by a grant from Allied-Signal Corporation.

REFERENCES

1. J. S. Voronov and N. B. Delone, Sov. Phys. JETP **23**, 54 (1966).
2. P. Agostini, G. Barjot, J. F. Bonnal, G. Mainfray, and C. Manus, IEEE J. Quant. Electron. **QE-4**, 677 (1968).
3. D. Strickland and G. Mourou, Optics Commun. **56**, 219 (1985) and P. Maine, D. Strickland, P. Bado, M. Pessot and G. Mourou, IEEE J. Quant. Electron. **24**, 398 (1988).
4. See, for example, S. Augst, D. Strickland, D. D. Meyerhofer, S. L. Chin and J. H. Eberly, Phys. Rev. Lett. **63**, 2212 (1989); M. Ferray, L. A. Lompré, O. Gobert, A. L'Huillier, C. Manus, A. Sanchez and A. S. Gomes, Optics Commun. **75**, 278 (1990); J. Squier, F. Salin, G. Mourou and D. Harter, Optics Lett. **16**, 324 (1991); and J. D. Kmetec, J. J. Macklin and J. F. Young, Optics Lett. **16**, 1001 (1991).
5. A recent book contains accounts of results in these areas: Atoms in Strong Radiation Fields, edited by M. Gavrila (Academic Press, Orlando, in press 1992).
6. Earlier attempts to study ionization by purely numerical methods are due to Geltman, et al.: S. Geltman, J. Phys. B **10**, 831 (1977), and *ibid.* **13**, 115 (1980); see also E. J. Austin, *ibid.* **12**, 4045 (1979).
7. K. C. Kulander, Phys. Rev. A **35**, 445 (1987); *ibid.* **36**, 2726 (1987); and *ibid.* **38**, 778 (1988).
8. Q. Su, J. H. Eberly, and J. Javanainen, Phys. Rev. Lett. **64**, 862 (1990), Q. Su and J. H. Eberly, J. Opt. Soc. Am. B **7**, 564 (1990), and Q. Su and J. H. Eberly, Phys. Rev. A **43**, 2474 (1991).
9. Early predictions of stabilization were obtained from analytic studies by J. I. Gersten and M. Mittleman, J. Phys. B **9**, 2561 (1976) and M. Gavrila and J. Kaminiski, Phys. Rev. Lett. **52**, 613 (1984) and based on high field-strength and frequency requirements that are now recognized to have been too restrictive.
10. See J. Javanainen, J. H. Eberly and Q. Su, Phys. Rev. A **38**, 3430 (1988), J. H. Eberly, Phys. Rev. A **42**, 5750 (1990), and Q. Su and J. H. Eberly, Phys. Rev. A **44**, 5997 (1991).
11. For confirmations of stabilization (decrease of ionization rate with increasing laser intensity) see M. Dörr, R. M. Potvliege and R. Shakeshaft, Phys. Rev. Lett. **64**, 2003 (1990) and M. Pont and M. Gavrila, *ibid.* **65**, 2362 (1990).

12. Confirmation of the first report [8] of electron localization accompanying ground state ionization has come so far only from the 3-D results of K. C. Kulander, K. J. Schafer, and J. L. Krause, Phys. Rev. Lett. **66**, 2601 (1991).

13. See W. C. Henneberger, Phys. Rev. Lett. **21**, 838 (1968).

14. For example, see M. Gavrila, in Fundamentals of Laser Interactions, edited by F. Ehlotzky (Springer-Verlag, Berlin, 1985), p. 3; and T. P. Grozdanov, P. S. Krstic and M. H. Mittleman, Phys. Lett. A **149**, 144 (1990).

15. C. K. Law, Q. Su and J. H. Eberly, Phys. Rev. A **44**, 7844 (1991).

QUANTUM OPTICS OF STRONGLY DRIVEN TWO-LEVEL ATOMS

Thomas W. Mossberg

Department of Physics
University of Oregon
Eugene, Oregon 97403

BASIC CONCEPTS

Introduction. Strong driving fields can add a new richness to the quantum optical behavior of atoms. This follows from the fact that in driven systems the relevant energy levels become those of the coupled atom-field system rather than those of the atom alone. Also, since the energies of the coupled atom-field states depend on the intensity of the driving field(s), one gains an additional control parameter over the quantum optics of the system. We briefly touch on two different situations in which driven atoms display unique quantum optical properties. First, we consider modifications in driven-atom dynamics consequent to coloring of the electromagnetic vacuum. These modifications follow from the reservoir-dependent nature of spontaneous emission – the primary relaxation mechanism in the systems considered. Second, we discuss the use of driven two-level atoms as one- and two-photon gain media.

Dressed States. As a preliminary, we first explore the nature of the coupled atom-field system. In Fig. 1, we show the energy levels of a two-level atom, a single mode of the radiation field, and two adjacent doublets of the coupled atom-field states (dressed-states). The dressed states, like the harmonic oscillator states, have a ladder structure except that each rung on the dressed-state ladder consists of a doublet split by the generalized Rabi frequency $\nu_s = (\Delta^2 + \Omega^2)^{1/2}$, where $\Delta = \nu_l - \nu_a$, Ω is the resonant Rabi frequency, and $\nu_l(\nu_a)$ is the driving-field (atomic) frequency. In Fig. 1c, the dressed-states (center labels) are shown to asymptotically approach atom-field product states (extreme right and left) for large values of Δ. If we denote the population in dressed state i (summed over n) by Π_i and the spontaneous emission rate between the dressed levels i and j by A_{ij}, it is straightforward to show that in steady state

$$A_{12}\Pi_1 = A_{21}\Pi_2. \tag{1}$$

Spontaneous Emission. The spontaneous emission rate between two levels e and g can be written as

$$A_{eg} = \alpha_o \sigma(\nu_{eg}, r_a) d_{eg}^2(\Delta, \Omega), \tag{2}$$

Recent Developments in Quantum Optics, Edited
by R. Inguva, Plenum Press, New York, 1993

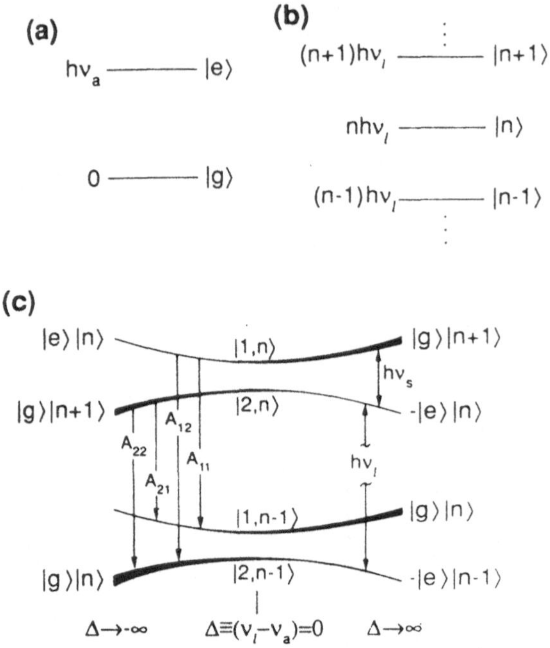

Figure 1. Energy levels of (a) two-level atom, (b) single-mode electromagnetic field, (c) coupled atom-field system (dressed states). Line thickness is representative of relative population.

where α_o is a constant $\sigma_{eg}(\nu_{eg}, r_a)$ is the effective spectral energy density associated with the electromagnetic vacuum, r_a is the spatial location of the atom, and d_{eg} is an electric-dipole matrix element associated with transition *eg*. Following Einstein, $\sigma(\nu, r)$ can be computed by populating each field mode with a single real photon, determining the mode's spectral energy density at (ν, r), weighting the mode according to the square of its field-dipole coupling coefficient, and summing. In the present case, the argument r is needed because cavity environments are generally non-isotropic and mode functions (e.g. standing-waves) have amplitudes that are spatially varying. Frequently, the "density of electromagnetic modes" is written in place of $\sigma(\nu, r)$. Strictly speaking, however, mode density is appropriate only in isotropic environments where all modes can be considered equivalent in terms of volume and spectral width. In such cases, σ is independent of position and is proportional to the mode density.

It is interesting to provide estimates of how much a single cavity model can enhance the free-space spontaneous emission rate. Following Eq. 2, the spontaneous emission rate into a cavity mode, A_{cav}, is given by

$$A_{cav} = A_{fs}\left(\frac{3\sigma_{cav}(\nu_a)}{\sigma_{fs}(\nu_a)}\right) = A_{fs}\eta_{cav}$$

where $A_{fs}(\sigma_{fs})$ is the free-space spontaneous emission rate (free-space vacuum spectral energy density) and σ_{cav} is the spectral energy associated with the cavity-mode vacuum. The factor of 3 arises from isotropic character of the free-space vacuum in contrast to the cavity-mode vacuum which is assumed optimally coupled to the atom. In the case of a closed cavity of quality factor Q and volume V_{cav}, $\eta_{cav} \approx (3Q\lambda^3/8\pi V_{cav})$, while in the case of an open optical resonator of length l, finesse F, and mirror radius R ($R >> l$), $\eta_{cav} \approx (\lambda F/\sqrt{lR})$. Finally, we consider mode-degenerate open cavities

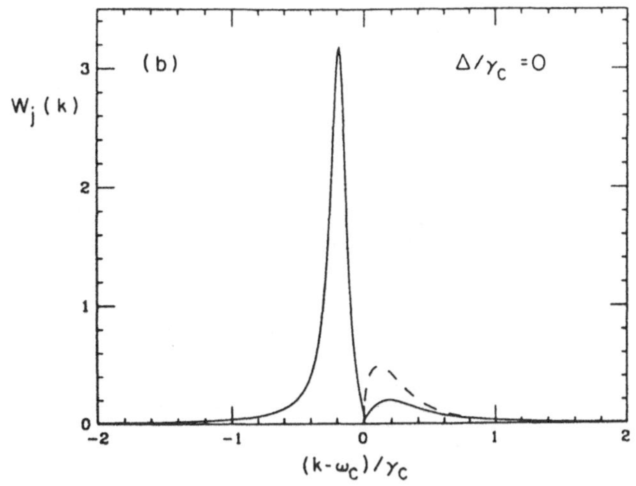

Figure 2. Spontaneous emission spectrum in a colored vacuum. Ref. 3.

(e.g. concentric, confocal). In these cavities, it is possible to utilize "supermodes" having a very small diameter at the cavity's center. In the case of optimal (nearly 2π steradians) degeneracy the central spot diameter may approach λ and $\eta_{cav} \approx F/(4\pi^2)$. In all cases, a spatial average over λ^3 is assumed.

DRIVEN TWO-LEVEL ATOMS IN A COLORED VACUUM

Controlling Spontaneous Emission Rates. The electromagnetic vacuum can be perturbed by the imposition of boundary conditions. It has been pointed out that this fact can be utilized to suppress or enhance the rate of spontaneous radiative decay.[1] It follows that suppression will occur in cavities having an effective vacuum spectral energy density at frequency ν_a small compared to that found in free space. Enhancement occurs in the opposite situation. Modification of spontaneous radiative decay rates has indeed been observed.[2]

In cases where σ has a fairly constant value over the spectral width of the atomic resonance, variations of σ effect the atomic decay dynamics only through a relatively trivial change in time scale. The fundamental dynamics and accessible steady-states available to a system are unchanged. In the presence of a weak driving field ($\Omega < A_{eg}$), for example, all steady-state conditions achievable by varying σ are also achievable by simply varying the driving field intensity. The situation is entirely different if $\sigma(\nu)$ exhibits spectral variation over the resonance profile of the atom. In such cases, atomic decay can become non-exponential and totally new steady-state behavior observable.[3] Spontaneous emission lineshapes assume interesting non-Lorentzian character. In Fig. 2, we show the spontaneous emission lineshape for an atom coupled to an open cavity having a sharp cut on at $(k - \omega_c)/\gamma_c$. The solid (dashed) line represents the spectrum of light emitted into out the cavity sides (into the cavity modes). Experimental work in this regime has been limited because of the difficulty associated with introducing significant spectral structure into $\sigma(\nu)$ on frequency scales as fine as typical atomic resonance widths.

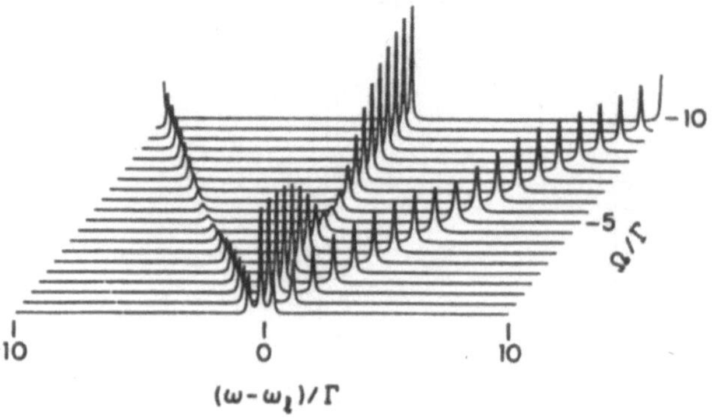

Figure 3: Strong-field resonance fluorescence spectrum in a colored vacuum. Ref. 5.

Spontaneous Emission of Strongly Driven Atoms. In the presence of a strong driving field ($\Omega > A_{eg}$), the atomic resonance splits and assumes the form of the three-component Mollow spectrum.[4] The overall width of the resonance becomes of order $2\nu_s$. If $\sigma(\nu)$ is constant over this entire spectral profile, decay dynamics (save for time scale) and allowable steady-states are no different from those found in free space. However, if $\sigma(\nu)$ varies in magnitude from one Mollow component to the other, phenomenologically new behavior emerges.[5] In Fig. 3, we show the Mollow spectrum as a function of Ω/Γ in the presence of a vacuum reservoir displaying a sharp spectral density peak of width Γ at $(\omega - \omega_l)/\Gamma = -5$. This peak is generated through the effect of an open cavity surrounding the radiating, driven atom. The spectrum is observed out the side of the cavity. When the lower frequency sideband (lower frequency dressed-state transition) becomes resonant with the vacuum spectral energy peak, it is dramatically broadened and reduced in amplitude.

A very well-known result related to the behavior of two-level atoms is that a driving field, no matter how strong, cannot produce a steady-state atomic inversion. While it is seldom state explicitly, the equations used to derive this fundamental result depend critically on the presence of a white (spectrally flat) vacuum. In the presence of a colored vacuum, i.e. one in which $\sigma(\nu)$ exhibits significant variations within the spectral region of interest, the usual equations of driven atomic motion do not apply and steady-state inversion is indeed a possibility.[6] A simple way of seeing this follows from Eq. 1 and Fig. 1c. The steady-state dressed-state populations depend on the ratio of A_{12} and A_{21}. For $\Delta \neq 0$, one can thus enhance, through appropriate changes in $\sigma(\nu)$ and hence the A_{ij}, the population in the dressed-state sublevel corresponding predominantly to the atomic excited state. In appropriate limits, one obtains a steady-state atomic inversion. In Fig. 4, we plot the excited-state population as a function of Δ/Ω under the condition that σ differs by a factor of 10 at the frequencies of the Mollow sidebands. An experimental demonstration of the manipulation of excited-state population through control of the vacuum is shown in Fig. 5. The top three traces represent fluorescence intensity emitted by strongly driven atoms out the side of a confocal cavity as a function of the cavity resonance frequency (where σ exhibits a cavity-induced peak). The driving laser, incident through the side of the cavity, is tuned above (top trace), to, and below the atomic resonance. The side fluorescence intensity is simply proportional to the number of excited-state atoms. The tuning of the cavity (which contains no real photons) clearly affects the excited-state population.

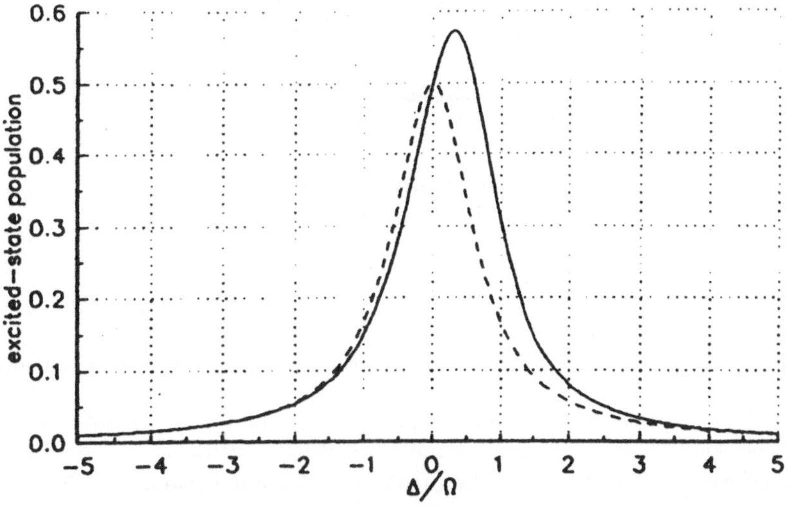

Figure 4. Excited-state population of a driven atom as a function of driving-field frequency in colored and normal (dashed) vacua. Ref. 6.

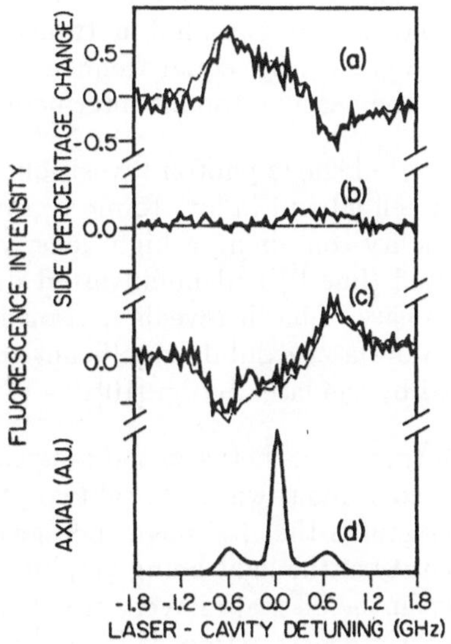

Figure 5. Experimental observation of vacuum controlled atomic excitation. Ref. 6.

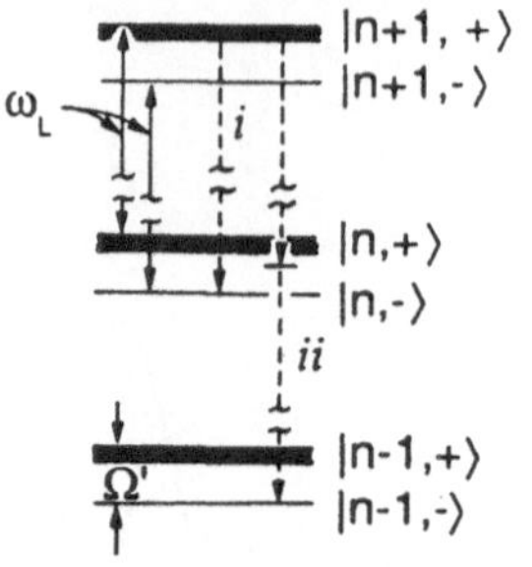

Figure 6. Three consecutive dressed-state doublets. Here ω_L and Ω' represent the laser and generalized Rabi frequencies, respectively.

Under the specific experimental conditions employed an inversion was not expected.[7]

The analysis of the previous paragraph, though adequate to provide a basic explanation of the inversion effect, does not indicate the richness of the non-Markovian equations of motion that follow in the presence of a colored vacuum. The full implications of these equations remain to be explored.

DRIVEN TWO-LEVEL ATOMS AS ONE- AND TWO-PHOTON GAIN MEDIA

In Fig. 1c, it is shown that the populations of the sublevels within each dressed-state doublet are imbalanced for $\Delta \neq 0$. For a particular sign of Δ, one of the dressed-state transitions is inverted, and one-photon gain is present at the corresponding transition frequency. Similarly, if one considers a sequence of three consecutive dressed-state doublets, see Fig. 6, one notes that two-photon transitions between non-adjacent doublets are also inverted. Significantly, equal frequency two-photon transitions are enhanced by the presence of the nearly resonant intermediate dressed-state doublet.

Laser action based on inverted single-photon transitions between adjacent doublets has been demonstrated in a cell [8] and in an atomic beam.[9] In Fig. 7, we show the emission out the end of a cavity containing a high density of driven atoms when the cavity is tuned to an inverted (line b) and noninverted dressed-state transition as a function of atomic beam density. Line b reveals a transition to lasing. It has been calculated that this one-photon laser exhibits significant squeezing arising from shifts in the dressed-levels induced by the laser action.[10]

The inverted two-photon transitions between the dressed-levels have been employed in the realization of the first continuous-wave optical two-photon laser.[11,12] An interesting aspect of two-photon gain is that its magnitude grows linearly (in the absence of saturation) with the intensity of the light being amplified. Thus the gain within the cavity of a two-photon laser in the off state is very small. As a result, the two-photon laser cannot self-start. To initiate laser action, it is necessary to inject a field of sufficient intensity to bring the intracavity gain up to a value that exceeds the cavity losses. Once this occurs, the two-photon laser is self-sustaining and the injected signal can be discontinued. The need for a trigger is a characteristic signature of the two-photon laser. In Fig. 8, we show an experimental demonstration of the switch-on behaviour of the two-photon dressed-atom laser. We plot power out the end of the cavity as a

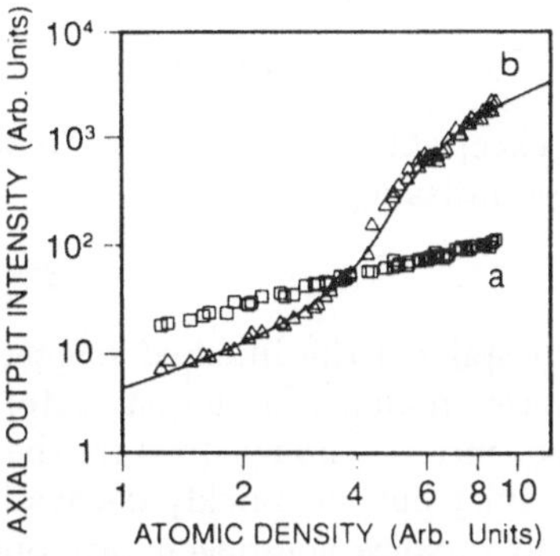

Figure 7. Power output from a cavity containing driven atoms as a function of atomic density. For trace a (b) the cavity is tuned to a dressed-state transition without (with) gain. Ref. 9.

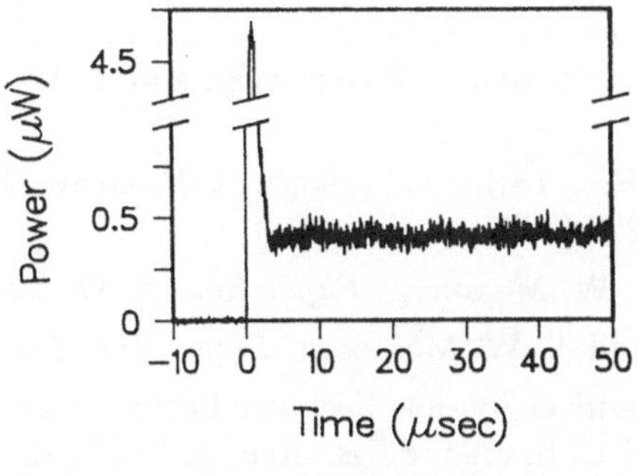

Figure 8. Power output from a cavity, tuned to an inverted two-photon dressed-state transition and containing driven atoms, as a function of time before and after application of a trigger field (sharp spike). Ref. 12.

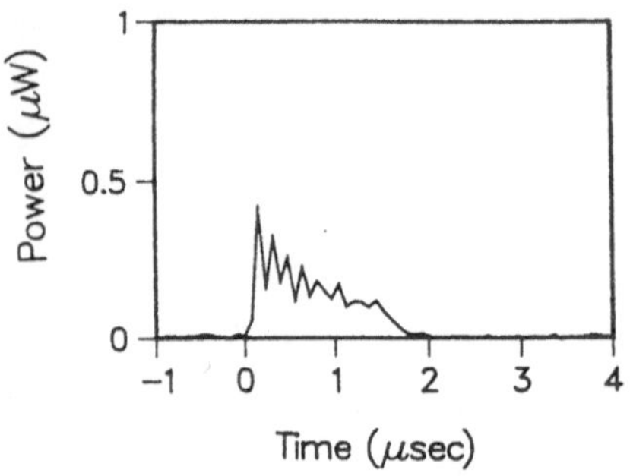

Figure 9. Same as figure 8 except that the trigger field was of insufficient intensity to initiate two-photon lasing.

function of time. The sharp spike is the injected trigger field. When the trigger is turned off, the two-photon laser remains in the on state. In Fig. 9, we show a case where the trigger field was not intense enough to start the two-photon laser. After the 1.4 μsec trigger is over the cavity output quickly decays to zero. Future work in this system will include attempts to realize non-degenerate operation and detailed studies of the switch-on characteristics.

REFERENCES

1. D. Kleppner, *Phys. Rev. Lett.* **47**, 233 (1981).
2. D. J. Heinzen and M. S. Feld, *Phys. Rev. Lett.* **59**, 2623 (1987); R. G. Hulet, E.S. Hilfer, and D. Kleppner, *Phys. Rev. Lett.* **55**, 2137 (1985); W. Jhe, A. Anderson, E. A. Hinds, D. Meschede, L. Moi, and S. Haroche, *Phys. Rev. Lett.* **58**, 666 (1987); P. Goy, J. M. Raimond, M. Gross, and S. Haroche, *Phys. Rev. Lett.* **50**, 1903 (1983); K. H. Drexhage, Prog. Opt. XII, (E. Wolf, Ed.) North-Holland, New York (1974).
3. M. Lewenstein, J. Zakrzewski, T. W. Mossberg, and J. Mostowski, *J. Phys.* B **21**, L9-14, (1988); M. Lewenstein, J. Zakrzewski, and T. W. Mossberg, *Phys. Rev.* **A 39**, 808 (1988).
4. B. R. Mollow, *Phys. Rev.* **188**, 1969 (1969); R. E. Grove, F. Y. Wu, and S. Ezekiel, *Phys. Rev.* **A 15**, 227 (1977).
5. M. Lewenstein and T. W. Mossberg, *Phys. Rev.* A **37**, 2048 (1988).
6. Y. Zhu, A. Lezama, and T. W. Mossberg, *Phys. Rev. Lett* **61**, 1946 (1988).
7. Note that steady-state inversion has been predicted under quite different conditions by M. Lindberg and C. Savage, *Phys. Rev.* A **38**, 5182 (1988).
8. G. Khitrova, J. F. Valley, and H. M. Gibbs, *Phys. Rev. Lett.* **60**, 1126 (1988).
9. A. Lezama, Y. Zhu, M. Kanskar, and T. W. Mossberg, *Phys. Rev.* A **41**, 1576 (1990).
10. J. Zakrzewski, M. Lewenstein, and T. W. Mossberg, *Phys. Rev.* A 44, 7717; 7732; 7746 (1991).
11. M. Lewenstein, Y. Zhu, and T. W. Mossberg, *Phys. Rev. Lett.* **64**, 3131 (1990); Y. Zhu, Q. Wu, S. Morin, and T. W. Mossberg, ibid. **65**, 1200 (1990).
12. D. Gauthier, Q. Wu, S. Morin, and T. W. Mossberg, *Phys. Rev. Lett.* 68, 464 (1992).

THE OPTICAL RING RESONATOR AS A MODEL OF A TWO-LEVEL ATOM

R.J.C. Spreeuw and J.P. Woerdman

Huygens Laboratory, Leiden University
P.O. Box 9504, 2300 RA Leiden
The Netherlands

1. INTRODUCTION

In recent years optics has become a popular playground for testing of basic quantum mechanical concepts. Here we refer to questions dealing with quantum mechanical measurement theory and with the transition between quantum physics and classical physics [1]-[5]. In this paper we discuss classical optical experiments displaying features which are usually associated with quantum physics. Our favorite optical system in this context is the optical ring resonator. The static properties of such a resonator are reviewed in Section 2, in the language of coupled-mode theory. Dynamical properties are reviewed in Section 3; this includes phenomena such as Rabi oscillations, Zener tunneling, Bloch oscillations, multiphoton transitions and the Bloch-Siegert shift. The experimental results reviewed in Sections 2 and 3 establish in a sense the *bona fide* character and practical convenience of our model system. An important advantage of the macroscopic nature of the system is that all parameters are accessible to direct experimental control, including parameters that in the equivalent microscopic quantum system would be difficult or impossible to vary. Finally, in section 4 we give an outlook on further possibilities.

2. THE OPTICAL RING RESONATOR AS A COUPLED-MODE SYSTEM

In this Section we discuss an optical resonator in the language of coupled-mode theory. A *ring* resonator will turn out to be particularly useful from out point of view; later in the Section we will briefly comment on the merits of the Fabry-Perot type resonator.

Consider a ring resonator with clockwise (*cw*) and counterclockwise (*ccw*) travelling waves propagating along the ring (Fig. 1). The eigenfrequencies of the *cw* and *ccw* modes can be lifted by varying a control parameter S which acts in a nonreciprocal way on the modes. For this we may use the Sagnac effect which is produced by rotation of the ring. Crossing of the eigenfrequencies as a function of S is shown in Fig. 2a; it represents an optical level crossing. The *cw* and *ccw* modes are coupled due to partial reflection by a dielectric perturbation (amplitude reflectivity r) somewhere along the ring; this can be realized by introducing a glass plate in a conventional ring

Recent Developments in Quantum Optics, Edited
by R. Inguva, Plenum Press, New York, 1993

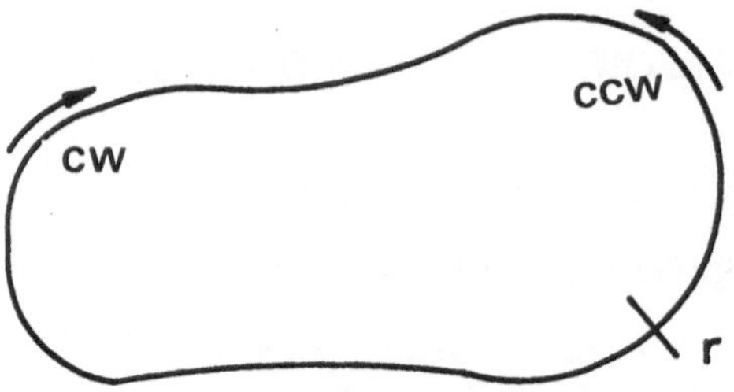

Figure 1. Ring resonator with clockwise and counterclockwise travelling-wave modes. The modes are coupled by a dielectric perturbation with amplitude reflectivity r.

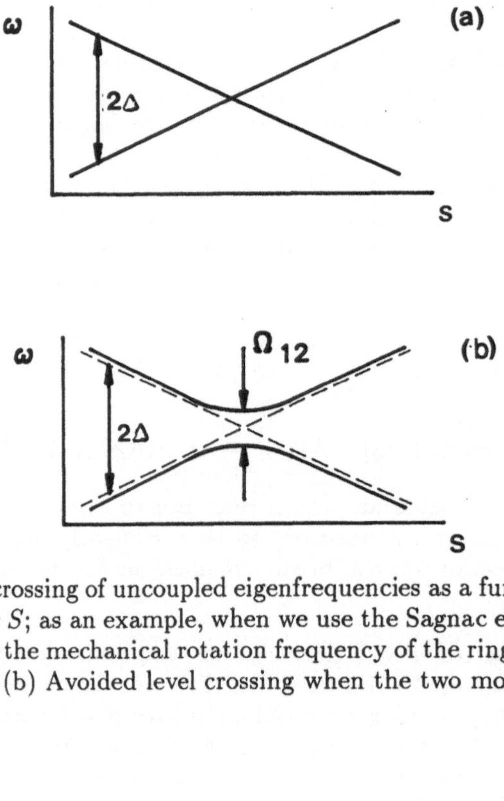

Figure 2. (a) Level crossing of uncoupled eigenfrequencies as a function of the control parameter S; as an example, when we use the Sagnac effect the control parameter is the mechanical rotation frequency of the ring resonator sketched in Fig. 1. (b) Avoided level crossing when the two modes are coupled.

cavity or an air gap in a fiber-optic ring cavity [6,7]. The equations of motion of the complex travelling wave amplitudes a_1 and a_2 in the presence of coupling follow from elementary optical considerations and can be written in matrix notation [8],

$$\frac{dA}{dt} = -iHA \tag{1}$$

with A the state vector (a_1, a_2) and H the dynamical matrix

$$H = \begin{pmatrix} \Delta & W_{12} \\ W_{21} & -\Delta \end{pmatrix}. \tag{2}$$

For the lossless configuration sketched in Fig. 1 the matrix H is Hermitian, with the mode detunings $\pm\Delta$ as its diagonal elements and the couplings $W_{12} = W_{21}^*$ as its off-diagonal elements. In this Hermitian case the coupling is called conservative, which refers to the fact that $|a_1|^2 + |a_2|^2$, i.e. the total energy of the two modes, is conserved. The coupling rate is given by $|W_{12}|^2 = W = r(c/L)$, where L is the circumference of the cavity.

Note that Eq. (1) is analagous to the Schrödinger equation of a two-level system. As an alternative for the complex state-vector A one may choose to work with the real three-dimensional "Bloch vector" $\vec{r} = (r_1, r_2, r_3)$, with

$$r_1 = 2Re\rho_{12}$$

$$r_2 = -2Im\rho_{12}$$

$$r_3 = \rho_{11} - \rho_{22} \tag{3}$$

where $\rho_{ij} = a_i a_j^*$ are the components of the density matrix. Note that the components of the Bloch vector can be considered as the expectation values of the spin components of the spin-$\frac{1}{2}$ particle with spinor A, i.e. $r_i = <\sigma_i> = A^\dagger \sigma_i A$ [9]. This description leads to the optical Bloch equations (see Eq.(6) below).

If several dielectric perturbations are present along with the ring their amplitude reflectivities ($\propto W_{12}$) should be added. Thus, such perturbations may interfere constructively or destructively depending on their separation along the ring. A simple example of this is the interference between the reflections from the front and rear dielectric interfaces when a glass plate is used as coupling element.

Note that Eq.(1) is not restricted to *optical* modes. In fact Eq(1) represents the canonical form of the equations of motion of any pair of linearly coupled harmonic oscillators, if the variation of the oscillator amplitudes is negligible on the time scale of the oscillator period [10]. Of course, in our case the harmonic oscillators are the *cw* and *ccw* modes of the electromagnetic field in the ring resonator.

Diagonalization of H (Eq.(2)) yields two normal modes with eigenfrequencies

$$\omega_\pm = \pm\sqrt{\Delta^2 + W^2}. \tag{4}$$

A plot of Eq.(4), as given in Fig. 2b, shows that the coupling leads to an avoided level crossing: the eigenfrequencies are pushed apart, opening up a frequency gap

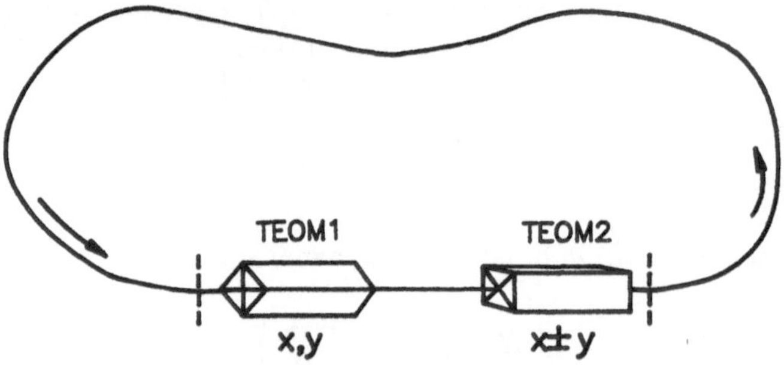

Figure 3: Polarization implementation of two-mode system. TEOM stands for transverse electro-optic modulator. In this case the ring configuration is not essential, also a linear configuration (Fabry-Perot cavity) can be used, with mirrors at position indicated by the dashed lines.

$\Omega_{12} = 2W$. Instead of using the *cw* and *ccw* travelling waves as a basis for the field in the ring resonator we may equally well use the standing waves $cw \pm ccw$. In that case the dynamical matrix becomes

$$H = \begin{pmatrix} W & \Delta \\ \Delta & -W \end{pmatrix} \tag{5}$$

so that coupling and detuning are seen to have reversed their roles [8].

As has been demonstrated experimentally [6,11] we may play essentially the same game by using the polarization modes (i.e. x, y or σ^+, σ^-) instead of the propagation modes (cw, ccw). In this case the Bloch-vector approach would lead to a description of polarization in terms of the Poincaré sphere. The two polarization modes may be coupled by means of a birefringent element, either linear or circular, and detuned (relative to each other) by means of a controllable birefringence, such as an electro-optic or magneto-optic (Faraday) modulator. The control parameter S is then an electric voltage or a magnetic field. Although many configurations are possible, they are all described by Eqs. (1,2). using $x \pm y$ instead of x, y as a basis, the same birefringent element may be interpreted as a coupling instead of a detuning element, and vice versa. Also, when dealing with two polarization modes a ring resonator is not really necessary; a Fabry-Perot type resonator will do as well (see Fig. 3).

Interesting possibilities arise if we consider the case of *four* coupled modes in a ring resonator: two propagation modes (*cw* and *ccw*), coupled by partial reflection, and two polarization modes (x and y) coupled by birefringence. Recently the four-level case has been studied theoretically and experimentally [11]. The complexity of this case is such that a full analysis requires group-theoretical methods; the group involved is U(2,2). However, the useful properties of some high-symmetry configurations of the four-level case can also be explored heuristically. This applies for instance to configurations which allow to simulate Sagnac-detuning of two counterpropagating modes (cw, ccw) by means of the Faraday effect, as demonstrated in Ref.[7].

So far, the discussion has been restricted to a single, be it four-fold degenerate resonance of the ring resonator. Clearly, the ring has a series of resonances, separated along the frequency axis by the free spectral range c/L. When varying the detuning parameter S this gives rise to a manifold of level crossings, instead of just a single crossing, so that, when introducing a coupling W a manifold of avoided crossings results, a so-called optical band structure. The analogy with electronic band structure has been extensively discussed elsewhere [7,12,13].

3. DYNAMICAL ASPECTS OF THE OPTICAL RING RESONATOR

The optical ring resonator can be driven by varying one of the control parameters (coupling or detuning) as a function of time. As discussed in Section 2, coupling and detuning can be implemented by means of a variety of optical elements. In this Section we will restrict the discussion to the two mode case as described by Eqs. (1,2) and consider only harmonic driving. As an example, when we substitute $W_{12} = W_{21} = W_o sin\Omega t$ in Eq.(5) we have again the Hamiltonian of an optically driven two-level atom, this time with transition frequency 2Δ, the coupling strength with the optical field being described by the Rabi frequency W_o. As another example, when we substitute $\Delta = \Delta_o sin\Omega t$ in Eq.(5) we have again the Hamiltonian of an optically driven two-level atom, this time with transition frequency $2W$ and Rabi freqency Δ_o. Thus by a suitable choice of the basis the Hamiltonian of the driven optical resonator may always be mapped on that of a two-level atom driven by a classical optical field. In this sense the driven ring resonator may be called an "optical atom."

Harmonic variation of W or Δ can be accomplished by harmonically driving an electro-optic or magneto-optic (Faraday) modulator inside the ring. For instance, the partial backreflector which couples a *cw* and *ccw* mode can be modulated by implementing it as a Fabry-Perot with an electro-optic phase modulator between its mirrors. Note that we have now three harmonic oscillators in the problem: the two optical mode oscillators, referred to in Section 2, and one electrical oscillator, e.g. an LC circuit, which drives the intracavity modulator. In practice, the mode oscillators have an optical frequency ($10^{14} - 10^{15}$ Hz) and the electrical oscillator a radio frequency ($10^6 - 10^7$ Hz).

We have demonstrated experimentally a number of effects which are well known for driven two-level atoms [6]. For a classification of the effects it is useful to distinguish the regimes $\Omega << \Omega_{12}, \Omega \sim \Omega_{12}$ and $\Omega >> \Omega_{12}$, where Ω_{12} is the transition frequency of the equivalent two-level atom. Assume now that in Fig. 4 the system is prepared, at $t = 0$, in an eigenstate (top of the mountain hyperbola) and that, for $t > 0$, the control parameter S becomes a harmonic (Ω) function of time. The response of the system becomes adiabatic in the limit $\Omega << \Omega_{12}$ (keeping the amplitude of the driving field constant); the representative dot then sweeps back and forth along the mountain hyperbola in Fig. 4. In atomic physics this corresponds to the AC polarization set up in an atom when drive by a low-frequency electromagnetic field; in solid-state physics it is closely related to so-called Bloch oscillation in an electronic band structure and for this reason the same name is sometimes used for the adiabatic response upon (harmonic) driving of the optical resonator. Upon increasing the value of Ω the response acquires some nonadiabaticity: Zener tunneling of light occurs across the gap Ω_{12}. In the limit $\Omega >> \Omega_{12}$ Zener tunneling dominates the response. Experimentally we have observed both Bloch oscillation and Zener tunneling, see Fig. 5 [6].

The resonant condition ($\Omega \sim \Omega_{12}$) takes a special place. For a driven two-level atom this condition leads to Rabi oscillation, i.e., periodic population transfer between

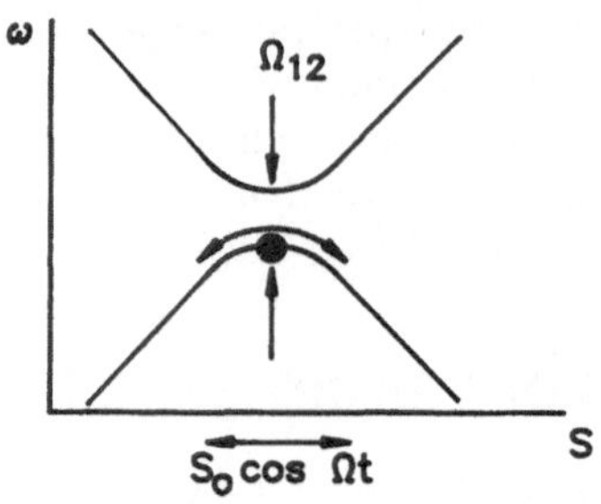

Figure 4. Adiabatic response of optical two-level system upon an harmonic driving field. The dot represents the instantaneous state of the system and 'rolls' back and forther over the mountain hyperbola.

levels and 1 and 2 at the Rabi "flipping" frequency. We have observed this Rabi oscillation for the resonantly driven optical resonator, in the time domain as well as in the spectral domain (Autler-Townes doublet splitting); Fig. 6 shows an example [6]. An important point to make is that for our optical implementation of the driven two-level atom the Rabi frequency (i.e. either W_o or Δ_o) can be easily made as large as the transition frequency Ω_{12}, or even larger, which is rather unusual in two-level atom physics. This implies that we may grossly violate the so-called rotating-wave approximation (RWA), which is an exceptional situation in quantum optics; see Sections 5, 6 for further discussion. We have already observed consequences of violation of the RWA in our experiments [6]; we aim to demonstrate in the near future other consequences such as the Bloch-Siegert shift and multiphoton transitions (subharmonic resonance), i.e. resonant behavior at $\Omega = \Omega_{12}/(2n+1)$.

4. FUTURE POSSIBILITIES

So far the driven coupled-mode systems which we have discussed were fully linear (see Eqs. (1)(2)). Interesting possiblities arise if we extend the discussion to nonlinear systems, i.e. to cases where either the detuning Δ or the coupling W_{12} (or both) is a function of the mode amplitudes A_1 and A_2. The generic equation of motion for this case can be written as

$$\frac{d}{dt}\begin{pmatrix} A_1 \\ A_2 \end{pmatrix} = -i\begin{pmatrix} \Delta(A_1, A_2) & W_{12}(A_1, A_2)cos\ \omega t \\ W_{21}(A_1, A_2)cos\ \omega t & -\Delta(A_1, A_2) \end{pmatrix} \begin{pmatrix} A_1 \\ A_2 \end{pmatrix}. \tag{6}$$

Eq. (6) generalizes several popular model Hamiltonians of quantum optics; a detailed discussion may be found in Refs. [14,15]. We give now two examples.

As a first example, the case $\Delta = constant$ and $W_{12} = W_{12}(A_1, A_2)$ corresponds to the well known Jaynes-Cummings model [16,17], which describes the coupling of a two-level atom and a single (cavity) model of the electromagnetic field. In our case the driving field is the radiofrequency (RF) field applied to an intracavity electro-optic or Faraday modulator. Since a RF field is classical under all practical circumstances we deal with a semiclassical version of the Jaynes-Cummings model. It has been

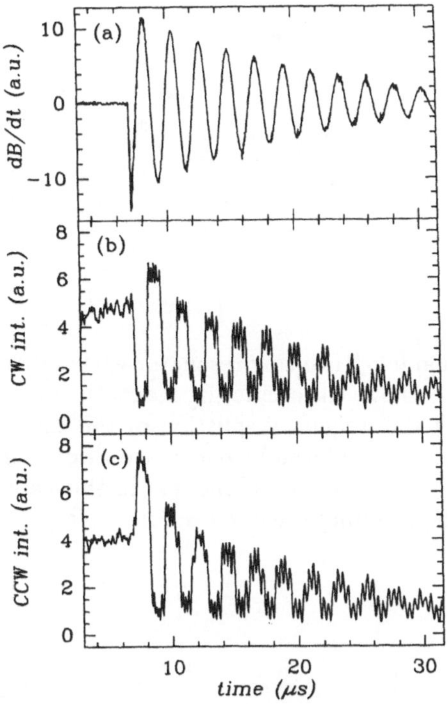

Figure 5. Bloch oscillations and Zener tunneling in optical ring resonator with a driven Farady modulator. The *cw* and *ccw* waves in the ring are coupled by a backscattering element (glass plate). Box (a) shows the oscillating magnetic field, box (b) the *cw* intensity in the ring and box (c) the *ccw* intensity in the ring. The *cw* and *ccw* intensities are seen to oscillate in antiphase, in synchronism with the magnetic field; this represents the diabatic response (Bloch oscillation). A high-frequency ripple is superimposed on this slow, adiabatic response. The ripple betrays that, starting from a pure state, *both* normal modes have become populated. Thus the ripple shows that Zener tunneling of lilght is taking place.

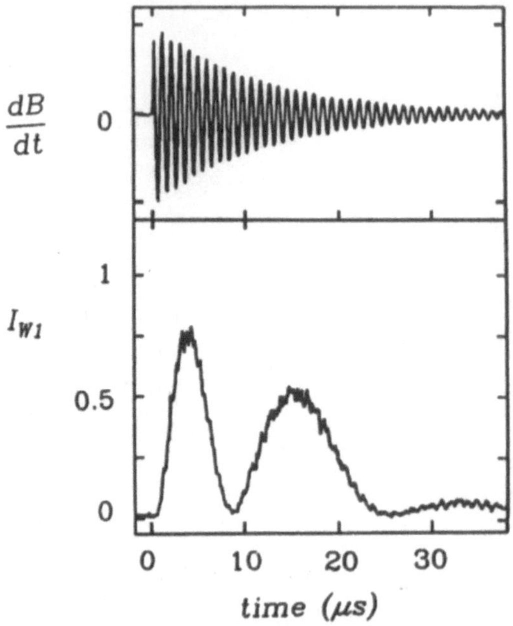

Figure 6. Rabi oscillation in optical ring resonator with a driven Faraday modulator. The upper box shows the oscillating magnetic field, which is approximately resonant with the frequency splitting of the *cw* and *ccw* waves (this frequency splitting is produced by backscattering). The lower box shows Rabi oscillation of one of the normal modes. In this case the normal modes correspond to the standing waves $cw \pm ccw$.

predicted that chaos may occur in the semiclassical Janes-Cummings model when the driving field is so strong that the RWA is violated [18]. Possibly, our system offers the first possibility of an experimental verification of this prediction [14,15].

As a second example, the case $\Delta = \Delta(A_1, A_2)$ and W_{12} = constant may be mapped on the mechanical problem of a driven three-dimensional top [21]. The dynamics of a driven top may be chaotic for certain regions of the parameter space. If, in such a region, one varies the angular momentum of the top from $j >> \hbar$ to $j \sim \hbar$ one may study the transition from classical chaos to "quantum chaos" [19,20]. In the optical implementation the angular momentum of the top (j) corresponds to the number of photons in the ring resonator (N), more precisely: $N = 2j$. Thus optical experimentation with weal light fields in the ring may possibly allow us to enter the regime of quantum chaos.

Experimentally, the most convenient way to realize a nonlinear detuning or coupling seems to be adding an electrical feedback loop to the optical ring resonator [14,15]. Bulk-optic nonlinearities are too weak for this purpose, certainly in the quantum regime

ACKNOWLEDGEMENTS

We gratefully acknowledge useful discussions with E.R. Eliel,, M.P. van Exter, F. Haake and F.T. Arecchi.

REFERENCES

1. Z. Y. Ou, X. Y. Zou, L. J. Wang and L. Mandel, "Observation of nonlocal interference in separated photon channels," *Phys. Rev. Lett.* **65** (1990) 321.
2. L. J. Wang, X. Y. Zou and L. Mandel, "Experimental test of the de Broglie guided-wave theory for photons," *Phys. Rev. Lett.* **66** (1991) 1111.
3. P. G. Kwiat, W. A. Vareka, C. K. Hong, H. Nathel and R. Y. Chiao, "Correlated two-photon interference in a dual-beam Michelson interferometer," *Phys. Rev.* A**41** (1990) 2910.
4. W. M. Itano, D. J. Heinzen, J. J. Bollinger and D. J. Wineland, "Quantum Zeno effect," *Phys. Rev.* A**41** (1990) 2295.
5. A. Aspect, P. Grangier and G. Roger, "Experimental tests of realistic local theories via Bell's theorem," *Phys. Rev. Lett.* **47** (1981) 460.
6. R. J. C. Spreeuw, N. J. van Druten, M. W. Beijersbergen, E. R. Eliel and J. P. Woerdman, "Classical realization of a strongly driven two-level system," *Phys. Rev. Lett.* **65** (1990) 2642.
7. R. J. C. Spreeuw, J. P. Woerdman and D. Lenstra, "Photon band structure in a Sagnac fiber-optic ring resonator," *Phys. Rev. Lett.* **61** (1988) 318.
8. R. J. C. Spreeuw, R. Centeno Neelen, N. J. van Druten, E. R. Eliel and J. P. Woerdman, "Mode coupling in a HeNe ring laser with backscattering," *Phys. Rev.* A**42** (1990) 4315.
9. L. Allen and J. H. Eberly, *Optical Resonance and Two-Level Atoms*, (Dover, New York, 1987).
10. G. Weinreich, "Coupled piano strings," *J. Acoust. Soc. Am.* **62** (1977) 1474.
11. R. J. C. Spreeuw, M. W. Beijersbergen and J. P. Woerdman, "Optical ring cavities as tailored four-level systems: an application of the group U(2,2)," submitted to *Phys. Rev.* A.
12. J. P. Woerdman and R. J. C. Spreeuw, *Optical level crossings*, in: *Analogies in Optics and micro-electronics*, edited by W. van Haeringen and D. Lenstra, Kluwer, Dordrecht, 1990, p. 135.
13. D. Lenstra, L. P. J. Kamp and W. van Haeringen, "Mode conversion, Bloch oscillations and Zener tunneling with light by means of the Sagnac effect," *Opt. Commun.* **60** (1986) 339.
14. R. J. C. Spreeuw and J. P. Woerdman, "The driven optical ring resonator as a model system for quantum optics," to be published in *Physica* B(1991).
15. R. J. C. Spreeuw and J. P. Woerdman, *"Optical atoms"*, to be published in: *Progress in Optics,* E. Wolf, editor, Elsevier, Amsterdam (1992).
16. E. T. Jaynes and F. W. Cummings, *Comparison of quantum and semiclassical radiation theories with application to the beam maser*, Proc. IEEE **51** (1963) 89.
17. R. Graham and M. Hr$\tilde{m}$ohnerbach, "Quantum chaos of the two-level atom," *Phys. Lett.* **101A** (1984) 61.

18. P. W. Milonni, J. R. Ackerhalt and H. W. Galbraith, "Chaos in the semiclassical N-atom Jaynes-Cummings model failure of the rotating-wave approximation," *Phys. Rev. Lett.* **50** (1983) 966.

19. F. Haake, *Quantum signatures of chaos*, Springer, Berlin, 1991.

20. F. Haake, M. Kuś and R. Scharf, "Classical and quantum chaos for a kicked top," *Z. Phys.* B **65** (1987) 381.

21. F. Haake, G. Lenz and F. Puri, "Optical tops," *J. Mod. Opt.* **37** (1990) 155.

TAILORING ATOMIC RESPONSE*

Surya P. Tewari

School of Physics
University of Hyderabad
Hyderabad-500 134, India

The nonlinear generation of radiation by atomic vapors depends critically on the dispersion and absorption properties of the medium. The absorptive media are of little use. The requirement on the dispersive properties of the medium depends on the geometry of the incident radiation. For example, for plane wave fields one needs the medium to be dispersion free while for focused beams, and say for third harmonic generation, negatively dispersive media are needed. It has been demonstrated[1] that, by the use of an extra resonant laser it is possible to tailor the atomic response in the atomic vapors. For example, a control over the phase-matching was predicted by the use of the intensity and the detuning of the resonant laser radiation. The experiments[2,3] confirmed the predictions, in that the generation of radiation in the forbidden range of frequencies was made possible by the use of an extra resonant laser.

Recently, the use of intense laser radiation under the simultaneous condition of two-photon resonance has revealed a new limit for increasing the efficiency of the generating process over a fairly wide range of frequencies.[4,5] In the following, we consider tailoring the atomic response to achieve optimal generation of radiation. We begin briefly, with the problem of defining an optimal generation, by identifying the conditions that are required to achieve it. For simplicity, we shall consider only the plane wave approximation of the incident fields.

Consider the intensity of the generated field which is given by

$$|\mathcal{E}_g|^2 = \frac{4\pi^2}{c^2}\omega_g^2|\chi^{NL}(\omega_g)|^2|F(\overline{\Delta k}\ell)|^2|\mathcal{E}_f^n|^2\ell^2 \tag{1}$$

$$F(\overline{\Delta k}\ell) = \frac{1 - exp(i\overline{\Delta k}\ell)}{-i\overline{\Delta k}\ell} \tag{2}$$

$$\overline{\Delta k} = \frac{2\pi\omega_g}{c}\chi^{(1)}(\omega_g) = \Delta k + i\alpha \tag{3}$$

$$\omega_g = n\omega_f \tag{4}$$

*Work done in collaboration with G. S. Agarwal

Recent Developments in Quantum Optics, Edited
by R. Inguva, Plenum Press, New York, 1993

here Δk, α are real, $\omega_g(\omega_f)$ is the frequency of the generated (fundamental) field, $\mathcal{E}_g(\mathcal{E}_f)$ is the slowly varying envelope of the generated (fundamental) field, $\chi^{NL}(\omega_g)$ $[\chi^{(1)}(\omega_g)]$ is the non-linear [linear] susceptibility of the atomic medium at the frequency ω_g, c is the velocity of light and ℓ is the length of the interaction region.

In deriving Eq. (1), the plane wave approximation and the assumption of the collinearity of propagation of the incident and generated radiation, along the positive z-direction, has been made. For high output intensity of the generated radiation we need maximum values of all factors on the right hand side. The behavior of the function $|F(\overline{\Delta k\ell})|^2$ for a medium nearly resonant at the generated frequency is shown in Fig. 1, for different values of the linear absorption coefficient α. The function has its peak at zero values of Δk and α. When ω_g becomes resonant with atomic transition (ω_o), $\Delta k \to 0$ and $\alpha >> 1$, depending upon the atomic density. The $F(\overline{\Delta k\ell})$ is then very small and hence minimal generation in absorptive medium occurs. It is clear from Fig. 1, that the largest value of $|F(\overline{\Delta k\ell})|^2$ occurs for $\Delta k = 0, \alpha = 0$. One gets such a region approximately by choosing $|\omega_g - \omega_o| = \delta >> \gamma$ (γ being the width of the atomic transition at frequency ω_o). For, then $\Delta k \approx 1/\delta$ and $\alpha \approx 1/\delta^2$, demanding that generated frequency should be far away from any resonance. In this situation, the largest value of length, ℓ, in the right hand side of Eq. (1) turns out to be $1/\alpha$. Consequently, $|\mathcal{E}_g|^2$ can become proportional to δ^4 in Eq. (1), for large detunings. There is a limitation to the largest detuning, however, as detuning with one level may result in resonance with some other level in the atom. For a large value of $\chi^{NL}(\omega_g)$, in Eq. (1), one must have a resonance in one of the intermediate energy levels. Usually a two-photon resonance is attempted[6] as the absorption due to two-photon process is weak for the fundamental field. Thus, the condition required for optimal generation may be enumerated as (i) $\Delta k = \alpha = 0$, (ii) $\ell \approx 1/\alpha$, (iii) two-photon resonance.

We shall see that the tailoring of the atomic response shall allow to generate these conditions even in a medium which is on resonance with ω_g.

Consider an atom with four energy levels $|1>, |2>, |3>$ and $|4>$ with their energies $E_4 > E_3 > E_2 > E_1$ interacting with fields at frequencies ω_a, ω_b, ω_c, ω_d and ω_ℓ according to the following Hamiltonian (see the level structure in Fig. 2)

$$H = \sum_{i=1}^{4} E_i|i><i| - \Big[\hbar K_{12}e^{i(\omega_a+\omega_b)t}|1><2| + \hbar G_{23}e^{i\omega_c t}|2><3|$$

$$+ \hbar G_{13}e^{i\omega_d t}|1><3| + \hbar G_{34}e^{i\omega_l t}|3><4| + h.c.\Big] \quad (5)$$

$$\omega_d = \omega_a + \omega_b + \omega_c, \hbar K_{12} = \hbar K_{21}^*, \quad (6)$$

$$\hbar K_{21} = \sum_j \Big[\frac{<2|ex|j><j|ex|1>}{\hbar(\omega_j - \omega_a - \omega_l)} + \frac{<2|ex|j><j|ex|1}{\hbar(\omega_j - \omega_b - \omega_l)}\Big]\mathcal{E}_a\mathcal{E}_b \quad (7)$$

here $\hbar K_{12}$ is the two-photon resonant part of the Hamiltonian, and G_{23}, G_{13}, and G_{34} are the Rabi frequencies $(d_{mn} \cdot \mathcal{E}_q/\hbar, q = c, d, l)$ of the relevant transitions. d_{mn} being the dipole matrix element for the dipole allowed transition between the levels $|m>$ and $|n>$. The fields ω_a and ω_b are assumed to be on exact resonance with the two-photon transition between levels $|1>$ and $|2>$. We shall also assume that there is exact resonance between the field frequencies ω_c, ω_d and ω_l with respect to the transitions 2⇔3, 3⇔1, and 4⇔3 respectively. In the absence of the intense

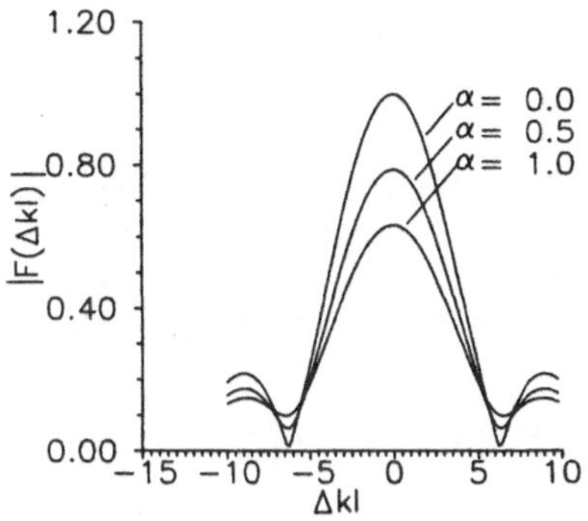

Figure 1. The values of $|F(\overline{\Delta k \ell})|$ versus $\overline{\Delta k \ell}$ for different values of α.

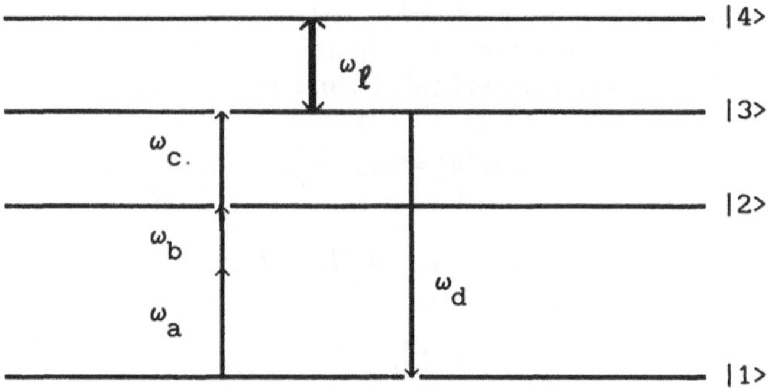

Figure 2. The energy level diagram considered in the text.

radiation, ω_l, the generation of the radiation, ω_d (resonant with $3 \Leftrightarrow 1$, ω_o) is negligible for length α^{-1}, of a medium which is an assembly, with density, n, of the four level atoms assumed in (5). Here, α is given by

$$\alpha = \frac{2\pi n \omega_d |d_{13}|^2}{\hbar c \Gamma_{31}} \tag{8}$$

where d_{31} is the dipole matrix element, and $\Gamma_{31}(\Gamma_{mn})$ is the transverse relaxation rate of the density matrix element $\rho_{31}(\rho_{mn})$. Note that in the absence of ω_l, and the presence of exact resonance ($\omega_d = \omega_O$), all but one condition of optimal generation are met with. Consequently, negligible generation would occur in the absence of ω_l. On switching the intense beam at ω_l, the response of the atomic medium changes in two ways. The $\chi^{(1)}(\omega_d)$ as well as the $\chi^{(NL)}(\omega_d)$ get modified. These modifications for weak fields at ω_a, ω_b, ω_c can be obtained perturbatively by solving the density matrix equations including in them phenomenologically the relvent relaxation rates. Working to the lowest order in the fields $\mathcal{E}_a$, $\mathcal{E}_b$, $\mathcal{E}_c$ and $\mathcal{E}_d$ but to all orders in $\mathcal{E}_l$ one has

$$\chi^{(1)}(\omega_d) = \frac{n}{\hbar} \frac{|d_{31}|^2}{\Delta_{31} - \frac{|G_{34}|^2}{\Delta_{41}}} \tag{9}$$

$$\chi^{(NL)}(\omega_d) = \frac{n}{\hbar^2} \frac{d_{13}d_{32}K_{21}}{\Delta_{21}\left[\Delta_{31} - \frac{|G_{34}|^2}{\Delta_{41}}\right]} \tag{10}$$

$$\Delta_{31} = \omega_{31} - \omega_d - i(\Gamma_{31}/2) \tag{11a}$$

$$\Delta_{21} = \omega_{21} - \omega_a - \omega_b - i(\Gamma_{21}/2) \tag{11b}$$

$$\Delta_{41} = \omega_{41} - \omega_d - \omega_l - i(\Gamma_{41}/2), \tag{11c}$$

Expressions (9) and (10) differ from similar expressions in Ref. (1), in the explicit presence of the two-photon resonance denominator (Δ_{21}) in $\chi^{(NL)}(\omega_d)$. In Ref. (1) this denominator was a part of the effective three-photon matrix element and consequently was assumed to be far detuned. For the homogeneously broadened system Eqs. (9) and (10) give the same results as discussed in Ref. (1). Here we consider the variation in susceptibilities for atoms moving with different velocities, a case generally found in atomic vapors. The velocity dependent detunings are then given by

$$\Delta_{31}(v) = \Delta_{31} + k_d v \tag{12a}$$

$$\Delta_{21}(v) = \Delta_{21} + (k_a + k_b)v \tag{12b}$$

$$\Delta_{41}(v) = \Delta_{41} + (k_d + k_l)v \tag{12c}$$

$$k_j = \omega_j/c, j = a, b, c, d, l \tag{13}$$

v is the z-component of the velocity vector, $\underline{v}$, along the direction of propagation of the collinear radiations. In Figs. 3, 4, we have plotted the real and imaginary parts of the susceptibilities. We have multiplied each term with the corresponding Dopplerian weight factor $exp(-v^2/v_T^2)$ such that the graphs represent the behavior of the integrand of all terms needed in the propagation equation for the generation of the field, ω_d, in the resonant medium.

There are two groups of atoms which are independently in resonance with the components of the Autler-Townes doublet. Their velocities are given approximately by

$$k_d v_{1,2} \cong \pm\left[|G_{34}|^2\left(1 + \frac{k_l}{k_d}\right) - \frac{1}{4}\left\{\frac{\Gamma_{31}}{2}\left(1 + \frac{k_l}{k_d}\right) - \frac{\Gamma_{41}}{2}\right\}^2\right]^{\frac{1}{2}} X\left(1 + \frac{k_l}{k_d}\right)^{-1} \tag{14}$$

Now these velocities are real numbers if G_{34} is such that the term under the square root sign in Eq. (14) is positive. Further, for $\nu_{1,2} < \nu_T$, the symmetry of the graphs suggests that the integrated Δk and α (represented by double bar above the respective symbols and which are proportional to the $\text{Re}\chi^{(1)}$ and the $\text{Im}\chi^{(1)}$ of Fig. 3, respectively), have the values such that

$$\overline{\overline{\alpha}} >> 1, \qquad \overline{\overline{\Delta k}} \approx 0. \tag{15}$$

Thus the system continues to behave like a strongly absorbing homogeneously broadened medium. While for $v_{1,2} >> v_T$, one deduces that

$$\overline{\overline{\Delta k}} \approx 0, \qquad 0 << \overline{\overline{\alpha}} << 1 \tag{16}$$

This region is of interest as it creates the conditions of optimal generation, even though the incident fields and the generated field are apparently on exact resonance with the atomic transitions.

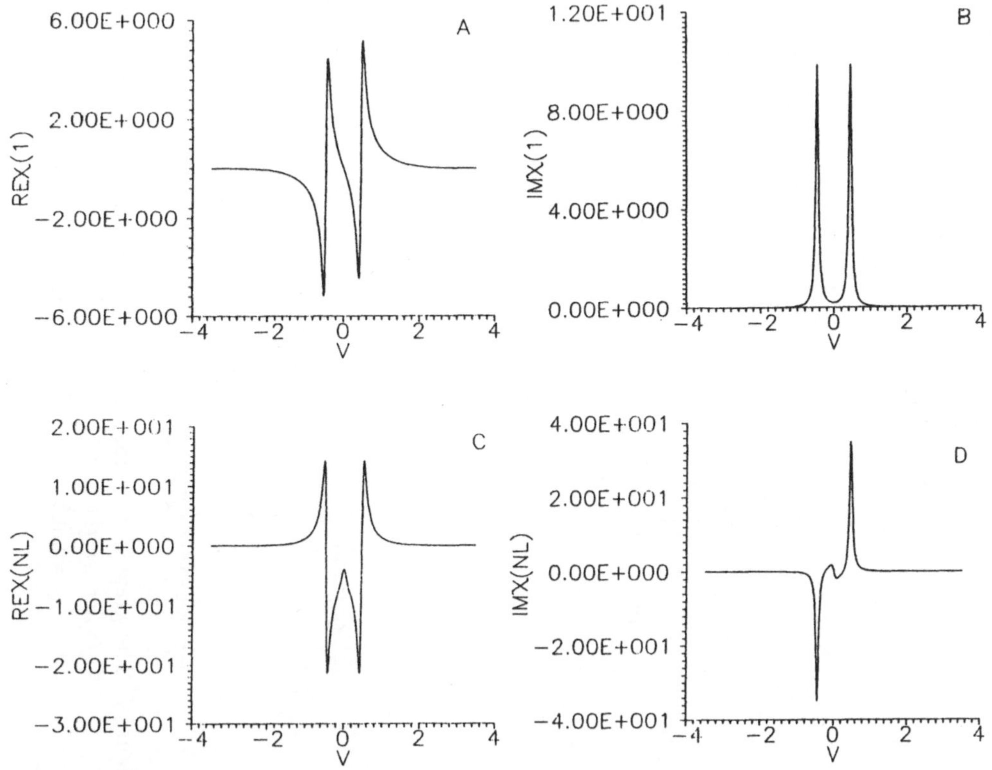

Figure 3. The graphs show the variation of the susceptibilities, within the Doppler profile, multiplied by the Doppler weight factor $exp(-v^2/v_T^2)$, for $\Delta_{21} = \Delta_{41} = \Delta_{31} = 0$, $\Gamma_{21} = \Gamma_{41} = 0.1\Delta\omega_D$, and $G_{34} = 0.5\Delta\omega_D$. All frequencies are normalized to $\Delta\omega_D$ and v is normalized to $v_T(= \Delta\omega_D/k_d)$.

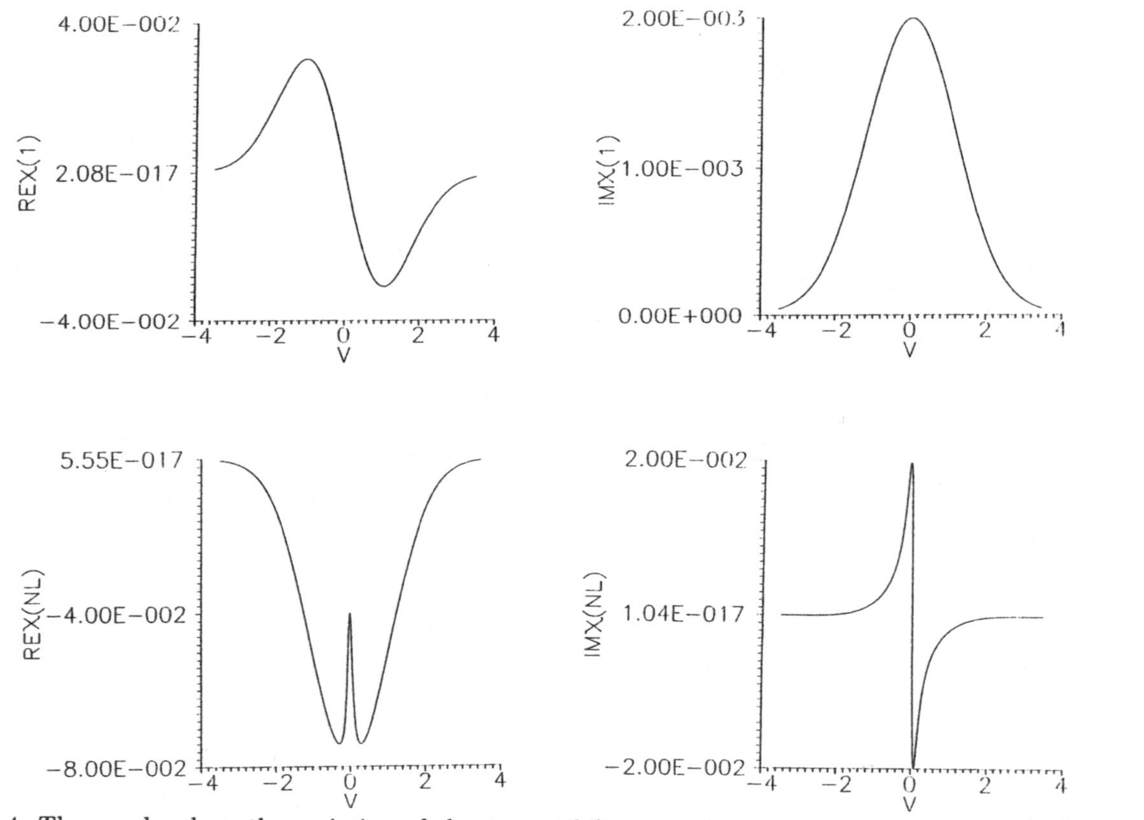

Figure 4. The graphs show the variation of the susceptibilities, within the Doppler profile, multiplied by the Doppler weight factor $exp(-v^2/v_T^2)$, for $\Delta_{21} = \Delta_{41} = \Delta_{31} = 0$, $\Gamma_{21} = \Gamma_{41} = 0.1\Delta\omega_D$, $\Gamma_{31} = 0.1\Delta\omega_D$ and $G_{34} = 5.0\Delta\omega_D$. All frequencies are normalized to $\Delta\omega_D$ and v is normalized to v_T $(=\Delta\omega_D/k_d)$.

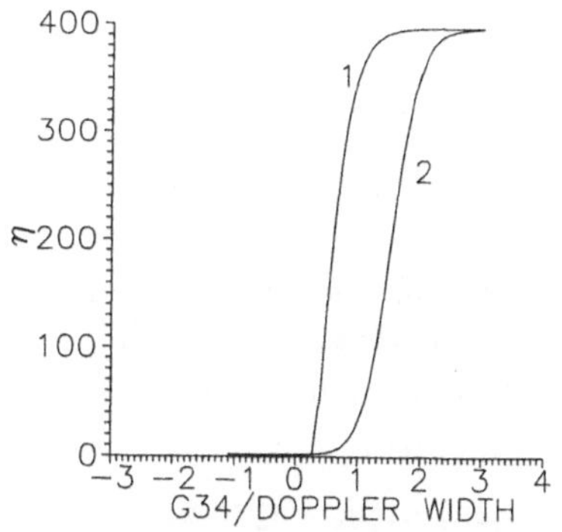

Figure 5. The efficiency factor η is plotted against the Rabi frequency, G_{34}, normalized to the Doppler width, for two values of $\Gamma_{31}(1) = 0.1\Delta\omega_D$ and $(2) = 10\ \Delta\omega_D$. Other parameters for this graph are $\Gamma_{21} = \Gamma_{41} = 0.01\Delta\omega_D$.

Note that optimal conditions are useful only for non-zero value of the integrated non-linear susceptibility of the resonant situation.† From Fig. 4 one notes, on the basis of the symmetry and the shape of the integrands, that $\mathrm{Im}\chi^{(NL)}(\omega_d)$ is negligible while the $\mathrm{Re}\chi^{(NL)}(\omega_d)$ is non-zero, and that it has enhanced value because of the two-photon resonance denominator in Eq. (10). The effect of two-photon resonance denominator is evident by the structure in the integrands of the non-linear susceptibility in the $v \approx 0$, region.

The generated intensity is obtained by solving the propagation equation using the integrated susceptibilities discussed above. The generated intensity is found to be proportional to the efficiency factor, η^2, given by

$$\eta = \frac{|\overline{\overline{Re\chi^{\mathrm{NL}}(\omega_d)}}|}{|\overline{\overline{\alpha}}|}$$

In Fig. 5, we have plotted the values of the efficiency factor, η, for different values of the Rabi frequency, G_{34}. As expected from the above analysis, η remains small for values of Rabi frequency, G_{34} smaller than the Doppler width, $\Delta\omega_D$. For larger values of G_{34}, η first increases sharply and then saturates to a value inversely proportional to the relaxation rate, Γ_{41}. It is interesting to note that at large G_{34} the efficiency factor is independent of the relaxation rate Γ_{31}, which corresponds to the transition with which the generated field is in resonance.

Similar effects have recently been discovered by Harris, Field and Imamoglu for an another level scheme.[4] The detailed comparison of the two level schemes has been made and the efficiency factors, η, in the two schemes are found to be comparable.[5]

We have thus shown that the tailoring of atomic response in presence of two-photon resonance is possible even in Doppler broadened atomic vapors, allowing to obtain efficient generation of radiation even at exact resonance between the generated field and the bare atomic transitions.

†Note that $\chi^{(NL)}(\omega_d)$ is negligibly small for $v_{1,2} << v_T$. As may be inferred from the symmetry of the graphs in Fig. 3c,d. Thus giving an additional reason for the generated field to be negligible in the so called Doppler limit, i.e., $\Delta\omega_{D(=k_d v_T)} >> \Gamma_{31}, \Gamma_{41}, \Gamma_{21}, G_{34}$. Note also the difference between the Doppler limit and the limit $v_T << v_{1,2}$, which requires $G_{34} >> \Delta\omega_D >> \Gamma_{31}, \Gamma_{41}, \Gamma_{21}$.

REFERENCES

1. Surya P. Tewari & G. S. Agarwal, *Phys. Rev. Lett.*, **54**, 1811 (1986).
2. R. P. Blacewiz, M. G. Payne, W. R. Garret & J. C. Miller, *Phys. Rev* **A34**, 5171, (1986).
3. R. N. Compton, J. C. Miller, *J. Opt. Soc. Am.* **B2**, 355, (1985); D. Normand, J. Morellec & J. Reif, *J. Phys.* **B16**, L227, (1983).
4. S. Harris, J. E. Field & A. Imamoglu, *Phys. Rev. Lett.* **64**, 1107, (1990).
5. Surya P. Tewari & G. S. Agarwal, *Phys. Rev. Lett.* **66**, 1797 (1991).
6. S. E. Harris & D. M. Bloom, *Appl. Phys. Lett.* **24**, 229 (1974).

GENERIC INSTABILITIES AND CHAOS IN STIMULATED SCATTERING PHENOMENON

J.S. Uppal, Weiping Lu, A. Johnstone and R.G. Harrison

Physics Department
Heriot-Watt University
Edinburgh EH14 4AS, UK

INTRODUCTION

Pulsating instabilities and deterministic chaos are known to be common features of many nonlinear processes. These implications are being strongly felt in a broad range of physical sciences. Lasers, in particular, have been found to exhibit a rich variety of nonlinear dynamical behaviour which, in addition to regular periodic oscillations, includes unstable or even chaotic solutions. The observation of deterministic instabilities in lasers is particularly significant as they provide ideal systems for quantitative investigations, owing to their simplicity and the mathematics that describe them, enriched by the possibility of a quantum description. Furthermore the timescale of these instabilities (nanosec. to microseconds) provides simplicity in maintaining stable operating conditions over many periods. From the prolific activity of this field in the area of lasers, attention is now being focussed on other important areas of nonlinear optics. In this paper, we try to explore the generic nature of nonlinear dynamical behaviour in fundamental nonlinear processes. The requirement of feedback for the emergence of instabilities is normally provided externally either by optical cavities as in the lasers or by counter-propagating beams [1-6]. There are, however, many optical phenomena in which feedback is an implicit and integral feature for their interaction. Significantly a broad range, if not the majority, of nonlinear optical interactions satisfy this criteria of which general three-wave mixing and parametric processes merit special attention as cornerstones of the field. Through recent major advances in optical fibre technology we are now able to interrogate these phenomena under ideal cw and for low power pulsed optical excitation by utilizing the long interaction length of this medium to readily induce nonlinearity under these conditions. Furthermore plane wave analysis is an excellent approximation in describing beam propagation in such media. In addition, however, fibre systems are critical components to an entire future telecommunications technology and any qualitative insight into the nonlinear dynamical behaviour of light propagation in such media has potentially important applications.

In this paper we consider in particular stimulated scattering phenomena and provide, to our knowledge, first evidence of chaotic dynamics in one of these basic processes, namely stimulated Brillouin scattering (SBS). Single mode optical fibre is used

Recent Developments in Quantum Optics, Edited
by R. Inguva, Plenum Press, New York, 1993

to generate SBS under cw single mode pump conditions, resulting in first order Stokes emission only [7]. We find both the transmitted pump and back scattered SBS to exhibit chaotic behaviour under all operating conditions investigated, including those close to the threshold for SBS; the SBS exhibiting massive instabilities with modulation depths ~100%. This fundamental and conceptually simple nonlinear wave interaction contrasts with those reported in which feedback either by a reflecting surface or involving counter propagating pump beams is used as a necessary condition for generating dynamical instabilities. Indeed we find that the inclusion of external optical feedback modifies but does not suppress the chaotic dynamics and also gives rise to a wide range of quasi-oscillatory and bursting instabilities. The generality of these observations has been established for fibre lengths ranging from 38 to 300 metres and for launch powers down to ~ 100 mW (threshold for SBS in 200 m fibre).

Subsequent mathematical analysis, based on coupled nonlinear field (pump and Stokes) and material equations, has established the central role of nonlinear refraction and its interplay with the nonlinear gain of the SBS interaction as being responsible for the dynamical behaviour for both cavityless and cavity operation, omission of coexisting dispersive effects, as in conventional treatments, resulting in stable solutions except for the case of low ratio of SBS intensity to input intensity where stable oscillations appear. Further the generality of this treatment points to the likelihood of similar dynamical behaviour in other stimulated scattering phenomena and 3-wave mixing interactions. Significantly the good agreement of our experimental findings with these theoretical results shows the dynamics to be deterministic in origin and not to be due to, or unduly influenced by, the noise structure of the initial spontaneous Brillouin scattering.

EXPERIMENT

Our experimental arrangement is schematically shown in Fig. 1. The cw emission of a single-mode argon-ion laser at 514.5 nm, with an instantaneous ($\leq$1 msec) linewidth of ~15 kHz (Coherent Innova 100) was used as a pump source providing variable output power stabilized to $\pm$2%. Two 10 x microscope objectives L_1 and L_2 were used to couple the light into and out of the optical fibre, respectively. The fibre comprised a pure SiO_2 core of diameter 4.8 μm with a B_2O_3-doped SiO_2 cladding. It was optically isolated from the argon laser using a Faraday isolator (OFR Model IO-5- 532) giving an isolation factor of 35 dB between them. The pump signal and backscattered signal, comprising the SBS signal together with residual scatter, were sampled via the beam splitter in Fig. 1. These signals together with the signal transmitted through the fibre were detected using photodiodes (HP-type BPX65) D_1, D_2 and D_3 which have a rise time of $\leq$0.5 nsec. A transient digitizer (LeCroy TR 8828C) having a data sampling rate of 5 nsec interfaced with a Masscomp Computer (MC5600) was used for temporal recording and subsequent signal processing to provide power spectra and phase-portrait reconstruction. Results were cross checked from independent recordings on a Tektronix 7104 oscilloscope for both temporal measurements and direct phase-portrait reconstruction of the SBS versus transmitted pump signals (oscilloscope in the x-y mode) and on a Hewlett-Packard spectrum analyzer (HP8590A) for corresponding spectral analysis. Feedback from the end faces of the fibre was eliminated using an index-matching liquid for the cavityless system. Single-mode propagation in the fibre was ensured by looping a small section of the front end of the fibre, thereby suppressing higher-order modes and was confirmed by observation of the spatial form of the transmitted signal. A Fabry-Perot

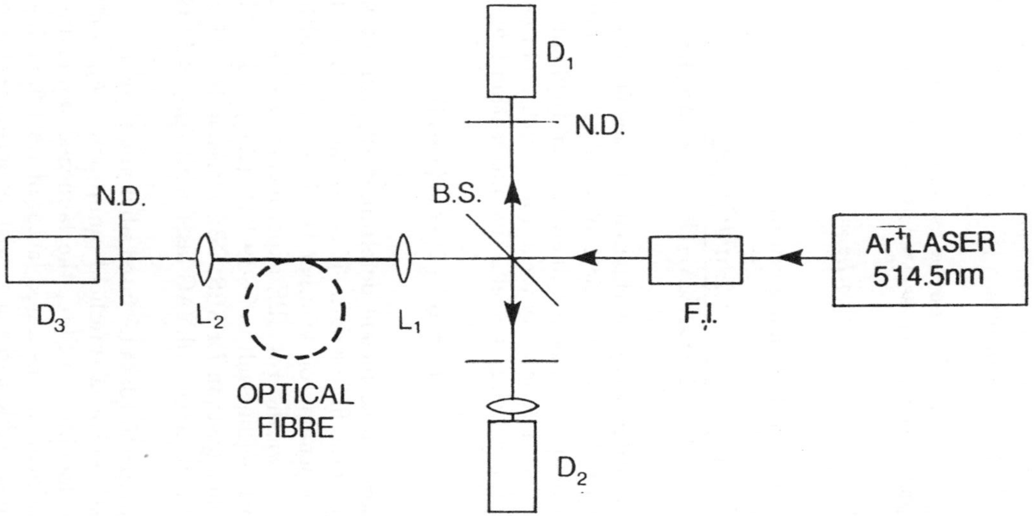

Figure 1. Schematic set-up of experimental arrangement. F.I. is Faraday isolator, N.D. neutral density filters and D_1, D_2, D_3 are photodetectors.

interferometer with variable spacer confirmed throughout only first-order Stokes SBS emission.

At low input power the backscattered signal increased linearly with input power due to Fresnel reflection up to the threshold launch power for SBS (~100 mW for a 200-m fibre). The single thereafter increased rapidly and nonlinearly with launch power, the conversion efficiency to SBS being ~40% for a launch power of ~300 mW, accounting for losses arising from looping the fibre. Corresponding fibre transmission measurements showed a nonlinear response for input power ≥100 mW and for pump powers ≥200 mW gave a saturated maximum of 20 mW. These general features are in agreement with the findings of others [8] and are shown in Fig. 2.

The threshold power for SBS can be estimated from small signal steady state coupled wave analysis of stimulated scattering and is given as

$$g.Le = 21$$

where g is the SBS exponential gain given by [8]

$$g = \frac{2\pi n^7 P_{12}^2 K.P_L}{C\lambda^2 \rho_o V_A \Delta\nu_b (1 + \Delta\nu_L/\Delta\nu_B) A}$$

where n is the refractive index, ρ_o is the material density, V_A is the acoustic velocity and P_{12} is the longitudinal elasto optic coefficient, A is an effective cross sectional area of the guided mode, $\Delta\nu_B$ is the linewidth for Brillouin scattering, $\Delta\nu_L$ is the laser linewidth and factor K is unity for a fibre which maintains polarization, and is equal to the one-half otherwise. The effective interaction length L_e is given by

$$L_e = \alpha^{-1}(1 - exp(-\alpha 1))$$

where α is the absorption coefficient determined experimentally to be $4.6\text{x}10^{-3}$ m^{-1} (20dB/Km), and L is the fibre length. It is pertinent to note that the SBS gain can be quite different depending on whether the laser bandwidth $\Delta\nu_L$ is much less or greater than SBS linewidth. For SBS experiments, we have used therefore a single mode argon laser having linewidth of ~19 KH_2 ($\Delta\nu_B$ = 145 MH). Since the gain for SRS is two orders smaller than for SBS, SBS emission is preferentially suppressed by using a broadband multimode Nd:YAG laser with $\Delta\nu_L$ = 40 GHz.

Using physical properties data for fused silica and an estimated value of A=$0.9\text{x}10^{-11}$ m^2, the above relation gives a threshold pump power P_L ~ 82 mW. Accounting for losses arising from looping the fibre, the launched power for obtaining SBS is increased by an experimentally measured factor of 1.6-1.8, giving a threshold power for SBS~130 mW in reasonable agreement with experimental findings.

Representative recordings of the chopped SBS signal and transmitted pump signal are shown in the traces (a) and (b) of Fig. 3 which demonstrate the magnitude of the instabilities. Trace (a) shows sustained instabilities with modulation depth of ~100%. Similar results were obtained for all input powers, limited to a maximum value of ~700 mW by the laser output power. From trace (b) of the transmitted pump the ratio of the unstable to the underlying stable portion of the transmitted signal provides a measure of the SBS conversion efficiency of 40%, consistent with the measurement above. These general features were confirmed for various fibre lengths

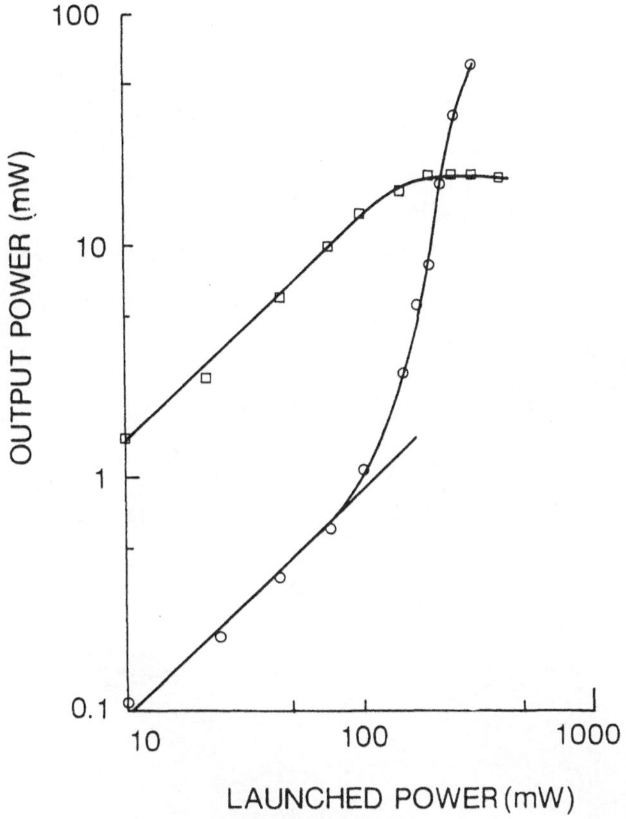

Figure 2. Outward power of backward Brillouin wave and forward transmitted laser wave as a function of input power

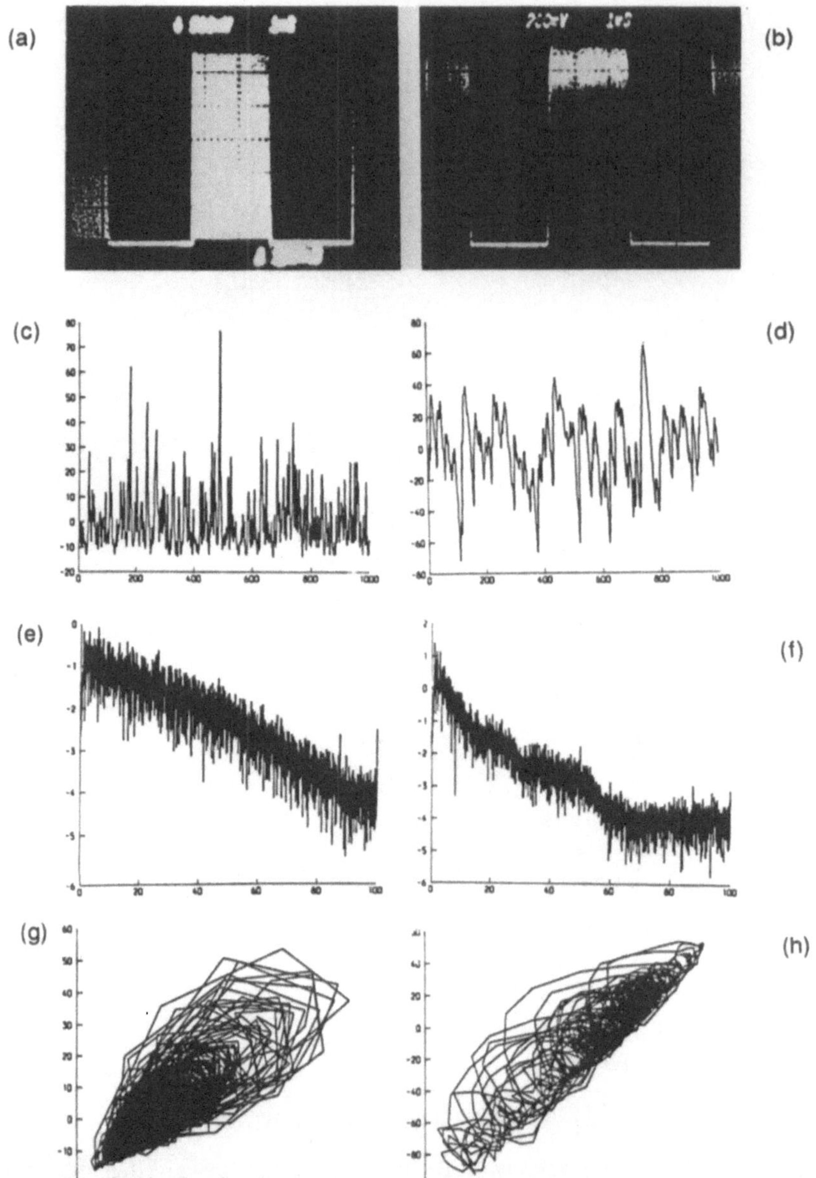

Figure 3. Characteristics of dynamical outputs of back scattered (left column) and transmitted (right column) waves in the cavityless condition. (a),(b) backscattered and transmitted signals detected respectively by D_2 and D_3 for a chopped launch power of 0.33 W where the time scale is 1 msec/division throughout and fibre length is 200 m. (c)-(h) sample time series, power spectrum and phase portraits taken from 16K digitized recordings of backscattered and transmitted signals without chopper: The time series (traces (c) and (d)) comprise 1000 data points (step interval 5 nsec); corresponding power spectra (traces (e) and (f)) are for 16K data points with a vertical scale gain of 10 dB/division; traces (g) and (h) are the corresponding phase portraits comprising 2000 data points constructed with a delay of 15 and 3 step intervals respectively, where fibre length is 100 m.

from 38 to 300 m with an expected increase in the SBS threshold for reduced fibre length; ~500 mW for the 38-m fibre. Representative forms of the dynamical features from the SBS and the transmitted pump signals, for a 100 m length fibre taken without the chopper and for a launch power of 330 mW, are displayed in the digitized recording of the trace (c)-(h) of Fig. 3 showing the time series, power spectrum and reconstructed phase portraits. The latter were obtained by a two-dimensional plot of data points $x_i = 1,2...16,000$ versus x_{i-K} for the appropriate value of delay K.

The time recording of the SBS (trace (c)) shows erratic and aperiodic emission, over a broad time scale down to ~50 nsec, the corresponding power spectrum (trace(e)) exhibiting pure broad band features consistent with fully developed chaos with a spectral bandwidth ~30 MHz which may be compared with the Brillouin gain bandwidth of ~ 145 MHz. The reconstructed phase portrait (trace(g)) exhibits motion on an outward spiralling and folding trajectory particularly evident from observing the temporal evolution of the trajectory. The corresponding recordings of the pump signal transmitted through the fibre (traces (f) to (h)) exhibit similar chaotic features though over a reduced spectral bandwidth. The phase portrait essentially mirrors that of the SBS due to the parametric nature of the nonlinear interaction providing further evidence of the deterministic rather than stochastic nature of this process, also corroborated by the continuous form of the trajectories spiralling and folding motion within the digitization resolution.

More generally the spectral broadening was found to depend upon the input pump power, varying from ~10 MHz close to threshold to ~20 MHz at twice the input power. The general features of traces (c)-(d) are typical of those obtained for a wide range of pump powers and were found to be highly reproducible. Interestingly no evidence of precursor routes to chaos were found.

In further investigations of these phenomena we also considered [9] the influence of external feedback on the dynamics of the SBS process. Feedback was provided by removal of the index-matching fluid from the fibre ends to give a nominal reflectivity of ~4% from each of the end faces.

Here we highlight sample results from an extensive experimental investigation of the nonlinear dynamics of SBS. Sustained quasi- regular oscillations at twice the fibre transit time ($2T_r$), similar to that reported earlier by Bar-Joseph et al. [3], are found to exist for only limited operating conditions and at low pump power. More generally the emission is found to comprise more than one frequency component as in Fig. 4(a) (fibre length 25 m). The power spectra of this data shows two independent frequencies and their harmonics along with a weak broad-band background which increases with pump power from the SBS threshold, indicative of the emergence of weak chaotic structure underlying the dominant quasi-periodic features. The number of such coexisting components increases with pump power, resulting in more complex time structure and random modulation of the emission. For further increase in pump power, this emission ultimately degenerates into bursts of oscillations as shown in Fig. 4(b) (fibre length of 40 m), the duration and separation of which appears random.

Some sample phase-portraits for a 100 m long fibre are shown in Fig. 5. Limit-cycle behaviour [trace (a)], dominant at the higher launch powers (500 mW sec), ranged in period from 200-300 nsec (cavity round-trip time $t_c \sim 1$ μsec) with evidence of correlation with t_c for the 38 m fibre. The torroidal phase portrait [trace (b)] provides evidence of two-frequency oscillation while period-two and period-three bifurcations of the basic limit cycle [trace (a)] have also been observed. Traces (c) and (d) suggest

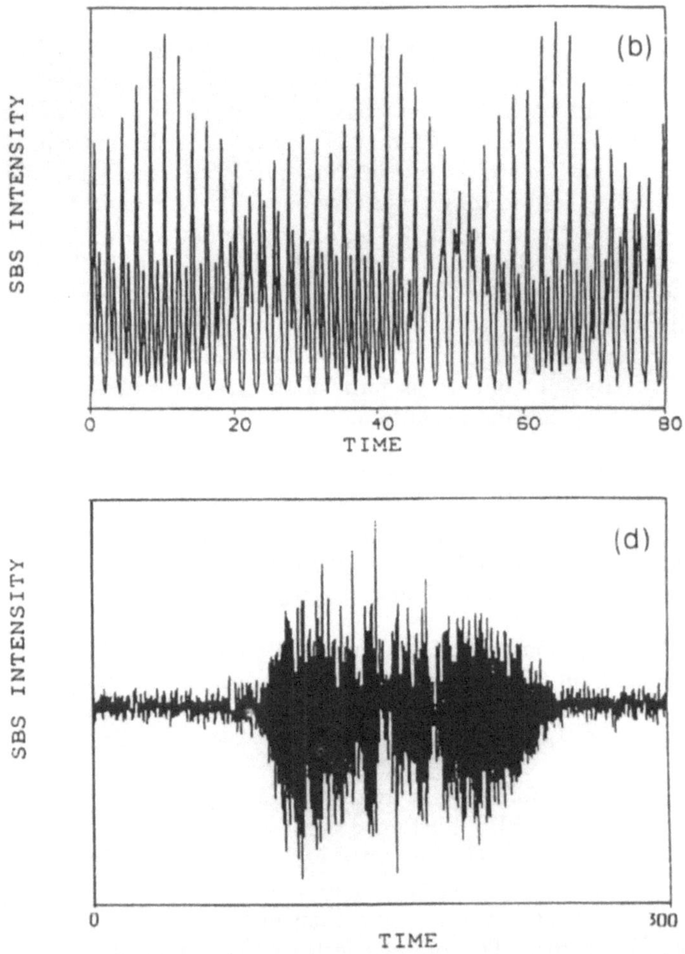

Figure 4. Two distinctive modes of oscillatory emissions of backward SBS obtained from experimental data for natural boundary conditions. (a) Quasi-periodic emissions for a fibre of 25 metres and pump power ~150 mW. (b) Bursting oscillations from a fibre of 40 metres and pump power ~200 mW. Here the time is normalized to the trip time of the medium T_r; typical duration of bursting is hundreds of T_r. Note data of trace (b) is taken at low resolution (320 ns digitization interval) resulting in loss of high frequency information.

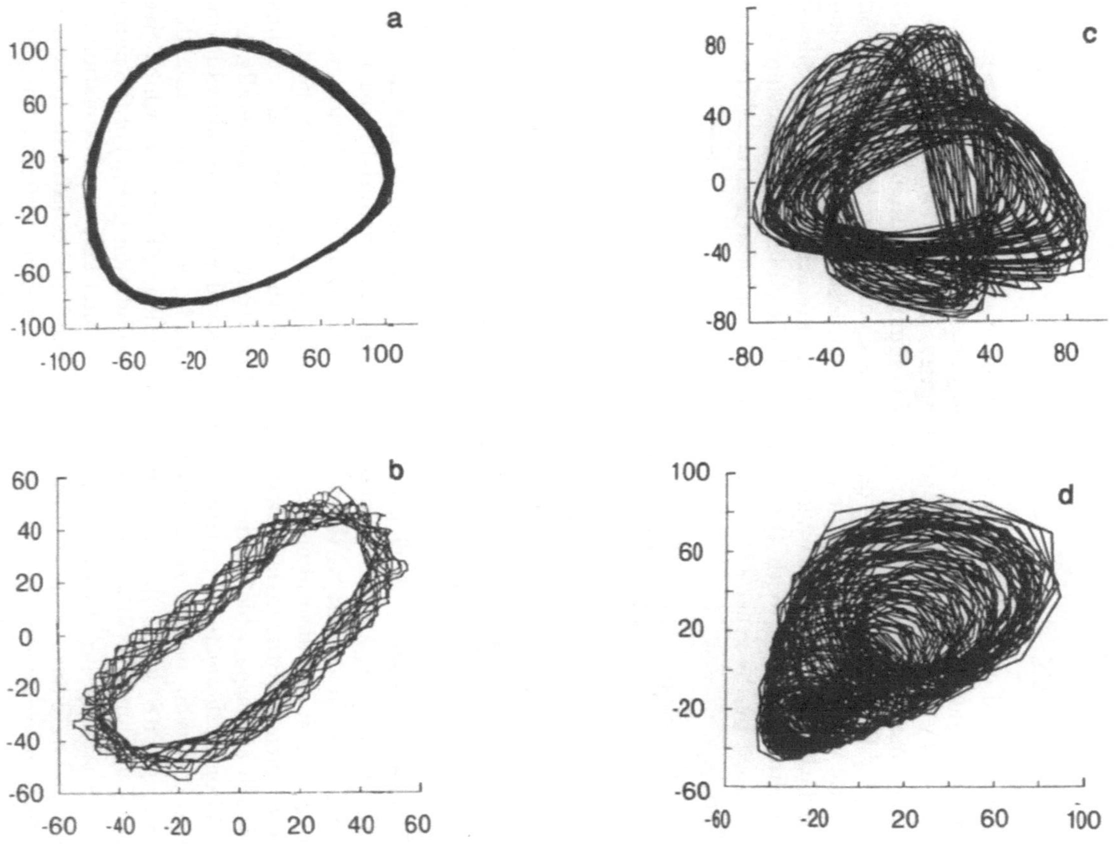

Figure 5. Some representative phase portraits of the transmitted pump signal [traces (a)-(c)] and SBS signal [trace (d)] from the fiber in the presence of optical feedback (a) limit cycle and (b) torroidal motion. Traces (c) and (d) are examples of chaotic attractors. 2000 data points of 16-K recordings are shown throughout. Fiber length is 100m.

a weakly chaotic motion indicative of the breakdown of a two torus which is further supported by the observed lifting of the broad background in their power spectra relative to (a) and (b). Significant further analysis of the experimental data for fibres with and without index matching will be required in order to resolve this question. This work is currently in progress. The attractor in trace (d) is similar to those obtained without optical feedback.

THEORETICAL ANALYSIS

Pure Stimulated Scattering

Our analysis is based on a full description of stimulated interaction scattering in which account is taken of the gain of the scattering processes as well as the nonlinear refraction induced by the pump and scattered signals. For clarity we restrict our attention to the common case of counter-propagating pump and stimulated scattered signals. Stimulated Brillouin scattering is described as a parametric coupling between light and acoustic waves [10]. The nonlinear interaction results from electrostriction which for a sufficiently high pump intensity or long interaction length stimulates growth of the Stokes signal. With the inclusion of the nonlinear dispersion this process is described by the general coupled first order nonlinear partial differential equations [9,11]

$$\frac{\partial E}{\partial t} + \frac{c\partial E}{n\partial z} - \frac{in_2\omega}{n}(|E|^2 + 2|E_s|^2)E + \frac{\alpha}{2}E = iV_1 E_s A$$

$$\frac{\partial E_s}{\partial t} - \frac{c}{n}\frac{\partial E_s}{\partial z} - \frac{in_2\omega}{n}(|E_s|^2 + 2|E|^2)E_s + \frac{\alpha}{2}E_s = iV_1 E A^*$$

$$\frac{\partial A}{\partial t} + c_s\frac{\partial A}{\partial z} + \alpha_A A = iV_3 E E_s^* \qquad (1)$$

where E, E_s and A are slowly varying amplitudes of pump, Stokes and material fields respectively. c and c_s are the velocities of the light and material respectively, n the linear refractive index, and n_2 the nonlinear refractive index coefficient which is related to $\chi^{(3)}$. α is the absorption coefficient of the pump and Stokes field, α_A the damping coefficient of the material wave and V_1, V_3 the opto-material coupling coefficients which are responsible for the stimulated scattering gain. Generally, adiabatic elimination of the material equation may be invoked when the bandwidth of the stimulated scattering process is sufficiently large as in, for example, stimulated Raman scattering (SRS) [(10,11)]. The nonlinear refraction or dispersion which give rise to self phase modulation (SPM) and cross phase modulation (XPM) of the pump and Stokes fields are described by the third terms in the first and second equations of Eqn. (1) and lead to changes in phase of both pump and Stokes fields in space and time. Here the response time of n_2, which may involve contribution from a number of physical mechanisms, e.g. electronic, electrostrictive, electrocaloric etc., is considered fast such that it adiabatically follows the electric field amplitude. These dispersive terms are generally omitted in conventional treatments since they have been considered to have no direct effect on the scattering processes [10]. Exceptions to this are in consideration of the influence of transverse effects such as self focussing on the gain characteristics of SRS [10,11] and the role of nonlinear dispersion on the evolution of ultrashort pulsed SRS [11]. Returning to eqn. (1), in the usual case the material attenuation coefficient α_A/c_s is much larger than the gain coefficient of the stimulated scattering. Consequently the material wave can be considered as highly damped and the term $\frac{\partial A}{\partial Z}$ generally omitted [10]. Furthermore for theorectical and numerical analysis we normalize the above variables as follows:

$$\varepsilon = E/E_o,\ \varepsilon_s = E_s/E_o,\ B = -\frac{iA}{E_o^2}\frac{\alpha_A}{V_3}$$

Eqn. (1) then becomes

$$\frac{\partial \varepsilon}{\partial \tau} + \frac{\partial \varepsilon}{\partial \xi} - iu(|\varepsilon|^2 + 2|\varepsilon_s|^2)\varepsilon + \frac{\beta}{2}\varepsilon = -g\varepsilon_s B$$

$$\frac{\partial \varepsilon_s}{\partial \tau} - \frac{\partial \varepsilon_s}{\partial \xi} - iu(|\varepsilon_s|^2 + 2|\varepsilon|^2)\varepsilon_s + \frac{\beta}{2}\varepsilon_s = g\varepsilon B^*$$

$$\frac{1}{\beta_A}\frac{dB}{d\tau} + B = \varepsilon\varepsilon_s^* \qquad (2)$$

where the parameters

$$\tau = \frac{ct}{nL},\ \xi = z/L,\ u = \frac{n_2\omega L}{c}|E_o|^2,\ \beta = \alpha\frac{nL}{c},\ \beta_A = \alpha_A\frac{nL}{c}\ \text{and}\ g = \frac{V_1V_3nL}{\alpha_A c}|E_o|^2,$$

where L is the interaction length, E_o the pump field amplitude at $\xi = 0$ (entrance end). Here g is the small signal amplitude gain of the stimulated scattering. We note that the nonlinear refractive effect does not change the threshold for stimulated scattering which remains [11] $G \equiv 2gl_c \sim 20$, where l_c is given as $l_c = (1 - e^{-\alpha L})/\alpha$ is the effective length.

First we demonstrate the essential role of nonlinear refraction in promoting dynamic behaviour in these processes. To do so we have numerically integrated the coupled nonlinear eqn. (2) in both space and time by using the method of characteristics. Figure 6 shows two sets of time-dependent solutions of the Stokes intensity at $\xi = 0$ without and with the nonlinear refractive coefficient u while other parameters remain unchanged. When u is zero or small the Stokes wave is stable and constant, approached through a transient relaxation oscillation with period $2T_r$ (trace(a)) where T_r is the single transit time through the medium. This behaviour is in accord with the early work of Johnson and Marburger [12]. On increasing u new features appear. While the amplitude of oscillation with period $2T_r$ decays, a new oscillation develops with period T_r (trace (b)) to form a stable sustained oscillation. Such behaviour is also mainfested in the material along the direction of propagation resulting in spatial dynamical features. In contrast we find no conditions of operation for which the truncated description of stimulated scattering, in which dispersive effects are neglected, gives rise to sustained dynamical behaviour. Consequently while such conventional treatments may adequately describe the general steady state features they do not account for the dynamics of these processes. It is the interplay between the nonlinear gain and nonlinear refraction as described by eqn. (2) which give rise to the spatio-temporal phase changes of the pump and Stokes signals and provide the basis for the dynamical instabilities.

The dependence of dynamical behaviour of stimulated scattering on the nonlinear refractive contribution is further detailed in Fig. 7, which shows a quasi-periodic

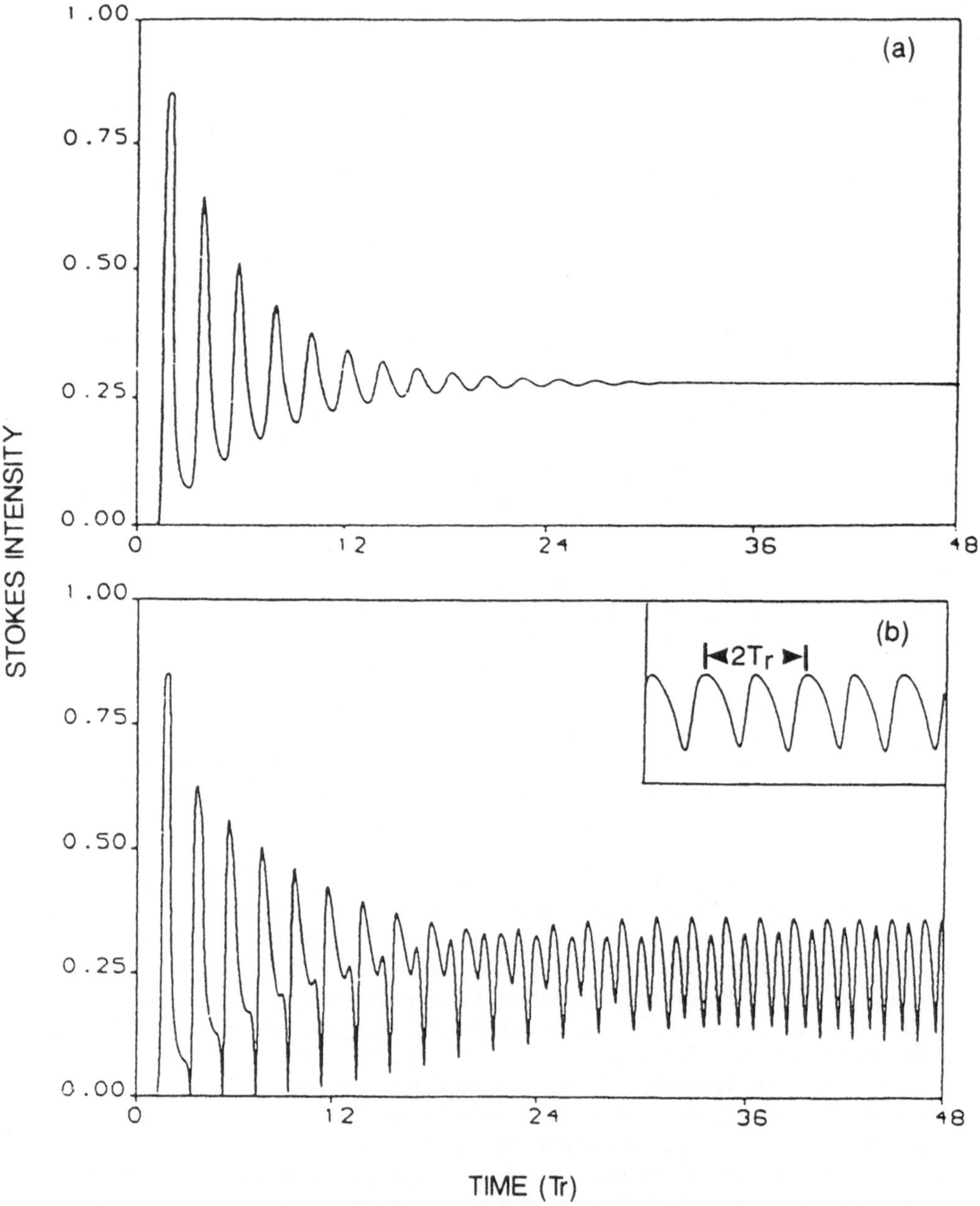

Figure 6. Time series of the SBS signal intensity demonstrating the essential role of nonlinear refraction in promoting dynamical behaviour, (a) $u = 0.0$, relaxation oscillations, (b) $u = 6.4$, sustained oscillations. The other parameters are fixed at $\beta = 0.175$, $\beta_A = 250$, $g = 22.5$.

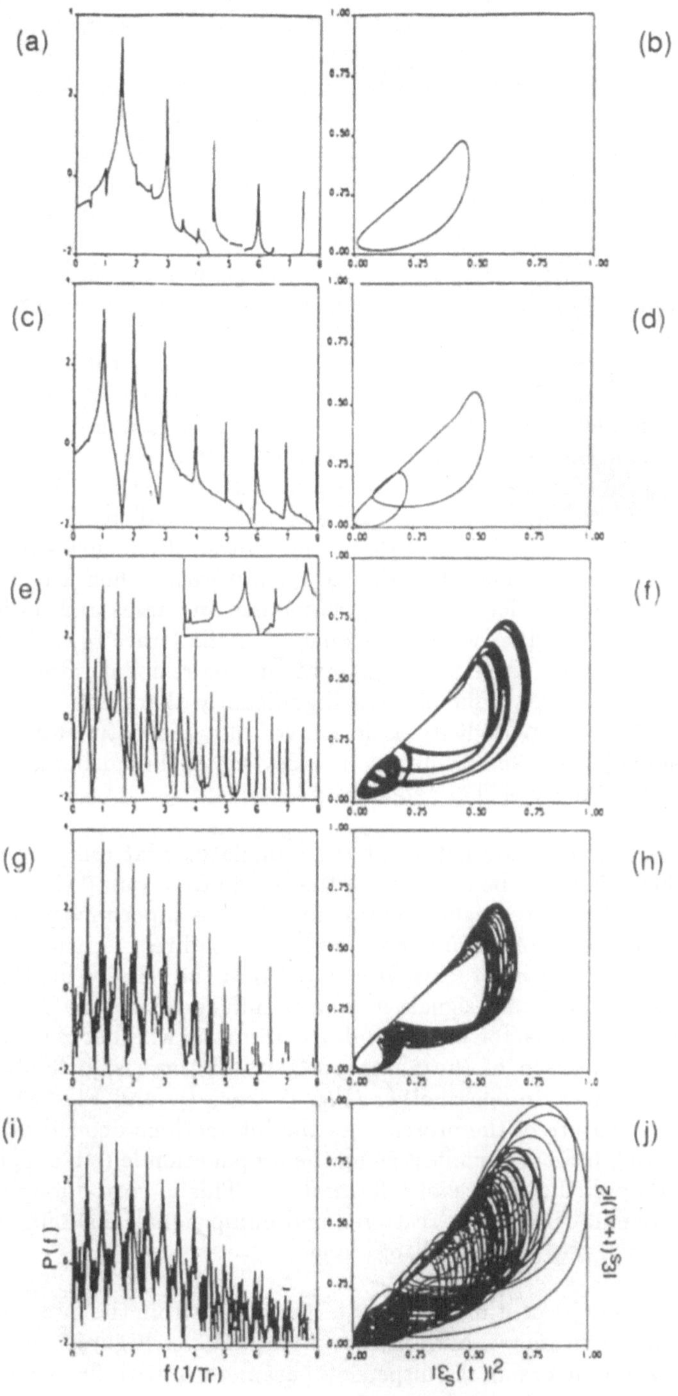

Figure 7. A quasi-periodic route to chaos for Stokes emission on increasing the nonlinear refraction coefficient u shows in both power spectra (left column) and reconstructed phase portraits (right column). u = 10.5, 14.3, 15.0, 16.5 and 18.0 from top to bottom, respectively. The other parameters are same as those in Fig. 1.

sequence to chaos on increasing u. All other parameters are as in Fig. 6. The power spectra and phase portraits shown comprise 40,000 data points and are small time window samples ($\sim 100T_r$) after the transient time, of a total computer simulation over $\sim 1000T_r$. The observed sequence of instabilities on increasing u from 6.4 (as in Fig. 6 trace (b)), is as follows. A bifurcation leads to a new periodic oscillation with frequency $3/2(1/T_r)$ along with its harmonics, as shown in the power spectra (Fig. 7 trace (a)). The corresponding phase portrait shows a single closed loop limit cycle (trace (b)). On increasing u the oscillatory behaviour bifurcates to two rational frequencies of comparable strength at $f = 1, 2(1/T_r)$ along with their harmonics and combinations (Fig. 7 trace (c)). The corresponding phase portrait (trace (d)) exhibits a double loop limit cycle. More generally further new frequencies can appear on increasing u with values $(\frac{n}{2} + 1)1/T_r$ $(n = 0, 1...)$ as well as their harmonics and sub- harmonics resulting in increasingly complex but still limit cycle behaviour. However such limit cycle behaviour breaks when a new low frequency component incommensurate with those already present emerges as u approaches 15.0. Since the component is initially quite weak the combination frequencies of this with the other components is not clearly visible in the power spectra (trace (e)). Its influence is however clearly evident in the emergence of quasiperiodicity in the phase portrait (trace (f)). Due to the presence of the commensurate frequencies the phase portrait is significantly more complex than that of a two torus. When u is increased further the strength of the new low frequency component increases and its frequency shifts to a higher value. Whether we observe quasi periodic (trace (g) and (h)) or complex limit cycles behaviour depends on the irrational or rational ratio of this frequency with the other frequency components. Significantly there appears to be a gradual break up of the quasi-periodicity as it evolves into a chaotic state on increasing u (traces (i) and (j)), the spectra lifting and developing broad band structure though retaining some memory of sharp spectral peaks.

Through the parametric nature of the stimulated scattering interaction similar dynamical behaviour can be expected in the transmitted pump signal. This has been confirmed, the transmitted signal showing instabilities superimposed on a steady state background, the depth of modulation providing an indirect measure of the conversion efficiency of the scattering process. It is noted that the relative depths of modulation of the scattered and pump signals is mainly influenced by the dispersion terms in eqn. (2) through the factor of two and according to whether ε is larger or smaller than ε_s. A comparison of the phase portraits of these signals show them to be inverted with respect to one another (Fig. 7 trace (j) and Fig. 8) consistent with the parametric nature of the process. As the interactions depend on both time and space, the instabilities are manifest along the propagation length as spatial, temporal, periodic, quasiperiodic and chaotic formations. This is corroborated by the lack of correlation we find between the scattered and pump signals emerging from either end of the interaction region in the chaotic region.

Since the dynamics of stimulated scattering results from the interplay between gain and dispersion there are two critical control parameters. Figure 7 showed the dynamical behaviour on increasing the dispersion parameter u with fixed gain which may be conveniently interpreted as variation of n_2 with the pump intensity held constant. On the other hand, it is of more practical value to consider the dependence of dynamical behaviour on pump intensity as the main control parameter. We find only in the limit of small n_2 that the solutions are stable for all pumping conditions. The dynamics occurring at higher n_2 are initially dominated by limit cycle behaviour which, for fixed n_2, becomes progressively more complicated with an increase in pumping field

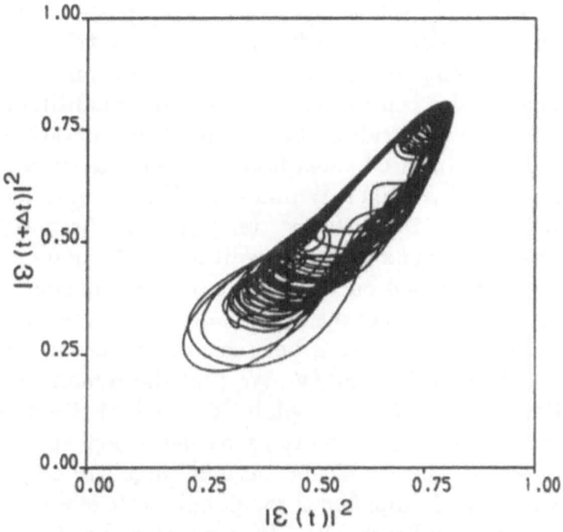

Figure 8. The phase portrait for transmitted pump signal with the same parameters as those in Fig. 2(j). The parametric form of the trajectories is evident from comparing this trace with Fig. 2(j).

strength. For larger n_2 these features are replaced by the emergence of complex quasi-periodicity and chaos in increasing the pump field. The threshold pump field strength for onset of dynamical instabilities (second threshold) is found to rapidly converge to the first threshold marking the onset of stimulated scattering with increase of n_2. In general we find that rich dynamical behaviour emerges when the magnitude of u, defining nonlinear refraction, is comparable to that of the gain g.

Comparing these results with our experimental observations of chaotic SBS Stokes emission (see Fig. 3) under single mode conditions, we see that qualitative agreement is quite good in confirming: a) the dominance of chaotic dynamics with modulation depths ~100% for all pump intensities including those close to the threshold for SBS, the parametric form of these instabilities manifesting themselves in the transmitted pump signal superimposed on a stable dc background, b) broad-band spectra of these signals with spectral fine structure at T_r^{-1} and relative spectral narrowing of the transmitted pump signal and c) phase portrait reconstruction showing motion on an outward spiralling and folding trajectory, the pump signal phase portrait being an inverted form of the SBS. While limit cycle and quasiperiodic precursors have so far not been experimentally observed, this may be attributable to unavoidable spontaneous noise in breaking the scenario. The inclusion of noise in our model description is currently under considerations. Towards quantifying the dynamics of SBS in a practical system, we consider, as examples, the generation of these processes under cw pump conditions in fused silica fibres at 514 nm. Pump power thresholds for SBS and SRS are typically ~300 mW and ~30 W for a fibre length of ~50 m and effective cross-section $A_{eff} = 10 \sim 20\mu m^2$, respectively. As noted above the detail structures of the dynamics depend on the magnitude of the nonlinear refractive index n_2. Reported values for n_2 based on the electronic contribution only vary over ~2-3.5x10^{-22}m^2/V while the inclusion of the other contributions to n_2 e.g. electrostriction etc. and their enhancement by beating between the pump and Stokes signals may increase this value up to ~6x10^{-22} m^2/V. We find the dynamics of SBS in the low limit of n_2 gives limit cycle behaviour while in the high limit of n_2 it comprises periodic, quasiperiodic and chaotic solutions as described above and is consistent with experimental observations. Details of these findings will be published elsewhere. For the case of SRS analysis if simplified through adiabatic elimination of the material equation due to the large gain bandwidth. We also note that the nonlinear refractive coefficient u is much higher than that for SBS owing to the requirement of higher pump power to promote SRS. Consequently interplay between the nonlinear refraction and gain is enhanced, providing favoured conditions for rich dynamical behaviour. Indeed we find chaotic dynamics to prevail almost from the onset of SRS with no evidence of precursor dynamical scenarios. An example is shown in Fig. 9 for a pump intensity imperceptibly close to the threshold (trace (a)) and for a slightly higher value (trace (b)).

Stimulated Scattering in Presence of External Feedback

As shown in our experimental data the influence of external feedback is to considerably simplify the dynamical behaviour of the SBS emission so providing an excellent basis for quantitative tests of our mathematical description. To simulate experimental conditions we consider Eqn. (1) accounting for weak feedback (~4% natural reflectivity of the fibre) namely $R_1R_2 << 1$, where R_1, R_2 are the reflectivities of intensity of the entrance and exit faces of the medium. Since the reflected field intensities from both ends are weak we neglect their interaction with the initial pump and Stokes fields already present in the interaction length in Eqn. (1). Furthermore the re-

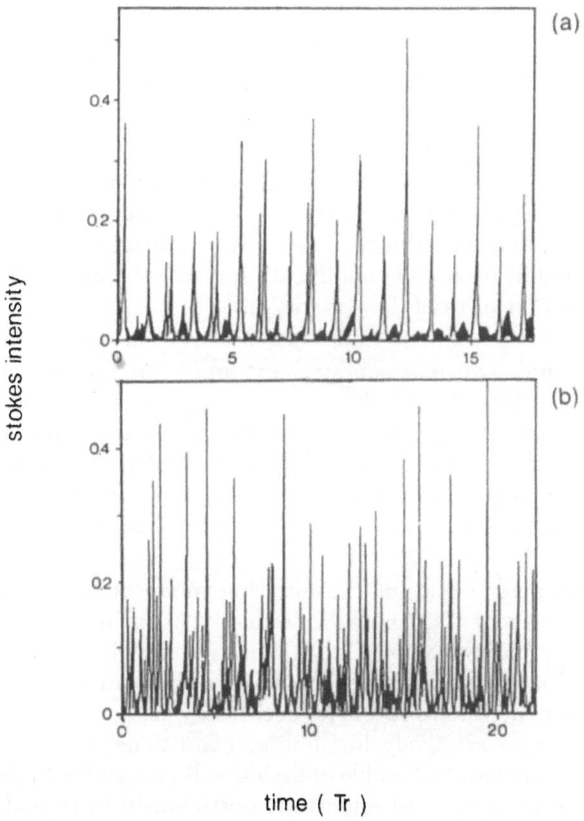

Figure 9. Sample time series of Stokes intensity on increasing the pumping parameter: (a) $g = 19.6$, $u = 168$, $\beta = 0.175$, (b) $g = 21$, $u = 180$, $\beta = 0.175$, here $\beta_A >> 1$ has been assumed. These parameters correspond to those for fused silica at $\lambda = 514$ nm with pump powers (a) ~35W, (b) ~38 W, and fibre length ~50 m.

flected pump, collinear with the incident pump, undergoes two reflections and is also neglected. For these conditions the boundary conditions become [9]

$$\varepsilon_{S,R}(t,0) = \sqrt{R_1}\varepsilon_s(t,0),$$

$$\varepsilon_S(t,L) = \sqrt{R_2}\varepsilon_{S,R}(t,L), \tag{3}$$

where $\varepsilon_{S,R}$ is the forward SBS signal induced by the cavity effect, the evolution of which is governed by the equation

$$\frac{\partial \varepsilon_{S,R}}{\partial \tau} + \frac{\partial \varepsilon_{S,R}}{\partial \xi} = -\frac{1}{2}\beta\varepsilon_{S,R} + iu(2|\varepsilon|^2 + |\varepsilon_S|^2)\varepsilon_{S,R} \tag{4}$$

In contrast with cavityless operation the role of the cavity is to provide additional feedback of the SBS signal to that indigenous to the fundamental SBS interaction. It is the relative phase and amplitudes of these contributions which we conclude leads to the observed modification and simplification of the dynamics of the SBS emission when compared to that of cavityless operation.

In Fig. 10 we show two representative examples for direct comparison with the experimental data of Fig. 4. Results are in good agreement with our experimental findings both regarding their temporal features and the pump scaling required to promote the change in dynamics from sustained to bursting modes of operation. Our theoretical results further confirm the eventual suppression of bursting features at the highest pump powers and the emergence of complex sustained oscillations.

We further find a tendency for the bursting mode of operation to predominate for larger fibre length and also for higher reflectivity on increasing the pump power, the duration of the individual profiles of bursting oscillation increases while their separation increases. In general we find the presence of external cavity feedback suppresses chaotic dynamics in favour of oscillatory behaviour. While these findings have been restricted to relatively low finesse conditions we have yet to determine whether regions of conventional stable emission will prevail for higher values of cavity finesse. These considerations are important particularly in regard to the utilization of stimulated scattering as lasing systems.

CONCLUSION

In conclusion, we have shown through our experimental and theoretical findings that nonlinear dynamics is generic to stimulated scattering phenomena. It is exhibited in both the Stokes and transmitted pump signals with period, quasiperiodic and chaotic behaviour over a broad range of experimentally controlled parameters, including those close to the threshold for stimulated scattering. The inclusion of external optical feedback modifies the form of the chaotic dynamics and results in a rich variety of classified precursor dynamical features. The interplay between the nonlinear refraction and the gain of stimulated scattering process is shown to be responsible for the dynamical characteristics. We have demonstrated the unique merit of a single-mode optical fibre in establishing these features under cw conditions of operation.

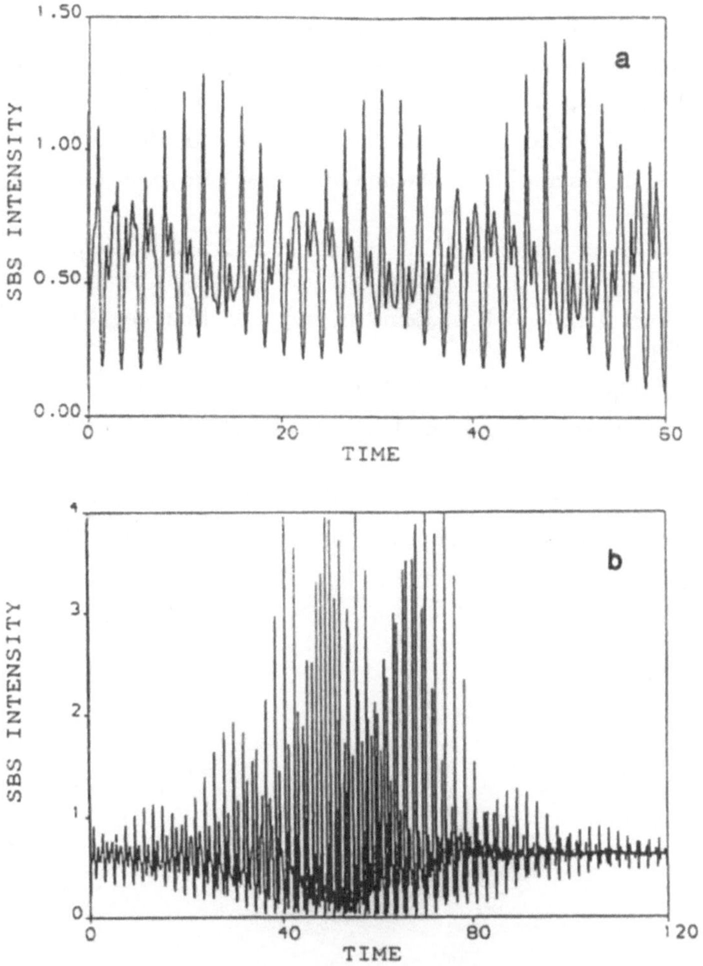

Figure 10. Two distinctive modes of oscillatory emissions of SBS with cavity obtained from theoretical solutions. (a) Quasi-periodic emissions for a fibre of 25 metres and pump power ~150 mW. (b) Bursting oscillations from a fibre of 40 metres and pump power ~200 mW. Here the time is normalized to the trip time of the medium T_r.

ACKNOWLEDGEMENTS

We acknowledge the Science and Engineering Research Council for support of the work.

REFERENCES

1. K. Baumgartel, U. Motschmann and K. Sauer, Opt. Commun. 51, 53 (1984).
2. C. J. Randall and J. R. Albritton, Phys. Rev. Lett. 52, 1887 (1984).
3. I. Bar-Joseph, A. A. Friesem, E. Lichtman and R. G. Waarts, J. Opt. Soc. Am. B, 2, 1606 (1985); Y. Silberberg and Bar-Joseph, J. Opt. Soc. Am. B. 1 (1984).
4. C. Montes and J. Coste, Laser and Particle Beams, 5, 405 (1987).
5. P. Narum, A. L. Gaeta, M. D. Skeldon and R. W. Boyd, J. Opt. Soc. Am. B., 5, 623 (1988).
6. A. L. Gaeta, M. D. Skeldon, R. W. Boyd and P. Narum, J. Opt. Soc. Am. B., 6, 1709 (1989).
7. R. G. Harrison, J. S. Uppal, A. Johnston and J. V. Moloney, Phys. Rev. Lett. 65, 167 (1990).
8. D. Cotter, J. Opt. Commun. 4, 10 (1983).
9. A. Johnstone, Weiping Lu, J. S. Uppal and R. G. Harrison, Opt. Commun. 81, 222 (1991).
10. R. Shen, The Principles of Nonlinear Optics (Wiley, New York 1984).
11. G. P. Agarwal, Nonlinear fibre optics (published by Academic Press Inc.) (1989).
12. R. V. Johnson and J. H. Marburger, Phys. Rev. A4, 1175 (1971).

POPULATION FLUCTUATIONS IN TWO-LEVEL ATOMS INDUCED BY LASER NOISE

Gautam Vemuri, M.H. Anderson, R.D. Jones,* J. Cooper and S.J. Smith

Joint Institute for Laboratory Astrophysics
University of Colorado
Boulder, CO 80309-0440

ABSTRACT

This paper describes the effects of laser noise on the fluorescence from an ensemble of two-level atoms. The first half of the paper reports theoretical and experimental results when a phase diffusing optical field interacts with two-level atoms. Measurements on the variance of the fluctuations in the fluorescence intensity and the power spectra of these intensity fluctuations are shown to be very sensitive to the pump field strength and the statistical parameters of the field. The second half of the paper presents theoretical and numerical results when the pump field is assumed to be a combination of a coherent and a chaotic field. Langevin equations describing the time dependent behavior of the dipole moment and population inversion are numerically integrated using Monte-Carlo methods. Results on fluctuations in fluorescence intensity, power spectra of intensity fluctuations and the spectra of scattered light are presented.

INTRODUCTION

There has been a great deal of interest in the past few years in the study of the interaction of light with two-level atoms.[1] In particular, the problem of the light having statistical fluctuations in amplitude, in phase or in both has received considerable attention from both theorists and experimentalists.[2] The intrinsic quantum fluctuations in the system of two-level atoms can be probed by studies of the fluorescence produced by the system.[1] The characteristics of the fluorescence are sensitive to the nature of the field exciting the atoms. As a matter of fact, the influence of pump statistics on the fluorescence spectra and other nonlinear optical processes has been studied at length. The effect of these pump fields on atomic populations is fairly easy to study and understand and hence most previous theoretical works have concentrated on calculation of the mean values of the atomic populations interacting

* Present address: N.I.S.T. Boulder, CO 80309

Recent Developments in Quantum Optics, Edited
by R. Inguva, Plenum Press, New York, 1993

with resonant or near resonant laser light. It is, however, expected that fluctuations in the exciting field will cause fluctuations in the fluorescence signal which can often impose limits on experimental resolution. It is thus important to understand how the fluorescence signal depends on the pump field statistics.

Several stochastic models of the pump field have been put forth to elucidate the fluctuations in the laser field. The three primary models are the Phase Diffusion Model (PDM), the Chaotic Field Model (CFM) and the real Gaussian Field Model (GFM). In the PDM, the amplitude of the pump field is assumed to be a time dependent stochastic quantity. Most lasers operating in a single axial mode, high above threshold can be described by the PDM. In the CFM, the amplitude of the pump field is assumed to be a two-dimensional, complex Gaussian process. In contrast, the GFM assumes that the real part of the electric field is a stochastic quantity. In addition to the models mentioned here,there also exist jump models where the amplitude, phase or frequency is taken to be a discontinuous Markov process.

Recently, Zoller and co-workers[3] have done extensive studies on laser noise induced population fluctuations in two-level atoms, using the PDM and the PJM. They demonstrated that the behavior of the variance of the population fluctuations provides a better signature of the statistics of the pump field than the mean signal itself. They showed that while both models predict the same behavior for the fluorescence intensity as a function of detuning the laser from the atomic transition frequency (Δ), the variance of the fluctuations in the intensity as a function of time showed dramatic differences. These authors hence concluded that the nature of the variance of the fluorescence intensity fluctuations could be used to predict the statistical nature of the pump field.

The original work of Zoller and co-workers[3] assumed that the atoms were stationary and that the spatial intensity profile of the laser was uniform across the interaction region. Later, this work was extended to account for residual Doppler broadening caused by non-stationary atoms and for the non-uniform intensity profile of the laser.[4] These modifications were necessary to achieve quantitative agreement with the experiments described below.

One of the important problems in high resolution spectroscopy is achieving better resolution by suppressing various line broadening mechanisms. Recently, Agarwal and Singh[5] reported a new mechanism that leads to the suppression of line broadening effects. They showed that a strong competition between the effects of coherent and incoherent pumps can lead to the narrowing of spectral profiles, which can have important consequences in spectroscopic studies where there may be overlapping lines. They demonstrated that line narrowing is possible if the strength of the coherent pump is such that many Rabi oscillations are possible within the coherence time of the incoherent pump. This is the regime where the usual optical Bloch equations cannot be used; more general equations are required, along with specific models for the pump field. Agarwal and Singh used a two-state frequency model for the pump field which made it possible to obtain analytical expressions for the spectra of scattered light.

Motivated by the works of Zoller, et al. on "noise spectroscopy"[3] and Agarwal's work on line narrowing,[5] we recently introduced a new model based on Monte-Carlo techniques, which makes it possible to study a two-level atom interacting simultaneously with a coherent and a chaotic field.[6] This model is of relevance to multimode lasers where chaotic fields occur almost universally. The model we have developed is used to study fluctuations in the fluorescence from an ensemble of two-level atoms and calculate the spectra of scattered light. This model allows one to consider a finite correlation time for the stochastic part of the field. In this paper, we present results for two distinct cases; one where the stochastic fluctuations of the laser field are on a very fast time scale compared to the radiative decay rate of the two-level atoms (γ) and one where the time scale is comparable to γ. The latter corresponds to an Ornstein-Uhlenbeck or colored noise process. We characterize the stochastic component in terms of a Gaussian random process. Our results are found to be sensitive to the ratio of the Rabi frequency to the bandwidth of the chaotic field.

PHASE DIFFUSING FIELDS

In this section we describe our work on population fluctuations in two-level atoms pumped by phase diffusing optical fields. To produce the phase diffusing laser field we use the method described by Elliott and Smith[7] in which a Gaussian white noise source is shaped with an RC filter network to produce noise with a known power spectrum. The field thus produced has a frequency correlation function of the form

$$< \omega(t)\omega(t+\tau) >= b\beta \, exp(-\beta|\tau|)$$

where $1/\beta$ is the time scale of fluctuations and b is related to the laser full width at half maximum (FWHM). A schematic of our experiment is shown in Fig. 1. The two level system was prepared by optical pumping in a beam of atomic sodium (see below). The sodium $3S_{1/2} \rightarrow 3P_{3/2}$transition has a natural linewidth of 10 MHz and is convenient for our noise modulation system which can produce laser linewidths int he range of 1 to 20 MHz. This linewidth also insures that the residual Doppler broadening of $\approx$7.5 MHz did not dominate over the natural linewidth.

To provide a two state system and control the laser scan, a portion of the laser was split off prior to noise modulation and modulated with an electro-optic modulator (EOM) to produce variable frequency modulated (FM) sidebands. This beam was circularly polarized and directed into the atomic beam ahead of the interaction region and the upper sideband was tuned into resonance with the atoms, thereby optically pumping the atoms into the $F = 2$, $m_f = 2$ hyperfine state. In addition, the FM sidebands were frequency modulated (dithered) at a much lower frequency of 1 kHz. The fluorescence from the pumping region was detected and analyzed at the dither frequency with a lock-in amplifier. The lock-in amplifier generated an error signal which, when input to the scan control of the ring dye laser, regulated the laser carrier frequency, ω_L, such that the upper sideband of the optical pumping beam remained locked on resonance. The acousto-optic modulator (AOM) in the noise modulation system produces a net positive 400 MHz offset of the input laser carrier frequency, so, when the upper FM sideband of the optical pumping beam was tuned to $\omega_L + 400$ MHz, the noise modulated laser beam was at zero detuning (small deviations from zero detuning exist due to skewness of the two beams in the interaction region). Scanning of the noise modulated laser was achieved by scanning the frequency of the FM signal to the EOM(Fig. 1b). A heterodyne signal, obtained from portions of the noise modulated beam and optical pumping beam, provides a measurement of the detuning of the noise modulated laser. The noise modulated laser was also circularly polarized to couple the prepared ground state to only the $f = 3$, $m_f = 3$ hyperfine level of the $3P_{3/2}$ upper state, thus providing a two level system.

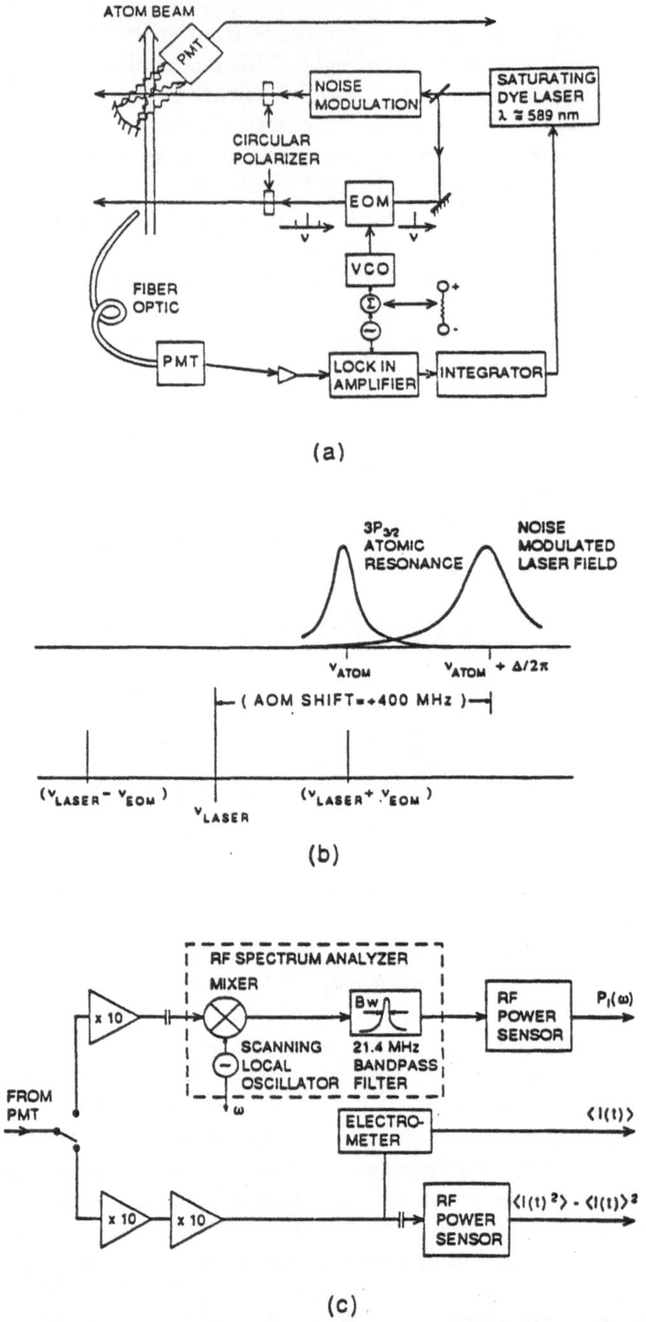

Figure 1. Schematic diagram of the experiment. (a) Details of the apparatus. The upper sideband from the optical pumping beam is locked on resonance by feeding a signal derived from the pumping fluorescence back to the dye laser. The dye laser frequency at $\nu_L = \nu_{atom} - \nu_{EOM}$ is offset by an acousto-optic modulator in the noise modulation system. The noise modulated laser frequency at $\nu_{noise} = \nu_{atom} - \nu_{EOM} + 400$ MHz is scanned by scanning ν_{EOM}. (b) Illustration of the frequencies of the optical pumping and the noise modulated lasers. The detuning of the noise modulated laser is $\beta/2\pi = 400$ MHz $-\nu_{EOM}$. (c) The detection electronics.

We conclude this section with a discussion of the methods by which the variance and spectrum of the intensity fluctuations are obtained from the photomultiplier tube (PMT) signal (Fig. 1c). To measure the variance of the fluorescence intensity the signal from the PMT was amplified by a factor of 100 with two 50-ohm input broadband amplifiers. As the noise modulated laser was scanned across the transition the dc component of the signal was determined with an electrometer and the RF power was measured with a power sensor consisting of a capacitively blocked 50-ohm resistor thermally coupled to a monolithic silicon thermocouple. The power sensor measures the true mean of the square of the ac component. The RF power and the dc voltage are proportional to the variance $(\Delta I)^2$ and the time average $< I(t) >$ of the intensity, respectively. These quantities were then logged by a computer as functions of the laser detuning. The ac power in the signal is dissipated across 50 ohms in the power sensor and is converted to an RMS voltage. The RMS voltage, proportional to ΔI, is thus put on the same scale as the dc voltage level which is proportional to $< I(t) >$.

Spectra of the intensity fluctuations were taken with the laser at a fixed detuning and one broadband amplifier removed. A commercial spectrum analyzer was used as a tunable bandpass filter with the bandwidth typically set at 100 kHz and the center frequency tuned from 0-100 MHz. In the spectrum analyzer, the signal is mixed with the output of a tunable local oscillator and the resulting signal is passed through a 21.4 MHz bandpass filter. The power in the 21.4 MHz signal, proportional to the spectrum of the intensity fluctuations $P_I(\omega)$, is measured with the square-law power sensor. The statistics of the intensity fluctuations are not known and could conceivably be a function of the laser detuning and intensity if the laser is saturating, so it was necessary to measure the true power in the 21.4 MHz signal. Our spectrum analyzer is equipped with an envelope detector that measures the average of the absolute amplitude, not the squared amplitude, which can result in systematic errors.

Variance of the Intensity Fluctuations

In Fig. 2 we present data collected with $\beta/2\pi$ = 100 MHz and values for the spectral density, b, which give laser linewidths of ≈ 4, 10, and 15 Mhz, and spatially averaged Rabi frequencies of $\approx$2, 10, and 25 MHz. Measurements of ΔI and $< I(t) >$ are shown in the lower double peaked curve and upper curve, respectively. The theoretical calculations are shown as circles or squares plotted over the data.

A number of parameters are necessary for the calculations shown in Fig. 2. A value of β is known from the shaping filters in the noise generation system. The value of b is obtained from fits to heterodyne measurements of the laser spectrum. The value of residual Doppler broadening is obtained from low intensity scans with the noise off and ranged from 7.5 to 8.0 MHz (giving 14 MHz total linewidth for sodium).[4]

As can be seen in Fig. 2 the agreement between theory and experiment for the low power scans is quite good with discrepancies on the order of only a few percent. As the power level is increased the agreement is worse (discrepancies of up to 15%), most likely due to a lack of knowledge of the precise intensity distribution across the interaction region. A number of trends are evident in Fig. 2. Within a column with uniform Rabi frequency Ω, the extent of the central dip in ΔI is reduced as the bandwidth of the laser spectrum is increased (FWHM=2b). The two peaks of ΔI are no longer resolved in the low power scans when the FWHM of the laser becomes comparable to the natural linewidth, γ. For a given value of b, the extent of the dip increases and the overall magnitude of the fluctuations is reduced as the power level of the laser is increased. For a detector of infinite bandwidth, Ref. 3 predicts a

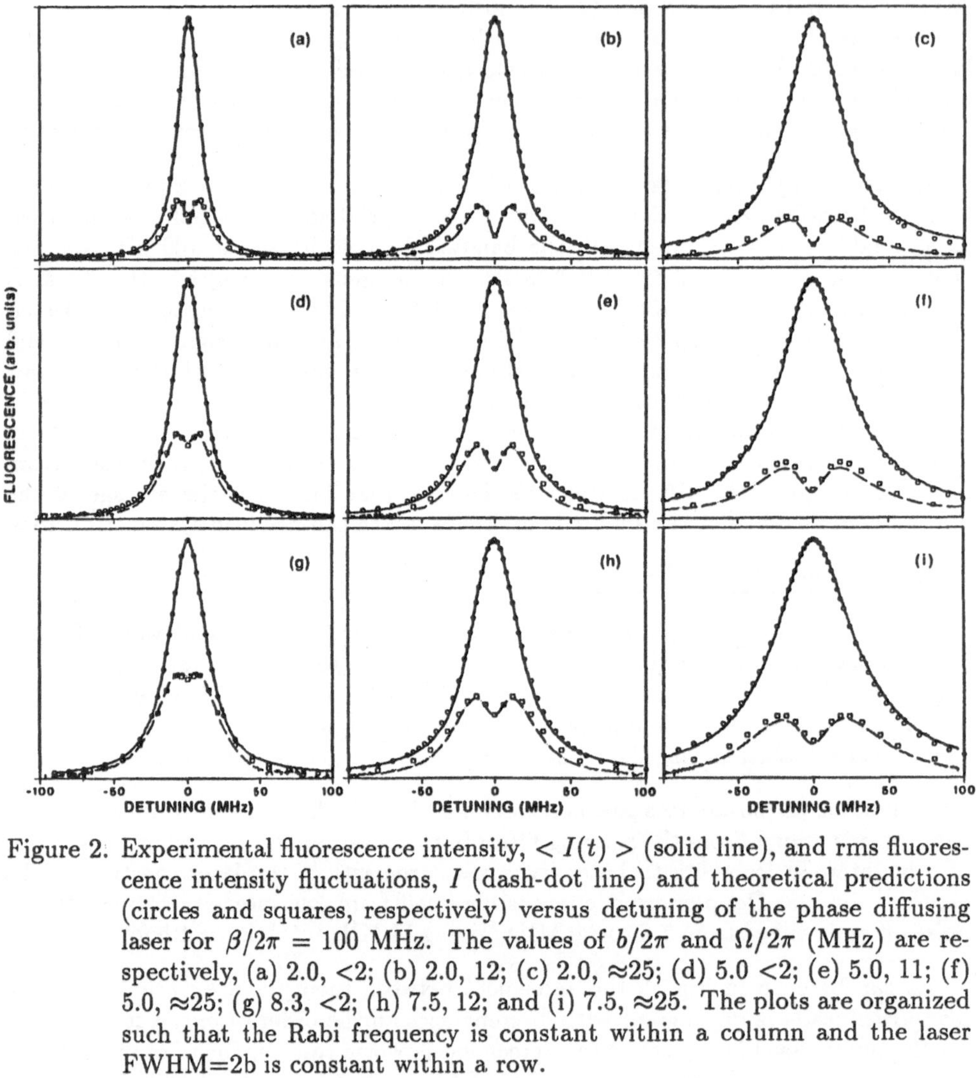

Figure 2. Experimental fluorescence intensity, $< I(t) >$ (solid line), and rms fluorescence intensity fluctuations, I (dash-dot line) and theoretical predictions (circles and squares, respectively) versus detuning of the phase diffusing laser for $\beta/2\pi = 100$ MHz. The values of $b/2\pi$ and $\Omega/2\pi$ (MHz) are respectively, (a) 2.0, <2; (b) 2.0, 12; (c) 2.0, ≈25; (d) 5.0 <2; (e) 5.0, 11; (f) 5.0, ≈25; (g) 8.3, <2; (h) 7.5, 12; and (i) 7.5, ≈25. The plots are organized such that the Rabi frequency is constant within a column and the laser FWHM=2b is constant within a row.

much higher noise ration $\Delta I / < I >$ in the wings of the ΔI profile than is observed in Fig. 3. The bandwidth of the intensity fluctuations increases with the detuning due to contributions to $P_I(\omega)$ at the generalized Rabi frequency $\Omega' = (\Omega^2 + \Delta^2)^{\frac{1}{2}}$. Consequently, as the laser is detuned more of the power in the spectrum of the intensity fluctuations is shifted beyond the bandwidth of the detector, resulting in a reduction in the wings of the profile of ΔI. The effect of finite β is qualitatively similar to that of a finite detector response and becomes appreciable if Ω approaches the value of β. For $\beta/2\pi = 100$ MHz and these intensity levels, finite β effects are quite small except in the region of the far wings.

Spectra of the Intensity Fluctuations

Spectra of the intensity fluctuations for the parameters of Fig. 2(a) and (c) are shown in Fig. 3(a) and (b), respectively. The calculated spectra, with the same corrections as discussed for Fig. 2 are shown in Fig. 3(c) and (d). The data at low frequencies are obscured by local oscillator effects of the spectrum analyzer and are not shown. In Fig. 3(a), a discontinuity in the spectra at 6 MHz is due to imperfect matching of the individual shaping filters for the AOM and EOM which shape the spectrum of the frequency fluctuations of the input field. In addition, the spectra in Fig. 3(b) with $\Omega/2\pi = 25$ MHz contain discontinuities at 15, 25, and 35 MHz which are due to spurious resonances of the $LiTaO_3$ crystal of the EOM. These artifacts are not apparent in the variance of the fluorescence or in the heterodyne measurements of the laser spectrum. Clearly, the spectrum of the intensity fluctuations is very sensitive to details in the spectrum of the laser frequency fluctuations. Since the variance is proportional to the integral of the $P_I(\omega)$, these artifacts are effectively averaged out in the data plots of variance vs. detuning.

Despite the discrepancies mentioned above, the data in Fig. 3 are qualitatively in agreement with the calculations. In Fig. 3(a), for $\Omega/2\pi < 2$ MHz (below saturation) and $\Delta = 0$, it can be seen that the fluctuations roll off at frequencies of $\approx \gamma/2\pi$. The spectra become broader as the laser is detuned from resonance, as predicted in Ref. 3, due to the excitation of transients at frequencies given by the laser detuning. For laser detunings of $\Delta/2\pi > 0$ MHz and $\Omega/2\pi < 2$ MHz, Ref. 3 predicts a peak located at Δ that is resolved from the central peak, but the overall magnitude of the fluctuations is much too small for it to be observed in this experiment. When the laser power is increased, as in Fig. 3(b) where $\Omega/2\pi \approx 25$ MHz, the features of the spectra are broadened considerably, in this case due to excitation of transients which contribute to $P_I(\omega)$ at the generalized Rabi frequency Ω'. For the curve with $\Delta/2\pi = 30$ MHz, the peak at the generalized Rabi frequency is clearly evident. The arrows in Fig. 3(b) locate the value of Ω' and thus the positions at which the peaks are expected to occur. The peak at Ω' in the spectra of $\Delta/2\pi = 15$ MHz, is just beginning to become resolved.

Finite Correlation Time of the Phase Fluctuations

Spectra of the intensity fluctuations for $\Omega/2\pi \approx 25$ MHz, $\Delta/2\pi = 30$ MHz, $\beta/2\pi =$ 100, 30, and 10 MHz, and $b/2\pi = 2.0$, 2.1, and 2.5, respectively, are shown in Fig. 4. The values of b are chosen to maintain the measured laser FWHM between 4 and 5 MHz. The area under the data plots is proportional to the variance at $\Delta/2\pi = 30$ MHz. Figure 4(b) shows the calculated spectra for the same parameters, using the methods of Ref. 3 for finite β, but with no corrections for spatial intensity distribution or Doppler broadening. The overall scale of Fig. 4(b) is reduced relative to Fig. 4(a). Again, the effects of the spatial intensity distribution are clearly evident in the data.

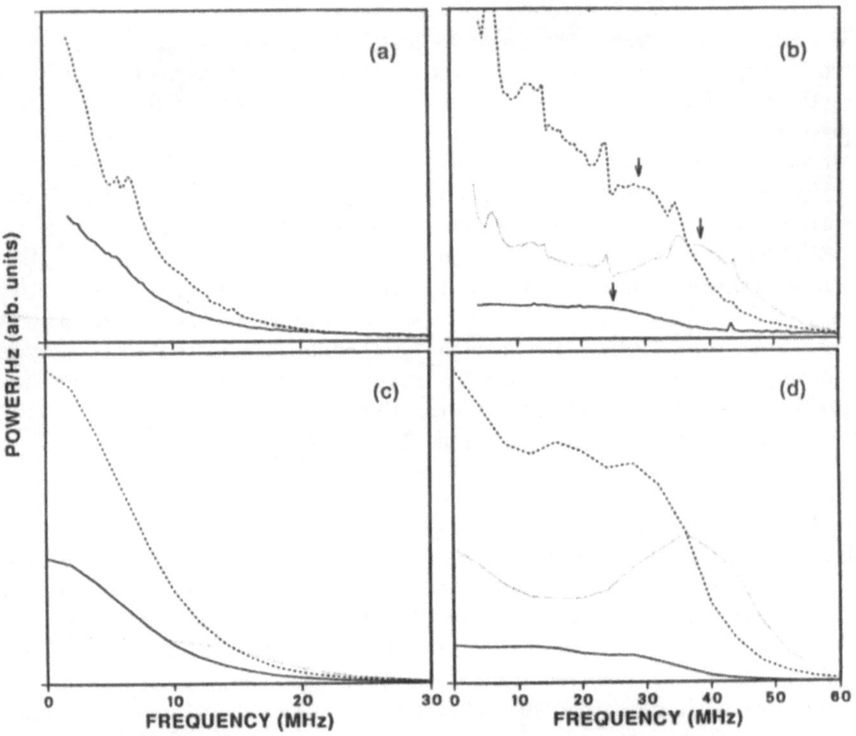

Figure 3. Spectra of the fluorescence intensity fluctuations, $P_I(\omega)$, with $\beta/2\pi = 100$ MHz obtained experimentally, (a) and (b); and theoretically, (c) and (d). In (a) and (c) $b/2\pi = 2.0$ MHz and $\Omega/2\pi < 2$ MHz at detunings of $\Delta/2\pi =$ 0 (solid line), 5 MHz (dashed line), and 15 MHz (dotted line). In (b) and (D) $b/2\pi = 2.0$ MHz and $\Omega/2\pi \approx 25$ MHz at detunings of $\Delta/2\pi = 0$ (solid line), 15 MHz (dashed line), and 30 MHz (dotted line). Note the change of scales.

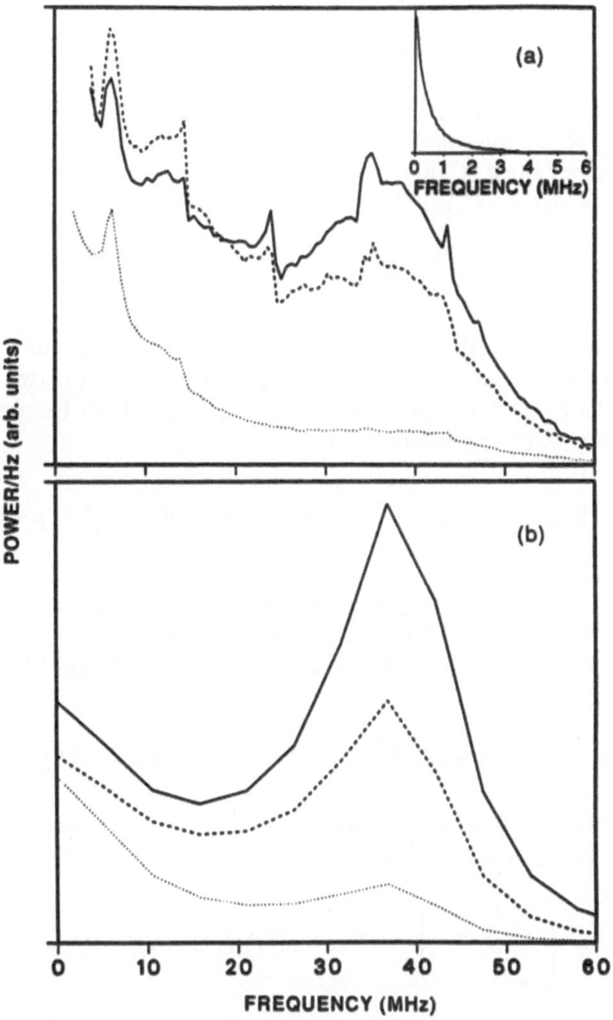

Figure 4. Spectra of fluorescence intensity fluctuations, $P_I(\omega)$, obtained experimentally (a) and theoretically (b) for $\Omega/2\pi \approx 25$ MHz and $\Delta/2\pi = 30$ MHz. The laser FWHM is constant at $\approx$4.5 MHz and the statistical parameters of the field are $\beta/2\pi = 100$ MHz and $b/2\pi = 2.0$ MHz (solid lines), $\beta/2\pi = 30$ MHz and $b/2\pi = 2.1$ MHz (dashed lines), $\beta/2\pi = 10$ MHz and $b/2\pi = 2.5$ MHz (dotted lines), and $\beta/2\pi = 1.0$ MHz and $b/2\pi = 4.5$ MHz (inset). The spectra are normalized to the same fluorescence intensity at line center.

The general effect is a reduction of the spectra at frequencies greater than $\beta/2\pi$. This continues as β is reduced below γ, and for $\beta/2\pi = 1$ MHz (shown in the inset with $b/2\pi = 4.5$ MHz) the spectrum is devoid of structure resulting from the dynamics of the atom. At $\beta/2\pi = 10$ MHz only a trace of the peak at Ω' is visible.

Profiles for ΔI, for the same parameters as in Fig. 4, are shown in Fig. 5. There is a general trend to reduce the magnitude of the fluctuations as β is decreased. This effect is more pronounced at high power, as in Fig. 4, than at low power. At high power the PDM ($\beta \to \infty$) predicts an increase in the bandwidth ($\approx \Omega'$) of the fluctuations as the laser power is increased. If β is comparable to Ω' then a reduction of the magnitude of the fluctuations results (as compared to the case of $\beta \to \infty$). As previously discussed, since Ω' increases with detuning, a finite β will also have more effect in the wings of the ΔI profile.

COHERENT AND CHAOTIC FIELDS

In this section we describe our work on the interaction of an ensemble of two-level atoms with coherent and chaotic fields. The dynamics of a two-level atom with resonance frequency ω_o, in the presence of a pump field (with frequency ω_c) is given by (in the rotating wave approximation)[6]

$$\partial\rho/\partial t = -i/\hbar\{[\vec{d}\cdot\hat{\epsilon}/\hbar(\epsilon_o+\epsilon_1(t))S^+ + H.C.],\rho\} - i\Delta[S^z,\rho] - \gamma(S^+S^-\rho - 2S^-\rho S^+ + \rho S^+S^-) \tag{1}$$

where S^+, S^- and S^z are the dipole moment and inversion operators, γ is the radiative decay rate, ϵ_o is the coherent part of the field, ϵ_1 is the chaotic component, and Δ is the detuning parameter $(\omega_o - \omega_c)$. The stochastic part of the field has the correlation functions

$$< \epsilon_1(t)\epsilon_1^*(t') >= \beta\Gamma/2 \; exp(-\Gamma|t-t'|); \quad < \epsilon_1(t)\epsilon(t') >= 0 \tag{2}$$

where $1/\Gamma$ is the time scale of fluctuations and $\beta\Gamma/2$ is the variance of ϵ_1. The optical Bloch equations can be easily written down as

$$\dot{\psi} = \begin{pmatrix} < \dot{S}^+ > \\ < \dot{S}^- > \\ < \dot{S}^z > \end{pmatrix} = C_o\psi - ix(t)C_+\psi - ix^*(t)C_-\psi + g \tag{3}$$

where

$$C_o = \begin{pmatrix} -\gamma^{+i\Delta} & 0 & -i\Omega \\ 0 & -\gamma - i\Delta & i\Omega \\ -i\Omega/2 & i\Omega/2 & -2\gamma \end{pmatrix}; C_+ = \begin{pmatrix} 0 & 0 & 0 \\ 0 & 0 & -2 \\ 1 & 0 & 0 \end{pmatrix}; C_- = \begin{pmatrix} 0 & 0 & 2 \\ 0 & 0 & 0 \\ 0 & -1 & 0 \end{pmatrix};$$

$$g = \begin{pmatrix} 0 \\ 0 \\ -\gamma \end{pmatrix}; \quad \Omega = 2\vec{d}\cdot\hat{\epsilon}/\hbar; \quad x(t) = (\vec{d}\cdot\hat{\epsilon}/\hbar)\epsilon_1(t)$$

The fluctuating fluorescence intensity is given by

$$I = \sum(\frac{1}{2} + \psi_3) \tag{4}$$

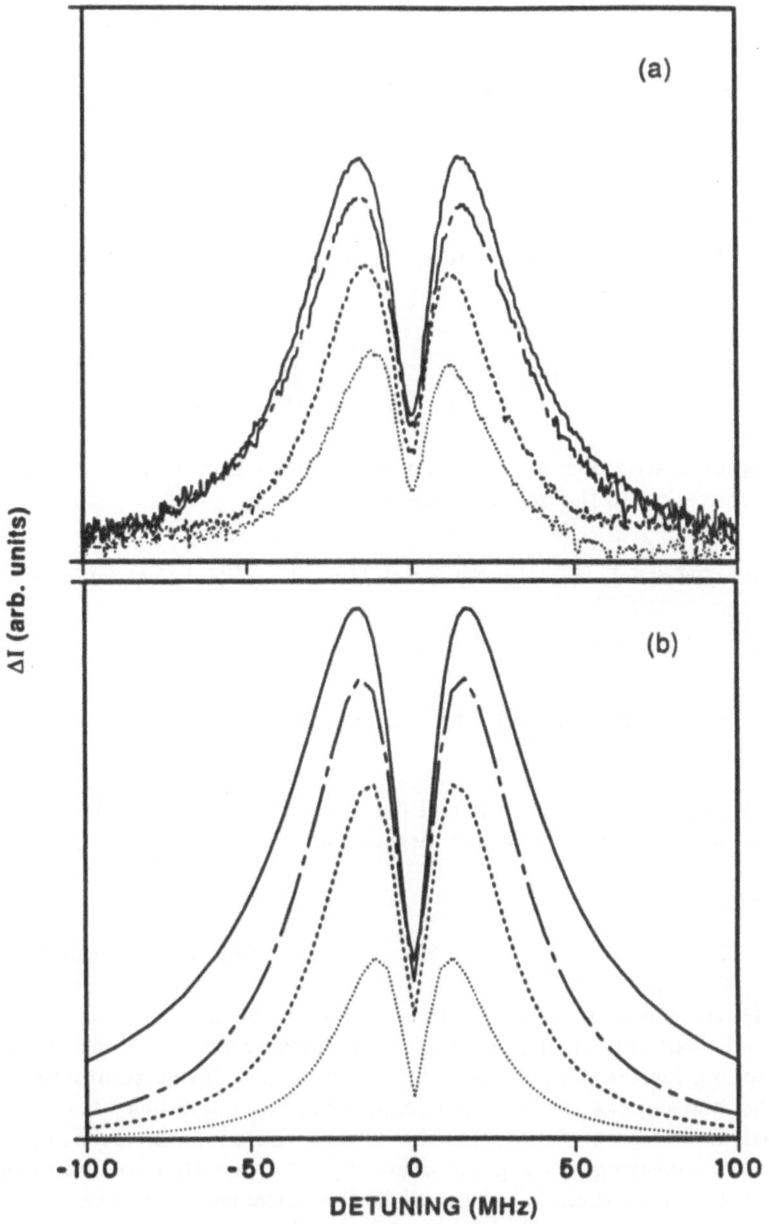

Figure 5. Relative rms fluorescence intensity fluctuations, ΔI, obtained experimentally (a) and theoretically (b) for $\Omega/2\pi \approx 25$ MHz with constant laser FWHM ≈ 4.5 MHz. The statistical parameters of the field are $\beta/2\pi = 100$ MHz and $b/2\pi = 2.0$ MHz (solid lines), $\beta/2\pi = 30$ MHz and $b/2\pi = 2.1$ MHz (chain-dash lines), $\beta/2\pi = 10$ MHz and $b/2\pi = 2.5$ MHz (dashed lines), and $\beta/2\pi = 1.0$ MHz and $b/2\pi = 4.5$ MHz (dotted line). The plots have been normalized to the same fluorescence intensity at line center.

where the sum is over all atoms in the ensemble. The variance of the intensity fluctuations is given by

$$(\Delta I)^2 = < \psi_3(t)\psi_3(t) > - < \psi_3(t) >< \psi_3(t) > \tag{5}$$

and the power spectra of the intensity fluctuations by

$$S_I(\omega) = \int_o^\infty \{< \psi_3(t+\tau)\psi_3(t) > - < \psi_3 >< \psi_3 >\} \, exp(i\omega\tau) \tag{6}$$

To calculate the spectra of scattered light we introduce the correlation matrix $R(t+\tau,t)$ defined as

$$R(t+\tau,t) = \begin{pmatrix} < S^+(t+\tau)S^-(t) > \\ < S^-(t+\tau)S^-(t) > \\ < S^z(t+\tau)S^-(t) > \end{pmatrix} \tag{7}$$

The correlation matrix obeys an equation that follows from the Bloch equations and the quantum regression theorem and is given by

$$dR/d\tau = C_o R(t+\tau,t) - i[x(t+\tau)C_+ + x^*(t+\tau)C_-]R(t+\tau,t) + g < S^-(t) > \tag{8}$$

The spectrum of the scattered field is then defined as

$$S(\omega) = Re \int_o^\infty d\tau \; exp(-i(\omega - \omega_c)\tau[< R_1 > - < \psi_1 >< \psi_2 >] \tag{9}$$

In our work, we solve the Eqs. (3) and (8) using Monte-Carlo methods and obtain the desired quantities. Details of this method can be found in Ref. 6.

Fluctuations in Fluorescence Intensity

In Fig. 6(a) is shown the mean intensity as a function of detuning for $\Gamma = 100\gamma$, $\beta = 20\gamma$, and three Ω's of 40γ, 100γ, and 400γ. We note that with an increase in Ω, there is a narrowing of these curves. The limit of this narrowing was reported and dealt with in Ref. 6. Fig. 6(b) shows the fluctuations in fluorescence versus detuning for the parameters of 6(a). There is a dip at zero detuning due to the fact that the slope of the inversion versus detuning curve has a zero slope at this point. With an increase in Δ on either side of zero, there is an increase in the variance with the peak appearing at the point where Fig. 6(a) has its maximum slope. With further increase in Δ, there is a decrease in the variance due to the flattening tails in Fig. 6(a). If the chaotic field is treated as a colored noise or Ornstein-Uhlenbeck process, qualitatively similar results are obtained.

Fig. 7(a) shows the power spectra of the intensity fluctuations when $\Gamma = 100\gamma$, $\beta = 20\gamma, \Delta = 0$ and Ω's of 40γ, 100γ, and 400γ. Clearly there are peaks in the power spectra at precisely the values of the Rabi frequency. We find that with an increase in the Ω, the relative height of the Rabi frequency component increases with respect to the zero frequency component. In Fig. 7(b) are shown spectra for $\Gamma = 100\gamma$, $\beta = 20\gamma$, $\Omega = 40\gamma$ and Δ's of 10γ, 40γ, and 100γ. Once again the spectra exhibit peaks at the generalized Rabi frequencies. With an increase in detuning, the zero frequency

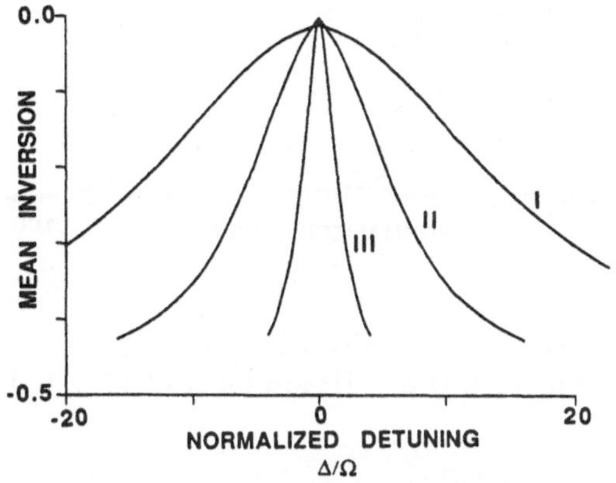

Figure 6(a). Mean inversion as a function of the normalized detuning, Δ/Ω for $\Gamma = 100\gamma$ and $\beta = 20\gamma$. Curves I, II and III are respectively for Rabi frequency, Ω, of 40γ, 100γ and 400γ.

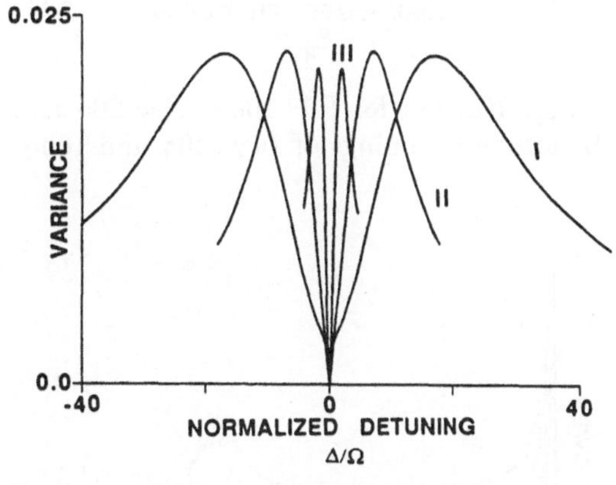

Figure 6(b). Variance of population fluctuations, i.e. intensity fluctuations $(\Delta I)^2$, as a function of Δ/Ω for the same parameters as Fig. 6(a).

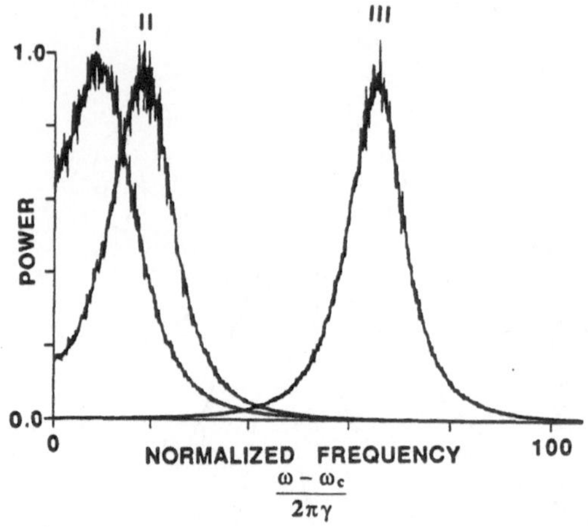

Figure 7(a). Power spectra of intensity fluctuations for $\Gamma = 100\gamma$, $\beta = 20\gamma$ and $\Delta = 0$. Curves I, II and III are for Ω of 40γ, 100γ, and 400γ.

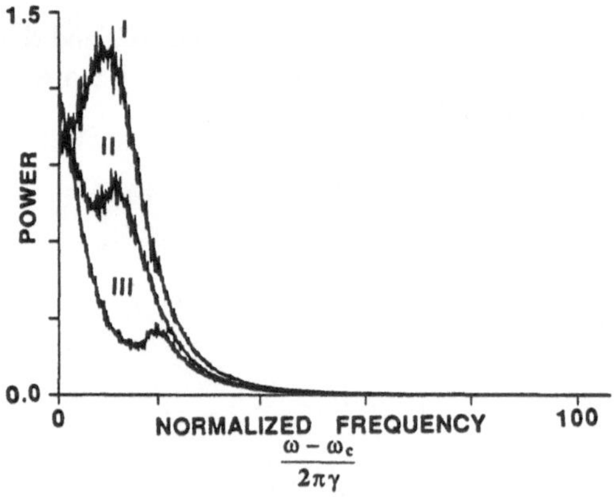

Figure 7(b). Same as Fig. 7(a) but for $\Gamma = 100\gamma$, $\beta = 20\gamma$ and $\Omega = 40\gamma$. Curves I, II and III are for detunings of 10γ, 40γ and 100γ.

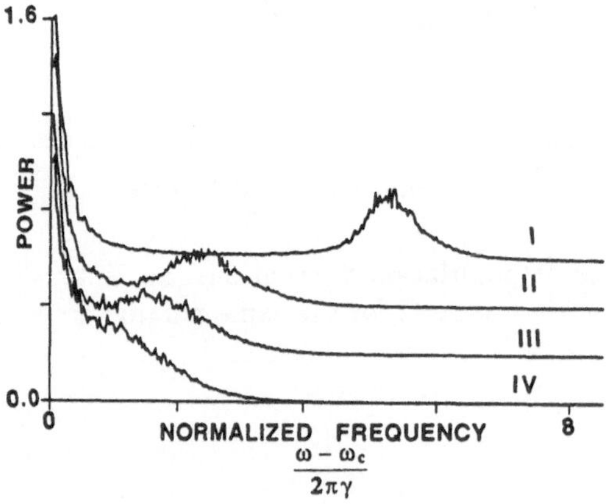

Figure 7(c). Same as Fig. 7(a) but for $\Gamma = \gamma$, $\beta = 20\gamma$ and $\Omega = 10\gamma$. Curves I, II, III and IV are for Ω of 30γ, $10\gamma, 5\gamma$ and 0 respectively.

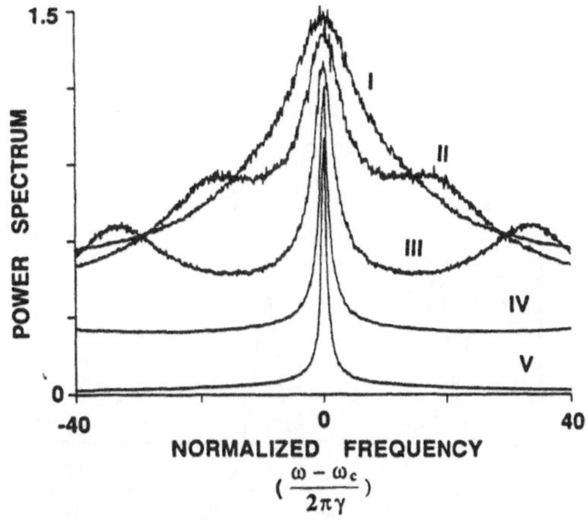

Figure 8(a). Spectrum of resonance fluorescence for $\Gamma = 100\gamma$ and $\beta = 20\gamma$. Curves I, II, III, IV and V are for Ω of 40γ, 100γ, 200γ, 400γ and 1000γ respectively.

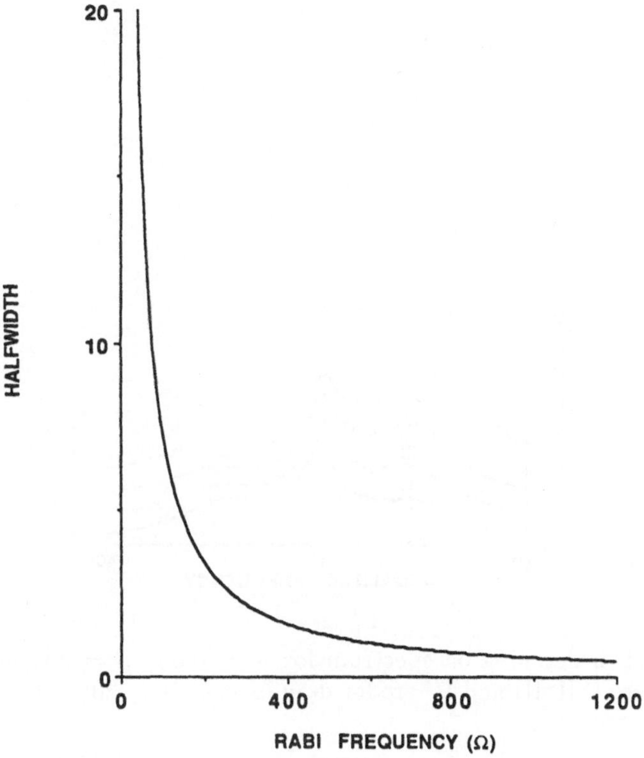

Figure 8(b). Halfwidth of central component as a function of Ω for $\Gamma = 100\gamma$ and $\beta = 20\gamma$.

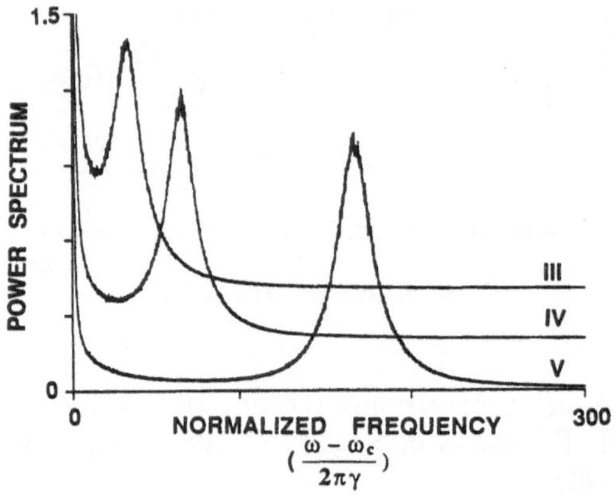

Figure 8(c). Sidebands of fluorescence spectrum of Fig. 8(a) for Ω of 200γ, 400γ and 1000γ.

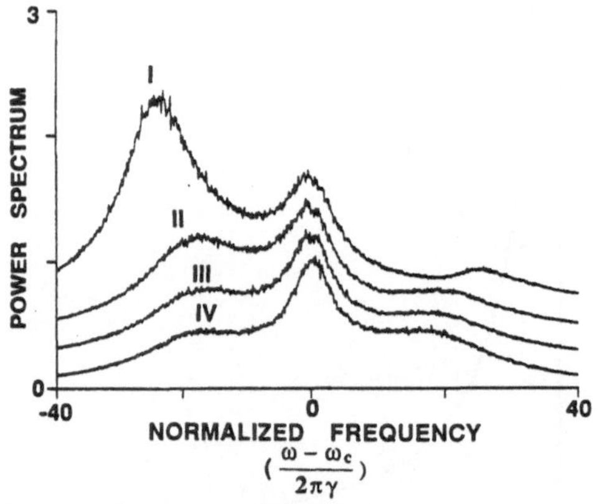

Figure 9. Effect of detuning on spectrum for $\Gamma = 100\gamma$, $\beta = 20\gamma$ and $\Omega = 100\gamma$. Curves I, II, III and IV are for detunings of 50γ, 20γ, 10γ and 0.

component becomes more dominant and the Rabi frequency component gets relatively suppressed. In Fig. 7(c) are spectra for $\Gamma = \gamma$, $\beta = 20\gamma$, $\Omega = 10\gamma$ and Δ's of 0, 5γ, 10γ and 30γ. In this case, we find that the zero frequency component is always the dominant component. With increasing detuning, the generalized Rabi frequency components become more prominent and well resolved from the zero frequency component.

Spectra of Scattered Light

In this section we present the results of spectra of scattered light when a two-level atom interacts with a coherent and a chaotic field. In Fig. 8(a) is the spectra for $\Gamma = 100\gamma$, $\beta = 20\gamma$, $\Delta = 0$ and Ω of 40γ, 100γ, 200γ, 400γ, and 1000γ. For values of Ω low compared to Γ but large compared to $(\gamma + \beta)$, we recover Mollow's three peak spectrum, i.e. we find peaks centered at $\omega = \omega_c$ with a width of $(\gamma + \beta)$ and two sidebands at $\omega = \omega_c \pm \Omega$ with widths of $3/2\ (\gamma + \beta)$. As Ω becomes larger than Γ, the usual Bloch description can no longer be used. This is because within the correlation time of the stochastic field, several Rabi oscillations due to the coherent component are possible. This leads to extreme narrowing of the central component. In fact the central component narrows from a halfwidth of $(\gamma + \beta)$ to γ (see Fig. 8(b)). The sidebands of Fig. 8(a) are shown in Fig. 8(c) and these show marginal narrowing with increasing Ω.

In Fig. 9 is shown the effect of detuning the pump from the transition frequency of the two-level atoms. We choose values of $\Gamma = 100\gamma$, $\beta = 20\gamma$, $\Omega = 100\gamma$ and Δ's of 0, 10γ, 20γ, and 50γ. We find that the three-photon sidebands become more and more pronounced with increasing detuning. On reversing the sign of the detuning, there is a reversal in the asymmetry of the spectra.

Qualitatively similar behavior is observed when the stochastic component is treated as a colored noise process. We refer the reader to Ref. 6 for further details on this work.

ACKNOWLEDGEMENTS

This work was supported by the U.S. Department of Energy, Office of Basic Energy Sciences, Chemical Sciences Division, G.S. Agarwal, R. Roy, P. Zoller, H. Ritsch and D.S. Elliott, have collaborated with us on various aspects of the work described here and their input is gratefully acknowledged.

REFERENCES

1. B. R. Mollow, *Phys, Rev.* **188**, 1969 (1969).
2. H. J. Kimble and L. Mandel, *Phys. Rev.* **A15**, 689 (1977); G. S. Agarwal, *Phys. Rev. Lett.* **37**, 1383 (1976); *Phys. Rev.* **A18**, 1490 (1978); J. H. Eberly, *Phys. Rev. Lett.* **37** 1387 (1976); K. Wodkiewicz, *Phys. Rev.* **A19**, 1686 (1979).
3. Th. Haslwanter, H. Ritsch, J. Cooper, and P. Zoller, *Phys. Rev.* **A38**, 5652 (1988); H. Ritsch, P. Zoller, and J. Cooper, *Phys. Rev.* **A41**, 2653 (1990).
4. M. H. Anderson, R. D. Jones, J. Cooper, S. J. Smith, D. S. Elliott, H. Ritsch, and P. Zoller, *Phys. Rev.* **A42**, 6690 (1990).
5. G. S. Agarwal and S. Singh, *Phys. Rev.* **A39**, 2239 (1989).
6. G. Vemuri, R. Roy, and G. S. Agarwal, *Phys. Rev.* **A41**, 2749 (1990); G. Vemuri and G. S. Agarwal, *Phys. Rev.* **A42**, 1687 (1990).
7. D. S. Elliott and S. J. Smith, *J. Opt. Soc. Am.* **B5**, 1927 (1988).

LASER-PHASE-NOISE INDUCED STOCHASTIC RESONANCE FLUORESCENCE

A.A. Rangwala*, K. Wo'dkiewicz†, and C. Su

Center for Advanced Studies and Department of
Physics and Astronomy
University of New Mexico,
Albuquerque, New Mexico 87131 USA

ABSTRACT

Stochastic fluctuations of coherent and incoherent components of resonance fluorescence intensity induced by Wiener-L'evy laser-phase noise are investigated. Statistical properties of the atomic dipole moment for different values of laser linewidth and different strength of the driving field are calculated. The stochastic Mollow spectra of atomic dipole fluctuations are derived. It is shown that these spectra exhibit a triplet structure which is purely classical and entirely laser-noise dependent.

INTRODUCTION

In a recent experiment by Anderson et. al.[1] fluorescence from a large number of atoms ($\sim 10^5$) has been observed by a detector with an area much larger than the coherence of the fluorescence light with fluctuating phase. This fluorescence was purely classical and the measured fluctuations of light intensity have been induced by classical laser phase fluctuations.

The aim of the present investigation[2] is to extend the above results to the forward component of the scattered light which is governed by atomic dipole moments.

STOCHASTIC BLOCH EQUATIONS AND RESONANCE FLUORESCENCE INTENSITY

We restrict our analysis to two-level atoms described by the stochastic optical Bloch equations[3]

$$\mathring{d} = \left(-i\Delta - \frac{A}{2}\right) d - \frac{i\Omega}{2} e^{-i\phi(t)} \mathrm{w} \qquad (1)$$

* Permanent address: Department of Physics, University of Bombay, Bombay 400 098, INDIA

† Permanent address: Institute of Theoretical Physics, Warsaw University, Warsaw 00-681, POLAND

$$\overset{\circ *}{d} = \left(i\Delta - \frac{A}{2}\right)d^* + \frac{i\Omega}{2}e^{i\phi(t)}\mathrm{w} \tag{2}$$

$$\overset{\circ}{\mathrm{w}} = -A(1+\mathrm{w}) + i\Omega\left(d^* e^{-i\phi(t)} - d e^{i\phi(t)}\right) \tag{3}$$

where d = dipole moment of the two-level atomic system, w = atomic population difference of the levels, Δ = detuning frequency $= w_o - w_L$, w_o = atomic transition frequency, w_L = laser light frequency, A = Einstein A-coefficient $= \frac{1}{T_1}$, T_1 = Lifetime for spontaneous decay, Ω = Rabi frequency, $\phi(t)$ = fluctuating phase of the laser field.

We suppose that the phase fluctuation of laser light is described by the Wiener-L'evy process:

$$<\overset{\circ}{\phi}(t)\,\overset{\circ}{\phi}(t')>= 2\Gamma\delta(t-t') \tag{4}$$

where Γ is the Lorentzian laser linewidth.

The resonance fluorescence intensity, I, consists of two parts, I_{coh} and I_{incoh}

$$I = p(t) = \frac{(1+w(t))}{2} = I_{coh} + I_{incoh}, \tag{5}$$

where I_{coh} = coherently (Rayleigh) scattered intensity $= d^*(t)d(t)$, I_{incoh} = incoherent intensity $= p(t) - d^*(t)d(t)$.

Use of Eqs. (1), (2), (3) and (4) leads to the following steady-state averages[2]:

$$\psi_1 =< I_{coh}(\infty) >= \frac{A\Omega^2}{2}\frac{Re[(\mathcal{L}_1^{-1}\mathcal{L}_2\mathcal{L}_3)^{*-1}((\mathcal{L}_2\mathcal{L}_3)^{-1} + \frac{\Omega^2}{2})]}{D} \tag{6}$$

$$\psi_2 =< d\mathrm{w}e^{i\phi}(\infty) >= -\frac{2i\mathcal{L}_3^{*-1}}{\Omega} < d^2 e^{2i\phi}(\infty) > \tag{7}$$

$$\psi_3 =< d^2 e^{2i\phi}(\infty) >= \frac{A^2\Omega^2}{2}\frac{[(1 - A\mathcal{L}_1^*)((\mathcal{L}_2\mathcal{L}_3)^{-1} + \frac{\Omega^2}{2}) - 2i\Omega^2\mathcal{L}_3^{-1}Im\mathcal{L}_1]}{D} \tag{8}$$

$$\psi_4 =< de^{i\phi}(\infty) >= \frac{iA\Omega}{2}\frac{\mathcal{L}_1^*}{\Omega^2(Re\mathcal{L}_1) - A} \tag{9}$$

$$< I(\infty) >= \frac{1}{2}\frac{\Omega^2 Re(\mathcal{L}_1)}{\Omega^2 Re(\mathcal{L}_1) - A} \tag{10}$$

where

$$D = [\Omega^2 Re(\mathcal{L}_1) - A]\left\{A\left|(\mathcal{L}_2\mathcal{L}_3)^{-1} + \frac{\Omega^2}{2}\right|^2 - 2\Omega^2 Re\left[\mathcal{L}_3^{*-1}\left((\mathcal{L}_2\mathcal{L}_3)^{-1} + \frac{\Omega^2}{2}\right)\right]\right\} \tag{11}$$

and the complex Lorentzian functions $\mathcal{L}_i(\Delta)$ are defined by

$$\mathcal{L}_1(\Delta) = \left(i\Delta - \frac{A}{2} - \Gamma\right)^{-1}, \mathcal{L}_2(\Delta) = \left(i\Delta - \frac{3A}{2} - \Gamma\right)^{-1}, \mathcal{L}_3(\Delta) = \left(i\Delta - \frac{A}{2} - 2\Gamma\right)^{-1} \tag{12}$$

Two typical numerical results are depicted in Figs. 1 and 2.

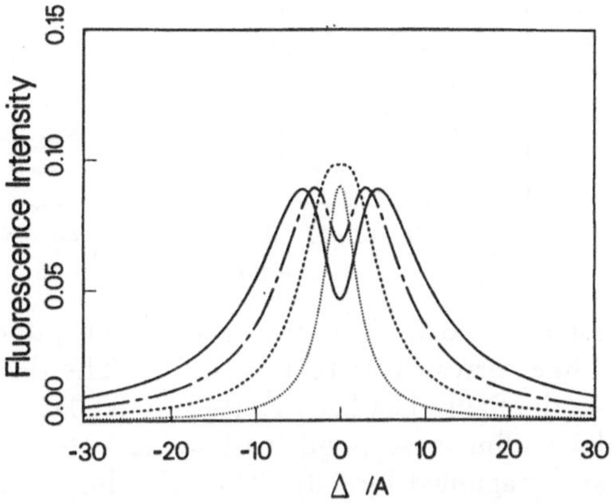

Figure 1. Coherent fluorescence $< I_{coh} >$ as a function of laser detuning Δ with $\Gamma = 1.2A$. The dotted line corresponds to $\Omega = 0.8A$. The dashed line corresponds to $\Omega = 2.4A$. The dash-dotted line corresponds to $\Omega = 3.2A$. The solid line corresponds to $\Omega = 1.6A$.

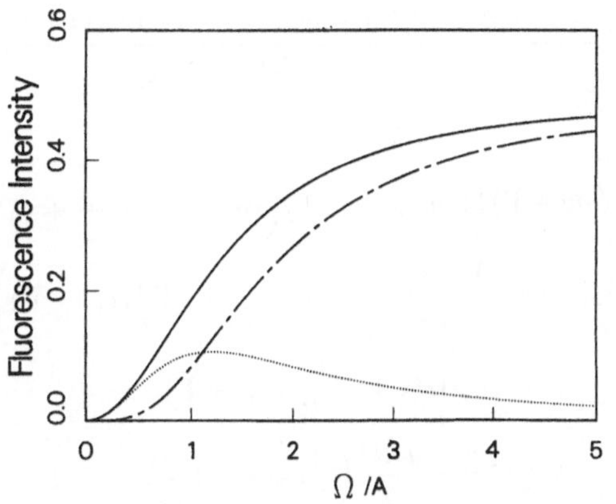

Figure 2. Intensity fluorescence as a function of Rabi frequency at exact resonance $\Delta = 0$ and $\Gamma = 1.2A$: total intensity $< I >$ (solid line), coherent intensity $< I_{coh} >$ (dotted line), incoherent intensity $< I_{inc} >$ (dash-dotted line).

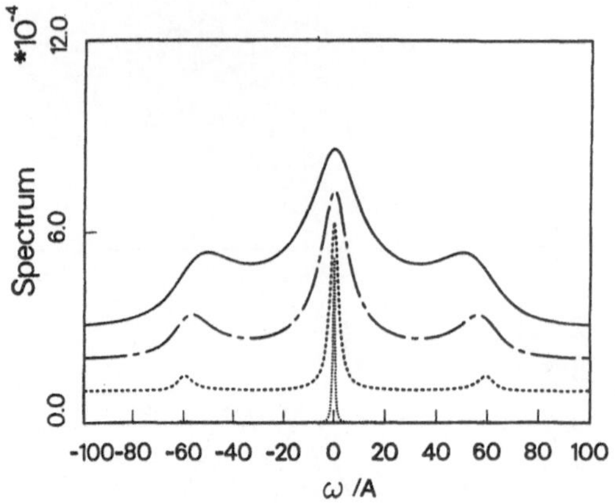

Figure 3. The SMS for $\Omega = 60A$, $\Delta = 0$ and different values of laser linewidth. The dotted line corresponds to $\Gamma = 0.5A$. The dashed line corresponds to $\Gamma = 2.0A$ (the figure has been shifted by 0.0001 and magnified by 3.0). The dash-dotted line corresponds to $\Gamma = 6.0A$ (the figure has been shifted by 0.0002 and magnified by 6.0). The solid line corresponds to $\Gamma = 10A$ (the figure has been shifted by 0.0003 and magnified by 7.5).

POWER SPECTRA OF DIPOLE FLUCTUATIONS

The steady-state autocorrelation of dipole moment is given by

$$S_d(w) = \lim_{t\to\infty} 2Re \int_0^\infty e^{-i(w-w_0)\tau} < d^*(t+\tau)d(t) > d\tau \tag{13}$$

This leads us to the following stochastic Mollow spectrum (SMS) in presence of laser phase fluctuations[2]

$$S_d(w) = 2Re\left\{\frac{N_1(w-w_0)\psi_1 + N_2(w-w_0)\psi_2 + N_3(w-w_0)\psi_3 + N_4(w-w_0)\psi_4}{N(w-w_0)}\right\} \tag{14}$$

where

$$N_1(w) = (iw+\Gamma)\left\{(iw+A+\Gamma)\left(iw+i\Delta+\frac{A}{2}+4\Gamma\right)+\frac{\Omega^2}{2}\right\} \tag{15}$$

$$N_2(w) = \frac{i\Omega}{2}\left\{\left(iw+i\Delta+\frac{A}{2}+4\Gamma\right)(iw+\Gamma)\right\} \tag{16}$$

$$N_3(w) = \frac{\Omega^2}{2}(iw+\Gamma) \tag{17}$$

$$N_4(w) = -\frac{iA\Omega}{2}\left(iw+i\Delta+\frac{A}{2}+4\Gamma\right) \tag{18}$$

and

$$N(w) = (iw+\Gamma)\left\{\left(iw-i\Delta+\frac{A}{2}\right)\mathrm{x}\right.$$
$$(iw+A+\Gamma)\left(iw+i\Delta+\frac{A}{2}+4\Gamma\right)$$
$$\left.+\frac{\Omega^3}{2}\left(iw+\frac{A}{2}+2\Gamma\right)\right\} \tag{19}$$

with $\psi_i^{\prime}s$ given by Eqs. (6), (7), (8) and (9).

It is important to note that the expression for $S_d(w)$ in Eq. (14) involves four complex Lorentzians: $\mathcal{L}_1$, $\mathcal{L}_2$, $\mathcal{L}_3$ and $\mathcal{L}_4 = (i\Delta + \frac{A}{2} + 4\Gamma)^{-1}$.

An important numerical result for the SMS is displayed in Fig. 3 which exhibits a triplet structure.

CONCLUSION

The single main result of the present investigation is that for strong laser excitations the SMS exhibits a purely classical and noise-dependent triplet structure.

REFERENCES

1. M.H. Anderson, R.D. Jones, J. Cooper, S.J. Smith, D. S. Elliot, H. Ritsch and P. Zoller, *Phys. Rev. Lett.* **64**, 1346 (1990).
2. A.A. Rangwala, K. Wo‘dkiewicz and C. Su. *Phys. Rev.* **A42**, 6651 (1990).
3. See for example, L. Allen and J.H. Eberly, *“Optical Resonance and Two-Level Atoms”(Wiley, New York, 1975).*

A DOUBLE RESONANCE MODEL FOR LASERS WITHOUT INVERSION†

G. Bhanu Prasad

School of Physics
University of Hyderabad
Hyderabad - 500 134, India

1. INTRODUCTION

Recently several mechanisms have been proposed for achieving laser action without population inversion.[1,2] Some of these mechanisms depend on interference effects arising from different pathways while others make use of additional pumping by coherent fields.[3] In this work we study the fairly standard laser model. However we assume that the excited state of the laser transition is coupled to a high lying state with a coherent field. We show that under suitable conditions such a pumping scheme can produce gain on the laser transition even if there is no inversion among the bare states of the atom.

2. LADDER SYSTEM FOR GAIN WITHOUT INVERSION

We consider a three level atom with non-equidistant, non- degenerate energy levels $|1>$, $|2>$, and $|3>$ (Fig. 1) with energies $\hbar\omega_{23}$, $\hbar\omega_{13}$ and 0 respectively. The transition $|1> \leftrightarrow |2>$ is driven by an intense laser beam. Level $|1>$ can decay to level $|2>$ and level $|2>$ to level $|3>$ at the rate of $2\gamma_{21}(\equiv 2\gamma_1)$ and $2\gamma_{32}(\equiv 2\gamma_2)$ respectively, where $2\gamma_{fi}$ being the probability per unit time that the system makes a transition from $|i>$ to $|f>$ due to spontaneous emission. The level $|2>$ is being pumped (incoherent pumping) at the rate of 2Λ. Here $|2> \leftrightarrow |3>$ is the laser transition, which is being probed by a weak laser. Generally the transition from $|1>$ to $|3>$ is forbidden due to parity considerations.

The electric field at the atom can be written as

$$E = \mathcal{E}_1 e^{-i\omega_1 t} + \mathcal{E}_2 e^{-i\omega_2 t} + C.C. \tag{2.1}$$

where $\mathcal{E}_1$ is the field on $|1> \leftrightarrow |2>$ transition and $\mathcal{E}_2$ is the field on $|2> \leftrightarrow |3>$ transition.

In the dipole approximation the total Hamiltonian of the system without counter rotating terms can be written as

† Work done in collaboration with G.S. Agarwal

$$H = (\Delta_1 + \Delta_2)|1><1| + \Delta_2|2><2| - \{G_1|1><2| + G_2|2><3| + C.C.\} \quad (2.2)$$

The parameters

$$G_1 = \frac{\vec{d}_{12} \cdot \vec{\mathcal{E}}_1}{\hbar}, \qquad \frac{\vec{d}_{23} \cdot \vec{\mathcal{E}}_2}{\hbar} \quad (2.3)$$

are the Rabi-frequencies associated with the coupling of laserfields $\mathcal{E}_1$ and $\mathcal{E}_2$ to the transitions $|1> \leftrightarrow |2>$ and $|2> \leftrightarrow |3>$ respectively, while the detunings between the laser and atomic frequencies are given by

$$\Delta_1 = \omega_{12} - \omega_1 \qquad \Delta_2 = \omega_{23} - \omega_1 \quad (2.4)$$

The equation of motion for the electric field amplitude $\mathcal{E}_2$ can be written as

$$\dot{\mathcal{E}}_2 = -\kappa\mathcal{E}_2 + 2\pi i \omega_2 \mathcal{P}(\omega_2) \quad (2.5)$$

where $\mathcal{P}(\omega_2)$ is the induced polarization at the probe frequency ω_2 and κ is the cavity losses. The induced polarization is related to the density matrix element by

$$\mathcal{P}(\omega_2) = n\rho_{23}d_{32} \quad (2.6)$$

where n is the number density of atoms.

In (2.5), we can see that the first term is responsible for the losses in the cavity, and also if $Im(\mathcal{P}) < 0$ the second term can lead to gain. We can define the gain coefficient to be

$$\mathcal{G} = -\frac{4\pi n\omega_2|d|^2}{\hbar c} Im\Big(\frac{\rho_{23}}{G_2}\Big) \quad (2.7)$$

i.e. for positive $\mathcal{G}$ we have amplification of the probe and for negative $\mathcal{G}$ we have probe absorption.

Using the Hamiltonian (2.2), the density matrix equations for the ladder system can be written and these equations are solved in the steady state. From this solution the gain coefficient has been computed and is plotted in Fig. 2 as a function of Δ_2 for different values of Δ_1. We can find that in some regions of frequencies of the probe field, $\mathcal{G}$ is positive and thus one can have gain. Note that we have gain even when the pumping is such that there is no inversion in terms of the bare states of the atom.

3. ORIGIN OF GAIN-DRESSED ATOM ANALYSIS

If the laser field G_1 is strong i.e. $|G_1| >> \gamma_1, \gamma_2$ then we can solve the density matrix equations in a representation in which the resonant part of the Hamiltonian is diagonal. The weak field G_2 is treated perturbatively. The total Hamiltonian is written in the form

$$H = H_d + H_1 \quad (3.1)$$

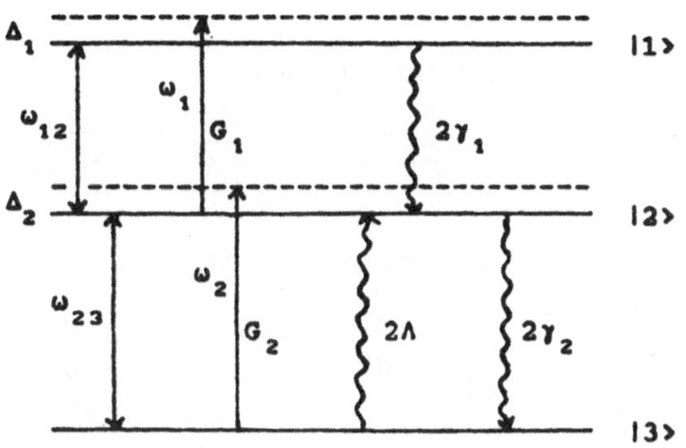

Figure 1. Schematic diagram of the energy levels of the ladder system showing various interactions and relaxations.

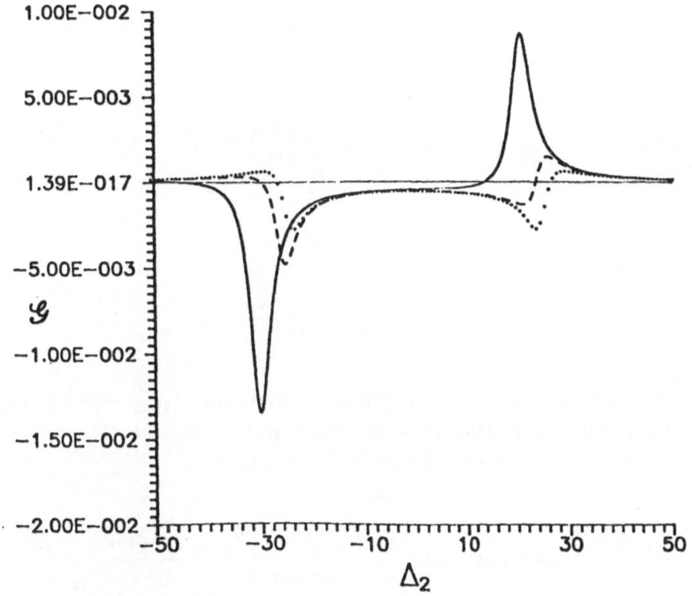

Figure 2. Gain coefficient $\mathcal{G}$ as a function of Δ_2 for $G_1 = 25$, $G_2 = 1, \gamma_1 = \gamma_2 = 1$, $\Lambda = 0.97$ and $\Delta_1 = 0, 2$ and 10 for the dotted, broken and solid lines respectively.

where

$$H_d = \begin{pmatrix} \Delta_1+\Delta_2 & -G_1 & 0 \\ -G_1^* & \Delta_2 & 0 \\ 0 & 0 & 0 \end{pmatrix}, \quad H_1 = \begin{pmatrix} 0 & 0 & 0 \\ 0 & 0 & -G_2 \\ 0 & -G_2^* & 0 \end{pmatrix} \tag{3.2}$$

the eigenvalues of H_d begin λ_+, λ_- and 0, where

$$\lambda_\pm = \frac{\Delta_1 + 2\Delta_2 \pm \Omega}{2}, \quad \Omega = \sqrt{\Delta_1^2 + 4|G_1^2|} \tag{3.3}$$

The eigenstates of the Hamiltonian H_d can be written as

$$\begin{pmatrix} |\psi_1> \\ |\psi_2> \\ |\psi_3> \end{pmatrix} = \begin{pmatrix} cos\vartheta & -sin\vartheta & 0 \\ sin\vartheta & cos\vartheta & 0 \\ 0 & 0 & 1 \end{pmatrix} = \begin{pmatrix} |1> \\ |2> \\ |3> \end{pmatrix} \tag{3.4}$$

where ϑ is determined from the expression

$$\vartheta = tan^{-1}\sqrt{\frac{\Omega - \Delta_1}{\Omega + \Delta_1}} \tag{3.5}$$

Using the transformation defined by (3.4) the density matrix equations are written in this representation and these are solved for the case $\gamma_1 = \gamma_2$ to the first order in the probe field. The gain coefficient in terms of $(\hat{\rho}_{33}^{(0)} - \bar{\rho}_{11}^{(0)})$, $(\bar{\rho}_{33}^{(0)} - \bar{\rho}_{22}^{(0)})$, $\bar{\rho}_{12}^{(0)}$ and $\bar{\rho}_{21}^{(0)}$ is found to be

$$\mathcal{G} = -\frac{4\pi n\omega_2|d|^2}{\hbar c} cos\vartheta sin\vartheta \; X \; Im\left[\frac{tan\vartheta(\bar{\rho}_{33}^{(0)} - \bar{\rho}_{11}^{(0)}) + \bar{\rho}_{12}^{(0)}}{\lambda_+ - i(\gamma_1 + \Lambda)} + \frac{cot\vartheta(\bar{\rho}_{33}^{(0)} - \bar{\rho}_{22}^{(0)}) + \bar{\rho}_{21}^{(0)}}{\lambda_- - i(\gamma_1 + \Lambda)}\right] \tag{3.6}$$

Next we write $\mathcal{G}$ as

$$\mathcal{G} = \mathcal{G}_p + \mathcal{G}_c, \tag{3.7}$$

where $\mathcal{G}_p$ is the contribution due to population inversion $(\bar{\rho}_{33}^{(0)} - \bar{\rho}_{11}^{(0)})$, $(\bar{\rho}_{33}^{(0)} - \bar{\rho}_{22}^{(0)})$ i.e. the inversion between one of the dressed states and $|3>$, and $\mathcal{G}_c$ is the contribution due to the coherence between the dressed states i.e.

$$\mathcal{G}_p = -\frac{4\pi n\omega_2|d|^2}{\hbar c} cos\vartheta sin\vartheta \; Im\left[\frac{tan\vartheta(\bar{\rho}_{33}^{(0)} - \bar{\rho}_{11}^{(0)})}{\lambda_+ - i(\gamma_1 + \Lambda)} + \frac{cot\vartheta(\bar{\rho}_{33}^{(0)} - \bar{\rho}_{22}^{(0)})}{\lambda_- - i(\gamma_1 + \Lambda)}\right] \tag{3.8a}$$

$$\mathcal{G}_c = -\frac{4\pi n\omega_2|d|^2}{\hbar c} cos\vartheta sin\vartheta \; Im\left[\frac{\bar{\rho}_{12}^{(0)}}{\lambda_+ - i(\gamma_1 + \Lambda)} + \frac{\bar{\rho}_{21}^{(0)}}{\lambda_- - i(\gamma_1 + \Lambda)}\right] \tag{3.8b}$$

These quantities are computed numerically and are plotted as functions of Δ_2 in Figs. 3a,b,c for different values Δ_1. From Fig. 3a we see that for $\Delta_1 = 0$, the gain is completely due to the coherence between the dressed states $|\psi_1>$ and $|\psi_2>$.

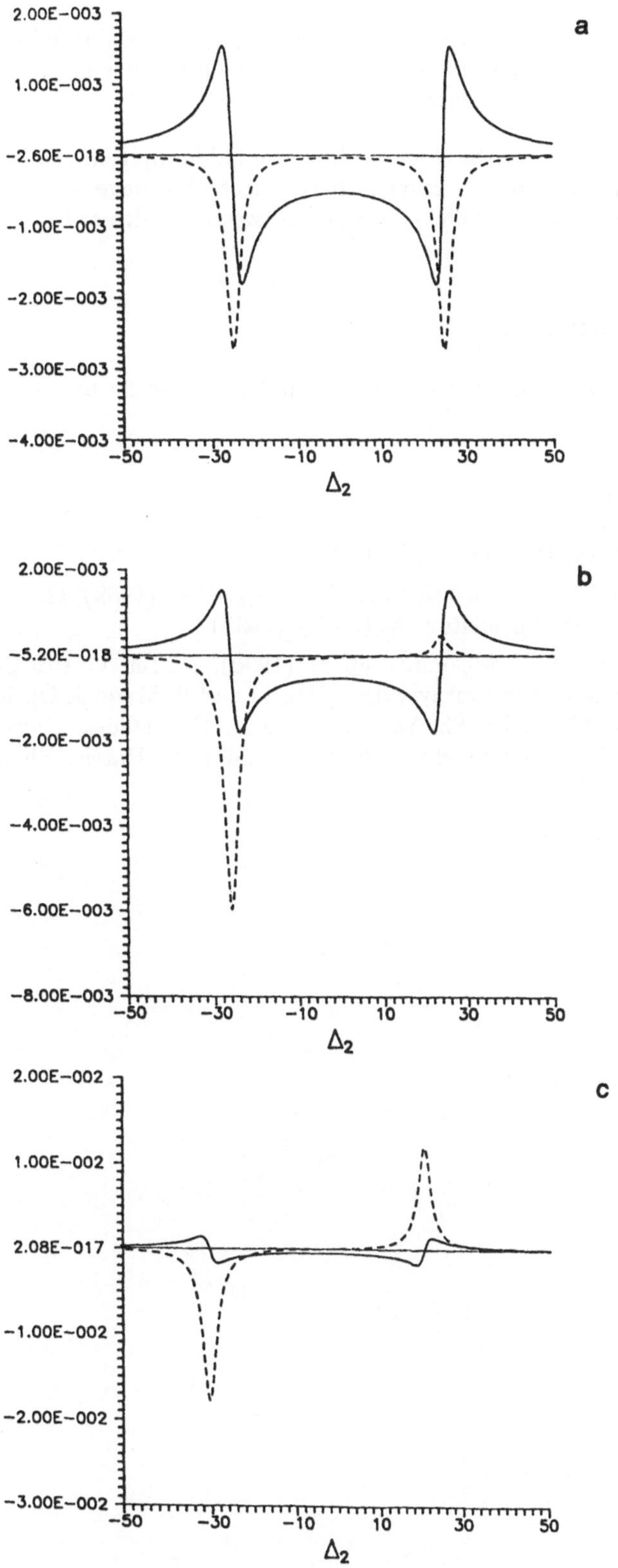

Figure 3. - - - = $\mathcal{G}_p$, ___ = $\mathcal{G}_c$ as functions of Δ_2 for a) $\Delta_1 = 0$, b) $\Delta_1 = 2$, c) $\Delta_1 = 10$ and rest of the parameters are same as in Fig. 2.

As Δ_1 is increased, the contribution of the population inversion between dressed states to the gain becomes important, even though there is no inversion between the bare atomic states for the parameters of interest.

Thus in conclusion we have shown how the ladder system can be used to produce gain without population inversion in bare states. We have also analyzed the origin of gain in such a system. The coherence between the dressed states is shown to be important for gain.

ACKNOWLEDGEMENTS

This work was supported by the Council of Scientific and Industrial Research, India.

REFERENCES

1. S. Harris, Phys. Rev. Lett. 62, 1033 (1989).
2. R. Ghosh and G. S. Agarwal, Phys. Rev. A39 1582 (1989), G. S. Agarwal, S. Ravi and J. Cooper, Phys. Rev. A41, 4721 (1990).
3. G. S. Agarwal, Optics Comm. 80 37 (1990), N. Lu, Optics Commun. 73 479 (1989), Olga Kocharovskaya, Ruo-Ding Li and P. Mandel, Optics Comm. 77 215 (1990), M. O. Scully, Shi-Yao Zhu, and A. Gavrielides, Phys. Rev. Lett. 62 2813 (1989), A. Imamoğlu, J. E. Field and S. E. Harris, Phys. Rev. Lett. 66 1154 (1991).

SQUEEZING OF CLASSICAL SOLITONS

D. Ranganathan

Physics Department
Indian Institute of Technology
New Delhi - 110 016 India

The absence of a superposition principle makes it difficult to find complex solutions to the nonlinear Schröedinger equation,

$$[i\frac{\partial}{\partial z} + \frac{\sigma}{2}\frac{\partial^2}{\partial T^2} + |u|^2]u = 0, \tag{1}$$

which describes the propagation of solitons in an optical fibre [1,2]. Here, following Hasegawa's notation,

$$z = \frac{\varepsilon^2 x}{\lambda}, \quad T = \frac{\varepsilon(t - k'x)}{(\lambda|k''1)^{\frac{1}{2}}}, \quad u = \frac{(2\pi\alpha\lambda n_2)^{\frac{1}{2}}E}{\varepsilon}, \tag{2}$$

and $\sigma = \pm 1$, for positive and negative group velocity dispersion respectively. One way to superimpose is to use Backlund transformations, which connect one solution u of equation (1) to another solution v of the same equation. The Backlund transformation of the nonlinear Schröedinger equation is [3].

$$(u \pm v)_T = (u \mp v)(4\eta^2 - |u \pm v|^2)^{\frac{1}{2}}, \tag{3}$$

where η is a real free parameter. Of most interest is the case when u and v differ by the addition of a single soliton, which is,

$$(u - v) = \eta \mathrm{Sech}[\eta(T - \xi z - \xi_o)]exp[i(\frac{\eta^2 - \xi^2}{2}z - \xi T - \eta_o)] \tag{4a}$$

for $\sigma = +1$, the case of bright solitons, and

$$(u - v) = \sqrt{\rho}e^{i\phi}, \ \rho = \rho_o[1 - a^2\mathrm{Sech}^2(\sqrt{\rho_o}a\{T - \xi z - \xi_o\})]^{\frac{1}{2}},$$

$$\phi = \int \frac{\rho_o^3(1 - a^2)}{\rho}dT - \frac{\rho_o}{2}(3 - a^2)z - \eta_o, \tag{4b}$$

Recent Developments in Quantum Optics, Edited
by R. Inguva, Plenum Press, New York, 1993

for $\sigma = -1$ the case of dark or 'gray' (a < 1) solitons.

So the Backlund transformation can be interpreted as an annihilation or creation operator for solitons ; as successive applications of equation (3) change the soliton content of the solution by one. As the solitons are classical, the annihilation and creation operators are identical and commute with each other.

Setting $v = 0$ and exponentiating the operator in eq. (3) we get the soliton displacement operator. If we square and exponentiate we get the squeezing operator for solitons. Such a solution is not of much use as the solitons thus generated will disperse rapidly due to the forces between them. For instance, the force between two bright solitons is attractive and varies exponentially with the distance [4], while the force between dark solitons is repulsive.

A better method of soliton superposition is given below. It uses the fact that while solutions of the nonlinear Schröedinger equation are complex, the 'potential' occurring in it is always real. Consider a linear superposition of the solutions u of equation (1),

$$U = \sum^{i} c_i u_i, \quad (|U|^2 = |u|_i^2 \text{ for all } i). \tag{5}$$

here the c_i are constants. We can easily see that U is a solution of equation (1) if all the u_i are solutions. Such a solution is quite stable, as each soliton feels its own potential and the effect of all the others on it coherently cancels out. In general, if the u_i''s are identical, then such superpositions are possible so the analogue of Fock and Squeezed states can always be constructed.

We illustrate this for the Fock state of occupation number N; a solution with N identical solitons. From equation (4), these solitons can still differ in their initial positions and phases. These 'labels' serve to distinguish the solitons from each other and are purely classical having no quantum mechanical equivalent. We choose the initial positions of all the solitons to be the same. Then from equation (5),

$$|N >= \sum^{i} |i > exp(i\delta_i) = |1 > \sum^{i} exp(i\delta_i). \tag{6}$$

$|1>$ is one of the solitons of Eq. (4), while the condition in Eq. (5) is

$$N + 2\sum_{i=2}^{N} \cos(\delta_i - \delta_1) + 2\sum_{i=2}^{N}\sum_{j=3}^{i} \cos(\delta_i - \delta_j) = 1 \tag{7}$$

An iterative solution to the above equation can always be found, such that given a n-1 soliton solution, the n soliton solution can be explicitly constructed from it. For instance, the first few phases are, $\delta_1 = 0°$, $\delta_2 = 120°$, $\delta_3 = 180°$, $\delta_4 = 60°, ...$ Such a solution is of interest for dark solitons Eq. (4b), where it has been suggested that there are no stable multisoliton solutions [5].

Many inequivalent sequences of arbitrary length can be constructed by the following rules (i) add the next soliton 120° or 240° away in phase from the phase vector formed by the rest, (ii) add a soliton 180° away from any existing one to cancel its effect.

The Fock state with phases obeying equation (7), can be generated in the following manner. Consider a fibre ring cavity or similar device which produces a train of solitons separated by the cavity transit time. If these are now introduced into a second ring with the same transit time, the successive solitons can be superposed. By adding a modulator between the two cavities, the state described in equation (7) can be realized. From this Fock state it is possible to generate squeezed states. For commuting operators, the squeezed state

$$|S> = exp(-\frac{s}{2}\hat{a}^2)|0> = \sum_{2n=0}^{\infty} \frac{s^n}{2n!}|2n> \tag{8}$$

However, unlike the quantum case, the normalization is not arbitrary. Thus the entire series cannot be generated, as an infinite number of solitons would require an infinite amount of energy. Cauchy sequences approximating equation (8) are possible. In a sequence containing n terms, the first soliton occurs n times, the second $n-1$ times and so on. To generate such a sequence, the Fock states must be stacked in a third ring cavity of twice the length. In doing so, the successive solitons must be amplified in the ratio of times they appear in the expansion (8).

Such superpositions can be detected by crosscorrelation with a fundamental soliton of the type given in equation (4); other methods have also been suggested for measuring the amplitude and phase of such short pulses [6].

Partial support of the Department of Science and Technology, Government of India is acknowledged.

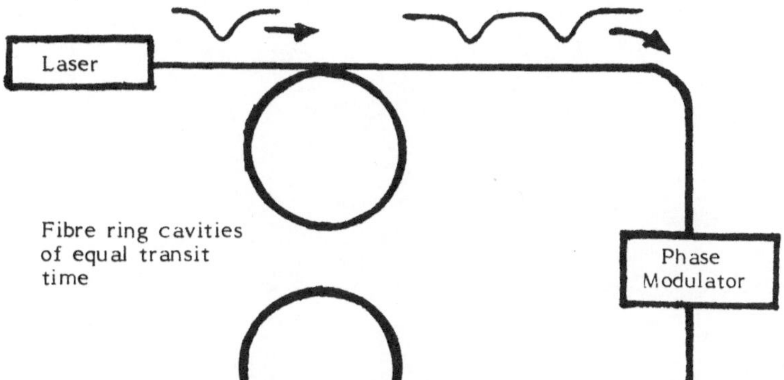

Figure 1. Fock State Generator

REFERENCES

1. A. Hasegawa and F. Tappert, *Applied Physics Letters*, 23 142 (1973); and 171 (1973).
2. A. Hasegawa, 'Optical solitons in fibres,' Springer New York (1989).
3. P. G. Drazin and R. S. Johnson, 'Solitons: an introduction,' Cambridge University press, Cambridge U.K. (1989).
4. J. P. Gordon, *Optics Letters*, 8 596 (1983).
5. K. J. Blow and N. J. Doran, *Physics Letters*, 107A, 55 (1985).
6. A. L. Lentine, L. M. F. Chirovsky, and L. A. D'Asaro, *Optics Letters*, 16, 36 (1991).

LADDER OF TRIPLETS IN THE SPECTRUM OF RESONANCE FLUORESCENCE FROM A TWO LEVEL ATOM

M. Krishna Kumari and Surya P. Tewari

School of Physics
University of Hyderabad
Hyderabad 500134, India

Recently, a great deal of interest has been shown in the study of the modifications in the resonance fluorescence spectrum from a two level atom (TLA) interacting with amplitude modulated intense laser radiation.[1-3] A striking observation has been made in the spectrum of a TLA interacting with a fully amplitude modulated (FAM) field. The spectrum consists of a series of peaks at the multiples of the modulation frequency and shows alternating linewidths and peak heights. These features differ from the Mollow's triplet observed in the case of strong monochromatic excitation. The observed features in the FAM case have been explained analytically in a recent communication.[1] In this paper, we show (i) the ladder of triplet structure of the multipeaked spectrum of a TLA in presence of partially amplitude modulated (PAM) field, (ii) how the singular structure for the FAM field arises from the ladder of triplets, (iii) minimum spontaneous emission noise at the inelastic coherent peaks.

Consider a two level atom with eigenstates $|1>$ and $|2>$, having energies E_1 and E_2 ($E_1 - E_2 = \hbar\omega_o$) respectively. We assume that the atom is driven by an amplitude modulated intense laser field, $\vec{F}(t)$, given by

$$\vec{F}(t) = \vec{\varepsilon}_o[1 + a cos(\Omega t + \phi_\Omega)] cos(\omega_L t + \phi_L). \tag{1}$$

Ω, ω_L and ϕ_Ω and ϕ_L are the frequencies and arbitrary initial phases of the modulation and the carrier vibration of the laser radiation, a is the modulation depth. We shall consider large modulation depth, $a >> 1$ (such that $a\vec{\varepsilon_o} >> \vec{\varepsilon_o}$). ε_o is the amplitude of the field. The field $\vec{F}(t)$ is equivalent to a superposition of three monochromatic and phase correlated fields having frequencies ω_L, $\omega_L \pm \Omega$. The Bloch equations for the mean atomic variables $< S^\pm >$ and $< S^z >$, in the rotating wave approximation are obtained from those in Ref. 1 by replacing $\chi(t)$ by Eq. 2. We take

$$\chi(t) = [\alpha_o + \alpha cos(\Omega t + \phi_\Omega)], \tag{2}$$

as the instantaneous Rabi frequency. α_o is the average Rabi frequency. It corresponds to the electric field $\vec{\varepsilon}_o(\alpha_o = 2\vec{d}_{12} \cdot \vec{\varepsilon}_o)$. α is the Rabi frequency of modulation and it corresponds to the envelope function $a\vec{\varepsilon}_o(\alpha = 2\vec{d}_{12} \cdot \vec{\varepsilon}_o a)$. *The two kinds of Rabi*

Recent Developments in Quantum Optics, Edited
by R. Inguva, Plenum Press, New York, 1993

frequencies can be varied independently. We examine the time averaged resonance fluorescence spectrum for $\Omega >> \alpha_o$. This limit correlates the results for the PAM field with the results for the FAM case.

TIME AVERAGED SPECTRAL RESPONSE IN A PARTIALLY MODULATED FIELD

The response of the TLA interacting with a PAM field has been the subject of several studies.[4] Most of the studies have been devoted to the demonstration of parametric resonance (PR) effects characterized by the condition $k\Omega - \alpha_o = 0$, k=integer. There is no PR in the limit, $\Omega >> \alpha_o$, which is of interest here. We give below the time averaged part of the steady state spectrum[5] calculated in the secular approximation using the Eberly-Wodkiewicz definition of spectrum for a finite value of α_o, $\phi_\Omega = \phi_L = 0$ and $\omega_L = \omega_o$.

$$\begin{aligned}
\bar{S}(\infty, \omega, \Gamma) &= \frac{C_{inc}}{[(\frac{\Gamma}{2}+\frac{\gamma}{2})^2 + D^2]} + \frac{C_{coh}}{(\frac{\Gamma}{2})^2 + D^2} \\
&+ \frac{C_{inc}^{L_o}}{[(\frac{\Gamma}{2}+\frac{3\gamma}{4})^2 + (D+\alpha_o)^2]} + \frac{C_{inc}^{R_o}}{[(\frac{\Gamma}{2}+\frac{3\gamma}{4})^2 + (D-\alpha_o)^2]} \\
&+ \sum_{n=1}^{\infty} \Big[\frac{S_{coh}^{n}}{[(\frac{\Gamma}{2}^2 + (D-n\Omega)^2]} + \Big\{ \frac{S_{inc}^{L_n}}{[(\frac{\Gamma}{2}+\frac{3\gamma}{4})^2 + (D-n\Omega+\alpha_o)^2]} \\
&+ \frac{S_{inc}^{R_n}}{[(\frac{\Gamma}{2}+\frac{3\gamma}{4})^2 + (D-n\Omega-\alpha_o)^2]} \Big\} + n \to -n \Big], \qquad (3)
\end{aligned}$$

where $C_{inc}, C_{coh}, C_{inc}^{L_o}, C_{inc}^{R_o}, S_{coh}^{n}, S_{inc}^{L_n}$ and $S_{inc}^{R_n}$ are the peak weights. Eq. 3 is plotted in Fig. 1a,b.

LADDER OF TRIPLETS

It is clear from (3) that the coherently scattered radiation from a TLA interacting with an amplitude modulated field has several inelastic components at $D \pm n\Omega = 0$ $(n \neq 0)$. The incoherent scattering also gets new frequencies at $D \pm n\Omega \pm \alpha_o = 0$. The spectral peaks may be grouped as $(a)(D + n\Omega) + 0 = 0, (b)(D + n\Omega) + \alpha_o = 0, (c)(D + n\Omega) - \alpha_o = 0$, for all n. Thus the multiple peak spectrum in (3) may be treated as a series of triplets occurring at each multiple of the modulation frequency, Ω. The central triplet (n=0) shows features similar to that of Mollow spectrum.

ORIGIN OF THE ALTERNATE STRUCTURE IN FAM CASE

We have examined a situation when α_o is very small. In this limit, the C_{coh} at D=0 (central peak) becomes very small and the peak has only incoherent contribution. The two Rabi side bands at $D \pm \alpha_o = 0$ merge with the central peak for small α_o. Thus the peak at D=0 has only incoherent contributions with widths $(\frac{\Gamma}{2}+\frac{\gamma}{2})$ and $(\frac{\Gamma}{2}+\frac{3\gamma}{4})$. The even n inelastic coherent peaks S_{coh}^{n} which are intensity (α_o) dependent also become very small for small α_o. As a result, the coherent peaks appear in the spectrum only at the odd n. The spectrum now displays a series of alternate triplets and doublets. This series of doublets and triplets observed for small α_o clearly indicate the origin

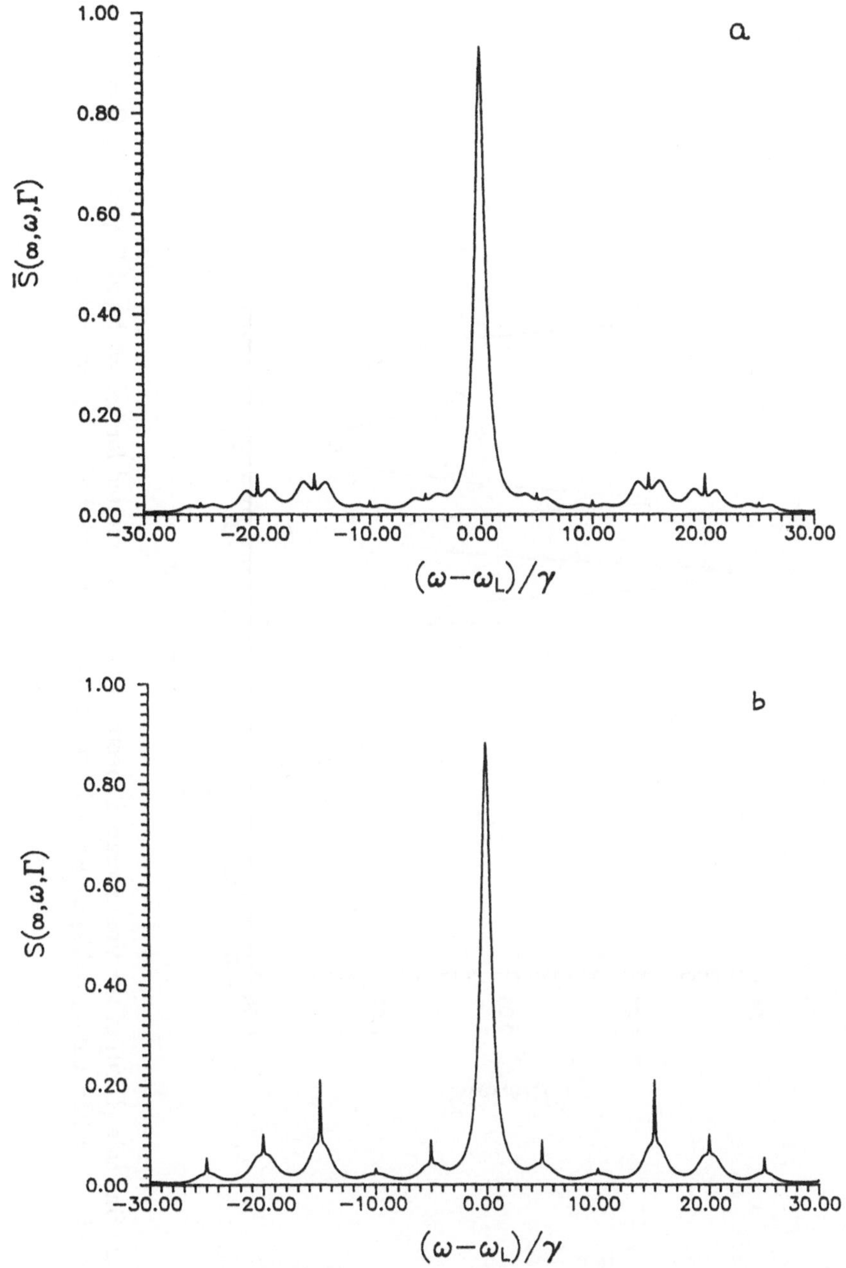

Figure 1. The time averaged spectrum $\bar{S}(\infty, \omega, \Gamma)$ for $\Delta = 0$ (a) shows the ladder of triplets for $\Gamma/\gamma = 0.1$, $\Omega = 5\gamma$, $\alpha = 23\gamma$ and $\alpha_o = \gamma$ (b) shows that for very small values of α_o and for $\Gamma/\gamma = 0.1$, $\Omega = 5\gamma$, $\alpha = 23\gamma$ and $\alpha_o = 0.5\gamma$ the peaks have finite coherent and incoherent contributions.

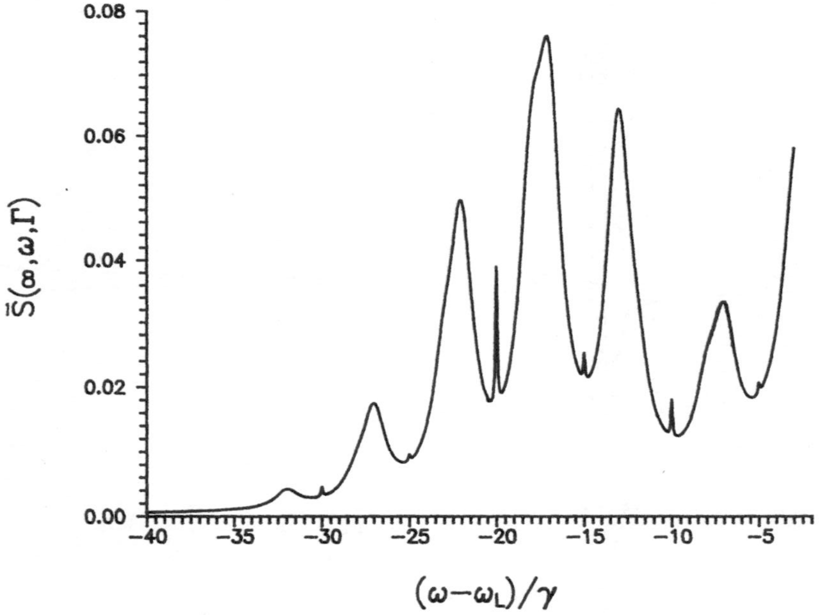

Figure 2. One side (D<0) of the time averaged spectrum $\bar{S}(\infty, \omega, \Gamma)$ plotted for $2\alpha_o = \Omega$ and $\Gamma/\gamma = 0.1$, $\Omega = 5\gamma$, $\alpha = 23\gamma$ and $\alpha_o = 2.5\gamma$ shows well separated coherent and the incoherent peaks.

of the singular alternating structure observed in the FAM case. In the limit when $\alpha_o = 0$ the central triplet merges to form a single peak at D=0 and all the doublets and triplets merge to give peaks at $D \pm n\Omega = 0$. The triplet structure at the odd multiple of n lead to higher and narrower peaks, while the doublets at even n give broad and short peaks. We have thus shown that it is the ladder of triplets in the PAM field that develops into its singular structure for the FAM field.

MINIMUM SPONTANEOUS NOISE AT THE INELASTIC COHERENT PEAKS

The inelastic coherent peaks at $D \pm n\Omega = 0, n \neq 0$, have only the instrument width $\frac{\Gamma}{2}$. The most remarkable feature to be noted here is that there is negligible incoherent contribution at the inelastic coherent peaks at $D \pm n\Omega = 0, n \neq 0$ and $0 \neq \alpha_o < \Omega$. The peaks at $D = \pm n\Omega$ can be said to be free of the spontaneous emission noise. To demonstrate this effect we plot in Fig. 2, only one side of the spectrum for $2\alpha_o = \Omega$ for which the coherent peaks at $D = n\Omega$ are well separated from the incoherent peaks at $D = (n + \frac{1}{2})\Omega$. Further separation between the two peaks, by increasing $(\alpha_o, \Omega/2)$ to larger values, reduces the coherent peaks to negligible height due to saturation effects.

REFERENCES

1. Surya P. Tewari and M. Krishna Kumari, *Phys. Rev.* **A 41**, 5273 (1990).
2. Y. Zhu, Q. Wu, A. Lezama, Daniel J. Gauthier and T. W. Mossberg, *Phys. Rev.* **A 41**, 6574 (1990).
3. H. Freedhoff and Z. Chen, *Phys. Rev.* **A 41** 6013 (1990).
4. See G. S. Agarwal and N. Nayak, *J. Phys.* **B 19**, 3385 (1986), G. S. Agarwal and N. Nayak, *J. Opt. Soc. Am.* **B 1**, 164 (1984), B. Blind, P. R. Fontana, P. Thomann, *J. Phys.* **B 13**, 2717 (1980), Surya P. Tewari and M. Krishna Kumari, J. Phys. B22, L475 (1989) and references therein.
5. M. Krishna Kumari, Ph.D. Thesis, University of Hyderabad,India (1990).

PHOTON-STATISTICS OF STIMULATED HYPER RAMAN SCATTERING IN AN INHOMOGENEOUSLY BROADENED THREE-LEVEL GASEOUS SYSTEM

P.S. Gupta and J. Dash

Dept. of Applied Physics
Indian School of Mines
Dhanbad-826004 (Bihar) India

INTRODUCTION

Stimulated Hyper Raman Scattering (SHRS) has a potential for providing intense coherent radiation at diverse frequencies and is proving to be an excellent method for probing atomic excitation levels. Hence a study of quantum statistics of SHRS is of great interest. SHRS being essentially a multi-level problem it is of interest to study the phenomenon using a three or four-level system and subsequently a considerable interest has been focused in this direction.[1-4] In the present paper we apply the combined atom-field density-matrix technique, which has been applied successfully in the study of other nonlinear optical processes,[5-8] to study the quantum statistics of SHRS in a gaseous system described by three energy levels

MODEL AND EQUATIONS OF MOTION

A laser pump field of frequency ω_L is applied to a three-level system (fig. 1) with an upper-level $|a>$, intermediate level $|b>$ and the lower level $|c>$ with energy E_a, E_b and E_c respectively. If $2\omega_L \approx \omega_{ac}$ two pump photons are absorbed in first step taking the system from $|c>$ to $|a>$ and a hyper Raman stokes photon at frequency ω_s is emitted in the next transition, the atom settling in level $|b>$. The present model resembles that of Vrehen and Hikspoors[3] but in a simplified form. Initially all the atoms are assumed to be in the lower level $|c>$.

The unperturbed hamiltonian H_o and the interaction hamiltonian H_{int} of the atomic system with fields can be written as

$$H_o = (a_L^+ a_L + 1/2)\omega_L + (a_s^+ a_s + 1/2)\omega_s + \frac{1}{\hbar}(E_a C_a^+ C_a + E_b C_b^+ C_b + E_c C_c^+ C_c) \quad (1)$$

and

Recent Developments in Quantum Optics, Edited by R. Inguva, Plenum Press, New York, 1993

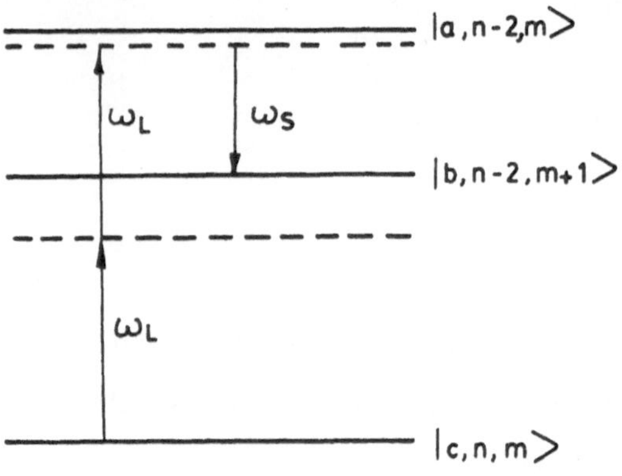

Figure 1

$$H_{int} = g_1(a_L a_L C_c C_a^+ + H.C.)\{\sqrt{2} sin K_L z(t)\}^2 + g_2(a_s^+ C_a C_b^+ + H.C.)\{\sqrt{2} sin K_s z(t)\} \tag{2}$$

where $a_L^+(a_L)$ and $a_s^+(a_s)$ are the creation (annihilation) operators of pump and hyper Raman stokes field respectively. $C_a^+(C_a), C_b^+(C_b)$ and $C_c^+(C_c)$ are the creation (destruction) operators of the three levels $|a>$, $|b>$ and $|c>$ respectively. g_1 and g_2 are coupling constants for the first and second transitions respectively, K_L and K_s are the wave vectors of pump field and hyper Raman field respectively.

The density matrix ρ of the coupled atom field system obeys the equation of motion:

$$i\dot{\rho} = [H_o + H_{int}, \rho] - \frac{i}{2}(\Gamma\rho + \rho\Gamma) \tag{3}$$

The relaxation mechanism is incorporated through the decay matrix

$$\Gamma = \begin{pmatrix} \gamma_a & 0 & 0 \\ 0 & \gamma_b & 0 \\ 0 & 0 & \gamma_c \end{pmatrix}$$

γ_a, γ_b, γ_c are the decay rates of the levels $|a>$, $|b>$ and $|c>$ respectively. Use of equations (1) and (2) in equation (3) gives the equations of motion for the elements of density-matrix. Then proceeding up to fourth order in the atom-field coupling constant and tracing over atomic variables[7,8] we obtain the equation of motion of the density-matrix for the radiation fields only. Evaluating time and velocity integrals under appropriate conditions[7,8] and incorporating the hyper-Raman stokes loss as a single photon process[8] we obtain

$$\frac{d\rho^{(n,m,t)}}{dt} = A(n+1)(n+2)\{U_1 - \frac{(n+1)(n+2)}{8}\frac{B}{A}f_1\}\rho(n+2,m,t)$$

$$-A(n-1)n\{U_1 - \tfrac{(n-1)n}{8}\tfrac{B}{A}f_1\}\rho(n,m,t)$$

$$-D(n+1)(n+2)(m+1)f_2\rho(n+2,m,t) + D(n+1)(n+2)mf_2\rho(n+2,m-1,t)$$

$$-Cm\rho(m,m,t) + C(m+1)\rho(n,m+1,t) \tag{4}$$

Where repetition of field variables are omitted by writing $\rho\ (n,m)$ for $\rho(n,m,n,m)$. Further

$$A = \frac{g_1^2 r_c}{4\gamma_c K_L u}, \quad B/A = 4g_1^2/\gamma_a\gamma_c, \quad D/A = g_2^2/2\gamma_a K_s u$$

r_c is the number the atoms entering $|c>$ per second.

$$f_1 = \left[U_1 + y_1\left(\frac{\partial V_1}{\partial x_1} - \frac{\partial U_1}{\partial y_1}\right) + \frac{\gamma_{ac}^2}{(\gamma_{ac}^2 + \Delta_1^2)}(U_1 + \frac{y_1}{x_1}V_1)\right]$$

$$f_2 = \Big[\frac{(y_2-y_1)(U_1-U_2)-(x_2-x_1)(V_1-V_2)}{(x_2-x_1)^2+(y_2-y_1)^2} + \frac{(y_2+y_1)(U_1+U_2)-(x_2+x_1)(V_1+V_2)}{(x_2+x_1)^2+(y_2+y_1)^2}$$

$$+\frac{(y_2+y_1)(U_1+U_2)+(x_2-x_1)(V_1-V_2)}{(x_2-x_1)^2+(y_2+y_1)^2} + \frac{(y_2-y_1)(U_1-U_2)+(x_2+x_1)(V_1+V_2)}{(x_2+x_1)^2+(y_2-y_1)^2}\Big] \tag{5}$$

$U_1(x_1,y_1)$ and $V_1(x_1,y_1)$ are the real and imaginary parts of the integral

$$\frac{i}{\pi}\int_{-\infty}^{\infty}\frac{e^{-t^2}}{x_1+iy_1-t}dt,$$

and $U_2(x_2,y_2)$ and $V_2(x_2,y_2)$ are the real and imaginary parts of the integral

$$\frac{i}{\pi}\int_{-\infty}^{\infty}\frac{e^{-t^2}}{x_2+iy_2-t}dt,$$

which are tabulated in Faddeyeva and Terentev[9], where

$$x_1 = \Delta_1/K_L u, \ y_1 = \frac{\gamma_{ac}}{K_L u}, \ x_2 = \frac{\Delta_2}{K_s u}, \ y_2 = \frac{\gamma_{ab}}{K_s u},$$

$K_L u$ and $K_s u$ are the broadening in the transition due to atomic motion (a Maxwell's velocity distribution is assumed). In the subsequent calculations we have taken $\Delta_1 = \Delta_2$.

PHOTON-STATISTICS

To solve, equation (4) without detailed balance we extend the slowly varying function technique of Gortz and Walls[10] to the present case of two variables.

We define

$$q_1(n,m) = \frac{\rho(n+1,m)}{\rho(n,m)}, \quad q_2(n,m) = \frac{\rho(n,m+1)}{\rho(n,m)} \tag{6}$$

which are characteristically slowly varying functions. From equation (6) we obtain

$$\dot{q}_1(n,m) = \frac{\dot{\rho}(n+1,m)}{\rho(n,m)} - q_1(n,m)\frac{\dot{\rho}(n,m)}{\rho(n,m)} \tag{7}$$

and

$$\dot{q}_2(n,m) = \frac{\dot{\rho}(n,m+1)}{\rho(n,m)} - q_2(n,m)\frac{\dot{\rho}(n,m)}{\rho(n,m)} \tag{8}$$

Use of equation (4) in equations (7) and (8) gives, for large values of n and m, in the steady state,

$$q_2(n,m) = \frac{G}{C}n(n+1)\frac{[1 - 2n^2\frac{F}{E} - \frac{C}{E}\frac{m}{n(n+1)}]}{[1 - 2(n+2)^2\frac{F}{E} - (m+1)\frac{G}{E}]} \tag{9}$$

where $E = Au_1$, $F = Bf_1$, $G = Df_2$ from which we obtain

$$\rho(n,m) = \rho(n,0)\prod_{j=0}^{m-1}\frac{G}{C}n(n+1)\frac{[1 - 2n^2\frac{F}{E} - \frac{C}{E}\frac{j}{n(n+1)}]}{[1 - 2(n+2)^2\frac{F}{E} - \frac{G}{E}(j+1)]} \tag{10}$$

Equation (10) gives the joint photon probability distribution of the pump and stokes field. It is seen to depend on the amplification parameter E, nonlinear terms F, G and loss parameter C.

RESULTS AND DISCUSSION

Since a single photon loss mechanism is adopted, all combinations of n and m are possible with the restriction $n+2m \leq n_o$, n_o being the initial pump photon number. But for a system without cavity loss states with $n + 2m = n_o$ are the only possible combinations for n and m. When we consider a system with low cavity loss, this condition may be used with reasonable accuracy[8] and this condition is imposed in the present calculations to avoid mathematical complexities.

The hyper-Raman stokes photon probability distribution at various values of nonlinear and loss parameters is illustrated in fig. 2, which shows a conversion efficiency varying from about 15 percent to 20 percent depending on parameters E, F, G and C. The conversion efficiency in the present case of three-level system is found to be greater than that in a two level system treated previously.[11] This may be attributed

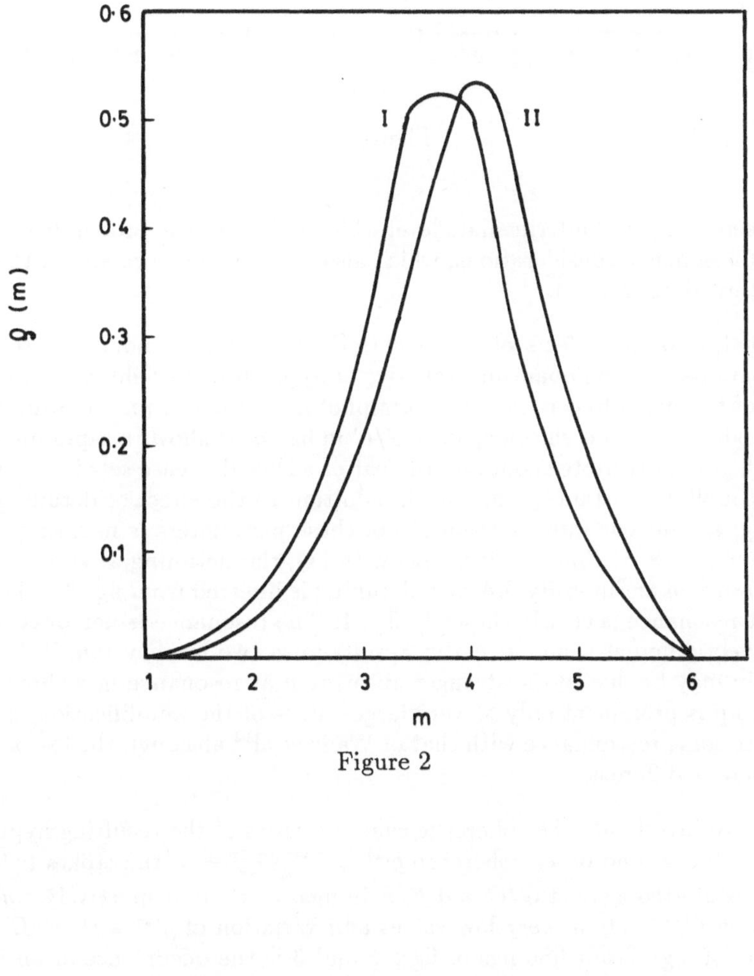

Figure 2

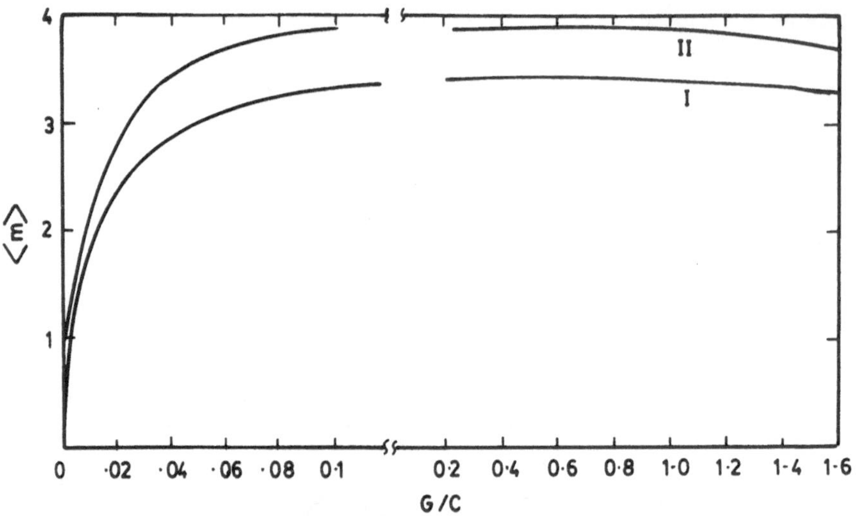

Figure 3

to the presence of a real intermediate level which enhances the probability of the process due to resonance considerations, which also justifies the necessity of the study of SHRS in three-level system.

Investigation of the effects of amplification parameter E, nonlinear terms F, G and loss parameter C on quantum statistics of hyper-Raman field is of interest. The variation of the mean hyper-Raman stokes photon number $< m >$ obtained numerically through eqn. (10) is shown against G/C in fig. 3. It shows a rapid increase with G/C which gradually flattens out and thereafter a slow decrease sets in. As G/C contains the coupling constants g_1 and g_2, in addition to the effect of detuning and line broadening, a judicious inter-relationship of these parameters is necessary to obtain the maximum $< m >$. Since G decreases with Δ_2, the detuning at stokes frequency, the appearance of an intensity dip with detuning is inferred from fig. 3. The presence of a dip at resonance is clearly shown in fig. 4. The resonance is narrower and dip is quantitatively different from our earlier results for a two level system.[11] The present value of dip may be due to the stronger absorption at resonance in a three-level system. The dip is prominent only at very large values of the amplification factor E/C. The dip has some resemblance with that of Walls et al[12] although the loss mechanism adopted here is different.

In order to investigate the coherence characteristics of the resulting hyper-Raman stokes field, the second order coherence $g^{(2)} = \frac{<m^2>-<m>}{<m>^2}$ of the stokes field is computed and is plotted against G/C and F/E in figs. 5 and 6, respectively. $g^{(2)}$ is seen to depend on G/C only at very low values and variation of $g^{(2)}$ with F/E is nonlinear. The most significant findings of figs. 5 and 6 is the occurrence of antibunching in the hyper-Raman stokes radiation since $g^{(2)} < 1$ for all values of G/C and F/E that we have considered. Interestingly SHRS from a three-level system appears to be dependent on the various parameters to a longer degree than the SHRS from a two

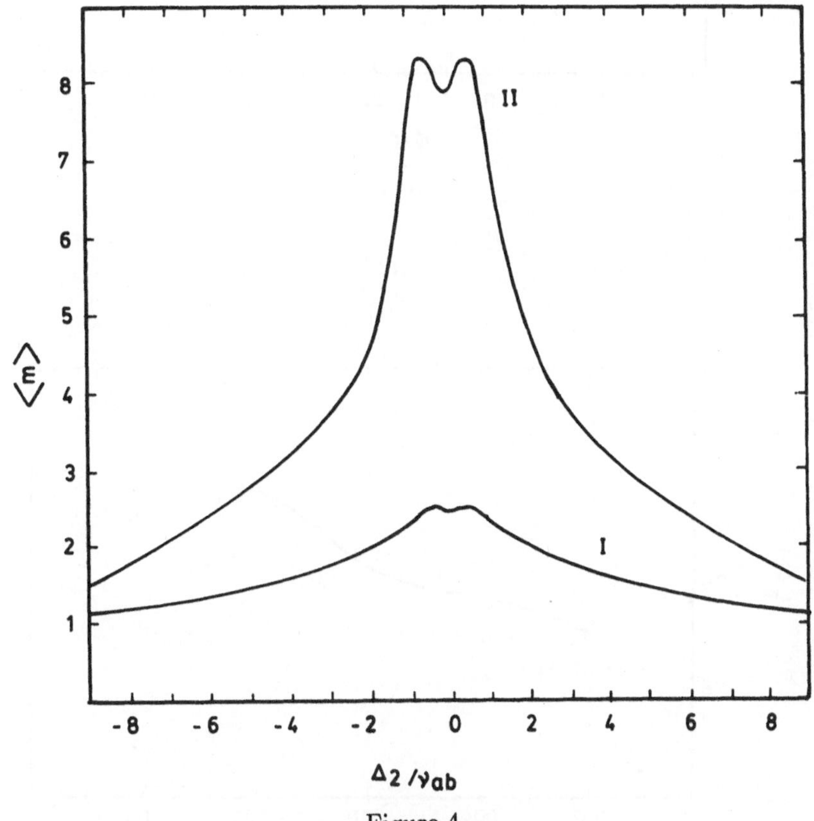

Figure 4

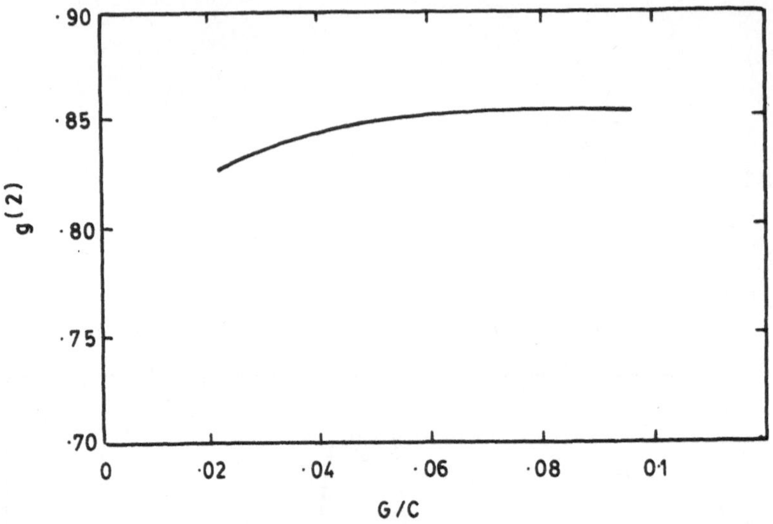

Figure 5

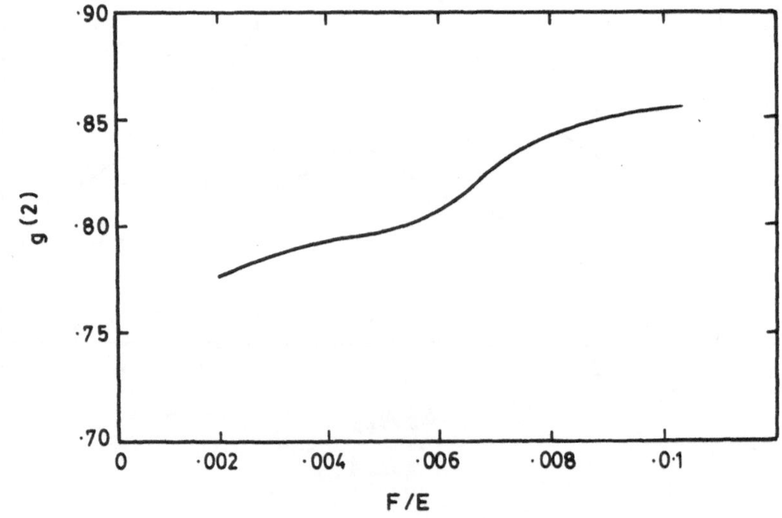

Figure 6

level system where detuning was of prime importance.[10] This is because three-level problem is more realistic. We are also trying to extend this study to a four-level system. The present model can also be applied to the study of other multiphoton processes in three-level systems.

REFERENCES

1. D. Krökel, K. Ludewigt and H. Welling, *IEEE J. Quant. Electron* **QE-22** (1986) 489.
2. D. J. Kim and P.D. Coleman, *IEEE J. Quant Electron* **QE-16** (1980) 300.
3. Q. H. F. Vrehen and H. M. Hikspoors. *Optics Commun.* **21** (1977) 127.
4. J. Perina, V. Perinova, C. Sibilia and M. Bertolotti, *Optics. Commun.* **49** (1984) 285.
5. K. J. McNeil and D. F. Walls, *J. Phys.* **A7** (1974) 617.
6. P. S. Gupta and B. K. Mohanty, *Optica Acta* **28** (1981) 521.
7. P. S. Gupta and J. Dash, *Optics Commun.* **79** (1990) 275.
8. P. S. Gupta and J. Dash, Czech. *J. Phys.* **B 40** (1990), 432.
9. Faddeyeva V. N. and Terentev N. M. *Tables of probability Integral for Comp. Argument.* Pergamon Press, New York, 1961.
10. R. Görtz and D. F. Walls, *Z. Physik* **B 25** (1976) 423.
11. B. K. Mohanty and P. S. Gupta, Czech. *J. Phys.* **B 31** (1981) 857.
12. D. F. Walls, P. Zoller and M. L. Steynross, *IEEE J. Quant. Electron* **QE 17** (1981) 380.

OPTOELECTRONIC EFFECT IN JUNCTION LASERS AND AMPLIFIERS AND THEIR APPLICATIONS

P.G. Eliseev and M.A. Man'ko

Lebedev Physics Institute
Leninsky pr. 53
Moscow 117924, USSR

We shall refer to electrical response of the laser diode in the oscillation regime to the light injection into the active region as "optoelectronic" response or signal. This response appears due to stimulated emission in the active region provoked by injected photons and by other photons generated by the noted ones. Stimulated emission leads to the decrease of the electron lifetime since its effective value is affected by rates of all recombination processes. As a result, the excess carrier concentration decreases with a consequent decrease of quasi-Fermi levels difference and the p-n voltage. Therefore, the response may be detected electrically in the pumping current circuits of the diode.

Such optoelectronic effect in junction lasers was firstly described in 1968 [1,2], where the saturation of the voltage applied to p-n junction of GaAs laser diode at the lasing threshold was demonstrated. The optoelectronic signal (OES) occurring in stripe GaAlAs/GaAs heterojunction lasers with a feedback provided by external mirror was investigated [3,4] and correlation between OES and the differential resistance of the laser diode was found. Practical application of the method with varying feedback was demonstrated in optoelectronic readout [5].

It is necessary to take into account that optoelectronic response is a property of semiconductor medium observed both in the laser regime and in the amplifying regime of the diode at subthreshold pumping condition [6]. As the current through the diode grows up from zero to the laser threshold, one may observe first the usual photoresponse due to photon absorption in the semiconductor medium, and then, the photoresponse is reduced to zero at the inversion threshold, and after the change in polarity the optoelectronic response appears to move to its peak value somewhere near the oscillation threshold. Thus, the optoelectronic response is an opposite effect by the sign compared to a regular photoresponse of photodiodes along with the stimulated emission being opposite to optical absorption process.

Semiconductor laser amplifiers (SLA) were studied, for example, in [4,7-9]. Such kind of an amplifier is a promising optoelectronic device to be applied in communication system as the preamplifier, modulator, regenerator, etc. It was also shown that due to optoelectronic effect SLA may be used as a monitor of the light power and optical signal in transmission line [3,6]. Our work on the development and investigation of SLA modules for single-mode fiber-optic system was presented in [10,14].

Recent Developments in Quantum Optics, Edited
by R. Inguva, Plenum Press, New York, 1993

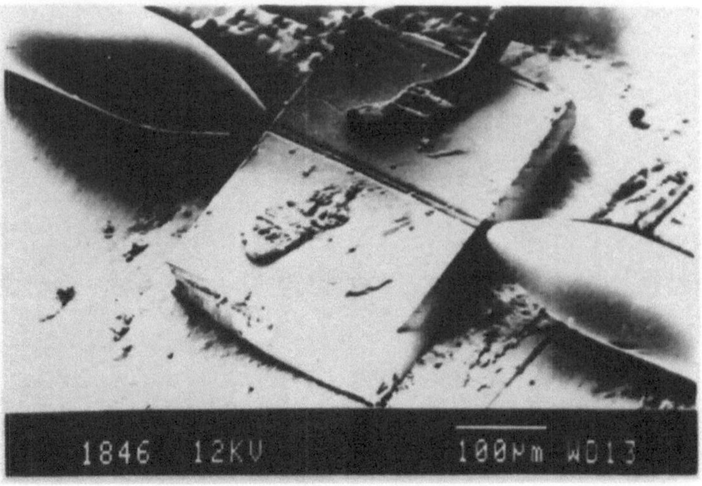

Figure 1. Raster electron microgram of SLA module.

The active element of SLA module is shown in Fig. 1. There is a diode chip with microlens-type fiber couplers attached to the chip facet at both sides. Hermetic module device contains the active element, fiber cable pigtails, the temperature sensor and the Peltier microcooler. The latter was used not only to stabilize the SLA temperature, but, if necessary, to change the spectral form of the gain curve. The module mass is only 6.5 g (with no cables). It is compatible to the standard IC panel. To produce the active element the double-heterostructure wafers of InGaAs/InP [15] had been grown by LPE process on the p-type InP substrates [16]. The second epitaxial process to bury the active stripe had been performed by MOCVD with a high-resistivity InP as a burying material. Both 1.3μm and 1.55 μm BH laser diodes have been prepared by the technique suitable for SLA applications. The active region of the diodes has the following parameters: thickness, 0.08-0.12 μm; width, 1.5-3 μm, length, 150-300 μm. The lasers readily operated with single transverse mode CW output up to ~10 mW with no kinks at room temperature. The threshold current 20-40 mA was quite reproducible, whereas the lower value was about 5 mA.

In order to get a travelling wave (TW) regime of the amplification the $\lambda/4$ antireflection coating on both facets of the diode has been performed by vacuum deposition technique. Residual reflectivity after AR coating was 0.005-0.03. The dielectric materials for the coating were $Si0$, $Zr0_2$, Eu_20_3. The increase in the threshold current after AR-coating was up to 1.5-1.6 times. Efficient coupling of the SLA diode to the fibers is obtained using hemispheric microlenses at the top of the tapered single-mode fiber. The carefully adjusted fiber couplers are fixed by special soldering.

The near TW and the resonant mode operation of the SLA are observed in the modules with fiber-to-fiber gaining the region 10-20 dB. The gain saturation behaviour is shown in Fig. 2. The diode used had the initial threshold current 25 mA, and the oscillation threshold ~39 mA after AR procedures. The saturation input power

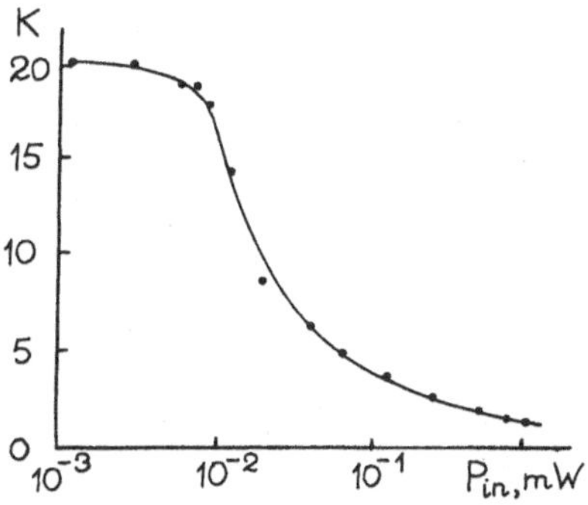

Figure 2. The dependence of the gain on the input optical power in SLA module with a laser diode 300 μm long at wavelength 1328 nm. The pumping current is 35 mA.

was near 16 μW. Output power versus pumping current dependence is shown in Fig. 3 together with OES curves at different input power. It is seen that the OES changes the polarity at ~18 mA, that is the inversion threshold (where the stimulated emission compensates the absorption). Above this point TW-amplification occurs which transforms to the resonant amplification regime as the current approaches the oscillation threshold. The OES magnitude passes negative extremum just at the threshold, where the injection locking regime begins to compete against the free-run oscillation. Due to this competition the OES magnitude decreases as the current grows. However, the sensitivity of the diode voltage to the injected light is seen even at twice threshold overdriving. The maximal amplitude of the OES voltage is found near 20 mV and seems to be quite large to easy detection. The diagnostics of the operation regimes of the SLA may be performed by these voltage measurements and the OES can be used for monitoring optical information passage in the regenerator device or for the distributed access to the data transmitted in the system. The magnitude and the sign of the electrical signals are dependent on the current bias in the SLA diode. It may be used to control the channel selection, or the decoder operation. When SLA module is in resonant mode of operation, OES response becomes selective to the carrier wavelength, and this principle may be used for spectral demultiplexing of the optical signal. The latter remains quite available for further transmission.

Another very useful application of optoelectronic regime of semiconductor injection laser operation takes place when one uses such a laser as a sensor. Injection lasers may be effectively used in a variety of sensors to detect, measure and control different parameters. Sensors of this type may have applications in mechanical equipment, tools, diagnostics apparatus, etc. One of the advantages of the sensors is a compact design of the device provided by very small dimensions of the laser source and also by original solutions of the interferometer scheme and of the problem of the fringe detection. We have investigated the problem of the external optical feedback in the diode laser as early as 1969 [17] in the simplest case of three mirror laser scheme, where

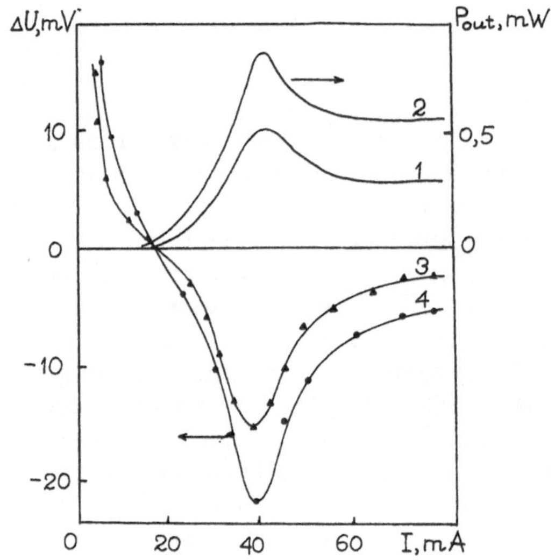

Figure 3. The output optical power (1,2) and the optoelectronic signal (3,4) vs d.c. current bias of the SLA module at different levels of the input power /pumping current of laser-generator is equal 60 mA (1,3), 80 mA (2,4)/.

the selectivity of such configuration had been demonstrated. In the papers [18,19] the composite cavity of the diode laser was realized by introducing the external plane mirror over the air gap to the one facet of the diode. The tunability of such a cavity is supplied by fine displacement of the external mirror. It was shown that oscillation regime, especially the spectral distribution of the laser emission over longitudinal modes is very sensitive to the external mirror position and this sensitivity may be observed even if the air gap is as large as 1-2 mm (feedback level 10^{-4}-10^{-5}). At a small distance the plane mirror can produce deep modulation of laser emission up to the breakdown of lasing [20]. In [21,22] it was shown that the diode voltage had a sine-shaped component of the dependence on the position of the external plane mirror. The spatial period of the sine wave is just equal to a half-wavelength of the laser emission. It exactly corresponds to the phase periodicity of reflected light as the latter produces enhancement or damping of average light intensity in the laser diode region.

Sensitivity of the laser oscillation regime to the external optical feedback in the diode laser may be considered as a kind of interferometric effect supplying the spatial resolution determined by the fraction of the wavelength. In [21,22] we reported the usage of commercial 1.3 νm diode laser ILPN-202 as a sensor of submicrometer displacement and vibrations with no additional optics and sensitivity 10 mV/μ. The diode package is coupled directly to the reflecting surface under investigation. The experiment was performed with no other photoreceiver except the laser diode itself. Therefore, this is an example of self-coupling effect applied to the interferometric measurements. It was shown that the device may be also effectively used for measuring other physical parameters which can be transformed into mechanical displacements. Adequate analysis of three-mirror cavity including two diode facets and one plane external mirror was given in [23], where the estimation of detection limit of the displacement sensor with laser diode as the detector of the reflected light and as the source of optoelectronic signal was done, and distant fiber sensor with detectivity limit about 4 nm was described.

REFERENCES

1. Eliseev P. G. and Man'ko M. A. Rept. III Polish Conf. on Radiospectroscopy and Quantum Electronics. Poznan, 1968. -Postepy Fiz., 1969, vol. 20, pp. 331-337.
2. Eliseev P. G., Ismailov I., Krasil'nikov A. I. and Man'ko M. A. Rept. IX Internat. Conf. on Semiconductor Physics. Moscow, 1968 -Proc. of the Conference, 1969, vol. 1, Leningrad, Nauka, pp. 519-523.
3. Vu Van Luc, Eliseev P. G., Man'ko M. A. and Mikaelyan G. IEEE J. Quantum Electron., 1983, vol QE-19, pp. 1080-1084.
4. Vu Van Luc, Eliseev P. G. and Man'ko M. A. Proc. of Lebedev Physics Inst., 1986, vol. 166, Moscow, Nauka, pp. 174-204. ["Nonlinear optics of semiconductor lasers," ed. by N. G. Basov, 1987, N.Y., Nova Science Publ., pp. 236-276].
5. Vu Van Luc, Eliseev P. G., Man'ko M. A. and Mikaelyan G. T. Kvantovaya Elektron. (Moscow) 1982, vol. 9, pp. 1825-1830. [Sov. J. Quantum Electron. 1982, vol. 12, pp. 1175-1178].
6. Vu Van Luc, Eliseev P. G., Man'ko M. A. and Mikaelyan G. T. Kvantovaya Elektron. (Moscow) 1982, vol. 9, pp. 1851-1854. [Sov. J. Quantum Electron. 1982, vol. 12, pp. 1194-1196].
7. Vu Van Luc, Eliseev P. G., Man'ko M. A. and Mikaelyan G. T. Kratkie soobshch. po fizike FIAN (Moscow) 1982, N 10, p. 31-35 [Lebedev Physics Inst. Reports, Arlegton Press, N.Y.].

8. Vu Van Luc, Eliseev P. G. and Man'ko M. A. Proc. of Lebedev Physics Inst., 1987, vol. 185, Moscow, Nauka, pp. 48-63 ["Injection lasers in optical communication and information processing systems," ed. by Ya. M. Popov, 1989, N.Y., Nova Science Publ., pp. 69-92].
9. Vu Van Luc, Eliseev P. G. and Man'ko M. A. Proc. of Lebedev Physics Inst., 1989, vol. 198, Moscow, Nauka, pp. 133-146 ["Laser Technology," ed. by V. A. Katulin, N.Y., Nova Science Publ. (in press)].
10. Bui Huy, Vu Van Luc. Duraev V. P. et al. Kvantovaya Elektron. (Moscow) 1989, vol. 16, pp. 1606-1608 [Sov. J. Quantum Electron., 1989, vol. 19, pp. 1034-1036].
11. Vu Van Luc, Duraev V. P., Eliseev P. G. et al. Technical Digest of XII Internat. Conf. on Integrated Optics and Optical Fiber Communication, vol. 3, (July 1989, Kobe, Japan) pp. 86-87.
12. Vu Van Luc, Duraev V. P., Eliseev P. G. et al. Preprint of Lebedev Physics Inst. N 47, 1989, Moscow [J. of Sov. Laser Research, 1989, Plenum Publ. Corp., N4, pp. 310-317].
13. Vu Van Luc, Eliseev P. G., Man'ko M. A. et al. Preprint of Lebedev Physics Inst. N137, 1989, Moscow [J. of Sov. Laser Research, 1990, Plenum Publ. Corp., N1, pp. 44-58].
14. Man'ko M. A. Proceedings 13th Internat. Annual School with Posters on Microelectronic Sensors and Semiconductor Lasers (Sozopol, Bulgaria, 14-18 May 1990) ed by Ph. Philipov, Technical University of Sofia Publ. pp. 213-219.
15. Bogatov A. P., Dolginov L. M., Druzhinina L. M. et al. Kvantovaya Elektron., (Moscow) 1974, vol. 1, pp. 2294-2295 [Sov. J. Quantum Electron., 1974, vol. 4, pp. 1281-1282].
16. Bezotosnyi V. V., Dolginov L. M., Eliseev P. G. et al. Kvantovaya Elektron. (Moscow) 1980, vol. 7, pp. 1990-1992 [Sov. J. Quantum Elektron., 1980, vol.. 10, pp. 1146-1148].
17. Eliseev P. G., Ismailov I., Man'ko M. A. and Strakhov V. P. Pis'ma Zh. Eksp. Teor. Fiz. (Moscow) 1969, vol. 9, pp. 594-595.
18. Eliseev P. G. and Man'ko M. A. Kratkie soobshcheniya po fizike, FIAN (Moscow) 1970, N4, pp. 47-52 [Lebedev Physics Inst. Reports, Arlegton Press, N.Y.].
19. Akermann D., Eliseev P. G., Keiper A. et al. Kvantovaya Elektron. (Moscow), ed. N. Basov, 1971, N1, pp. 85-90 [Sov. J. Quantum Electron., vol. 1, pp. 60-63].
20. Bogatov A. P., Eliseev P. G., Man'ko M. A. et al. IEEE J. Quantum Electron., 1973, vol. QE-9, pp. 392-394].
21. Vu Van Luc, Eliseev P. G., Man'ko M. A. and Tsotsoriya M. V. Preprint of Lebedev Physics Inst., N172, 1988, Moscow. 24p. (in Russian).
22. Vu Van Luc, Eliseev P. G., Man'ko M. A. and Tsotsoriya M. V. Kratkie soobshch. po fizike FIAN (Moscow) 1988, N4, pp. 42-44 [Lebedev Physics Inst. Reports, Arlegton Press, N.Y.].
23. Eliseev P. G., Tsotsoriya M. V. Proceedings 13th Internat. Annual School with Posters on Microelectronic Sensors and Semiconductor Lasers (Sozopol, Bulgaria, 14-18 May 1990) ed by Ph. Philipov, Technical University of Sofia Publ., pp. 61-68.

SOME PROPERTIES OF STIMULATED RAMAN SCATTERING EXCITED BY PICOSECOND LIGHT PULSES

G.L. Brekhovskikh, A.D. Kudryavtseva, A.I. Sokolovskaya
N.V. Tchernega

P.N. Lebedev Physical Institute
Leninsky Prospect
Moscow, USSR

G. Rivoire, R. Chevalier

IUI, Angers University
B^{d} Lavoisier
49045 Angers, France

For decision of some practical tasks it is interesting to get pulses of stimulated Raman scattering (SRS) radiation with good characteristics: high spatial coherency, the small divergence of beam etc. In nanosecond range of excitation it can be reached by selecting of suitable geometry of the active medium illumination. At the picosecond range the influence of geometry of illumination on the SRS characteristics is not studied well. In work [1] the experimental investigations of SRS energy and space characteristics at picosecond regime of excitation have been carried out. The parallel laser beam was used for excitation of SRS in number of liquids. It was shown that in spite of the good stability of the laser energy (~10%) the fluctuations of SRS pulses energy were rather large (till 100%). Statistic of backward SRS was independent upon the value of the active medium length. The divergency of SRS-beam was measured to be ≈100 times more than the laser beam divergency. In cross-sections of SRS-beams there were many maximums of intensity, distributed in a random way.

In the present work the SRS was excited with the help of focused beam, pulse duration being equal to 25 ps. The task was to study the influence of the active medium illumination geometry upon the energy and space characteristics of SRS. Experimental scheme is presented on fig. 1. SRS was produced in acetone with a single pulse of the second harmonic of YAG-laser radiation. The length of the sample (2) was equal to 6 and 2 cm. Maximum power of the laser pulse was $80 \cdot 10^{6}$W, divergence of radiation being $\approx 5 \cdot 10^{-4}$ rad. Deviations of laser energy from the mean value from shot to shot were not more than 10%. The laser beam was focused by the lens (1) f/10 cm. During the experiment the distance between the lens (1) and the cell filled with acetone (2) was changing. The different geometrical situations of excitation are displayed on fig. 2 (a,b,c). In the case of the laser beam focusing within the active volume (fig. 2b) three positions of the cell were investigated:

Recent Developments in Quantum Optics, Edited
by R. Inguva, Plenum Press, New York, 1993

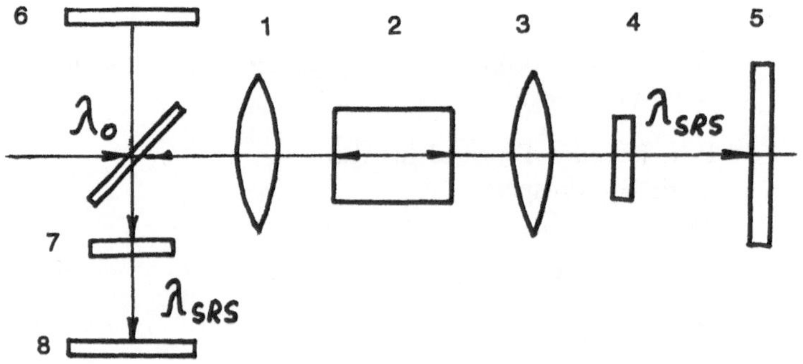

Figure 1. Experimental setup. 1,3 - lenses, 2 - cell with active material, 4,7, - color filters 5,6,8 - photoplates.

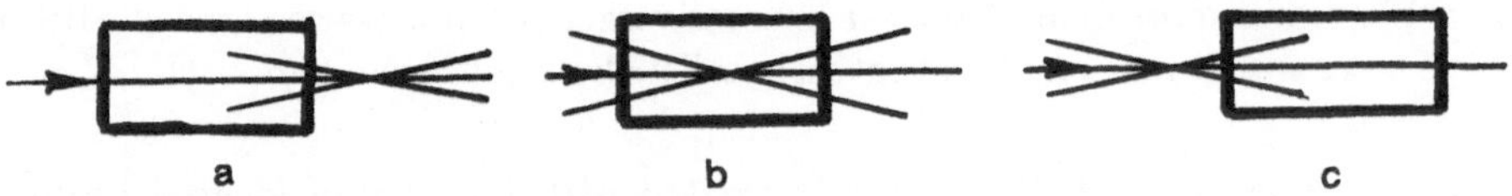

Figure 2. Different geometrical situations of excitation.

1) focal point was displaced near the entrance of the cell;
2) in the middle of the cell;
3) near it's exit at 1 cm distance up to the back window.

On the photoplates 6,8,5 (figure 1) we registered correspondingly intensity distribution in the beams cross-section of the laser light, backward SRS after passing through the lens (1) and forward SRS after passing through the lens (3) (f 10 cm), confocal; with the lens (1). Before photoplates (8) and (5) color filters (7) and (4), transmitting only light on the SRS wavelength, were placed. With the aim of divergence determination, the stokes beam cross-section has been registered by turns on the photoplates, placed at the different distances from the active material. For laser light and SRS energy measurement we used a method, elaborated in Angers University (France) [2], giving possibility to measure ultrashort pulses energy with the help of a system, consisting of the photoelectric receiver (photodiod SGD-100) and oscillograph Tektronix-7904 and having characteristic time more than measured light pulse duration.

Our measurements showed that when SRS is excited by focused pulse of picosecond duration, SRS energy fluctuations are rather large and reach 100%.

Spread value to a considerable degree is determined by the focal region position relative to the active volume, independently on the length L. The most spread is observed, when the focal region is placed inside the cell.

Forward SRS had been showed experimentally to come out of the active material by the wide beam with divergence, exceeding laser light divergence by 2 orders, independently on the illumination geometry, scattering layer thickness and pumping light power [3].

As concerns to backward SRS we have discovered that it is possible to achieve the Stokes radiation to be formed as a single, space-coherent beam with small divergence near to this of the laser beam. This effect has been reached by means of change both geometry and energy conditions of SRS excitation [3,4]. Fig. 3a presents the cross-section of the single Stokes beam, it's divergence was measured to be equal to $7 \cdot 10^{-4}$ rad. At fixed pumping energy formation of the single Stokes beam is more probable in a geometric condition when the focal point is out of the cell (fig. 2a,c). If the focus of the lens is inside the cell (fig. 2b) the probability of single Stokes beam formation increases as far as the distance from entrance window to focal point grows up. In the case of the focal point near the entrance of the cell the cross-section of backward Stokes radiation as a rule consists of some similar spots distributed in a random way (fig. 3b). Fluctuations of relative disposition of spots from shot to shot have been observed. The number of spots (N) was found to increase as far as the laser energy increases (fig. 3c). As a result we observed the rather difficult irregular distribution of the backward Stokes field.

We have ascertained the Stokes energy in each spot is limited, i.e. it cannot be more than 0.009 mJ corresponding to $\approx 10^{13}$ photons. Considerable growing of Stokes energy is possible by means of the appearance of new beams in spatial distribution with the same limit of energy (something like spatial quantization). Fig. 4 demonstrates the connection between backward Stokes energy (E_{SRS}) and the number of spots in it's cross-section (N). The strong fluctuations of SRS energy we observed are probably due to spread in numbers of spots (N) from shot to shot. Variations in spots distribution are also to be taken into consideration.

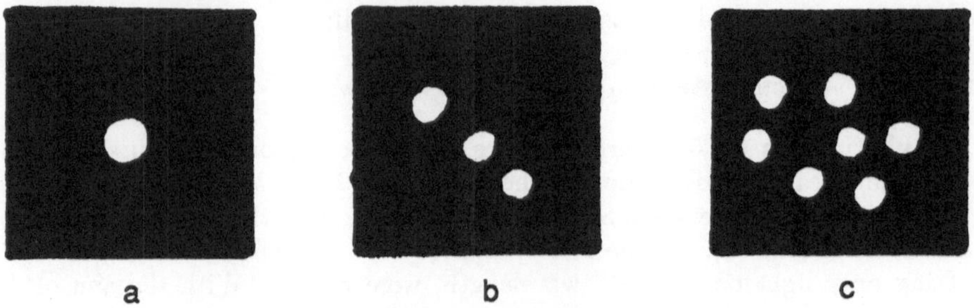

Figure 3. SRS beam spatial distribution.

The experimental results of this work are both of scientific importance and practical use. Producing of a single space-coherent Stokes beam with $\approx 10^{-11}$s time duration and diffractional divergence of radiation is an advance of a great significance for successful decision of many scientifical and practical problems. The physical nature

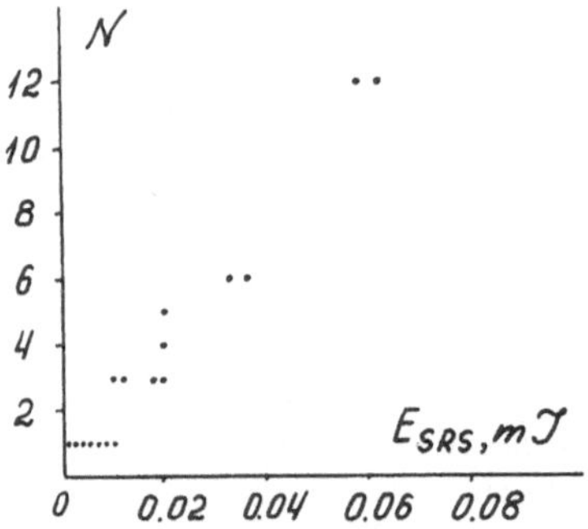

Figure 4. The spots number dependence in SRS spatial distribution on the SRS pulse energy.

of mechanisms giving rise to formation of discrete structure of the backward Stokes radiation being as a package of some space-separated portions limited both in energy and divergency is not clear yet. In future study of this problem is actual because of promotion to progress in existing conceptions of nonlinear phenomena developing in active media under exposure to ultrashort pulses of light.

REFERENCES

1. G. L. Brekhovskikh, A. I. Sokolovskaya, J. - L. Ferrier, Z. Wu, G. Rivoire. *Quantum Electronics Sov.* **10:** 622 (1983).
2. J. - L. Ferrier, N. Phun Xuan, G. Rivoire. *Mesures Regue Autom,* **45**, 53 (1980).
3. G. L. Brekhovskikh, A. I. Sokolovskaya, BN.Phu-xuan, G. Rivoire *Kratk.soobshch. phys.*, **1**, 12 (1990).
4. N. V. Tcherniega, A. I. Sokolovskaya, R. Chevalier, G. Rivoire, *Pisma VJTPh*, **16**, 23 (1990).

STATISTICAL DISTRIBUTION FOR LABORATORY GENERATED TURBULENCE

A.K. Razdan,* B.P. Singh, S. Chopra and M.B. Modi*

Physics Department
Indian Institute of Technology Delhi
New Delhi - 110016, India

*Defence Science Centre, Metcalfe House
Delhi - 110054, India

INTRODUCTION

It has long been realised that propagation of an optical beam through a turbulent medium will result in redistribution of energy of the beam leading to fluctuations in the beam intensity which can only be interpreted in statistical terms. The study of the form of probability distribution functions of these irradiance fluctuations is still a subject of controversy[1-3] and needs additional developments.

For the case of weak turbulence and relatively short propagation paths where the normalised variance of the irradiance $\sigma_I^2 << 1$, the method of smooth perturbations or Rytov theory[4] has been highly successful, but it cannot be applied for the longer paths where experimental results show σ_I^2 saturating to a value of the order of unity rather than increasing monotonically.[2,5] According to the Rytov approximation, the effect of turbulence is to perturb the propagating wave by a large number of independent, multiplicative events and a central limit theorem argument leads to a log-normal distribution. For the case of strong turbulence and large propagation paths the probability density function is expected to approach a negative exponential in the limit of infinite turbulence.[6,7]

THEORETICAL BACKGROUND

We calculate higher order skewness and excess coefficients using central moments which are then compared with those of some known distributions like log-normal, Rice-Nakagami, and gamma[8,9]. Higher order moments can give information concerning the contribution from the tails of the probability density functions. For practical purposes, moments up to eighth order are usually sufficient for the statistical characterization of the random process. Moments higher than eighth usually are difficult to extract and do not carry much statistical weight needed for characterization of the turbulence.

The nth order central moment n(n = 2,3,4....) of intensity 'I' can be written as:

Recent Developments in Quantum Optics, Edited
by R. Inguva, Plenum Press, New York, 1993

$$\mu_n = [I - < I >]^n$$

where <> is the ensemble average. The higher order non-dimensional coefficients are defined in terms of central moments as follows:

$$\Gamma_3\ (Skewness) = \mu_3/\mu_2^{3/2} \quad ; \ \Gamma_4\ (Excess) = \mu_4/\mu_2^2 - 3$$
$$\Gamma_5\ (Superskewness) = \mu_5/\mu_3\mu_2 - 10 \quad ; \ \Gamma_6\ (Superexcess) = \mu_6/\mu_2^3 - 15$$
$$\Gamma_7\ (Hyperskewness) = \mu_7/\mu_3\mu_2^2 - 105 \quad ; \ \Gamma_8\ (Hyperexcess) = \mu_8/\mu_2^4 - 105$$

From the experimental measurements the coefficients Γ_3 to Γ_8 are calculated. Measured higher order coefficients are then compared with those of model distributions like log-normal, Rice-Nakagami, and gamma to identify the best fit distribution.

EXPERIMENTAL ARRANGEMENT

A block diagram of the experimental set-up is shown in Fig. 1. Experiments have been carried out in the laboratory generated turbulence as well as in the open atmosphere. Turbulence was simulated in the laboratory by using two electric heater bars rated at 0.5 KW capacity each to heat the air between two aluminum plates along the path of the laser beam. A cool stream of air was supplied from above. The warm layer of the air rising up mixes with the cool descending layer creating a temperature fluctuation and thereby a refractive index fluctuation in the medium. By using mirrors the laser beam traverses a multipass pathlength of 6 meters before being allowed to fall on the detection system. For the case of open atmosphere the range consisted of a folded path length of 400 meters at a height of 15 meters above the ground.

A 7mW Melles Griot He-Ne laser was used as a source of radiation. After suitable attenuation and expansion the collimated beam of 1.5 cm cross-section was allowed to pass through a turbulent region. The resultant beam was detected by a photomultiplier tube (PMT model RCA 31034A) with a narrow band interference filter with transmission peak at 632.8 nm to eliminate background radiation. The PMT was illuminated through a pinhole of 50 μm diameter. The output of the PMT after suitable amplification and pulse shaping was fed to a multichannel analyzer (MCA model 256D). Before using the MCA the output of the detection system is fed to a photon correlator[10] (a home made R6501 microprocessor system) which works on the principle of directly registering the arrival time of the photoelectron pulses) to determine the sample time and the range of intensity correlation such that only independent events are used for our P(n,T) determinations. The data stored in various channels of the MCA is then used for constructing a frequency histogram from which higher order moments, cumulants and central moments up to eigth order are computed and analysed on a computer.

RESULTS AND DISCUSSION

The turbulence encountered here falls in the very weak turbulence region (as $\sigma_I^2 << 1$ in our case). From the measured data we calculate higher order moments Γ_3 to Γ_8 and compare them with those for certain well known model distributions.

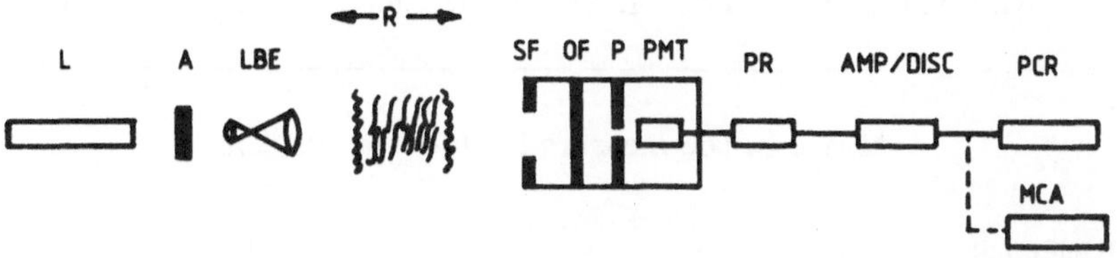

Figure 1. Block diagram of the experimental setup:L:He-Ne Laser(0.6328μm); A : Beam Attenuator; LBE : Laser Beam Expander; R : Simulated Turbulence Region; SF : Spatial Filter; OF : Optical Filter (0.6238μm); P : Pin Hole; PMT : Photomultiplier Tube; PR : Pre-Amplifier; AMP/DISC : Amplifier Discriminator; PCR : Photon Correlator; MCA : Multi Channel Analyser.

Table 1 shows the comparison of experimental data with those for standard model distributions viz. log-normal, Rice-Nakagami, and gamma. From the table it is clear that the log-normal is the best fit distribution in the very weak turbulence regime.

Because of a highly sensitive detection system used in our case, we have been able to characterize very weak turbulence simulated in the air itself as against liquids used by earlier workers. The results of open atmosphere experiments will be published elsewhere.[11]

Table 1. Comparison of theoretical and experimental values of higher order skewness and excess coefficients

	Γ_n (Theoretical) / Γ_n (Experimental)*					
	n=3	n=4	n=5	n=6	n=7	n=8
Log-normal	1.449	1.123	0.558	1.44	1.135	2.223
Gamma	1.873	0.841	0.314	1.066	0.67	1.532
Rice-Nakagami	5.779	4.498	1.407	6.858	3.633	14.164

*Error in the measured value is about 2%

REFERENCES

1. R. L. Phillips and L. C. Andrews, J.O.S.A. 71: 1440 (1981) and references therein.
2. J. R. Dunphy and J. R. Kerr, J.O.S.A. 63: 981 (1973)
3. J. H. Churnside and R. J. Hill J.O.S.A.'A' 4: 727 (1987)
4. V. I. Tatarskii, "Wave Propagation in a Turbulent Medium", McGraw-Hill, New York (1961).
5. J. R. Kerr, J.O.S.A. 62: 1040 (1972)
6. D. A. Dewolf, Proc. IEEE 62: 1523 (1974).
7. R. Dashen, J. Math. Phys. 20: 894 (1979).
8. A. K. Majumdar, J.O.S.A. 69: 199 (1979).
9. A. K. Majumdar and H. Gamo, Appl. Opt. 21: 2229 (1982).
10. B. P. Singh and S. Chopra, Rev. Sci. Instr. 59: 2096 (1988).
11. A. K. Razdan and S. Chopra, Proc. SPIE (1991) (to be published).

STATISTICAL PARAMETERS FOR NON-EXPONENTIALLY DECAYING SOURCES

Brahm Pal Singh and S. Chopra

Physics Department
Indian Institute of Technology Delhi
New Delhi - 110016 India

INTRODUCTION

One of the major problems in dynamic light scattering technique occurs when the scattered intensity time autocorrelation function (ACF) is known not to be a single exponential and also the exact theoretical form is unknown, e.g., polydisperse system. It is in fact important to determine whether a given measured ACF is a single exponential or, say, is the sum of two or more exponentials. For the later case we consider the intensity decay as nonexponential since only these will give rise to nonexponentially decaying autocorrection function.

Relevant fluctuation parameters can be investigated via photon counting distribution, energy variance (bunching parameter), and moments etc., which are dependent on the temporal behaviour of the intensity decay. Teich and Card[1] have calculated the photon counting fluctuation parameters for the exponentially decaying sources. However, it has been observed that there are a number of processes in which time dependent relaxation deviates from the exponential, e.g., relaxation function of polymers[5,6], certain neurobiological and psychophysical processes[7] and deviation from exponential law in spontaneous emission.[8] An attempt has been made to calculate the statistical parameters for nonexponentially decaying sources[2] but with a complicated form of analysis where the results are not easily amenable to the experiments.

THEORETICAL ANALYSIS

We have made an attempt to derive the exact photon counting distribution and related parameters for a pulse of light which decays nonexponentially. The results so obtained are valid for repeated and exhaustive sampling of the decaying light pulse or the sampling of an ensemble of such pulses of identical height, when the starting time (t_1) of sampling interval has a uniform distribution.

We describe the time dependent relaxation of the optical signal in its most general form as follows

$$I(t) = I_o \sum_i exp(-\mu_i t) \tag{1}$$

Recent Developments in Quantum Optics, Edited
by R. Inguva, Plenum Press, New York, 1993

where inverse μ_i are the relaxation times. The total energy illuminating the photomultiplier in time interval 'T' is given by

$$W(t,T) = \alpha \int_{t_1}^{t_1+T} I(t)dt \tag{2}$$

where α is the quantum efficiency of the detector. Substituting eq. (1) in eq. (2) and solving, we get

$$W(t,T) = \alpha I_o \sum_i exp(-\mu_i t_1)[\{1 - exp(\mu_i T)\}/\mu_i]$$

when the sampling interval is much less than the characteristic relaxation times the effective intensity can be given by

$$W(t,T) = W_o \sum_i exp(-\mu_i t_1) \tag{3}$$

where $W_o = I_o T$ and $\alpha = 1$ (for simplicity). If the starting time t_1 has a uniform probability of being anywhere between $t_1 = 0$ and $t_1 = t_2$, the probability density function can be calculated from the relation

$$P(W) = P(t_1) \mid dt_1/dW| \tag{4}$$

where

$$P(t_1) = -\frac{1}{t_2} \quad for\ 0 \leq t_1 \leq t_2 \tag{5}$$

$$P(W) = 1/\{W_o t_2 \sum_i \mu_i exp(-\mu_i t_1)\}$$

The mth order moments of $P(W)$ is defined by

$$< W^m >= \int_w W^m P(W)dW \tag{6}$$

The normalized variance of energy (bunching parameter) is given by

$$< (\Delta W)^2 > / < W >^2 = (< W^2 > - < W >^2)/ < W >^2 \tag{7}$$

The general expression for variance of energy for a sum of any number of exponentially sources can be given by

$$\frac{< (\Delta W)^2 >}{< W >^2} = t_2 \frac{\sum_i(\frac{1}{2}\mu_i)1 - exp(-2\mu_i t_2) + \sum_{i,j} \frac{1}{(\mu_i+\mu_j)}\{1 - exp(-(\mu_i + \mu_j)t_2)\}}{[\sum_i(\frac{1}{\mu_i})1 - exp(-\mu_i t_2)]^2} - 1$$

Photon counting distribution is given by the Mandel's formula

$$P(n,T) = \int_w \frac{W^n exp(-W)}{n!} P(W)\, dW \tag{8}$$

This integral can be solved for the sum of two exponentials as follows:

$$P(n,T) = (\frac{1}{t_2} n!) \int_w W_o^n (exp(-\mu_1 t_1) + exp(-\mu_2 t_1))^n$$

$$\text{x } exp(-W_o(exp(-\mu_1 t_1) + exp(-\mu_2 t_1)))\, dt_1$$

assuming $\mu_2 > \mu_1$ and $\mu_2 = p\mu_1$ and putting $exp(-\mu_1 t_1) = x$.

$$P(n,T) = (\frac{W_o^n}{t_2} n!) \int_L^U (x^p + x)^n\, exp(-W_o(x^p + x))(dx/(-\mu_1 x)). \tag{9}$$

where L and U are modified lower and upper limits.

$$P(n,T) = (\frac{W_o^n}{(-\mu_1 t_2 n!)}) \int_L^U \sum_k {}^nC_k x^{np} x^{n-k} x^{-1}\, exp(-W_o(x^p + x))\, dx.$$

$$= (\frac{W_o^n}{(-\mu_1 t_2 n!)}) \sum_k {}^nC_k \int_L^U x^{(np+n-k-1)} exp(-W_o(x^p + x))\, dx.$$

Putting $np + n - k - 1 = y$

$$P(n,T) = (\frac{W_o^n}{(-\mu_1 t_2 n!)}) \sum_k {}^nC_k \int_L^U x^y\, exp(-W_o(x^p + x))\, dx. \tag{10}$$

This is the form of a finite Laplace Transform and can be solved further by analytical or numerical methods.

DISCUSSION

These results are expected to be of interest for the interpretation of photon counting distribution generated by the polydisperse systems, laser near threshold of oscillation, laser light scattered from finite number of particles where central limit theorem is not applicable.

For laser oscillation near threshold, the intensity correlation function is given by

$$g(\tau) = \sum_m M_m\, exp(-\mu_m \tau)$$

symbols have their usual meanings. Photon counting distribution from such sources does not correspond to Bose-Einstein photon counting distribution as would be for a single exponential case. For polydisperse systems also the field ACF consists of sum or distribution of single exponentials given by[4]

$$g(\tau) = \int_{o}^{\infty} M(\lambda) exp(-\lambda\tau)\ d\lambda$$

Relaxation functions observed by dynamic laser light scattering near the glass transition are highly nonexponential. However, an empirical function of the form[6]

$$\phi_r(t) = exp\{-(t/\tau)^{\beta}\}$$

with $0 < \beta \leq 1$ has been found to fit very well. The parameter is a measure of the width of the distribution of relaxation time that would be required to fit the data to a sum of exponential relaxation functions.

Therefore, our procedure for nonexponentially decaying sources would be useful for the interpretation of these data. The similar nonexponential relaxation processes such as those stated above have also been observed in number of experiments, e.g., decay of photon echoes under certain conditions, relaxation of polymers under certain well defined conditions, some neurobiological and psychophysical process and the deviation from exponential decays in spontaneous emission.[3,4]

REFERENCES

1. M. C. Teich and H. C. Card, *Opt. Lett.* **4**, 146 (1979).
2. V. R. Subrahmanyan, V. Harwalkar and S. Chopra, *Opt. Commun.* **52**, 207 (1984).
3. R. Pecora, "Dynamic Light Scattering" (Plenum, New York, 1985).
4. Schmitz, K. S. An Introduction to Dynamic Light Scattering by Macromolecules. (Academic, New York, 1990).
5. A. Campaan, *Phys. Rev. B.* **5**, 4450 (1972).
6. G. D. Patterson, P. J. Carroll and J. R. Stevens, Jr. Polym. Sci., Polym. Phys., 21, 613 (1983).
7. Cartellucci and E. R. Kandel, Proc. Nat. Acad. Sci., USA, 71, 5004 (1974).
8. V. Harwalkar, Bohidar and S. Chopra, *Opt. & Quan. Elect.*, **15**, 241 (1983).

ON THE LINE SHAPES OF SATURATED ABSORPTION SPECTRAL PROFILES IN DISCHARGE PLASMAS

B.N. Jagtap, G.K. Bhowmick, S.A. Ahmad and V.B. Kartha

Multi-disciplinary Research Scheme & Spectroscopy Division
Bhabha Atomic Research Centre
Bombay 400085, India

INTRODUCTION

The coherence associated with an atomic transition driven by a laser, is affected by collisions in two ways: a) due to the phase interrupting effects and b) due to velocity relaxation of the level population owing to the velocity changing nature of collisions. THe latter effect which describes the extent of thermalization can be studied by saturated absorption spectroscopy (SAS) technique using a single axial mode tunable dye laser. The analyses of line shapes obtained by SAS technique [1-3] indicated that the thermal velocity changing collisions (VCC) lead to broad background superimposed on homogeneously broadened sharp resonance line. In fact the relative intensity of the broad Doppler pedestal gives a quantitative description of the velocity relaxation of the level populations owing to the velocity changing nature of collisions.

We have observed that the suppression of the Doppler pedestal is significantly affected by pump laser intensities and discharge current. In fact under specific experimental conditions, we have found that the spectral line exhibits marked deviation from the usual line shape of the saturated absorption signal [1] and shows additional features which are different from those reported so far.

EXPERIMENTAL SET-UP

We have performed a systemic study of line shapes of an atomic transition (λ 5915Å) in uranium using the SAS technique and the standard experimental set up used is shown in Fig. 1. The Doppler-free saturated absorption spectrum of UI λ 5915 Å line was recorded using a uranium hollow cathode filled with xenon as buffer gas at pressures 0.3- 1.0 torr. The output of a single-mode dye laser (Coherent Radiation model CR-599-21) with an effective line width of 5 MHz was split into a strong pump and a weak probe beam, and then both the beam are sent counter propagating through the hollow cathode lamp. The pump beam is chopped at 378 Hz, and the effect of interaction of uranium atoms with the pump beam on the probe beam transmission is measured as a function of frequency using a photomultiplier coupled to a lock-in amplifier and recorded on a strip chart recorder. The calibration of frequency is done by

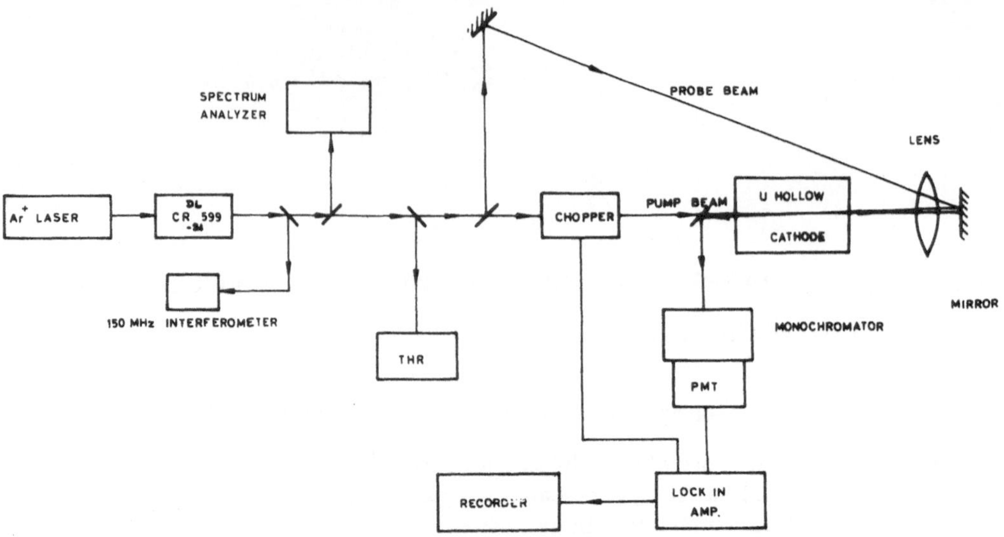

Figure 1. Experimental set up for saturated absorption spectroscopy.

simultaneously recording the transmission of a Fabry-Perot interferometer (FSR=150 MHz). The discharge current of hollow cathode (HC) was varied from 25 to 100 mA and laser power varied from 40 to 100 milliwatt.

RESULTS AND DISCUSSION

The saturated absorption signal recorded for 59 15.38 Å line of uranium with 0.95 Torr of Xe in HC and various discharge current showed that the Doppler pedestal is strongly influenced by the discharge current. When the buffer gas pressure is reduced to 0.44 torr and high discharge current (> 75 mA) the spectra develop a dip near the Lorentzian part resulting in two side wings symmetric with respect to the line centre (Fig. 2). The laser pump intensity used in these studies was 5 W/cm^2. When the pump intensity was reduced to 2.5 W/cm^2 under similar discharge conditions, the side wings observed at high current tend to disappear.

The strong suppression of the Doppler pedestal with increase in the discharge current has been interpreted in terms of shortening of life-time of the lower level through inelastic collisions with electrons in the hollow cathode discharge, resulting in corresponding reduction in the number of FCC that occur during this life-time.

For a two level atomic system undergoing "strong" collision and pumped by resonant photons, Smith and Hansch [1] have shown that the line shape for the saturated absorption signal under "strong collision" approximation is given by

$$\begin{aligned} S(\Delta\nu) &= C\ D(\Delta\nu_D)\Big\{\Big[\tfrac{1}{\Gamma_a+\gamma_a} + \tfrac{1}{\Gamma_b+\gamma_b}\Big] L(\gamma) \\ &+ \Big[\tfrac{\Gamma_a/\gamma_a}{\Gamma_a+\gamma_a} + \tfrac{\Gamma_b/\gamma_b}{\Gamma_b+\gamma_b}\Big] \tfrac{1}{2\pi^{1/2}\Delta\nu_D} D(\Delta\nu_D)\Big\} \ldots \end{aligned} \quad (1)$$

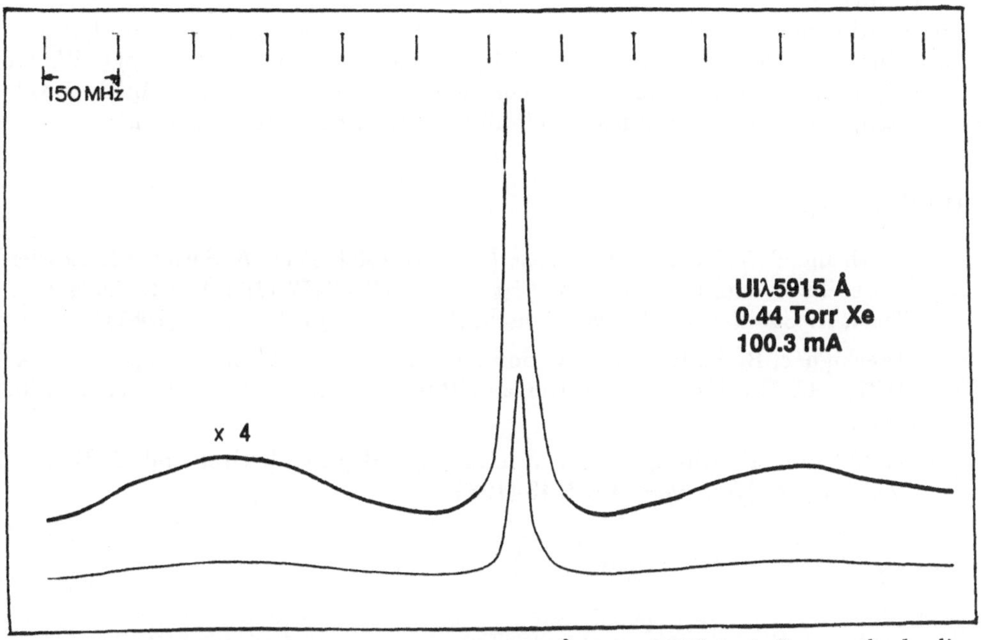

Figure 2. Saturated absorption signal of 5915 Å line of UII in hollow cathode discharge. The sidebands are emphasized by scaling up the spectrum by factor 4. The frequency marking correspond to 150 MHz.

Here, a and b are the upper and lower levels respectively. $D(\Delta\nu_D)$ and $L(\gamma)$ are the Doppler and Lorentzian line shapes with $\Delta\nu_D$ and γ as the respective widths. Γ_α, γ_α are respectively the velocity changing collision rate and decay rate from level α and C is the normalizing constant. We have found that under specific experimental condition the spectral line exhibits deviation from the usual line-shape (Eq. 1).

The additional features in saturated absorption signal (Fig. 2) appear when the buffer gas pressure is low, the current density and laser intensity are high. Thus we are in the region of low VCC constant rate, high decay rate caused by inelastic collision and non-linearity regime of pump laser absorption. The expression for line-shape given above (Eq. 1) cannot be used in this case. Tanebaum et al [3] have developed rate equation model for line-shape which is valid for all range of pump power but does not include the decay mechanism caused by discharge condition (a current). We are presently developing a generalized phenomenological model to at least qualitatively understand our new observed features in line-shape of saturated absorption.

REFERENCES

1. P. Smith and T.W. Hansch, *Phys. Rev. Lett.* **26**, 740 (1971). A. Siegel, J.E. Lawler, B. Couillaud and T.W. Hansch, *Phys. Rev.* **A23** 2457 (1981). J.E. Lawler, A. Siegel, B. Couillaud and T.W. Hansch, *J. Appl. Phys.* **52**, 4375 (1981).
2. C. Brechignac, R. Vetter and P.R. Berman, *J. Phys. B. At. Mol. Phys.* **10**, 3443 (1977). C. Brechignac, R. Vetter and P.R. Berman, *Phys. Rev.* **A17**, 1609 (1978).
3. J. Tenenbaum, E. Miron, S. Lavi, J. Liran, M. Strauss, J. Oreg and G. Erez, *J. Phys. B. At. Mol. Phys.* **16**, 4543 (1983).

EXCITONIC BISTABILITY IN MQW STRUCTURES UNDER GUIDED WAVE CONDITIONS

G.P. Bava*, F. Casteli+, P. Debernardi†, L.A. Lugiato‡

*Dip. di Elettronica, Politecnico di Torino
+Dip. di Fisica, Politecnico di Torino
†CESPA/CNR, c/o Politecnico di Torino, C.so Duca degli Abruzzi, 24, 10129 Torino, Italy
‡Dip. di Fisica, Università di Milano, Via Celoria, 16, 20133 Milano, Italy

ABSTRACT

In this paper a model for optical bistability in a ridge waveguide Fabry-Perot resonator is presented. The non-linearity is due to a Multiple Quantum Well structure inserted into the resonator. Near the excitonic resonance a bistable behaviour has been found, as shown by examples of steady-state response.

INTRODUCTION

In recent years a great interest has emerged about optical nonlinear effects in semiconductor materials, in particular in structures including multiple quantum wells (MQW) [1,2,3,4]. These materials are characterized by a very strong nonlinearity when the operating frequency is near an excitonic resonance, that is the resonance of the electromagnetic field with an electron- hole bound state due to the Coulomb interaction; the corresponding frequency appears just below the band gap (on the order of 10 meV). This strong nonlinearity, in MQW structures, persists also at room temperature, while in bulk materials it can be observed only at very low temperature [2,4,5]. This property makes MQW structures very attractive for the application to nonlinear photonic and/or electrooptic devices.

Experimental evidence of bistable behaviour in Fabry-Perot (FB) resonators, in which a MQW is inserted, was reported several years ago [1] and was the subject of several papers, as for instance [1,4,6,7,8,9]; recently, very interesting experimental results in a reflection configuration were reported in [10].

The investigation of bistable behaviour in FB resonators including a MQW has been carried out using simplified models which assume a cubic nonlinearity [9] or a saturable nonlinearity [7]. In order to identify the optimal conditions to achieve bistability in MQW structures we based on our analysis on a model which ensures an adequate description of the optical nonlinearity in these systems. This sophisticated

Recent Developments in Quantum Optics, Edited by R. Inguva, Plenum Press, New York, 1993

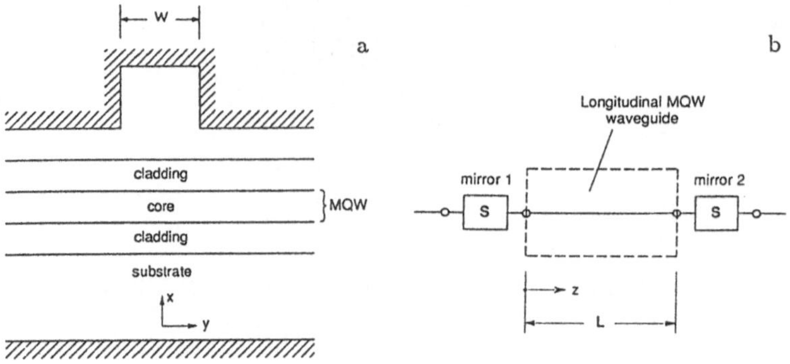

Figure 1. a) schematic waveguide cross section; b) longitudinal representation of the FP resonator.

model, which has been elaborated by the Frankfurt school [11,12], starts from a first principle microscopic description and allows for the evaluation of the contribution $\delta\varepsilon$ to the dielectric constant, which arises from both excitonic and band-to-band transitions, as a function of frequency, of the carrier density and of the parameters of the material. In this paper we use an extension of this model, which includes the effect of finite well thickness.

The system we analyze has the structure of an optical waveguide of the ridge type, containing a MQW in the core region (Fig. 1a), with the planes of the wells parallel to the direction of propagation; the waveguide, of length L, is connected to the exterior through mirrors and forms a FP resonator.

Examples of numerical computations will be reported by varying some significant parameters of the devices, such as mirror reflectivity, cavity length, detuning of the optical signal with respect to the unperturbed cavity resonance. For uniformity all the results will refer to a GaAs/AlGaAs/AlAs structure.

QW OPTICAL RESPONSE

The nonlinear dielectric response in MQW structures is due to a combination of many-body effects such as the coulombian interaction between carriers and the exclusion principle [2,5,11,13]; the most important aspects of the many-body behaviour, at room temperature, are represented by the renormalization of the band gap and the screening of the Coulomb interaction at increasing carrier density. As a consequence the optical nonlinearity is directly related to the carrier density generated by the optical absorption.

In the present paper the optical nonlinearity is evaluated by means of a model which accounts for only one conduction and valence subbands, but includes correctly the effects of the finite well thickness [14]. The contribution $\Delta\epsilon$ to the dielectric constant, due to the interaction of the electromagnetic field at the angular frequency ω with the carrier density per unit area N in a well of thickness d, is given by:

$$\Delta\varepsilon(\omega, N) = \frac{2}{V} \sum_{\underline{k}} \mu_k \chi_k \tag{1}$$

being V the volume under analysis ($=A \cdot d$, with A reference area), $\underline{k}$ the particle wavevector parallel to the QW layer, μ_k the dipole matrix element between conduction and valence band at a given wavevector k and for TE polarization. The microscopic polarization χ_k satisfies the equation [12]:

$$\chi_{\underline{k}} = \chi_k^o\left[1 + \frac{1}{\mu_k}\sum_{\underline{k'}} V_{s,eh}(\underline{k} - \underline{k'}|)\chi_{\underline{k'}}\right] \tag{2}$$

where χ_k^o is the corresponding quantity in the absence of Coulomb interaction which, adopting a spectral representation [12], can be written as

$$\chi_k^o(\omega) \simeq -\mu_k \frac{[1 - f_e(E_e) - f_h(E_h] \cdot (\hbar\omega - E_{cv}) - i\gamma_k[1 - f_e(\hbar\omega - E_g' - E_h) - f_h(E_h)]}{(\hbar\omega - E_{cv})^2 + \gamma_k^2}$$

In the previous equation f_i are the Fermi distributions of electrons (e) and holes (h), with $E_e(k)$ and $E_h(k)$ being the corresponding energies and $E_g' = E_g + \Delta E_g$ is the renormalized bidimensional bandgap. The quantity $E_{cv}(k) = E_e(k) + E_h(k) + E_g'$ denotes the energy difference between the conduction and the valence band states at the same k and includes the band gap renormalization $\Delta E_g; \gamma_k$ is a dynamical damping approximated by the following expression [12]:

$$\gamma_k(\omega) = \frac{2\gamma_o}{1 + e^{(E_{cv} - \hbar\omega)/KT}}$$

with γ_o properly chosen. Finally $V_{s,eh}(q) = V_{eh}(q)/\varepsilon_s(q,\omega)$ is the screened Coulomb potential for which a statical single plasmon pole approximation was adopted. $V_{eh}(q)$ is the Fourier transformed Coulomb potential, which takes the form [15]:

$$V_{eh}(q) = V_{2D}(q)\int_{z_e}\int_{z_h} \varepsilon_w \varepsilon^{-1}(z_e, z_h)|\Phi_e(z_e)|^2|\Phi_h(z_h)|^2 e^{-q|z_e - z_h|}dz_e dz_h \tag{3}$$

In the previous equation Φ_e and Φ_h are the electron and hole normalized wavefunctions in the potential well. In the limit case when the well thickness $d \to 0$ and potential well height $\to \infty$ the preceding expression reduces to the more familiar 2D potential $V_{2D}(q) = e^2/2qA\varepsilon_w$. In the above expression the static dielectric constant $\varepsilon(z_e, z_h)$ has been approximated by its well value ε_w when both coordinates z_e and z_h lie inside the well, by the barrier value when both coordinates z_e and z_h lie out of the well and by an appropriately weighted value in the other cases. The band gap renormalization, owing to the electron-hole plasma is computed as:

$$\Delta E_g = \sum_q \left\{\frac{1}{2}[V_{s,ee} - V_{ee}] + \frac{1}{2}[V_{s,hh} - V_{hh}] - [V_{s,ee}f_{eq} + V_{s,hh}f_{hq}]\right\}$$

where the quantities $V_{s,ee}$ and $V_{s,hh}$ are analogous to $V_{s,eh}$ and, in the same way, V_{ee} and V_{hh} are analogous to V_{eh}.

The choice of the well thickness plays a very important role; on one hand too small values of d give rise to a vanishing contribution to the MQW effect both due to the spreading of the wavefunctions and to the decreasing value of the overlap integral with the electric field of the guided mode; on the other hand, by increasing d the effect of the well becomes less and less effective as regards the excitonic interaction.

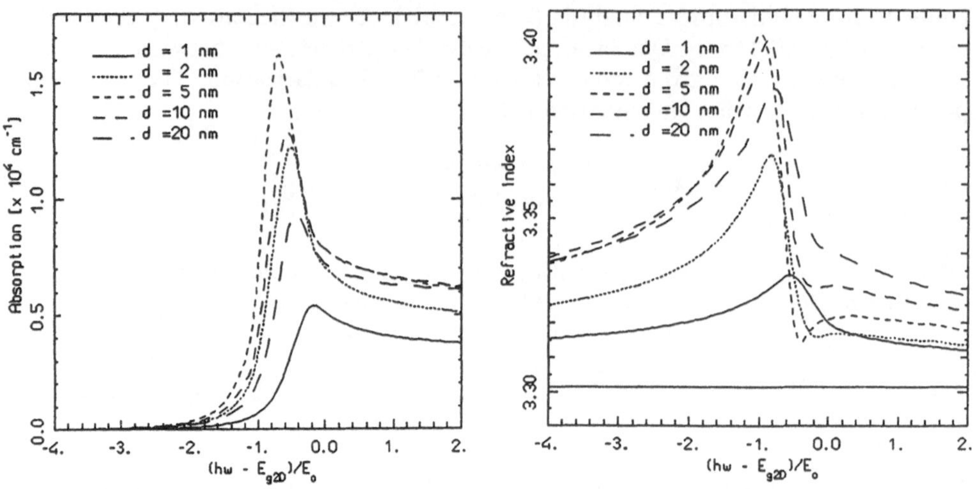

Figure 2. Absorption (left) and refractive index (right) spectra for a GaAs/AlGaAs QW at varying well width; E_{g2D} is the 2D bandgap and E_o the 2D exciton binding energy.

As a consequence an optimum value of d can be expected; for GaAs based structures and with respect to the absorption and refractive index variation, such an optimum corresponds approximately to $d \simeq 50\mathring{A}$; the typical behaviour is shown in Fig. 2. The equation (2) is solved by approximating the sum with an integral and by solving the resulting integral equation. For a GaAs/AlGaAs QW structure, using the numerical values for the material parameters given in ref. [12], the absorption (related to $Im\{\Delta\varepsilon\}$) and the refractive index (related to $Re\{\Delta\varepsilon\}$) spectra are shown respectively in Fig. 3a and in Fig. 3b, for several values of carrier density N (multiplied by the square of the 2D exciton Bohr radius a_o). The abscissa Δ is defined as:

$$\Delta = \frac{\hbar\omega - E_{g2D}}{E_o}$$

where E_{g2D} is the 2D bandgap and E_o the 2D exciton binding energy. In particular, in Fig. 3a, the exciton resonance peak and its bleaching at increasing carrier density are well evident.

MODEL FOR THE NONLINEAR INTERACTION

The non linearity in the field-carrier interaction will be introduced perturbatively through coupled mode theory, but considering strongly predominant only a single transverse mode (in our case the TE_{oo}) of the optical waveguide whose cross section is shown in Fig. 1a. This assumption is usually satisfactory since the coupling to other modes is very small and, moreover, we can suppose that the waveguide has been designed to work in monomodal conditions. In the framework of the coupled mode approach, if $C_F(z,t)$ and $C_B(z,t)$ denote the slowly varying (in z direction and in time) amplitudes of the forward and backward travelling waves respectively (normalized so that their squared moduli correspond to the travelling power), the following equations hold:

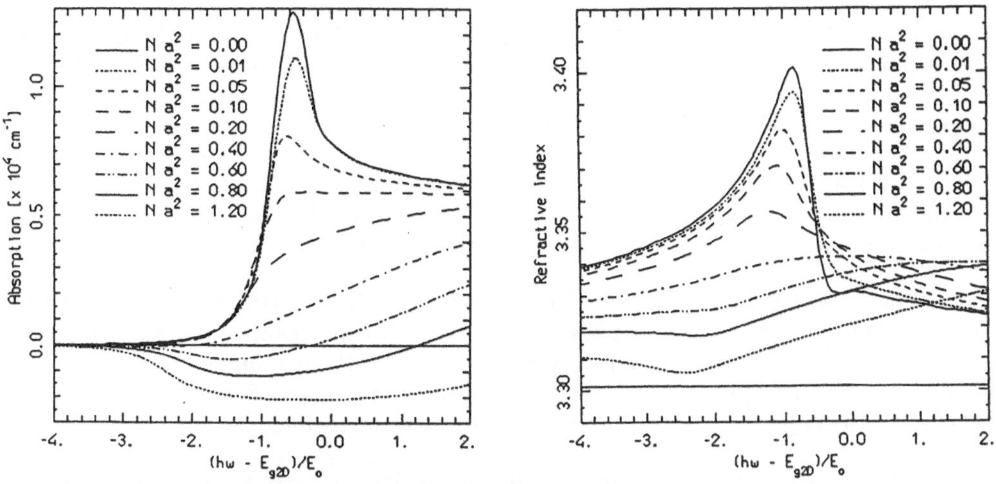

Figure 3. Absorption (left) and refractive index (right) spectra for a GaAs/AlGaAs QW at varying carrier density; E_{g2D} is the 2D bandgap and E_o the 2D exciton binding energy.

$$\begin{cases} \frac{\partial C_F}{\partial z} + \frac{1}{v_g}\frac{\partial C_F}{\partial t} = [h(z) - \alpha_i]C_F & (a) \\ \frac{\partial C_B}{\partial z} - \frac{1}{v_g}\frac{\partial C_B}{\partial t} = -[h(z) - \alpha_i]C_B & (b) \end{cases} \tag{4}$$

where v_g is the mode group velocity, α_i is the mode attenuation due to the material intrinsic losses (excluding the interaction with the carriers); the self-coupling coefficient $h(z)$, which accounts for the optical field coupling to the electrons and holes represented by $\Delta\varepsilon$ (see eq. (1)), is given by:

$$h = \frac{i\omega m}{2v_g}\frac{\int_{S_w}\Delta\varepsilon|\mathcal{E}_o|^2 ds}{\int_S \varepsilon|\mathcal{E}_o|^2 ds} \tag{5}$$

where m represents the number of wells of thickness d and $\mathcal{E}_o$ describes the mode field distribution.

Equations (4) must be coupled to the evolution equation for $N(y, z, t)$, on which $\Delta\varepsilon$ and, consequently, h depend. So we introduce the rate equation for the carriers in a well:

$$\frac{\partial N}{\partial t} = D\frac{\partial^2 N}{\partial y^2} - R(N) + Im\{\frac{\Delta\varepsilon}{\varepsilon}(N)\}(v_g\hbar)^{-1}\eta(y)(|C_F|^2 + |C_B|^2) \tag{6}$$

where D is the diffusion coefficient and $R(N) = AN + BN^2$ accounts for recombinations; the confinement factor $\eta(y)$ reads:

$$\eta(y) = \frac{\int_d \varepsilon|\mathcal{E}_o|^2 dx}{\int_S \varepsilon|\mathcal{E}_o|^2 ds} \tag{7}$$

and it is the measure of the amount of electrical power in a well with respect to the total mode power. The last term in the right hand side of eq. (6) accounts for optical carrier generation; more rigorously the carrier diffusion effects along z should be included and also the spatial grating due to the beat of forward and backward waves; for simplicity it has been assumed that diffusion compensates such a grating, since the diffusion length is larger than the grating period.

Equations (4) and (6) have been written including the dynamical aspects; however in the following of this paper only static solutions will be discussed and therefore the derivatives with respect to time will be dropped.

In the usual FP resonator (Fig. 1b) the resonant characteristics are connected to the multiple reflections between the mirrors. For the problem under consideration the boundary conditions can be expressed by the scattering parameters S_{ij} of the two mirrors, supposed symmetrical; $S_{11} = S_{22}$ are the mirror reflection coefficients at their two facets; $S_{12} = S_{21}$ are the mirror transmission coefficients.

For the problem under examination, in the stationary case, a partially analytical solution can be carried out, which can substantially reduce the computer time for a complete evaluation of the device performance. In fact eqn. (4) can be formally integrated as follows:

$$C_F(z) = C_F(0)\ exp[-\alpha_i z + \int_o^z h(\xi)d\xi]$$

$$C_B(z) = C_B(0)\ exp[+\alpha_i z - \int_o^z h(\xi)d\xi] \tag{8}$$

where, clearly, the complete solution requires the determination of the function $N(y,z)$ on which $h(z)$ depends. Taking into account eqn. (8) and the boundary conditions, it is rather easy to obtain the following relations between the transmitted power $|C_t|^2$ and the incident ($|C_i|^2$) and reflected ($|C_r|^2$) powers respectively:

$$|C_t|^2 = |C_{in}|^2 \frac{|S_{21}|^4 e^{2\int_o^L (\alpha+\alpha_i)dz}}{|1 - S_{11}^2 e^{-2L(j\beta_o+\alpha_i)+2\int_o^L h(z)dz}|^2} \tag{9}$$

$$|C_r|^2 = |C_t|^2 \frac{|S_{11}|^2}{|S_{21}|^4} e^{-2\int_o^L (\alpha+\alpha_i)dz} |1 - e^{-2L(j\beta_o+\alpha_i)+2\int_o^L h(z)dz}|^2 \tag{10}$$

where $\alpha = -Re\{h\}$ and β_o is the unperturbed waveguide propagation constant. At first sight, both the preceding relations appear completely similar to the well known formulas for FP resonators, when h is not dependent on z.

The diffusion equation (6) can be rewritten by expressing C_F and C_B by means of the solutions of Eqs. (4a) and (4b); the result is

$$D\frac{d^2N}{dy^2} = R(N) - \frac{1}{\hbar v_g}\frac{|C_t|^2}{|S_{21}|^2}\eta(y)\left[e^{2\int_z^L(\alpha+\alpha_i)dz} + |S_{11}|^2 e^{-2\int_z^L(\alpha+\alpha_i)dz}\right] Im\{\frac{\Delta\varepsilon}{\varepsilon}(N)\} \tag{11}$$

From the above formulation it appears that the strategy of solution of the nonlinear system is the following: first of all Eq. (11) is solved in order to obtain $N(y,z)$ as a function of the transmitted power. An iterative technique has been used for the integral equation along z starting from a constant value of $N(y,z)$ till the convergence of the solution; the differential equation along y was solved by a finite difference method.

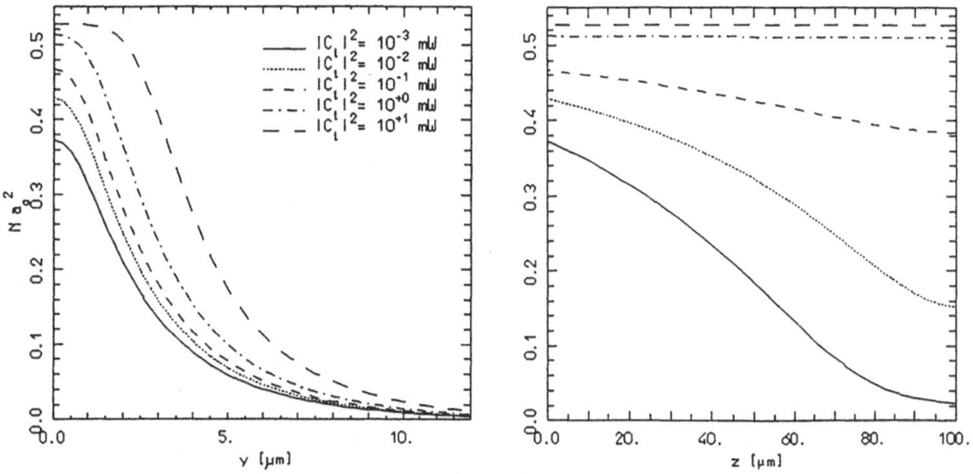

Figure 4. Transverse at $z = 0$ (left) and longitudinal at $y = 0$ (right) carrier distributions at varying output power. Parameter values: $W = 2\mu$m, $L = 100\mu$m, $|S_{11}|^2 = 0.9$, operating frequency corresponding to $\Delta = -2$; other parameters as in Sec. 4.

As an example, in Fig. 4, the transverse and longitudinal normalized carrier density distributions are reported for several values of the output power. It appears clearly the strong optical absorption at low power levels. Once, from the above procedure $N(y, z)$ is known, one can compute $|C_{in}|^2$ and $|C_r|^2$ as a function of $|C_t|^2$ from Eqs. (9) and (10), and so the transmission and reflection responses of the resonator as a function of the input power are evaluated. Such a strategy is numerically very convenient, since the first step gives a single valued relation between N and $|C_t|^2$; the bistable effects appear only in the final step. The more usual approach, i.e. starting from $|C_{in}|^2$, leads already in the first step to multi-valued functions, and thus to a much more complicated numerical procedure.

NUMERICAL RESULTS

The analysis of the FP behaviour was carried out with the following numerical values of the device parameters; for the material: $A = 0.2\,10^9 s^{-1}, B = 5\,10^{-4}\,cm^2 s^{-1}, D$ 16 $cm^2 s^{-1}$, $\alpha_i = 2.5\ cm^{-1}$; for the device: $d = 100\text{Å}$, $m = 6$, $W = 2\mu$m, $L = 50\mu$m, $|S_{11}|^2 = 0.9$, unperturbed cavity resonance frequency (ω_o) corresponding to $\Delta = -3$. The detuning, defined by $\delta = (\omega_o - \omega)/\omega$, is specified in each case. The parameters W, L, $|S_{11}|^2$ have the above values except when varied in a parametric way. The change of these parameters is very important because they affect strongly the bistable behaviour.

In the following figures the logarithm of the input, output and reflected powers (in mW) has been used in the horizontal and vertical axes in order to allow a better visualization of the results in a wide range of the parameters. In Fig. 5 some examples of numerical results for the transmission characteristics are reported at varying some of the above mentioned parameters. As a first comment, we can observe a rather nice bistable response in a certain range of each parameters. From Fig. 5 a) the great sensitivity to the detuning is evident, and from Fig. 5 b) it appears that the mirror reflectivity must be large enough (in our case approximately greater than 0.6) in order to allow hysteretic loops. With reference to Fig. 5 c) one must take into account

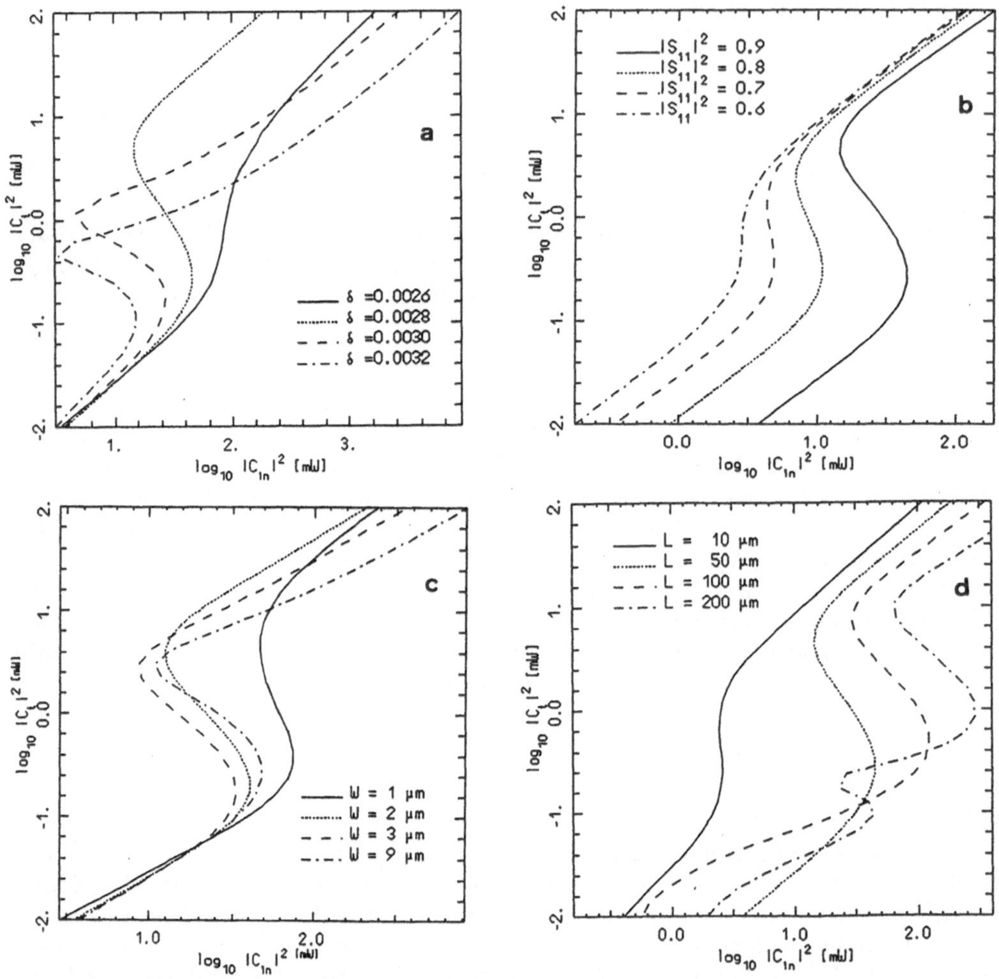

Figure 5. Transmission response of nonlinear FP resonator, a) at varying detuning δ, b) mirror reflectivity $|S_{11}|^2$ with $\delta = 0.0028$, c) ridge width W with $\delta = 0.00015$, d) resonator length L with $\delta = 0.0028$.

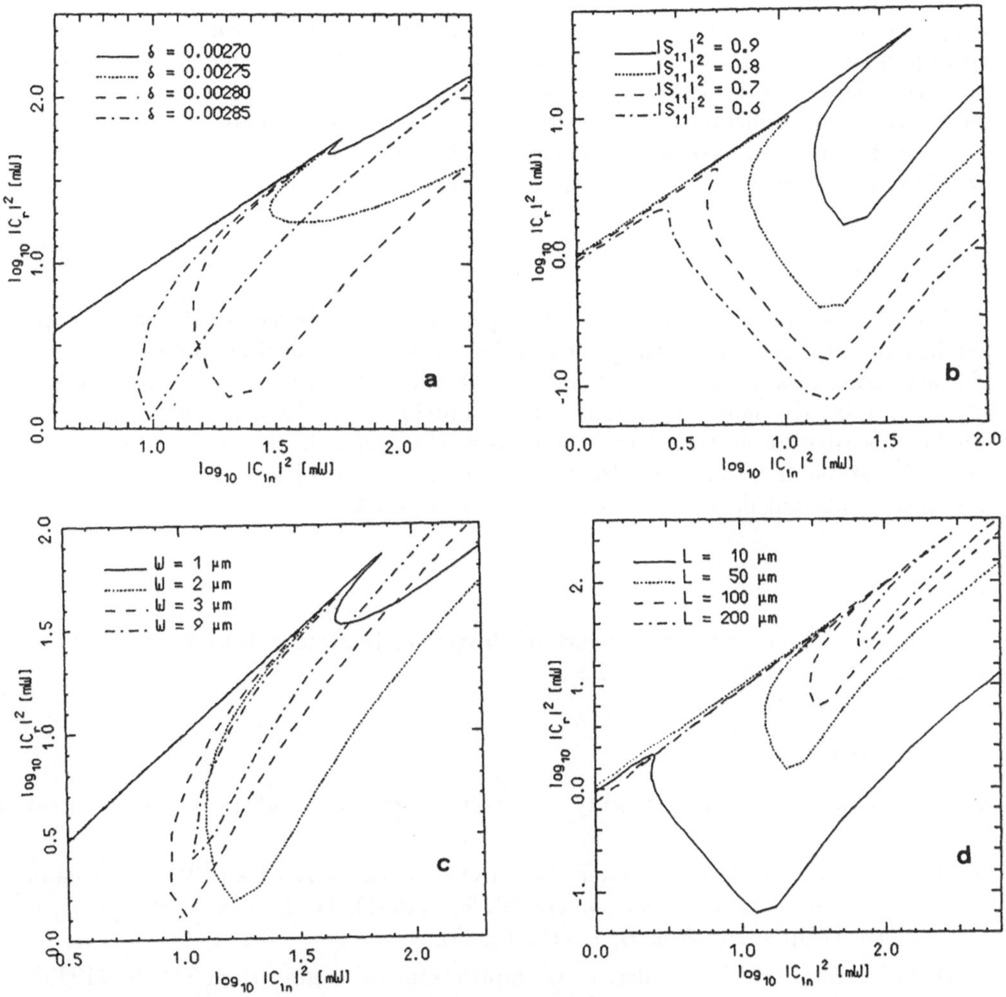

Figure 6. Reflection response of nonlinear FP resonator, a) at varying detuning δ, b) mirror reflectivity $|S_{11}|^2$ with $\delta = 0.0028$, c) ridge width W with $\delta = 0.00015$, d) resonator length L with $\delta = 0.0028$.

that for $W \geq 2.5\mu m$ the waveguide is no longer monomode; the reported results assume operation in the fundamental mode. For small values of W, a larger input power is needed to show bistability owing to the more important carrier diffusion effects, while for a W too large the total power needed to produce the required carrier density increases again. In Fig. 5 d) a clear minimum value of L appears, while for a certain length a multistable behaviour takes place; that is due, as a consequence of the nonlinear refractive index variation, to the tuning of the resonator to different longitudinal modes. In Fig. 6 examples of reflection characteristics of the FP resonator are reported. Clearly the input-output response for this configuration is complementary to the transmission case. All the diagrams of Fig. 6 are in strict correspondence with those of Fig. 5 and the same general comments apply; one must notice the more sensitive behaviour at the detuning δ, with respect to the transmission case. Recent experimental results referring to the reflection configuration, even with a different structure, have been reported in [10] and show a qualitative behaviour very similar to Fig. 6.

CONCLUSIONS

A model for the optical response of MQW structures, based on the microscopic semiconductor theory, has been applied to the study of the reflection and transmission characteristics of a nonlinear FP resonator. Owing to the very large absorption and change of refractive index, a rigorous approach including propagation effects has been used to describe the electromagnetic field inside the cavity. In the explored parameter range a bistable behaviour of refractive origin was observed and its dependence on the most important device parameters was investigated.

ACKNOWLEDGEMENTS

Work carried out in the framework of Progetto Finalizzato Telecomunicazioni of the Italian Research Council (CNR).

REFERENCES

1. H. M. Gibbs, "Optical bistability: controlling light with light," Academic Press, (1985).
2. D. S. Chemla, D. A. B. Miller, P. W. Smith, A. C. Gossard and W. Wiegmann, IEEE J. of Quantum Electron. QE-20, 265 (1984); D. S. Chemla and D. A. B. Miller, J. Opt. Soc. Am. B, 2, 1155 (1985).
3. D. S. Chemla, D. A. B. Miller, P. W. Smith, Optical engineering, 24, 556 (1985).
4. J. P. Pocholle et al., Revue Phys. Appl. 22, 1239 (1987).
5. S. Schmitt-Rink, D. S. Chemla, D. A. B. Miller, Phys. Rev. B, 32, 6601 (1985).
6. J. Goll and H. Haken, Phys. Stat. Sol. B, 101, 489 (1980).
7. E. Garmire, IEEE J. of Quantum Electron., 25, 289 (1989).
8. H. Yokoyama, IEEE J. of Quantum Electron., 25, 1190 (1989).
9. P. K. Milsom, A. Miller, Opt. and Quantum Electron., 21, 81 (1989).
10. J. L. Oudar, B. Sfez, R. Kuszelewica, J. C. Michel, R. Azoulay, Phys. Stat. Sol. B, 159, 181 (1990).
11. H. Haug and S. Schmitt-Rink, J. Opt. Soc. Am. B, 2, 1135 (1985).

12. C. Ell, R. Blank, S. Benner and H. Haug, J. Opt. Soc. Am. B, 6, 2006 (1989).
13. H. Haug and S. W. Koch, "Quantum Theory of the Optical and Electronic Properties of Semiconductors," World Scientific, 1990.
14. G. P. Bava and P. Debernardi, Electr. Lett., 27 603 (1991).
15. G. D. Sanders and Y. Chang, Phys. Rev. B, 32, 5517 (1985).

OPTICAL PHASE CONJUGATION IN MAGNETOACTIVE DOPED SEMICONDUCTORS

M. Bose, P. Aghamkar, and P. K. Sen*

Department of Physics
Barkatullah University
Bhopal - 462 026, India

INTRODUCTION

Optical Phase Conjugation (OPC) has already established itself as an active area of research in the realm of nonlinear optics due to a wide range of technological applications. Activities gained significant momentum with the development of the concept of 'transient' or 'real time' holography by Woerdman[1] and Stepanov et al.[2] A number of models have been proposed to deal theoretically with the occurrence of OPC in all kinds of matter including solids, liquids, gases, plasmas, liquid crystals and aerosols.

The most promising candidates amongst the various mechanisms are the three- and four-wave mixing as well as the stimulated scattering (SS) processes in the medium. Again, OPC-SS popularly known as "Self Phase Conjugation" has added advantage that it does not need any reference radiation. Out of the large number of SS processes like stimulated Raman scattering, stimulated Brillouin scattering (SBS), OPC-SBS is most preferred.[3] The role of semiconducting crystals in modern optical integrated circuits is unquestionable and sophisticated fabrication technology for this class of materials is already established. Hence, we have analytically investigated OPC-SBS in direct-gap III-V crystals which are nearly centrosymmetric and can be irradiated by high power pulsed lasers with photon energies in the vicinity of the crystal band-gap energies yielding large Brillouin susceptibility. We have studied OPC reflectivity of the n-type doped crystal immersed in a large magnetostatic field. Both doping as well as the magnetic field help in reducing the threshold value of the excitation intensity and in enhancing the reflectivity considerably. Recently, Ridley and Scott[4] performed an experiment using a high gain Brillouin amplifier in front of an SBS phase-conjugate mirror and obtained a reflectivity as high as 2×10^8. We have shown analytically that such complicated experimental alignments can be avoided by using a large magnetostatic field and a doped semiconductor.

---

* Present address: Shri G.S.I.T.S., 23, Park Road, Indore - 452 003 INDIA

OPC-SBS REFLECTIVITY

The crystal is irradiated by a slightly off-resonant pulsed laser. The pump electromagnetic radiation experiences SBS via its interaction with the acoustic photons. We consider the pump along the x-axis and the wave equation is given by

$$[\frac{\partial^2}{\partial x^2} + \nabla_T^2 + \frac{\omega_i^2}{c^2}]E(r_\perp, x, t) = -\mu_o \omega_i^2 P(r_\perp, x, t), \quad (1)$$

where $\nabla_T^2 = \partial^2/\partial y^2 + \partial^2/\partial z^2$. The electric field $E(r_\perp, x, t)$ varies as $exp[(\omega_i t - k_i x)]$. The induced polarization is assumed to comprise both linear (P_L) and nonlinear (P_{NL}) components given by $P = \epsilon_o[\chi^{(1)} + \chi^{(3)}|E(r_\perp, x)|^2 + ...]E(r_\perp, x, t)$.

We have neglected the even order optical susceptibilities like $\chi^{(2)}$, $\chi^{(4)}$. $\chi^{(1)}$ corresponds to the linear optical effects, while $\chi^{(3)}$ is responsible for the occurrence of OPC. $|E(r_\perp, x)|^2$ can be treated as a transmission function proportional to $(E_p + E_s)(E_p + E_s)^*$ with E_p and E_s being the electric fields associated with the pump and the Stokes' mode, respectively such that $P_{NL} = \epsilon_o \chi^{(3)}[|E_p|^2 + |E_s|^2 + E_p^* E_s + E_p E_s^*]E$.

We assume that the pump is applied along $-x$ direction while the backscattered Stokes' mode is along $+x$ with $\vec{k}_p = -\vec{k}_i$ and $\vec{k}_s = +\vec{k}_i$. Thus the wave equations under SVEA become

$$[\frac{\partial}{\partial x} - \frac{i}{2k_p}\nabla_T^2]E_p = [\alpha_p - i\alpha_{rp} - \frac{i\omega_p^2}{2k_p c^2}\chi_p^{(3)}|E_s|^2]E_p \quad (2a)$$

and

$$[\frac{\partial}{\partial x} + \frac{i}{2k_s}\nabla_T^2]E_s = [-\alpha_s + i\alpha_{rs} + \frac{i\omega_s^2}{2k_s c^2}\chi_s^{(3)}|E_p|^2]E_s. \quad (2b)$$

Here, $\alpha_{p,s}$ is the intensity dependent complex absorption coefficient of the crystal at frequency $\omega_{p,s}$ defined as

$$\alpha_{p,s} = \frac{\omega_{p,s}}{2c}(\chi_{p,s}^{(1)} + \chi_{p,s}^{(3)}|E_p|^2). \quad (3)$$

The parameters α_{rp} and α_{rs} in eqs. (2) account for material dispersion. The phase matching conditions for OPC-SBS as considered here are: $\hbar\omega_p = \hbar\omega_s + \hbar\omega_a$ and $\hbar\vec{k}_p = \hbar\vec{k}_s + \hbar\vec{k}_a$ with $(\omega_a, \vec{k}_a)$ representing the acoustic phonon mode. For $\vec{k}_p = -\vec{k}_s$, one gets $\vec{k}_a = 2\vec{k}_p$. Moreover, $\omega_p >> \omega_a$, $|\vec{k}_p| = |\vec{k}_s| = k$; $|\vec{k}_a| = 2k$ and $\omega_p \sim \omega_s = \omega$. Hence $\alpha_p = \alpha_s \approx \alpha$ and $\chi_p^{(3)} \sim \chi_s^{(3)} \approx \chi_B$, the Brillouin susceptibility (say). Following standard approach, one can get

$$E_s(x) = \beta(x)E_p^*(x) \quad (4)$$

where $|\beta(x)|^2$ can be defined as the OPC-SBS reflectivity. E_p is obtained as

$$E_p(x) = [E_p(L)exp(-\alpha(L - x))]exp(i\alpha_r(L - x)) \quad (5)$$

for $|E_s|^2 << |E_p|^2$, with $E_p(L)$ being the pump field at the entrance window $x = L \cdot \alpha_r$ has its origin in the dispersive properties of the crystal. Using (5) one finds

$$E_s(x) = E_s(o)exp[-(\alpha x + \frac{K}{2\alpha}(1 - exp(2\alpha x)))][(cos\alpha_r L + i\ sin\alpha_r L)exp(-i\alpha_r(L - x))]$$

where

$$K = \frac{\omega^2 \chi_B}{2kc^2}|E_p(L)|^2 exp(-2\alpha L). \tag{6}$$

Equation (6) manifests the phase conjugation property of back-scattered Brillouin mode with gain coefficient as $-\alpha x + K[1 - exp(2\alpha x)]/2\alpha$. Also, E_s (0) has its origin in the spontaneous scattering. It is evident from above that finite gain can be achieved only when $\alpha x - K[1 - exp(2\alpha x)]/2\alpha < 0$. For a material like optical fibre with very low loss or a bulk microcrystal, one finds $\alpha x << 1$, such that the gain condition is $\alpha < K$ while the threshold condition for the onset of SBS is $\alpha = K$. The threshold pump intensity is found to be

$$I_{th} = \epsilon_o n_o \alpha k c^3/(\omega^2|\chi_B|). \tag{7}$$

The OPC-SBS reflectivity at the entrance window ($x = L$) is obtained as

$$|\beta(x = L)|^2 = |E_s(o)/E_p|^2 exp[2(K - \alpha)L] \tag{8}$$

Since $|E_S(o)|^2 \sim 10^{-13}|E_p|^2$ (ref. 3), one has to achieve $(K - \alpha)L \sim 15$ for $|\beta(L)|^2 \sim 1$.

Hence K plays the most critical role. For low-loss systems with $\alpha L << 1$, large K can be obtained either at very high intensity or in a medium with large Brillouin susceptibility. We have shown that χ_B and hence $|\beta(L)|^2$ can be increased significantly by using a doped semiconducting microcrystal subjected to a large magnetostatic field.

We consider an n-type doped semiconductor acting as a single-component (electron) plasma and follow the hydrodynamic model under thermal equilibrium. The well established coupled-mode approach[5,6] for the magnetoplasma is employed to calculate the Brillouin susceptibility χ_B of the crystal. The electrostriction coefficient γ is responsible for the occurrence of SBS. Using the concept of density perturbations ($n_1 = n_{1l}$ and n_{1h}) oscillating at frequencies ω_a and ω_s, one can obtain the nonlinear current density $J_1(\omega_s) = -n_{1l} e v_o$ with v_o being the oscillatory electron fluid velocity. We find the nonlinear induced polarization using the relation

$$P_{NL} = \epsilon_o \chi_B |E_p|^2 E_s = \int J_1 dt \tag{9}$$

and the Brillouin susceptibility χ_B is obtained as

$$\chi_B = \gamma^2 \Omega_p^2 k^2/[2\epsilon_o \Gamma_a \omega_a(\omega^2 - \omega_c^2)]; \Omega_p^2 = n_o e^2/m\epsilon, \omega_c = -eB/m \tag{10}$$

for dispersionless acoustic wave propagation (i.e., $\omega_a^2 = k_a^2 v_a^2$). Ω_p and ω_c are the electron plasma and cyclotron frequencies, respectively. Equation (10) shows that the Brillouin susceptibility can be large at high doping concentration n_o. Also, a magnetic field yielding $\omega_c \sim \omega$ can enhance χ_B by a few orders of magnitude.

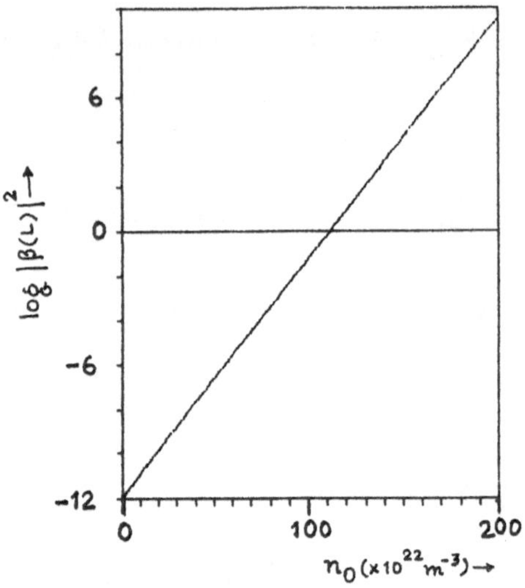

Figure 1. Variation of OPC reflectivity ($|\beta|^2$) with doping concentration n in n-InSb crystal at B=14 Teslas and I=10^{11} W cm^{-2}.

RESULTS AND DISCUSSIONS

We have studied the nature of the dependence of OPC-SBS reflectivity $|\beta(L)|^2$ on system parameters like the applied magnetostatic field B and the doping concentration n_o. The material chosen is n-type InSb crystal at 77 K, duly irradiated by a pulsed CO_2 laser. The material constants used are: $\gamma = 10^{-10}$ MKS units, $\rho = 5.8\text{x}10^3$ Kg m^{-3}, $\Gamma_a = 5\text{x}10^8 s^{-1}$, $\omega_a = 4.8\text{x}10^9 s^{-1}$, $k = 0.6\text{x}10^6 m^{-1}$, $\alpha = 100 m^{-1}$ and $L = 100$ micron. If B=14 Teslas, $\omega_c = 1.76\text{x}10^{14} s^{-1}$ and for $n_o = 10^{24} m^{-3}$, $\Omega_p = 1.2\text{x}10^{14} s^{-1}$, consequently, one finds I_{th} = 114 MW cm^{-2}. For efficient OPC-SBS, $(K - \alpha)L \sim 15$, hence pump intensity I >> 114 MW cm^{-2} will be essential to get $|\beta(L)|^2 \sim 1$. For I >> I_{th}, we can safely assume $K >> \alpha$ and $KL \sim 15$ yields $|\beta(L)|^2 \sim 1$.

The variation of OPC-SBS reflectivity of the semiconducting crystal with the doping concentration is shown in Fig. 1 and one may notice that the OPC-SBS reflectivity has a strong dependence on the doping concentration. The effect of magnetic field on the OPC-SBS reflectivity is plotted in Fig. 2. The numerical estimates made for the n-InSb crystal show that $|\beta(L)|^2$ can be remarkably enhanced, even at not-too-high excitation intensity and doping levels, by using a large magnetostatic field. Fig. 2 also manifest that the enhancement is considerably large only at high magnetic fields when ω_c becomes nearly resonant with the pump frequency ω.

ACKNOWLEDGEMENT

The financial support from the Council of Scientific and Industrial Research, India is gratefully acknowledged.

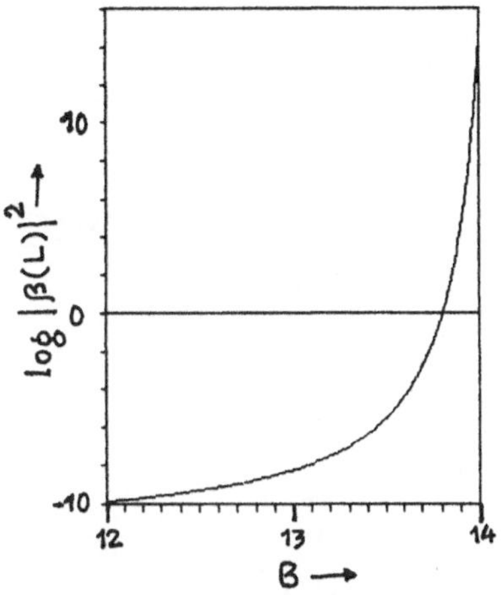

Figure 2. Dependence of OPC reflectivity ($|\beta|^2$) on the external magnetic field B(T) in n-InSb crystal for I=10^{11} W cm^{-2} and $n_o = 10^{24}m^{-3}$.

REFERENCES

1. J. P. Woerdman, *Opt. Commun.* **2**, 212 (1970).
2. B. I. Stepanov. E. V. Ivakin and A. S. Rubanov, *Sov. Phys. Dokl. Tech. Phys.* **15**, 46 (1970).
3. B. Ya. Zel'dovich, N. F. Pilipetskii and V. V. Shkunov, "Principles of Phase Conjugation" (Springer-Verlag, 1985).
4. K. D. Ridley and A. M. Scott, *Opt. Lett.* **15**, 777 (1990).
5. S. Guha and P. K. Sen, *Phys. Lett.* **A70**, 61 (1979).
6. P. K. Sen, *Phys. Rev.* **B21**, 3604 (1980).

RAMAN INDUCED PHASE CONJUGATION IN BENZENE*

Santosh Kumar Saha

Prachinmayapur (4th lane)
Nabadwip, Nadia, W.B.
741 302, India

Raman Induced Phase Conjugation (RIPC), a form of nondegenerate, backward four-wave mixing is used to measure the ratio of the resonant and non-resonant components of the third order nonlinearity in benzene from the 992 Cm^{-1} Raman mode. A computer code which models RIPC intensity is used to fit theoretical intensity to those obtained in experiment.

INTRODUCTION

The conventional phase conjugate backward wave geometry of frequency degenerate four-wave mixing can be used for Doppler- free high resolution laser spectroscopy.[1] Another sort of high resolution four wave mixing with folded boxcars geometry is to study a collision induced coherence or pressure induced extra resonance.[2] This latter scheme enforces a three dimensional phase matching geometry which enables spatial discrimination of the signal wave from the input beams even if the beams have nearly or exactly the same frequency. Good phase matching condition over a wide range of tunability is obtainable. Using the phase conjugate backward geometry a new Coherent Raman Spectroscopy (CRS) technique developed in which two beams (1st and 3rd) of frequencies ω and ν, nearly counter propagating, combine with a 2nd beam of frequency ω, with very small angle with pump beams, to generate an output beam of frequency ν. The output beam is little spatial separation from the probe beam. This new form of coherent, nondegenerate, backward four-wave mixing is Raman Induced Phase Conjugated Spectroscopy (RIPCS).[3] The superiority of this phase-conjugated backward wave geometry has been pointed out over the conventional coherent anti-Stokes Raman Scattering (CARS) due to phase matching over a wide range of tunability in RIPCS tecnique.[3]

We report here the first measurements of polarization dependence of of RIPC signal at resonance with a narrow band-width laser. The measurements are made in benzene at 992 Cm^{-1} Raman mode. The ratio of an imaginary part of resonant component ($\Delta''_{\Delta o}$) and a non-resonant component, ($\sigma+B_o$) of the third order nonlinear

*This experiment was done while author was at the Department of Physics, University of Southern California, Los Angeles, CA - 90089-0484, USA.

susceptibility has been determined. Finally, comparisons are made for measured value of this ratio using other techniques.

THEORY

The RIPC process is mediated by the third order nonlinear polarization $P(t)$. To calculate the polarization dependence of the RIPC spectra, it is convenient to consider $P(t)$ in any isotropic and nondispersive medium defined by[4]

$$P(t) = \frac{\sigma}{2}\vec{E}(t)\cdot\vec{E}(t)\vec{E}(t) + \int a(t-\tau)\vec{E}(\tau)\cdot\vec{E}(\tau)\vec{E}(t)d\tau + \int \mathcal{L}(t-\tau)\vec{E}(\tau)\cdot\vec{E}(t)\vec{E}(\tau)d\tau$$

where σ is electronic nonlinearity, a and b are isotropic and anisotropic part of nuclear response. Although third order nonlinear susceptibility (χ) is a 4th rank tensor containing 81 components only 3 of these are independent in isotropic medium. These are [3,5]

$$\chi_{1122} = \sigma + B_o + 2A_\mu$$

$$\chi_{1221} = \sigma + 2A_o + B_\mu$$

$$\chi_{1212} = \sigma + B_o + B_\mu$$

and $\chi_{1111} = \chi_{1122} + \chi_{1221} + \chi_{1212}$, the frequencies argumensts of χ is $\chi(-\nu;\omega,-\omega,\nu)$. Where A_μ and B_μ are the fourier components of a and b respectively (A_o and B_o, their dc counter parts) and related with Raman scattering cross section[3]

$$ImB_\Delta = \frac{\pi c^4}{\hbar\omega\nu^3}\frac{d^\nu\sigma_\perp}{d\Omega d\Delta}(1 - e^{-\hbar\Delta/KT})$$

similar relation with A_Δ and $\frac{\sigma_{11}}{2} - \sigma_\perp$, where $\hbar = \omega - \nu = -\Delta$

The nonlinear polarization density P expected in RIPC when the input beams are polarized as shown in figure 1 can be expressed in terms of the two components P_1, perpendicular to the polarization of the counter propagating beam (3rd beam) and P_{11}, parallel to the 3rd beam polarization.

$$P_\perp = \frac{1}{8}(\chi_{1212} - \chi_{1122})E_1E_2^*E_3 = \frac{1}{8}(B_\mu - 2A_\mu)E_1E_2^*E_3$$

$$P_{11} = \frac{1}{8}(\chi_{1212} + \chi_{1122})E_1E_2^*E_3 = \frac{1}{4}(\sigma + B_o + B_\mu/2 + A_\mu)E_1E_2^*E_3$$

The ratio $A''_\Delta/(\sigma+B_o)$ may be calculated by observing the polarization dependence of RIPC intensity.

EXPERIMENTAL PROCEDURE

The experimental arrangement is same as ref. 3. The following changes are made: 1) the spectrum of the probe laser is narrowed and its frequency can be varied linearly

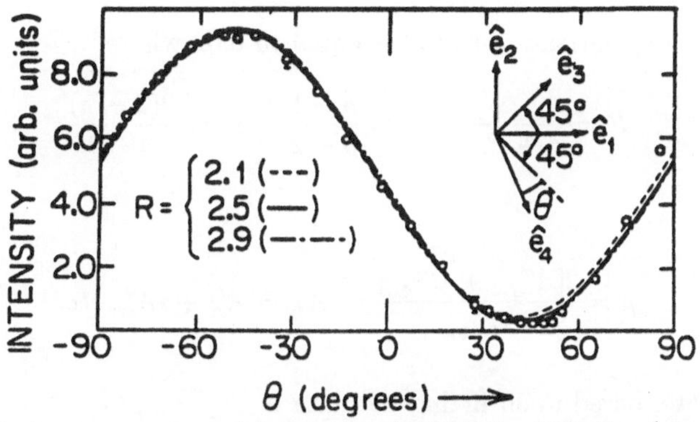

Figure 1. The circles are measured intensity of RIPC for 992 Cm^{-1} Raman mode of benzene for various polarizations of the signal beam. The solid lines are the calculated curves for $\rho = 0.02$ and different values of R.

with time. The line width (HWHM) of the laser limits the experimental resolution and its value in our case is 0.1 Cm^{-1}, 2) the line width of the pump beams is 0.01 Cm^{-1} with electronic line narrowing etalon (Model - Quanta Ray DCR-III) 3) the signal is then detected photoelectrically. A boxcar (PAR Model 162) average the signal pulses. The voltage is then plotted as a function of time by a chart recorder.

DATA ANALYSIS AND RESULTS

In our experimental configuration, the polarization of the RIPC intensity is measured at angle θ with respect to the cross-polarization of 3rd beam. Figure 1 shows the results of plotting the intensity as a function of θ at a constant frequency of $\omega - \nu = \Delta_o$. This frequency difference was found by locating the frequency of the absolute maximum intensity at angle θ=0, where the line shapes show no background level far from the resonance.[6] The lines are theoretical plot of the function.

$$(P_{\perp} cos\theta + P_{11} sin\theta)^2_{\mu=\Delta_o} = sin^2\theta + R^2(r_1 sin\theta + r_2 cos\theta)^2$$

by setting ρ, the depolarization ratio,[7] is equal to 0.02 where

$$R = \frac{A''(\Delta_o)}{\sigma + B_o}, \quad r_1 = \frac{\rho - 1}{\rho - 2}, \quad r_2 = \frac{2(3-\rho)}{\rho - 2}$$

and

$$\rho = \frac{2[A''_{-\Delta_o} + B''_{-\Delta_o}]}{B''_{-\Delta_o}}, \quad A_\Delta = A'_\Delta + \imath A''_\Delta, etc.$$

Our best estimate, based upon fitting the data is

$$R = 2.5 \pm 0.4$$

CONCLUDING REMARKS

We have employed the RIPCS to precise characterization of the Raman nonlinearity (A''_Δ) with respect to nonlinearity due to electronic (σ) and orientation contribution (B_o). The best fit value 2.5 ± 0.4 is different than the value of 3.6 ± 0.2 obtained by Levenson and Song[7] by the technique of Raman Induced Kerr Effect (RIKE). This inconsistency of the measurements may be coming from strain induced birefringence from the beam splitter in the case of RIPCS. This beam splitter is inserted in the signal beam in order to extract from the read beam. This effeect is minimized by extracting the signal in transmitted mode of the beam splitter. What was critical in the RIPCS experiments was non-vanishing minimum in figure. This spurious signal, although less than 2% (from figure 1), may be due to birefringence in the optics. On the other hand, this ratio measured by Owyoung and Peercy[8] using nonlinear interferometry technique of Stimulated Raman Scattering (SRS) is 1.8 ± 0.05. This is a close agreement with our result. Hence this technique (RIPCS) might be advantageous if one were examining Raman nonlinearity compared with the complicated nonlinear interferometry technique. The results obtained by various technique are given in the table.

Table 1. Comparison of the values for the ratio of Raman nonlinearity (A''_Δ) and non-resonant component or Optical Kerr component ($\sigma + B_o$) of the 992 Cm^{-1} line (line-width 1.2 Cm^{-1} (HWHM)) in benzene obtained from different measurements.[a]

Technique	Frequencies arguments ($-\nu; \omega$)	Beam profile with line width (HWHM)	$\frac{A''_\Delta}{\sigma+B_o}$
SRS	0.5616 μ; 0.532 μ	TEM_{001} (0.05 Cm^{-1}, ?)	1.8 ± 0.05
RIKE	0.555 μ; 0.587 μ	TEM_{001} (0.01 Cm^{-1}, 0.65 Cm^{-1})	3.6 ± 0.2
RPICS	0.5616 μ; 0.532 μ	Ary's function (0.1 Cm^{-1}; 0.01 Cm^{-1})	2.5 ± 0.4

[a]The accuracy of these nonlinear measurements have been severely restricted by the temporal and spatial characterization of the laser beams.

A new form of spectroscopy can be developed like a folded boxcars geometry[2] in which beam splitter problem to extract the signal from the read beam can be completely eliminated. Theoretical work on this route is in progress to enforce a three-dimensional phase-matching condition for Raman resonance like Zeeman resonance by Rothberg and Bloembergen.[9]

REFERENCES

1. P. F. Liao, N. P. Economou and R. R. Freeman, Phys. Rev. Lett. 39, 1473 (1977).
2. Y. Prior, A. R. Bogdan, M. Dagenais and N. Bloembergen, Phys. Rev. Lett 46, 111 (1981).
3. S. K. Saha and R. W. Hellwarth, Phys. Rev. A27, 919 (1983).
4. A. Owyoung, Ph. D. thesis (California Institute of Technology, 1971) (unpublished).
5. P. D. Maker and R. W. Terhune, Phys. Rev. A137, 801 (1965).
6. Details of our RIPC spectra will be published elsewhere.
7. M. D. Levenson and J. J. Song, J. Opt. Soc. Am. 66, 641 (1976).
8. A. Owyoung and P. S. Peercy, J. Appl. Phys. 48, 674 (1977).
9. L. J. Rothberg and N. Bloembergen, Phys. Rev. A30, 623 (1984).

NEW RESONANCES IN FOUR–WAVE MIXING DUE TO VELOCITY–CHANGING COLLISIONS

Suneel Singh and Charles M. Bowden

Weapons Sciences Directorate, AMSMI-RD-WS
Research, Development, and Engineering Center
U. S. Army Missile Command, Redstone Arsenal, AL 35898–5248

Effects of velocity–changing collisions (VCC's) between the active atoms of a vapor and heavy perturber atoms of a foreign gas, on four–wave mixing (FWM) in the vapor have been studied in the past. VCC's are known to give rise to such interesting effects as subnatural[1] narrowing and motional[2] narrowing of resonances and enhancement[3,4] of signals in FWM. Effects of VCC's in the presence of a saturating pump have also been studied[4] in two–level atomic vapors. The works dealing with multi-level atomic systems that model the atoms of the vapor as three–level systems of V or Λ configuration are, however, restricted either to the limit of nonsaturating[5] pump and probe fields or to the degenerate[3] case when the frequencies of the pump and probe fields are equal.

In this work we report the study of VCC's effects on nearly degenerate FWM in an inhomogeneously–broadened vapor consisting of effective three–level systems of Λ configuration, when one of the incident pumps is saturating. At high pump intensities we predict the existence of new resonance in FWM that occur due to VCC's.

We consider an atomic vapor consisting of four–level atoms. Three e. m. fields $\mathbf{E}_1(\mathbf{r},t)$, $\mathbf{E}_z(\mathbf{r},t)$ and $\mathbf{E}_3(\mathbf{r},t)$ given by

$$\mathbf{E}_j(\mathbf{r},t) = \vec{\varepsilon}_j \exp\left[i\left(\mathbf{k}_j\mathbf{r} - \omega_j t\right)\right] + \text{c. c.}, \quad j=1,2,3\,, \tag{1}$$

which couples the dipole transitions $|a\rangle \rightarrow |c\rangle$, $|b\rangle \rightarrow |c\rangle$, and $|b\rangle \rightarrow |d\rangle$, respectively (Fig. 1).

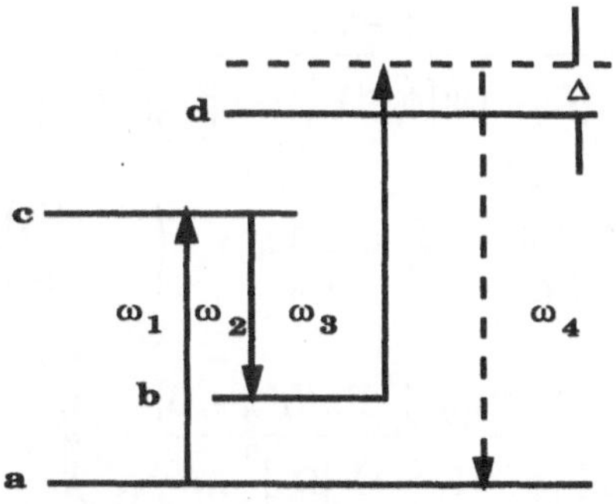

Figure 1. Energy level structure and the interaction scheme

Recent Developments in Quantum Optics, Edited
by R. Inguva, Plenum Press, New York, 1993

Such a situation may be realized by suitable choice of beam polarizations or in the near resonant case when each laser is tuned much closer to one particular resonance. The atoms of the vapor undergo collisions with the heavier perturber atoms of a foreign gas. Collisions with perturbers can be simultaneously velocity–changing (which results in a change in the velocity v or the vapor atoms) and phase–interrupting (i.e., collisions that dephase the dipole). The density matrix operator of the vapor atoms interacting with the external fields and undergoing collisions with the perturber atoms obey the equation[6]

$$\left(\frac{\partial}{\partial t}+\mathbf{v}\bullet\right)\rho=-\frac{i}{\hbar}[H_A+H_I,\rho]+\left(\frac{\partial}{\partial t}\rho\right)_{rel}+\Lambda+\left(\frac{\partial}{\partial t}\rho\right)_{coll}, \tag{2}$$

where H_A is the free–atom, and H_I is the interaction Hamiltonian given by $-\mathbf{d}\bullet\mathbf{E}$, $\mathbf{d}$ being the dipole matrix operator of the transitions. The second term on the right hand side describes the radiative relaxation and the diagonal matrix Λ describes the incoherent pumping of diagonal elements (populations). The last term on the right hand side accounts for collisional relaxation of the vapor atoms due to the dephasing and velocity–changing collisions. We adopt a strong collision model[7] to account for VCC's which assumes that the velocity distribution of active atoms is thermalized due to collisions with perturber atoms.

In the steady–state, we solve the density matrix equation (2) nonperturbatively to allow for an arbitrarily strong pump $\mathbf{E}_1$. The probe $\mathbf{E}_2$ and the other pump field $\mathbf{E}_3$ are assumed to be weak. Moreover, we consider the detuning $\omega_{db}-\omega_3$ of the field $\mathbf{E}_3$ from the level $|d\rangle$ to be much larger compared to other level–detunings, relaxation rates, and the inhomogeneous–broadening (Doppler width, γ_D). In this case we find that the nonlinear polarization $P(\omega_4)$ which generates the FWM signals at the frequency $\omega_4=\omega_1-\omega_2+\omega_3$ is proportional to the two–photon coherence ρ_{ba} between the two atomic levels (b and a) of the same parity and that we have to consider a reduced three–level system of Λ configuration interacting with a strong pump $\mathbf{E}_1$ and a weak probe $\mathbf{E}_2$.

We restrict our analysis to the near resonant case and consider a near collinear geometry in which the angle between the pump and the probe beams is very small. In the near resonant case the various detunings of the fields with the atomic levels, VCC's and dephasing collisional decay rates, and the Rabi frequency $2\Omega_1$ associated with the strong pump $\mathbf{E}_1$ are much smaller compared to the inhomogeneous broadening in the system (Doppler limit). The form of velocity–averaged FWM polarization thus obtained[8] in the Doppler limit is

$$P(\omega_4)\alpha\text{–}iZ\,\frac{\phi-i\Gamma_1}{[\phi-i(\Gamma_1-\Gamma)][\phi-i\Gamma_1]-\Gamma|\Omega_1|^2\sqrt{\pi}/\gamma_D}$$

$$\times\left\{\frac{\sqrt{\pi}}{\gamma_c}\bullet\frac{2|\Omega_1|^2}{\gamma_D}\bullet\frac{\Gamma}{\beta}\bullet\frac{\Gamma_2}{\Gamma_3}+1+\left(1+\frac{\Gamma_2}{\beta}\right)(\phi-i\Gamma_1)\right.$$

$$\left.\times\,\frac{\phi-i(\Gamma_2+\beta)+2i|\Omega_1|^2\Gamma_2/(\Gamma_2+\beta)}{[\phi-i(\Gamma_1-\Gamma)][\phi-i\Gamma_1]-\Gamma|\Omega_1|^2\sqrt{\pi}/\gamma_D}\right\}, \tag{3}$$

where

$$Z = \left[1 + \frac{\sqrt{\pi}}{\gamma_c} \bullet \frac{4|\Omega_1|^2}{\gamma_D} \bullet \frac{\Gamma}{\beta} \bullet \frac{\Gamma_2}{\Gamma_3} \bullet \left(\frac{\eta(\eta-\Gamma)+\xi^2}{\gamma_a(\gamma_a+\Gamma)}\right)\right]^{-1},$$

$$\beta = \Gamma_2\left[1 + \frac{2|\Omega_1|^2}{\Gamma_2}\left(\frac{1}{\gamma_a+\Gamma} + \frac{1}{\gamma_c+\Gamma}\right)\right]^{1/2},$$

$$\eta = \Gamma + (\gamma_c + \gamma_a)/2\ ,\quad \xi = (\gamma_c - \gamma_a)/2\ ,$$

$$\phi = \delta + \omega_{ba}\ ,\ \Gamma_1 = \gamma_{ba} + \Gamma\ ,\ \Gamma_2 = \gamma_{ca} + \Gamma, \text{ and } \Gamma_3 = \gamma_c + \Gamma\,. \tag{4}$$

In the above, $\gamma_{ij} = \frac{1}{2}(\gamma_i + \gamma_j) + \gamma_{ph}$, $i \neq j$, are the usual homogeneous widths associated with off–diagonal elements (level–coherences), γ_i is the population decay of level i, Γ is the VCC's decay rate, $\omega_{ij} = (\epsilon_i - \epsilon_j)/\hbar$ is the resonance frequency between levels i and j, ϵ_i being the energy of level i, $\delta = \omega_2 - \omega_1$ is the detuning of the probe field frequency from that of the pump, and $2\Omega_1$ is the Rabi frequency associated with the strong field $\mathbf{E}_1$.

From the first denominator in the above equation we note the existence of resonances at

$$\delta + \omega_{ba} = i\left(\gamma_{ba} + \frac{\Gamma}{2}\right) \pm \sqrt{-\frac{\Gamma^2}{4} + \frac{\sqrt{\pi}}{4}\frac{\Gamma}{\gamma_D}|2\Omega_1|^2}\,. \tag{5}$$

Figure 2 shows the behavior of FWM signals, $|P(\omega_4)|^2$ calculated numerically using Eq. (3) for a strong pump, $|2\Omega_1|^2 = 12\,\gamma_c$.

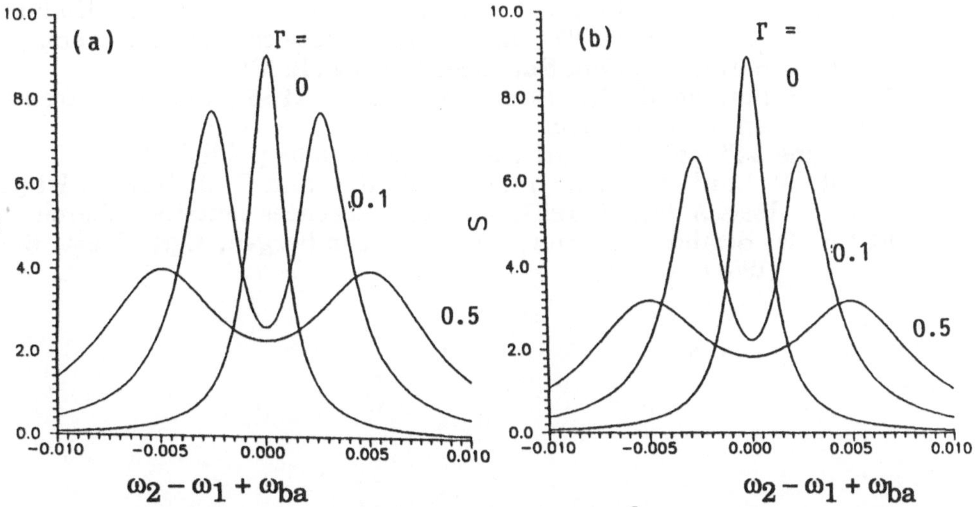

Figure 2. Four–wave mixing signals $S = |P(\omega_4)|^2$, as a function of relative detuning $\omega_2 - \omega_1 + \omega_{ba}$ for various values of the VCC's decay rate Γ, (a) in the absence of phase–interrupting collisions $\gamma_{ph} = 0$; (b) in the presence of dephasing collisions $\gamma_{ph} = 2\gamma_c$. All the parameters are scaled in terms of the radiative decay rate γ_c of the level c; $\gamma_a = \gamma_b = 0.1\gamma_c$, $\gamma_{ba} = 0.1\gamma_c$, and the Doppler width $\gamma_D = 100\gamma_c$. The coherence decay rate γ_{ba} is unaffected by dephasing collisions as the levels b and a belong to the same electronic manifold.

We observe from (5) that in the limit of strong pump field, $2\Omega_1 >> \gamma_c$, the first term under the square root is much less compared to the second, and hence two strong field–induced resonances occur at

$$\delta + \omega_{ba} = \pm |2\Omega_1| \sqrt{\frac{\sqrt{\pi}}{4} \frac{\Gamma}{\gamma_D}} . \qquad (6)$$

The origin of these resonances is in the strong–field dressing[9] of levels, due to which the levels a and b split in the presence of a strong pump field. The strong field splitting, however, is now dependent upon the VCC's and vanishes in the absence of the VCC's, i.e., when $\Gamma=0$. Thus these resonances can be termed as VCC–induced resonances in the presence of a strong field. It should also be noted here that the VCC's induced resonances in FWM occur even in the absence of the dephasing collisions ($\gamma_{ph} = 0$) and thus are different from the pressure–induced, extra resonances[10] discussed by Bloembergen and co–workers. As Eq. (5) shows, the width of these resonances goes as $(\gamma_{ba} + \Gamma/2)$ and thus is governed by the VCC's decay rate Γ.

REFERENCES

1. J. F. Lam, D. G. Steel, and R. A. McFarlane, **Phys. Rev. Lett.** 49:1628 (1982).
2. M. Gorlicki, P. R. Berman, and G. Khitrova, **Phys. Rev. A** 37:4340 (1988); S. Singh and G. S. Agarwal, **J. Opt. Soc. Am. B** 5:2515 (1988).
3. D. G. Steel and R. A. McFarlane, **Phys. Rev. A** 27:1217 (1983).
4. S. Singh and G. S. Agarwal, **Phys. Rev. A** 42:3070 (1990).
5. L. Rothberg, in "Progress in Optics," E. Wolf, ed., North–Holland, Amsterdam (1987), and the references contained therein.
6. P. R. Berman, **J. Opt. Soc. Am. B** 3:564 (1986).
7. P. R. Berman, **J. Opt. Soc. Am. B** 3:572 (1986), and the references contained therein.
8. Suneel Singh and Charles M. Bowden, to be published.
9. R. W. Boyd, M. G. Raymer, P. Narum, and D. J. Harter, **Phys. Rev. A** 24:411 (1981), and the references contained therein.
10. A. R. Bogdan, R. Prior, and N. Bloembergen, **Opt. Lett.** 6:82 (1981).

BISTABLE OSCILLATIONS IN PHASE CONJUGATE RESONATORS WITH $BaTiO_3$

P. Venkateswarlu, M. Dokhanian, P.C. Sekhar, H. Jagannath, M.C. George

Department of Physics
Alabama A&M University
Normal (Huntsville), AL 35762 USA

INTRODUCTION

A ring passive phase conjugator and several auxiliary semi-linear resonators can be formed with a $BaTiO_3$ crystal pumped by an Ar^+ laser. They can be made to function concurrently, or in bistable modes by adjusting their dimensions and mirror reflectivities. Several interesting results on bistabilities and hysteresis have been obtained.

Kwong and Yariv[1] used a single domain $BaTiO_3$ crystal to form a ring passive phase conjugator (RPPC) and two auxiliary resonators pumped by a multi-mode Ar^+ laser at 4880 Å. The auxiliary resonators were bistable. Jagannath et al[2] from this laboratory, reported the operation of a unidirectional ring resonator (UDRR) in addition to the RPPC and found bistability between the two. They also observed a linear passive phase conjugator (LPPC) which is bistable with the RPPC. It was found that auxiliary oscillations between the crystal and mirror are bistable separately with the UDRR and the LPPC.

EXPERIMENTAL RESULTS

The configuration of a ring passive phase conjugate resonator (RPPC) along with several semilinear resonators is shown in Fig. 1. The beam from an Ar^+ laser operating at 4880 Å, is incident on a single domain $BaTiO_3$ crystal with its c-axis in the horizontal plane. With the laser in a multimode condition, the RPPC (M_1M_2C) and the auxiliary resonators M_3C,..,M_6C are formed. It is found that any two of these resonators can be made to run either in a bistable mode or in a coexisting mode by changing the relative reflectivity of the mirrors or by introducing suitable attenuation in the oscillator. Fig. 2 shows the bistable oscillations between the RPPC and M_6C. In this case the RPPC dominates over the semilinear oscillator, in the sense that, when both shutters are open, only the RPPC oscillates and the other decays.

With the laser operating in single mode, the beam is incident on a face of the crystal suitable for self-pumped phase conjugation. The orientation of the crystal with respect to the incident light is kept in such a way that no self-pumped phase conjugation is

Recent Developments in Quantum Optics, Edited by R. Inguva, Plenum Press, New York, 1993

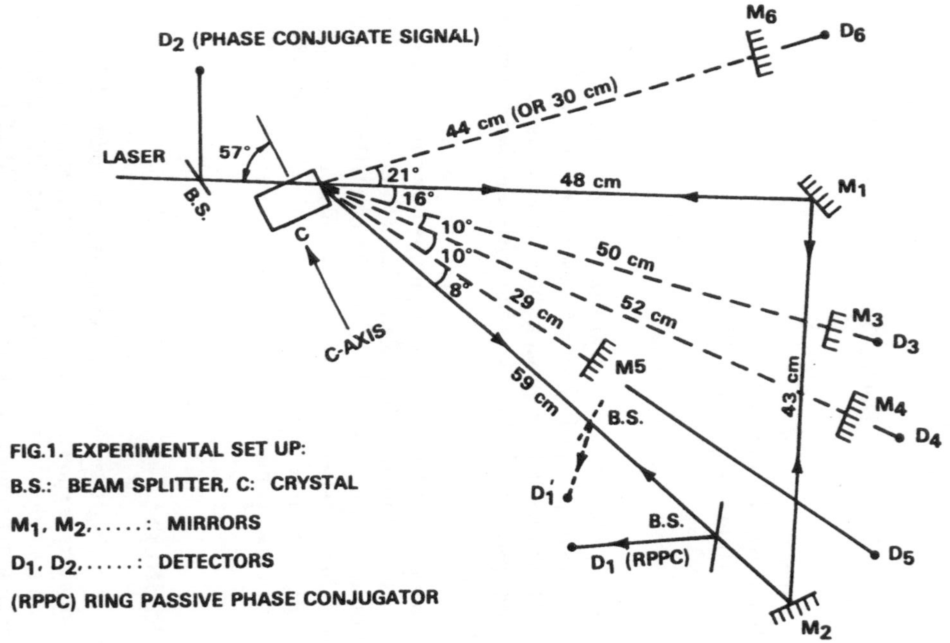

Figure 1. Experimental set-up

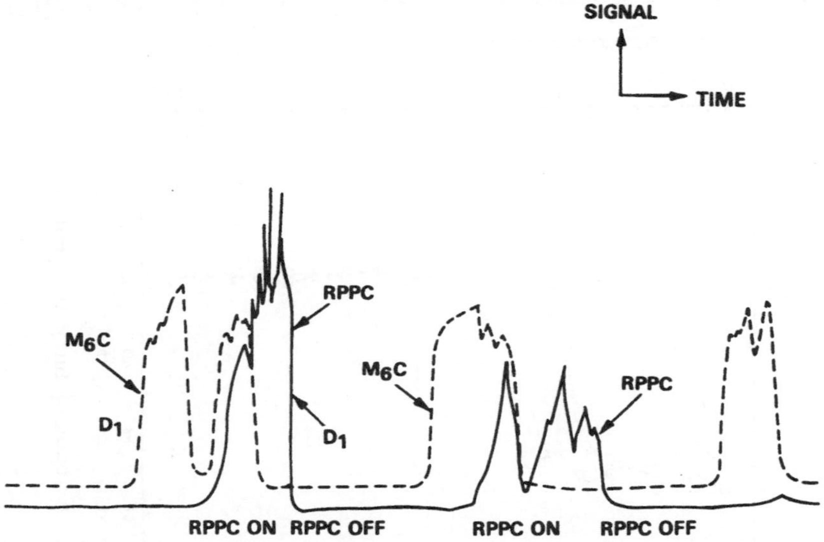

Figure 2. Bistable oscillations between RPPC and semilinear oscillator (M6C). M6C goes off when the RPPC is put on and is on when the RPPC is shut off. Total time of recording 6 minutes.

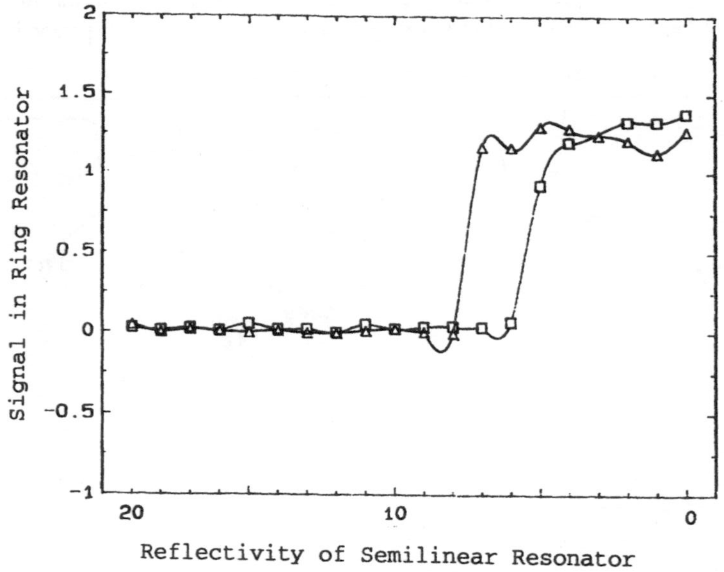

Figure 3: Hysteresis and bistability in ring resonator signal as a function of reflectivity of semilinear resonator mirror.

observable without either of the two semilinear resonators in operation. The beam, after passing through the crystal, generates a broad fan of amplified scattering toward an external variable beam reflector (V.B.R.) and the mirror M_1, which reflects the beam back to the crystal forming a ring resonator. A semilinear resonator, is also formed by reflecting the scattered light back to the crystal with another V.B.R.

Bistability and hysteresis have been observed in the signal intensity of the ring resonator by changing the reflectivity of the semilinear resonator mirror. This behavior has also been observed in the intensity of the signal of the semilinear resonator by changing its own reflectivity in a cycle. Fig. 3 shows the bistability and hysteresis observed in the signal intensity of the ring resonator as the reflectivity of the semilinear resonator is first decreased and then decreased in a cycle.

ACKNOWLEDGEMENTS

Work supported by ARO grant # DAAL03-87-G-0078 and NSF Grant # 8802971.

REFERENCES

1. S. K. Kwong, and A. Yariv, *Opt. Lett.*, **11**, 377 (1986).
2. H. Jagannath, P. Venkateswarlu, and M. C. George, *Opt. Lett.*, **12**, 1032 (1987).

RESONANT THIRD ORDER OPTICAL NONLINEARITY IN STYRYL 9

R. Vijaya, Y.V.G.S. Murti, K.R. Murali and T.A. Prasada Rao

Department of Physics
Indian Institute of Technology
Madras 600 036, India

INTRODUCTION

The third order nonlinear optical studies on the laser dye, Styryl 9 are discussed here. When an electromagnetic wave is incident on a material with its associated electric field being large, the induced polarization has terms proportional to the higher orders of the field. The ith cartesian component of the total polarization is[1]

$$P_i^{total} = \varepsilon_o(\chi_{ij}^{(1)} E_j + \chi_{ijk}^{(2)} E_j E_k + \chi_{ijkl}^{(3)} E_j E_k E_l + ...) \quad (1)$$

where $E_j, E_k, E_l, ...$ are the electric field components, $\chi^{(n)}$ are the nth order susceptibilities and ε_o is the permittivity of free space.

EXPERIMENTAL

In degenerate four wave mixing (DFWM), the frequencies of all the interacting fields are equal (ω) and the induced third order polarization at ω as a function of position $\vec{r}$ and time t is given by:

$$P^{(3)}(\vec{r}, t) = \varepsilon_o \chi^{(3)} A_1(\vec{r}) A_2(\vec{r}) A_3^*(\vec{r}) \ exp(i[(\omega + \omega - \omega)t - (\vec{k}_1 + \vec{k}_2 - \vec{k}_3) \cdot \vec{r}]) \quad (2)$$

A's are the amplitudes of the plane wave fields and $\vec{k}$'s are their wave vectors. This induced polarization generates a new field $\vec{E}_4$ of amplitude A_4, frequency ω and wavevector $\vec{k}_4$ which will emerge as a traveling wave from the medium. This output is the phase conjugate[2] of $\vec{E}_3(\vec{r})$. In the experimental arrangement used,[3,4] $\vec{E}_1$ and $\vec{E}_2$ are chosen to be intense laser fields (called pump beams) counterpropagating with respect to each other and $\vec{E}_3$ is called the probe field which intersects the beam 1 spatially at a small angle (θ). All the fields overlap within the medium. The output field with wave vector $\vec{k}_4$ will propagate in a direction opposite to that of the probe.

The nonlinear optical material Styryl 9 is a Lambdachrome laser dye from M/s Lambda Physik, FRG. This is studied in solution state in the solvents dimethyl

Recent Developments in Quantum Optics, Edited
by R. Inguva, Plenum Press, New York, 1993

sulphoxide (DMSO) and methanol taken in 1 mm-thick (L) quartz cuvettes. The laser wavelengths (λ) used in the studies are 532 nm from a frequency doubled Nd:YAG laser and 694 nm from a ruby laser, both operated in Q-switched single-shot mode. The absorption maximum of Styryl 9 lies at 565 nm, with very heavy absorption at 532 nm and quite appreciable values at 694 nm. The concentration range studied is 10^{-5} to 10^{-4} moles/liter. The maximum pump energies available at the sample at the two wavelengths are 50 mJ and 185 mJ respectively.

Phase conjugate efficiency is given by the parameter of reflectivity R defined as the ratio of the intensity of the phase conjugate beam to the intensity of the probe beam at the input face of the medium. R is measured in the experiment as a function of pump energy and concentration of the solution.

RESULTS AND DISCUSSION

To estimate the third order susceptibility $\chi^{(3)}$ of the material, we follow the model of Caro and Gower[5] for absorbing media. The expression for reflectivity is given as

$$R = \left| \frac{2Q sin((1/2)HL)}{H cos((1/2)HL) + \alpha sin((1/2)HL)} \right|^2 \tag{3}$$

where Q is the coupling constant defined by

$$|Q| = \left| \frac{3\omega I_1 e^{-\alpha L'} \chi^{(3)}}{4c^2 n^2 \varepsilon_o} \right| \tag{4}$$

α is the absorption coefficient, $L'(= L\ sec\theta)$ is the interaction length, I_1 is the intensity of first pump beam, c is the speed of light, n is the linear refractive index of the medium and $H = (4|Q|^2 - \alpha^2)^{\frac{1}{2}}$.

In table 1, the concentration corresponds to the optimum value where the reflectivity is a maximum. To fit the experimental data to the model, we have taken $\chi^{(3)} = \Gamma\alpha$, where Γ is a constant. This corresponds to the operating mechanism of thermal nonlinearity.[6] This is further confirmed by the results of our polarization discrimination studies. It is observed that the nonthermal contribution to the phase conjugate signal is small (about one fourth) compared to the thermal contribution.

Table 1. Phase conjugate reflectivity and third order susceptibility of Styryl 9

Solvent used	λ (nm)	Optimum Concentration (moles/liter)	Phase Conjugate Reflectivity	$\chi^{(3)}$ (m^2/V^2)
DMSO	532	7.80 x 10^{-5}	0.25	2.33 x 10^{-17}
Methanol	532	7.80 x 10^{-5}	0.23	3.48 x 10^{-17}
DMSO	694	2.33 x 10^{-4}	0.01	6.39 x 10^{-19}
Methanol	694	3.11 x 10^{-4}	0.03	3.77 x 10^{-18}

REFERENCES

1. N. Bloembergen, "Nonlinear Optics," Benjamin, New York (1965)
2. R. A. Fisher (Ed.), "Optical Phase Conjugation," Academic Press, New York (1983).
3. R. Vijaya, Y. V. G. S. Murti, G. Sundararajan and T. A. Prasada Rao, Degenerate four wave mixing in solutions of pure and iodine-doped polyphenyl acetylene, *Optics Commun.* **76**:256 (1990).
4. R. Vijaya, Y. V. G. S. Murti, G. Sundararajan and T. A. Prasada Rao, Nonresonant third-order optical response of polyphenyl acetylene, *J. Appl. Phys.*, **69**:3429 (1991).
5. R. G. Caro and M. C. Gower, Phase conjugation by degenerate four wave mixing in absorbing media, *IEEE J. Quant. Electr.* **QE-18**:1376 (1982).
6. B. Ya. Zeldovich, N. F. Pilipetski and V. V. Shkunov, "Principles of Phase Conjugation," Springer-Verlag, Berlin (1985).

INDEX